Blockchain Technologies

This book series aims to provide details of blockchain implementation in technology and interdisciplinary fields such as Medical Science, Applied Mathematics, Environmental Science, Business Management, and Computer Science. It covers an in-depth knowledge of blockchain technology for advance and emerging future technologies. It focuses on the Magnitude: scope, scale & frequency, Risk: security, reliability trust, and accuracy, Time: latency & timelines, utilization and implementation details of blockchain technologies. While Bitcoin and cryptocurrency might have been the first widely known uses of blockchain technology, but today, it has far many applications. In fact, blockchain is revolutionizing almost every industry. Blockchain has emerged as a disruptive technology, which has not only laid the foundation for all crypto-currencies, but also provides beneficial solutions in other fields of technologies. The features of blockchain technology include decentralized and distributed secure ledgers, recording transactions across a peer-to-peer network, creating the potential to remove unintended errors by providing transparency as well as accountability. This could affect not only the finance technology (crypto-currencies) sector, but also other fields such as:

Crypto-economics Blockchain
Enterprise Blockchain
Blockchain Travel Industry
Embedded Privacy Blockchain
Blockchain Industry 4.0
Blockchain Smart Cities,
Blockchain Future technologies,
Blockchain Fake news Detection,
Blockchain Technology and It's Future Applications
Implications of Blockchain technology
Blockchain Privacy
Blockchain Mining and Use cases
Blockchain Network Applications
Blockchain Smart Contract
Blockchain Architecture
Blockchain Business Models
Blockchain Consensus
Bitcoin and Crypto currencies, and related fields

The initiatives in which the technology is used to distribute and trace the communication start point, provide and manage privacy, and create trustworthy environment, are just a few examples of the utility of blockchain technology, which also highlight the risks, such as privacy protection. Opinion on the utility of blockchain technology has a mixed conception. Some are enthusiastic; others believe that it is merely hyped. Blockchain has also entered the sphere of humanitarian and development aids e.g. supply chain management, digital identity, smart contracts and many more. This book series provides clear concepts and applications of Blockchain technology and invites experts from research centers, academia, industry and government to contribute to it.

If you are interested in contributing to this series, please contact msingh@endicott.ac.kr OR loyola.dsilva@springer.com

Keshav Kaushik · Ishu Sharma
Editors

Next-Generation Cybersecurity

AI, ML, and Blockchain

Springer

Editors
Keshav Kaushik
University of Petroleum and Energy Studies
Dehradun, Uttarakhand, India

Ishu Sharma
Chitkara University Institute of Engineering and Technology
Chitkara University
Rajpura, Punjab, India

ISSN 2661-8338 ISSN 2661-8346 (electronic)
Blockchain Technologies
ISBN 978-981-97-1248-9 ISBN 978-981-97-1249-6 (eBook)
https://doi.org/10.1007/978-981-97-1249-6

This Springer imprint is published by the registered company Springer Nature Singapore Pte Ltd.
The registered company address is: 152 Beach Road, #21-01/04 Gateway East, Singapore 189721, Singapore

Paper in this product is recyclable.

This book is dedicated to my beloved Parents—Sh. Vijay Kaushik, Smt. Saroj Kaushik,
Wife—Priyanka, daughter Kashvi and son Harshiv. May god always bless us, Har Har Mahadev!!

Keshav Kaushik

This book is affectionately dedicated to my constant wellspring of strength, Dr. Jagdeep Sharma, my husband. His unwavering support, wisdom, and love have been the guiding lights that have illuminated my path. This dedication is a small token of appreciation for the endless moments of encouragement and shared joys.

Dr. Ishu Sharma

Preface

In a time when the digital world is changing at a never-before-seen rate, cybersecurity is a constantly changing battlefield. New frontiers are being opened by technological breakthroughs, but they also bring obstacles that call for creative answers. In this ever-changing world, "Next-Generation Cybersecurity—AI, ML, and Blockchain" shines as a lighthouse, providing an in-depth examination of how Artificial Intelligence (AI), Machine Learning (ML), and Blockchain technologies combine to strengthen cybersecurity defenses.

Our journey in the book covers up to 19 chapters and each chapter, written by top authorities in the subject, summarizes the most recent advancements and is complemented with case studies, real-world examples, and useful applications. This book presents a current and thorough assessment of the interaction between these transformational technologies and the field of cybersecurity, aiming to close a significant gap in the literature.

This book provides a route across the complex landscape of AI, ML, and blockchain in cybersecurity, whether your goal is to start cutting-edge research, strengthen your organization's defenses, or get a deeper understanding of the field.

Dehradun, India — Keshav Kaushik
Rajpura, India — Ishu Sharma

Contents

About the Editors

Keshav Kaushik is an experienced educator with around ten years of teaching and research experience in cybersecurity, digital forensics, and the Internet of Things. He is working as an Assistant Professor (Selection Grade) in the systems cluster under the School of Computer Science at the University of Petroleum and Energy Studies, Dehradun, India. He has published 110+ research papers in International Journals and has presented at reputed International Conferences. He is a Certified Ethical Hacker (CEH) v11, CQI and IRCA Certified ISO/IEC 27001:2013 Lead Auditor, Quick Heal Academy Certified Cyber Security Professional (QCSP), and IBM Cybersecurity Analyst. He acted as a keynote speaker and delivered 50+ professional talks on various national and international platforms. He has edited over twenty books with reputed international publishers like Springer, Taylor and Francis, IGI Global, Bentham Science, etc. He has chaired various special sessions at international conferences and also served as a reviewer in peer-reviewed journals and conferences. Currently, he is also serving as a Vice Chairperson of the Meerut ACM Professional Chapter and is also a brand ambassador for Bentham Science. Moreover, he is also serving as a guest editor in the *IEEE Journal of Biomedical and Health Informatics* (J-BHI) (IF:7.7).

Dr. Ishu Sharma is currently working as an Assistant Professor with Chitkara University Research and Innovation Network (CURIN), Chitkara University, Punjab. She has more than 14 years of experience in Academics and Management. Her research area includes Cybersecurity, Blockchain, Wireless Ad hoc Network, Wireless Sensor Network, and Artificial Intelligence. She completed Ph.D. in Computer Science and Engineering from Chitkara University, Punjab. She received a silver medal in Master of Engineering. She cleared the examination for Associate Member of the Institution of Engineers in the minimum duration. She was state-level topper in Diploma from the State Board of Technical Education, Haryana. She is associated with multiple EdTech Platforms and invited as a Guest Faculty in IIT Kanpur for e-Masters Course in Blockchain. She has published 40+ patents and 80+ Scopus-indexed research papers in various international journals and conferences. She has been invited as an Expert for Cybersecurity tool for various FDP and Guest Lectures in academic institutions and industries. She is guiding multiple Ph.D. and M.Tech. research scholars in the area of Cybersecurity and Blockchain.

Introduction to Cybersecurity with AI, ML, and Blockchain

Krishna Chaitanya Nosina and T. Swarna Latha

Abstract This chapter provides an introduction of cybersecurity with the aim of combining Artificial Intelligence (AI), along with Machine Learning (ML), as well as Blockchain technologies. In today's world, the need for robust cybersecurity measures is of utmost importance. The traditional approaches of security methods are often unable to defend the current evolving cyber threats, necessitating innovative solutions. AI and ML have emerged as the best technology in order to enhance the cybersecurity for processing of unlimited data, and also able to detect the regular patterns, and to identify the potential threats. Applications of AI and ML in cybersecurity include threat detection and analysis, intrusion detection and prevention systems, user and entity behavior analytics, and malware analysis. Blockchain, with its decentralized ability offers additional security benefits such as data integrity and secure identity management. With the proposed technologies, this chapter provides an opportunity for building more robust and intelligent security systems. However, the challenges such as privacy concerns and the evolving nature of cyber threats need to be addressed. This chapter serves as a basic level for understanding the essence of combining the technologies of AI, ML, and Blockchain in cybersecurity, setting the stage for further exploration of practical implementation and best practices in subsequent chapters.

Keywords Cybersecurity · Artificial intelligence (AI) · Machine learning (ML) · Blockchain · Threat detection · Data integrity · Security technologies

K. C. Nosina (✉)
Department of ECE, RSR Engineering College, Kadanuthala, Kavali 524142, India
e-mail: nosinakc@gmail.com

T. Swarna Latha
Department of CSE, PBR Visvodaya Institute of Technology and Science, Kavali 524201, India

K. Kaushik and I. Sharma (eds.), *Next-Generation Cybersecurity*, Blockchain Technologies, https://doi.org/10.1007/978-981-97-1249-6_1

1 Introduction

With the rise of digital technologies and the increasing sophistication of cyber threats, organizations and individuals need robust security measures to protect their sensitive data and systems. This chapter serves as an introduction to cybersecurity, wherein it focuses by combining the technologies of AI, ML, with Blockchain as powerful tools for enhancing security.

1.1 Understanding Cybersecurity

Cybersecurity [1] mainly aims to protect the computers, networks, as well as data from the illicit access, chance of theft, as well as damage, or even disruption. In the current internet world, where digital technologies [2] are deeply integrated into various aspects of our lives, the need for the better cybersecurity actions [3] has become very important.

A range of critical elements, as shown in Fig. 1, that are essential for the securement of Digital Assets and Data Against Cyber Threats will be covered by cyber security. Human beings play a vital role in addressing and managing the cyber environment, which requires ongoing user education so as to promote best practices and awareness. Application security is concerned with the safety of software and systems so as to avoid vulnerabilities and unauthorised access. The regulatory framework for cyber security consists of legislation and policies to guide organizations in ensuring their compliance with cybersecurity standards, as well as protecting user privacy. Implementing measures to ensure the safe transmission of data, as well as protection against unauthorised access and breaches are part of network security. In view of the cyber threat or disasters, disaster recovery and continuity planning ensure that unexpected events can be prepared for as quickly as possible in order to enable uninterrupted business operations. Together, these features contribute to creating an effective cyber ecosystem that fosters a secure and stable digital environment.

The primary objective of cybersecurity is to provide the privacy [4], veracity, and ease of use of digital assets. In particular, privacy safeguards important information. Integrity [5] ensures that data remains intact, unaltered, and free from unauthorized modifications. Availability [6] ensures that systems and data are accessible and usable when needed. Cyber security has a wide variety of strategies, technologies and processes. It involves identifying and assessing [7] vulnerabilities in computer systems and networks, implementing protective measures to prevent attacks, and developing incident response plans to effectively mitigate and recover from security incidents. Cyber threats exist in various forms, and they can have a serious impact. Malware [8, 9], such as viruses, worms, and ransomware, is designed to infiltrate and disrupt computer systems. To gain access to sensitive information, phishing attacks target individuals via misleading email or web sites. Social engineering [10, 11] exploits human vulnerabilities to manipulate those into disclosing secret

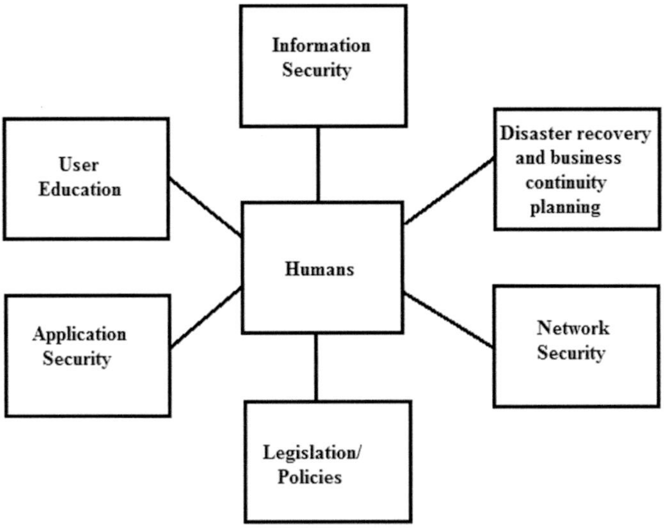

Fig. 1 Components in cybersecurity

information or compromise unlawful access. Other threats [12] such as denial-of-service (DoS) attacks, where attackers devastate a system with data transfer to make it unapproachable, and insider threats, where trusted individuals misuse their access privileges.

Cyber security professionals use a variety of techniques and technologies in order to effectively counter such threats [12]. The deployment of a firewall, intrusion detection system and antivirus software to eliminate the unauthorised access as well as detecting and combating any illegal activities [13–15] should also be undertaken in this context. In order to ensure confidentiality and security of this data, encryption shall be applied during transmission and storage. To verify the identity of individuals accessing systems [16] and data so that unauthorised access is prevented, access controls and authentication mechanisms are used. Cybersecurity is a constantly evolving field due to the rapid change with respect to the technology [17–19] and appearance of new threats. Cyber security experts need to keep up with the evolving attackers and constantly refresh their knowledge and skills in order to stay ahead. This includes staying informed about the latest threats, vulnerabilities, and security best practices, as well as engaging in regular training and professional development [20].

1.2 AI and ML in Cybersecurity

AI and ML have ruled as most effective tools [21–23] in the field of cybersecurity, revolutionizing the way threats are detected, analyzed, and mitigated. These technologies [24] enable cybersecurity professionals to effectively address the growing complexity and volume of cyber threats in today's interconnected world.

Artificial intelligence means the creation of machines capable of performing a job [25], such as learning to solve problems, making decisions and recognizing patterns, that requires high levels of Human Intelligence. The portion of AI dealing with algorithm [26] design is called ML. To increase the detection, analysis and response capabilities of threats, artificial intelligence and machine learning are being incorporated into cyber security. These technologies [27] can verify large amounts of information, capable of identifying the patterns, and to detect the incidents where human analysts miss. By continuously learning and adapting, AI and ML systems can improve their accuracy and also efficiency over time.

One of the significant contribution [28] of AI and ML in cybersecurity is detection of severe threats as well as analyze them efficiently. AI and ML algorithms have the ability to read the network traffic, to log files, and possess the behavior of the user to find suspicious activities and potential threats. They can detect known malware signatures [29] and also identify previously unknown threats using anomaly detection techniques. This helps in early detection of potential cyber-attacks and enables organizations to take proactive measures to prevent or mitigate the impact of such attacks. AI and ML techniques are also used in Intrusion Detection and Prevention Systems (IDPS). The systems [30] monitor network traffic in real time to detect and block malicious activities. Artificial Intelligence and Machine Learning algorithms can, by analysing patterns and behaviors, find out what's going on in the network as well as which kind of threat is likely to attack them; they can automatically perform mitigation measures.

User and Entity Behavior Analytics (UEBA) [31] is another area where AI and ML are extensively utilized. UEBA systems establish baseline behavior patterns for users and entities within an organization's network. They monitor and analyze user activities [32], network access, and data interactions to detect anomalous behavior of compromised user accounts. By identifying such threats in real-time [33], organizations can proactively respond and prevent potential data breaches. AI and ML also play a crucial role in malware analysis and antivirus solutions. Malware analysis involves studying malicious software [34] to understand its behavior, capabilities, and potential impact. This analysis may be automated by the use of artificial intelligence and machine learning techniques, helping to identify and classify threats much more quickly. By increasing the detection capabilities of ML models, they are continuously adapting to new threats and improving the effectiveness of antivirus solutions.

AI and ML have become obligatory in the field of cybersecurity [35]. These technologies offer superior capabilities for identification of threats, analysis, and response, allowing organizations to enhance their security posture in the face of evolving cyber threats. In today's increasingly complex digital environment, cyber security experts can use AI and machine learning to detect and mitigate risks more effectively, reduce the response times and remain at least one step ahead of those responsible for malicious activities.

1.3 Applications of AI, ML and Block Chain in Cybersecurity

These technologies have numerous applications across various cybersecurity domains, enabling organizations to enhance their security posture and effectively combat evolving cyber threats [36].

(a) **Threat Detection and Analysis**:

The threat detection and analysis is done by using AI with ML [37]. It is implemented with continuously learning from historical data, these algorithms can improve their accuracy and efficiency over time, enabling the early detection of cyber-attacks [38].

Using an artificial intelligence and blockchain methodology, Fig. 2 shows the detection and analysis of threats. The way that organizations approach the detection and defence of cyber threats has been transformed by these cutting edge technologies. Cybersecurity professionals can enhance their ability to detect threats, react more quickly and build resilience defenses through the unique capabilities of Artificial Intelligence, Machine Learning or Blockchain. Below, we delve into how each technology contributes to the process of threat detection and analysis:

(i). **Artificial Intelligence (AI) in Threat Detection**: AI in cybersecurity involves the application of intelligent algorithms that mimic human cognition to process vast amounts of data and detect for potential threats. The various purposes in threat detection using AI are [37].

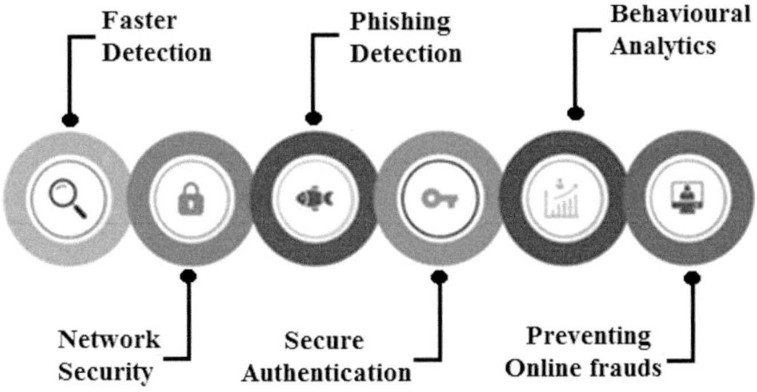

Fig. 2 Threat detection and analysis

- ***Behavior Analysis***: AI-powered systems can establish a baseline of normal behavior for users, systems, and networks. Any deviation from this baseline can be flagged as suspicious, potentially indicating an ongoing attack.
- ***Anomaly Detection***: AI algorithms can identify abnormal patterns or activities that do not match known signatures or established behavior. This helps detect zero-day attacks and novel threats that traditional signature-based systems may miss.
- ***Predictive Analytics***: The technology can analyze chronological data to predict potential future threats and vulnerabilities, enabling organizations to take proactive measures to prevent attacks.
- ***Natural Language Processing (NLP)***: This can be employed to analyze unstructured data.

(ii). **Machine Learning (ML) in Threat Detection**: The training algorithms are learned from the available data and also its performance is improved with ML. ML techniques play a vital task in cybersecurity threat detection as well as analysis [37]:

- ***Signature-less Detection***: ML algorithms can identify unknown threats by learning from past incidents and recognizing patterns that indicate malicious behavior.
- ***Classification and Clustering***: ML models can categorize threats into different classes based on their attributes, helping security teams prioritize and respond to incidents more effectively.
- ***Automated Response***: ML-powered systems can autonomously respond to low-level threats, reducing the burden on security personnel and enabling faster incident resolution.
- ***Data Preprocessing***: ML can assist in data cleaning, transformation, and feature extraction, which are crucial for preparing data for analysis.

(iii). **Blockchain in Threat Detection**: Blockchain technology, although initially associated with cryptocurrency, holds promise for enhancing threat detection and analysis in cybersecurity [38]:

- ***Immutable Logging***: Blockchain's decentralized and tamper-resistant nature can be utilized for secure and reliable logging of cybersecurity events and incidents.
- ***Distributed Threat Intelligence***: Blockchain can facilitate secure sharing of threat intelligence among organizations, enabling collective defense against cyber threats.
- ***Secure Identity Management***: Blockchain's cryptographic techniques can enhance identity and access management, reducing the risk of unauthorized access and identity-based attacks.

(iv). **Integration of AI, ML, and Blockchain**: The true power of threat detection and analysis in cybersecurity lies in the integration of AI, ML, and Blockchain technologies [38]:

- AI and ML can be combined to build dynamic and adaptive threat detection systems that learn from new data and adjust their models to stay ahead of evolving threats.
- AI-driven anomaly detection can be applied to Blockchain networks to identify suspicious activities, ensuring the integrity and security of distributed ledgers.
- Blockchain can be utilized to store and manage large-scale threat intelligence databases securely, while AI and ML can process and analyze this data for valuable insights.
- Smart contracts on a blockchain can autonomously trigger incident response actions based on predefined criteria, reducing response time and minimizing human error.

Cybersecurity practices have become radically altered by the integration of artificial intelligence, machine learning and Blockchain technology into threat detection and analysis. The convergence presents an opportunity for organizations to set up more responsive, adaptive and robust defense mechanisms in the face of rapidly emerging cyber threats which enable them to enhance their Digital Asset Protection. In the future, there is a promise that more advanced and efficient detection and analysis of threats will be possible, thanks to continuous advances in these technologies.

(b) **Intrusion Detection and Prevention Systems**:

IDPS is controlled [39] by AI and ML techniques to monitor network traffic in real-time. These systems can identify and block malicious activities by analyzing network packets and comparing them against known attack patterns [40]. ML models can also detect previously unseen attacks using anomaly detection algorithms, identifying deviations from normal behavior and raising alerts for further investigation.

Intrusion Detection and Prevention Systems (IDPS) play a crucial role in cybersecurity by monitoring network traffic and are shown in Fig. 3 and systems for signs of malicious activities and responding to potential threats. This technology in IDPS enhances their capabilities, making them more efficient, accurate, and resistant to tampering. Let's explore how these technologies contribute to IDPS:

(i) **AI in IDPS**: AI technologies provide IDPS with the ability to process large volumes of data and make intelligent decisions based on patterns and trends. Some ways where AI is utilized in IDPS [39] include:

- ***Pattern Recognition***: AI can identify complex patterns of network behavior associated with various types of attacks, enabling rapid threat detection.
- ***Predictive Analytics***: AI can analyze historical data and predict potential cyber threats, helping organizations take proactive measures to prevent attacks.
- ***Dynamic Rule Generation***: AI can autonomously generate and modify detection rules based on real-time data, adapting to new threats without manual intervention.

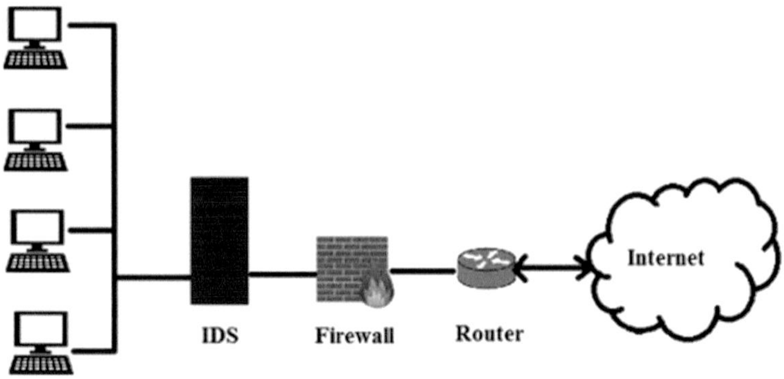

Fig. 3 Intruder detection system

(ii) **ML in IDPS**: ML complements AI in IDPS by enabling systems to learn from existing information and rise the performance w.r.t time. Some applications of ML in IDPS include [39]:

- ***Anomaly Detection***: ML algorithms can identify abnormal network activities that deviate from normal behavior, detecting zero-day attacks and novel threats.
- ***Traffic Analysis***: ML can classify network traffic and distinguish between legitimate traffic and suspicious or malicious activities.
- ***Signature-less Detection***: ML can detect threats without relying on predefined signatures, making IDPS more adaptive to new and unknown threats.
- ***Incident Severity Prediction***: ML models can predict the severity of detected incidents, allowing security teams to prioritize responses efficiently.

(iii) **Blockchain in IDPS**: Blockchain technology enhances the security and integrity of IDPS by providing a decentralized, immutable, and tamper-resistant platform. Key applications of Blockchain in IDPS include [39]:

- ***Secure Logging***: Blockchain's distributed and append-only nature ensures that logs of security events and incidents remain tamper-proof and trustworthy.
- ***Threat Intelligence Sharing***: Blockchain enables secure sharing of threat intelligence among different entities, fostering collaborative defense against cyber threats.
- ***Identity and Access Management***: Blockchain's cryptographic features can enhance identity management, reducing the risk of unauthorized access and insider threats.
- ***Smart Contract-based Response***: Blockchain's smart contracts can automate predefined responses to specific security events, improving incident response time and accuracy.

(iv) **Integration of AI, ML, and Blockchain in IDPS**: The integration of AI, ML, and Blockchain in IDPS creates a powerful defense system that is capable of providing more robust protection against cyber threats [40]:

- AI and ML can process large volumes of network data, identify patterns, and detect anomalies, enabling more accurate and timely threat detection.
- Blockchain ensures the integrity of IDPS logs, preventing attackers from tampering with critical data and covering their tracks.
- AI-driven anomaly detection and ML-based classification in IDPS can benefit from the immutability of Blockchain, which safeguards the integrity of the system's decision-making process.
- Threat intelligence sharing among different organizations via Blockchain can enhance the collective knowledge and improve the overall effectiveness of IDPS.

In order to counter sophisticated cyber threats, the integration of Artificial Intelligence, Machine Learning and Blockchain into intrusion detection and prevention systems such as IDNPS strengthens their capabilities. These technologies allow the Integrated Data Protection System to detect and respond more effectively to security incidents, provide accurate threat analyses as well as improve cooperation between cybersecurity stakeholders. By continually advancing these integrated solutions, organizations can bolster their cybersecurity posture and safeguard their digital assets in an ever-evolving threat landscape.

(c) **User and Entity Behavior Analytics (UEBA)**:

UEBA is another important application of AI and ML in cybersecurity [39]. UEBA systems establish baseline behavior patterns for users and entities within an organization's network. By monitoring and analyzing user activities, network access, and data interactions, AI and ML algorithms can detect anomalous behavior of user accounts [41]. It helps organizations identify suspicious activities and take proactive measures to prevent data breaches or unauthorized access.

UEBA is a cybersecurity approach that leverages the power of the technologies to detect and respond to insider threats, advanced persistent threats (APTs), and other sophisticated cyber-attacks. UEBA focuses on monitoring and analyzing the behavior of users and entities within an organization's network to identify abnormal or malicious activities. The integration of AI, ML, and Blockchain in UEBA enhances its capabilities, making it more effective and secure. Let's explore how these technologies contribute to UEBA:

(i) **AI in UEBA**: AI technologies in UEBA enable the system to process large data for patterns detection with anomalies in user and entity behavior. Some key applications of AI in UEBA include [40, 41]:

- ***Anomaly Detection***: AI algorithms can identify unusual activities or deviations from normal behavior, helping to identify potential insider threats or compromised accounts.

- ***Contextual Analysis***: AI can analyze the context of user actions and interactions with the network to distinguish between legitimate activities and suspicious behaviors.
- ***Dynamic Baselines***: AI can create dynamic baselines of user and entity behavior, adapting to changes in behavior patterns over time.
- ***Real-time Decision Making***: AI-driven UEBA systems can make real-time decisions, providing immediate alerts or responses to potential security incidents.

(ii) **ML in UEBA**: ML complements AI in UEBA by enabling the system to learn from data and improve its detection capabilities. Some applications of ML in UEBA include [41]:

- ***User Profiling***: ML algorithms can create profiles of user behavior based on historical data, helping to identify deviations from normal behavior.
- ***Pattern Recognition***: ML can recognize patterns associated with known attack methods, enhancing the system's ability to detect specific threats.
- ***Risk Scoring***: ML models can assign risk scores to users and entities based on their behavior, facilitating the prioritization of potential threats.
- ***Predictive Analytics***: ML can predict future behaviors based on historical data, allowing security teams to proactively address potential security risks.

(iii) **Blockchain in UEBA**: Blockchain technology strengthens UEBA by providing a secure and tamper-resistant policy for storing & sharing sensitive data. Key applications of Blockchain in UEBA include [40, 41]:

- ***Secure Data Storage***: Blockchain's decentralized and immutable nature ensures the integrity and security of user and entity behavior data.
- ***Identity Verification***: Blockchain's cryptographic features can enhance identity management, verifying the authenticity of user identities and reducing the risk of impersonation.
- ***Threat Intelligence Sharing***: Blockchain enables secure and anonymous sharing of threat intelligence among different organizations, enhancing the collective defense against cyber threats.

(iv) **Integration of AI, ML, and Blockchain in UEBA**: The integration of AI, ML, and Blockchain in UEBA creates a robust and efficient system to detect insider threats and other cyber-attacks [41]:

- AI and ML enable UEBA to process and analyze vast amounts of behavioral data, detecting subtle patterns and anomalies that may indicate potential threats.
- Blockchain ensures the integrity and security of user behavior data, preventing unauthorized access or modification of sensitive information.
- AI-driven UEBA systems can benefit from the tamper-resistant nature of Blockchain, safeguarding the accuracy and reliability of the system's decision-making process.

- Secure threat intelligence sharing via Blockchain enables organizations to gain insights into emerging threats and strengthen their overall cybersecurity posture.

These technologies, provide organizations with advanced capabilities to detect and respond to insider threats and sophisticated cyber-attacks. By continuously improving and advancing these integrated solutions, organizations can stay ahead of evolving threats and protect their critical assets from illicit access activities.

(d) **Malware Analysis and Antivirus Solutions**:

In analyzing malware and antivirus solutions, artificial intelligence and machine learning have also a key role to play. Malware analysis [42] is the study of malicious software, with a view to understanding its behavior, capabilities and potential impact. Such analysis is automated through the use of artificial intelligence and machine learning technologies [43], enabling faster detection and classification of computer viruses than conventional methods are able to do today. ML models can continuously learn from new threats, improving the effectiveness of antivirus solutions by enhancing their detection capabilities [44]. Additionally, AI and ML can be used in security information and event management (SIEM) [45] systems to comprehensive and investigate security logs. By applying AI and ML algorithms to SIEM data, organizations can gain insights into potential security incidents, identify patterns of attacks, and prioritize response efforts. Their applications span various domains, including threat detection and analysis, intrusion detection and prevention [46]. By leveraging AI and ML [47], organizations can bolster their cybersecurity defenses, detect and mitigate risks more effectively, and stay ahead of rapidly evolving cyber threats.

1.4 Blockchain and Cybersecurity

Blockchain technology [48], best known as the essential infrastructure for cryptocurrencies like Bitcoin for its potential to enhance cybersecurity. Blockchain's decentralized and immutable nature offers several security benefits, making it an attractive solution for various cybersecurity applications. The basic diagram for block chain in cyber security is shown in Fig. 4.

(a) **Data Integrity and Immutable Audit Trails**:

One of the key advantages [49] of Blockchain in cybersecurity is data integrity. Blockchain creates a tamper-proof and transparent ledger where data can be recorded in a decentralized manner. Every block consists of a cryptographic hash that links it to the prior block, in order to create a secure and verifiable record. This provides [50] assurance of data integrity. The diagram for blockchain audit trail mechanism is shown in Fig. 5.

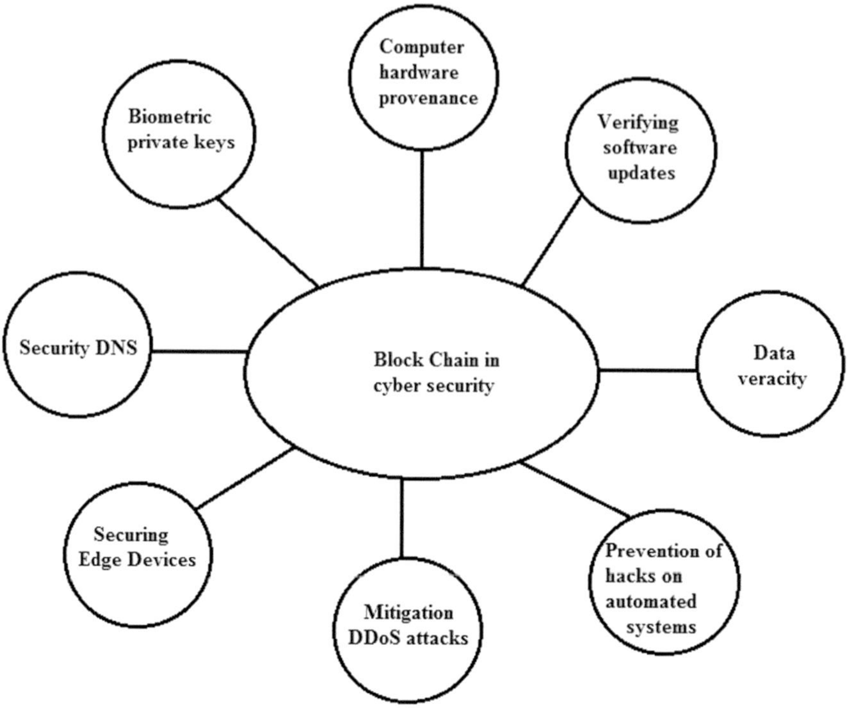

Fig. 4 Block chain in cyber security

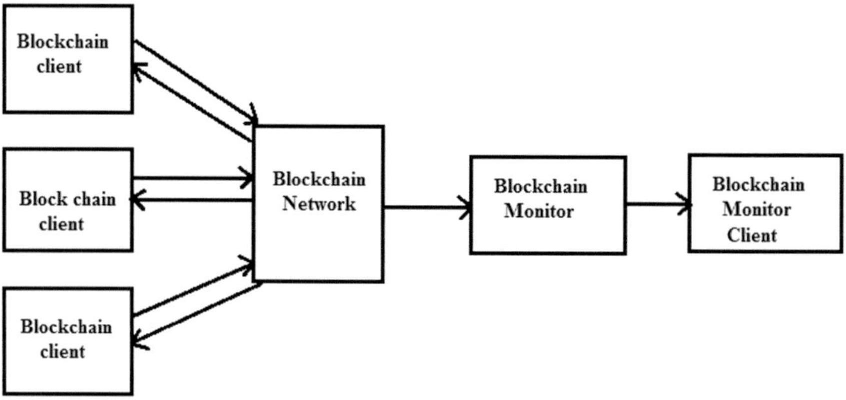

Fig. 5 Blockchain audit trail mechanism

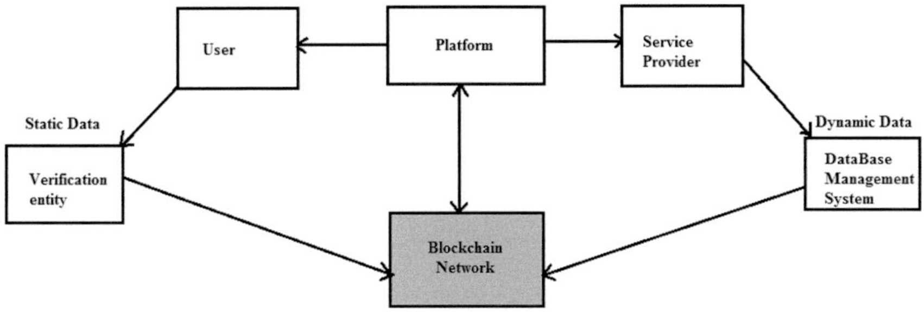

Fig. 6 Blockchain based secure identity and access management

(b) **Secure Identity and Access Management**:

In the area of security for identity and access management [51], bitcoin can also play an important role. A breach or identity theft is likely to occur in traditional centralized ID systems. Decentralized authentication mechanisms for the purpose of safeguarding identities are provided in Blockchain Based Identity Management Systems. With Blockchain, users can control their digital identities, avoids the risk of unauthorized access and theft of uniqueness [52] and is shown in Fig. 6.

(c) **Supply Chain Security**:

Furthermore, Blockchain technology can enhance supply chain security [52]. By leveraging Blockchain, organizations can establish transparent and verifiable tracking of goods and transactions. The diagram for supply chain security is shown in Fig. 7.

In terms of data sharing and verification [53], Blockchain provides a secure and transparent infrastructure. Blockchain-based smart contracts can enforce access control policies, facilitating secure data sharing among parties. Smart contracts ensure that data is accessed and shared based on predefined rules and conditions, avoiding the peril [54] of unauthorized information access and manipulation.

Blockchain technology also presents its own set of challenges [55, 56], including scalability, privacy concerns, and the potential for new vulnerabilities. Organizations must carefully evaluate the suitability of Blockchain for their specific cybersecurity needs and consider the trade-offs associated with its implementation [57].

Blockchain technology has the potential to enhance cybersecurity by providing data integrity [58], secure identity management, supply chain security, and transparent data sharing. It has a unique security advantage, which makes it an attractive solution for various cybersecurity applications because of its decentralized and immutable nature. However, organizations [59] must consider the specific requirements and challenges associated with Blockchain implementation to ensure its effective integration into their cybersecurity strategies.

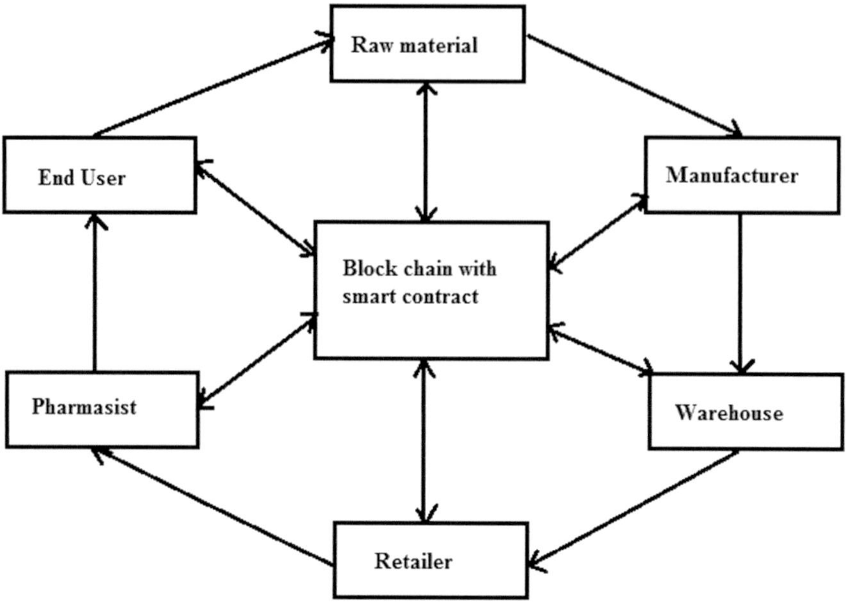

Fig. 7 Blockchain in supply chain security

1.5 Convergence of AI, ML, and Blockchain in Cybersecurity:

The convergence of these technologies [60] holds significant promise for strengthening cybersecurity. Each of these technologies brings unique capabilities to the table, and their integration offers new opportunities for enhancing security, detecting and mitigating threats, and ensuring data integrity. In this section, we will see the sights of the convergence of AI, ML, and Blockchain in the cybersecurity and discuss how they can complement and reinforce each other's strengths [61] as shown in Fig. 8.

AI and ML algorithms can play a crucial role [62] in analyzing the vast amount of data generated by Blockchain networks. Blockchain, with its decentralized and immutable nature, provides a transparent and secure infrastructure for storing data.

One area where the convergence of [63] AI, ML, and Blockchain can be particularly impactful is in threat detection and analysis. AI and ML algorithms can analyze data stored on the Blockchain, such as transaction records and network activity, to identify patterns and detect potential threats. By continuously learning from historical data and identifying abnormal behaviors or suspicious activities, AI and ML algorithms can provide early warnings and enhance threat detection capabilities. This can enable organizations to take proactive measures to prevent or mitigate the impact of cyber attacks. Blockchain provides a decentralized and secure [62] platform for managing digital identities, while AI and ML techniques can analyze user behavior and access patterns to detect unauthorized access attempts or unusual

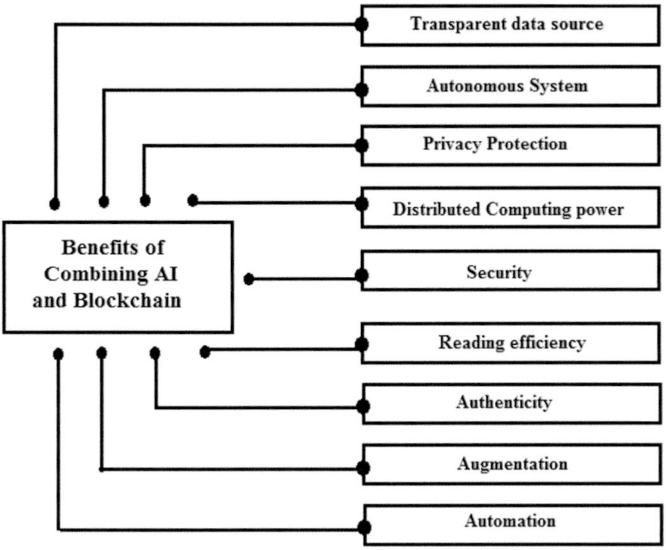

Fig. 8 Linking AI and blockchain

activities. In addition, organizations can build strong authentication mechanisms and detect potential fraud or misuse of access credentials through a combination of Blockchain's secure identity management with artificial intelligence and machine learning algorithms.

Additionally, the convergence of AI, ML, and Blockchain can improve the security and privacy of data sharing and collaboration [64]. AI and ML algorithms can analyze data shared through smart contracts to identify potential security risks or anomalies. By leveraging Blockchain's transparency and immutability, organizations can ensure the integrity and authenticity of shared data, while AI and ML algorithms can enhance data privacy by anonymizing or encrypting sensitive information. Blockchain can also endow with sheltered and transparent communications [65] for storing AI and ML models. AI and ML models can be vulnerable to attacks or tampering, as their accuracy and effectiveness rely on the integrity and confidentiality of the underlying data and algorithms. By storing AI and ML models on the Blockchain, organizations can ensure the immutability and integrity of these models, reducing the risk of unauthorized modifications or tampering. This can enhance the trustworthiness and reliability of AI and ML-based cybersecurity systems.

There are still many challenges to be tackled, despite the considerable opportunities for convergence of Artificial Intelligence, Machine Learning and Blockchain in cyber security. The privacy of the data stored in Blockchains is a particular challenge [66]. While it provides transparency and immutableness, the risk of disclosure of sensitive information is also raised by Blockchain. The privacy implications of protecting data need to be properly considered by enterprises and appropriate encryption or anonymization techniques should be implemented.

Another challenge is the need for robust governance frameworks and standards [67] for AI, ML, and Blockchain integration in cybersecurity. As such technologies evolve, it is essential to develop guidelines and best practices aimed at ethics and responsibility in their use. This includes addressing bias in AI algorithms, ensuring transparency in decision-making processes, and establishing accountability for the outcomes of AI and ML-based cybersecurity systems.

The convergence of these technologies in cybersecurity offers exciting opportunities [68] for enhancing security, threat detection, and data integrity. By leveraging the strengths of these technologies, organizations can build more resilient and intelligent security systems. However, addressing privacy concerns, establishing governance frameworks, and promoting responsible use are crucial for harnessing the full potential of the convergence of AI, ML, and Blockchain in cybersecurity.

1.6 Challenges and Future Directions

1.6.1 Challenges

The convergence of AI, ML, and Blockchain in cybersecurity presents several challenges that need to be addressed for successful implementation [69, 70]. These challenges include:

(a) **Privacy and Confidentiality Concerns**: The transparent and immutable nature of Blockchain raises concerns [71] about the exposure of sensitive information. Protecting privacy and ensuring confidentiality of encryption or anonymization techniques.
(b) **Ethical Considerations and Bias in AI**: Ensuring ethical use of AI and ML algorithms is crucial [72]. Bias in AI algorithms, whether due to biased training data or algorithmic design, can have serious implications. Organizations must address biases, establish governance frameworks, and adopt ethical AI principles to ensure fairness, transparency, and accountability.
(c) **Scalability and Performance**: As the volume of data continues to grow, organizations must ensure their infrastructure [73] can handle the computational demands of AI and ML algorithms. As far as transaction speed and duration are concerned, a blockchain network is facing challenges. In order to overcome these problems, R&D efforts should be focused on the development of scalable architectures and optimization algorithms.
(d) **Standardization and Interoperability**: Establishing standards and protocols to integrate Artificial Intelligence, Machine Learning or Blockchain into cybersecurity is essential in order for these technologies to be widely adopted and interoperable [74]. To facilitate the integration and ensure a seamless interchange between different systems, common frameworks, data formats and interoperable APIs are available.

(e) **Evolving Threat Landscape**: The threat landscape in cybersecurity [74] is dynamic, with new attack vectors and techniques emerging constantly. Research and development efforts should focus on staying ahead of emerging threats and developing proactive AI and ML algorithms that can detect and mitigate novel attack patterns.
(f) **Resilience and Adversarial Attacks**: Adversarial attacks pose a significant challenge [73] in the application of AI and ML in cybersecurity. Building robust and resilient AI models that can detect and defend against such attacks requires research into adversarial training, model interpretability, and anomaly detection techniques.

1.6.2 Future Directions

In addition to addressing the challenges, several future directions can enhance the convergence of AI, ML, and Blockchain in cybersecurity. These directions include:

(a) **Advancing Privacy-Preserving Techniques**: Research should focus on developing more sophisticated privacy-preserving techniques [75] to ensure confidentiality while leveraging the benefits of AI, ML, and Blockchain.
(b) **Trustworthy AI Systems**: Future efforts should prioritize the development of trustworthy AI systems [76] that uphold ethical considerations, fairness, transparency, and accountability in their operations.
(c) **Scalability and Performance Optimization**: Continued research is needed to enhance the scalability and performance [77] of AI, ML, and Blockchain systems to handle the increasing volume of data and transactions effectively.
(d) **Establishment of Standards and Protocols**: Industry-wide collaboration is necessary to establish standards and protocols [77] that enable interoperability and seamless integration of AI, ML, and Blockchain technologies across different cybersecurity systems.
(e) **Proactive Threat Mitigation**: Research should focus on proactively identifying and mitigating emerging threats [78] through advanced AI and ML algorithms that can detect and respond to novel attack patterns effectively.
(f) **Defense Against Adversarial Attacks**: Future directions should include strengthening defenses against adversarial attacks through robust AI models [79], adversarial training, model interpretability, and anomaly detection techniques.

2 Conclusion

We conclude here with the use of the above discussed technologies in cybersecurity holds immense potential for enhancing security, threat detection, and data integrity. With the strengths of these technologies and addressing the associated challenges, organizations can build robust and intelligent cybersecurity systems. AI and ML algorithms are capable of processing large volumes of data to identify patterns, anomalies, and potential threats. By continuously learning from historical data, these algorithms can improve their accuracy and efficiency, providing early warnings and enhancing threat detection capabilities. Blockchain technology offers unique security benefits, such as data integrity and tamper-proof records. This provides a transparent and secure infrastructure for various cybersecurity applications, including secure identity management, supply chain security, and transparent data sharing.

However, privacy and confidentiality concerns must be carefully considered, and appropriate encryption or anonymization techniques should be implemented to protect sensitive data. Ethical considerations, such as addressing bias in AI algorithms and ensuring transparency and accountability, are crucial for responsible AI usage.

Scalability and performance optimization are key areas of focus to handle the increasing volume of data and transactions in AI, ML, and Blockchain systems. Standardization and interoperability efforts are necessary to establish common frameworks and protocols for seamless integration across different cybersecurity systems. Looking ahead, future research and development efforts should aim to enhance privacy-preserving techniques, develop trustworthy AI systems, optimize scalability and performance, establish industry-wide standards and protocols, proactively mitigate emerging threats, and strengthen defenses against adversarial attacks.

By investing in these areas and exploring new avenues for innovation, organizations can yoke the full potential of these technologies in cybersecurity. This will enable them to build resilient, intelligent, and secure systems that effectively address the evolving cyber threat landscape. The integration of AI, ML, and Blockchain is not a standalone solution, but a powerful combination that can significantly enhance cybersecurity and pave the way for a more secure digital future.

References

1. Anderson R (2018) Security engineering: a guide to building dependable distributed systems. Wiley
2. Bishop M (2021) Computer security: art and science. Addison-Wesley
3. Schneier B (2015) Data and goliath: the hidden battles to collect your data and control your world. W. W. Norton & Company
4. Zhang H, Zhu S, Ghosh A (eds) (2021) Handbook of blockchain security: a technical guide to secure distributed ledgers. Springer
5. Buchanan W, Blyth A (eds) (2020) Cyber security handbook. Springer.
6. Cavoukian A, Castro J (2019) The rise of blockchain technology in securing the internet of things. Eur Data Prot Law Rev 5(2):226–241
7. Dhillon G, Moores T (2019) Principles of information security. Cengage Learning
8. Ford R, Sorell M, Traynor P (2019) Cybersecurity: protecting critical infrastructures from cyber attack and cyber warfare. Wiley.
9. Goodrich MT, Tamassia R (2021) Introduction to computer security. Pearson.
10. Janczewski L, Colarik A (2019) Cyber warfare and cyber terrorism. Springer

11. Kim D (2018) Cybersecurity and blockchain technology. In: The international conference on information networking (ICOIN).
12. Kouicem DE, Guesmi R (2019) Blockchain technology for secure cyber physical systems: a survey. Futur Gener Comput Syst 97:512–533
13. Liao CH, Lu YC, Huang CY (2020) Enhancing privacy and security in cloud-based medical systems using blockchain. Futur Gener Comput Syst 105:368–377
14. Mitnick KD, Simon WL (2017) The art of invisibility: the world's most famous hacker teaches you how to be safe in the age of big brother and big data. Little, Brown and Company.
15. Ross SM (2019) Computer security fundamentals. Pearson.
16. Samaniego M, Larrabeiti D, Díaz V (2019) Security in IoT communications based on blockchain technology. Sensors 19(6):1402
17. Schreiber P, Omerzu M, Kompara M (2020) Blockchain in cybersecurity: a systematic mapping study. Security and Communication Networks
18. Shariq S, Ahmad I (2021) Machine learning in cybersecurity: a survey. In: Machine learning for cybersecurity. Springer, pp 1–32
19. Vacca JR (2019) Computer and information security handbook. Morgan Kaufmann
20. Yampolskiy M (2018) Artificial superintelligence: a futuristic approach. CRC Press.
21. Chen M, Hao Y, Ma X (2020) Applications of machine learning in cybersecurity. IEEE Trans Netw Sci Eng 7(2):998–1010
22. Yadav P, Agarwal A (2020) A Comprehensive survey of deep learning in cybersecurity. J Inf Secur Appl 50:102419
23. Zhao Y, Ge L (2020) Machine learning-based cybersecurity solutions: a survey. Comput Secur 88:101670
24. Akhtar Z, Khan A, Mehdi S (2020) Applications of artificial intelligence in cybersecurity: a review. J Cybersecur 6(1):tyaa009.
25. Kambourakis G, Maragoudakis M (eds) (2020) Artificial intelligence applications in cyber security. Springer.
26. Thomas D (2021) Artificial intelligence in cybersecurity: a guide to detection, response, and prevention. Apress.
27. Nasseh S, Moallem A (2020) An overview of machine learning applications in cybersecurity. In: Proceedings of the 10th international conference on web intelligence, Mining and Semantics. pp 1–6.
28. Qin C, Cheng X (2020) Machine learning methods for cybersecurity. J Cybersecur 6(1):tyy027
29. Gandomi A, Haider M (2019) Beyond the hype: big data concepts, methods, and analytics. Int J Inf Manage 35(2):137–144
30. Rong C, Huang W (2018) A survey of advances in deep learning for cybersecurity. Inf Fusion 46:35–49
31. Qu Z, Zhang R (2020) Machine learning in cybersecurity: a comprehensive survey. IEEE Commun Surv & Tutor 22(3):1842–1872
32. Sitnikova E, Tulumenkov A, Ometov A (2019) Machine learning in network security: a comprehensive survey. IEEE Commun Surv & Tutor 21(3):2702–2733
33. Zhang X, Zhang Z, Wang H (2020) Machine Learning-based cybersecurity systems: a survey. Futur Gener Comput Syst 107:776–797
34. Sun Y, Liu X, Du X (2019) A survey of machine learning techniques applied to network security. Secur Commun Netw 2019:1–21
35. Bolon-Canedo V, Sanchez-Maroño N, Alonso-Betanzos A (2018) A review of feature selection methods in medical applications. Comput Biol Med 2018:279–291
36. Barbosa MA, de Souza RF, da Silva EP (2021) Artificial intelligence and machine learning for cyber security. Comput Secur 113:102363. https://doi.org/10.1016/j.cose.2021.102363
37. Choudhary KS, Kaushik A (2022) Artificial intelligence in cyber security: a review. ACM Comput Surv (CSUR) 55(3):1–33. https://doi.org/10.1145/3507490
38. Dash S, Mishra P, Rathore S (2022) Artificial intelligence in cyber security: a comprehensive survey. J Netw Comput Appl 186:103090. https://doi.org/10.1016/j.jnca.2022.103090

39. Dwivedi A, Jain R (2022) Artificial intelligence-based security solutions for cloud computing. Comput Secur 114:102469. https://doi.org/10.1016/j.cose.2022.102469
40. Garcia JC, Martinez JJ (2022) Artificial intelligence in cyber security: a systematic review. Comput Secur 115:102504. https://doi.org/10.1016/j.cose.2022.102504
41. Kaur A, Singh A (2022) Artificial intelligence and machine learning in cyber security: a review of literature. Secur J 35(1):145–166. https://doi.org/10.1057/s41284-021-00312-0
42. Kumar V, Kaur I (2022) Artificial intelligence and machine learning for cyber security: a state-of-the-art survey. Security and Privacy 10(1):3–19. https://doi.org/10.1002/sp.2168
43. Lakshmi V, Kumar S (2022) Artificial intelligence for cyber security: a survey. Comput Secur 116:102545. https://doi.org/10.1016/j.cose.2022.102545
44. Mishra S, Singh S, Sharma AK (2022) Artificial intelligence in cyber security: a comprehensive review. J Inf Secur 13(1):25–50. https://doi.org/10.1016/j.jis.2022.01.001
45. Patel M, Patel K (2022) Artificial intelligence in cyber security: a review. Int J Inf Secur 21(2):407–427. https://doi.org/10.1007/s10207-021-00463-y
46. Ranganathan S, Manjunath GK (2022) Artificial intelligence for cyber security: a survey and research directions. ACM Comput Surv (CSUR) 55(4):1–41. https://doi.org/10.1145/3507495
47. Saxena A, Jain R (2022) Artificial intelligence in cyber security: a state-of-the-art survey. IEEE Access 10:78105–78129. https://doi.org/10.1109/ACCESS.2022.3152332
48. Alam M, Ahad MA, Hossain MA (2022) Blockchain-based security solutions: a survey. Comput Secur 115:102505. https://doi.org/10.1016/j.cose.2022.102505
49. Bartoletti M, Lamberti F, Pompianu L, Satta G (2022) Blockchain security: a survey of current research. IEEE Commun Surv & Tutor 24(2):1603–1634. https://doi.org/10.1109/COMST.2021.3059155
50. Chen WY, Chen P, Lin LC (2022) A survey of blockchain security. ACM Comput Surv (CSUR) 55(3):1–33. https://doi.org/10.1145/3507489
51. Ellul J, Karame GO (2022) Blockchain security: a survey of current research and trends. Front Blockchain 5:68. https://doi.org/10.3389/fbloc.2022.787270
52. Han Z, Wang X, Zhou K, Wang L (2022) Blockchain security: a survey. AIEEE Access 10:6977–6992. https://doi.org/10.1109/ACCESS.2022.3152331
53. Jiang L, Zhang C, Wang X (2022) A survey on blockchain security: current status, challenges and opportunities. Secur Priv 10(1):20–36. https://doi.org/10.1002/sp.2170
54. Kamble VS, Patil S, Patil S (2022) Blockchain security: a comprehensive review. J Inf Secur 13(1):51–78. https://doi.org/10.1016/j.jis.2022.01.002
55. Khodaei A, Dehghantanha A (2022) A survey on blockchain security: challenges, solutions, and open problems. Comput Secur 116:102546. https://doi.org/10.1016/j.cose.2022.102546
56. Kumar V, Kaur I (2022) Blockchain security for industrial IoT: a systematic literature review. IEEE Syst J 16(2):815–832. https://doi.org/10.1109/JSYST.2021.3083220
57. Liu Y, Zhang W, Li R (2022) A survey on blockchain security: current status, challenges and opportunities. Secur Priv 10(1):37–52. https://doi.org/10.1002/sp.2169
58. Mishra S, Singh R (2022) Blockchain security: a comprehensive review. J Inf Secur 13(1):79–100. https://doi.org/10.1016/j.jis.2022.01.003
59. Nguyen TD, Chang H (2022) Blockchain security: a survey of current research and trends. IEEE Access 10:6969–6976. https://doi.org/10.1109/ACCESS.2022.3152330
60. Anoop R, Rathee A (2023) Blockchain and artificial intelligence: the convergence of two powerful technologies. Int J Inf Secur 22(1):3–19. https://doi.org/10.1007/s10207-022-00535-3
61. Bhardwaj A, Gupta A (2022) Convergence of artificial intelligence, machine learning, and blockchain: a survey of emerging security solutions. Comput Secur 116:102547. https://doi.org/10.1016/j.cose.2022.102547
62. Chang H, Lee J (2022) The convergence of artificial intelligence, machine learning, and blockchain for cybersecurity. IEEE Access 10:78129–78144. https://doi.org/10.1109/ACCESS.2022.3152333
63. Dwivedi A, Jain R (2022) Artificial intelligence and blockchain: a convergence for next-generation cybersecurity. Comput Secur 115:102503. https://doi.org/10.1016/j.cose.2022.102503

64. Gupta R, Kumar S (2022) Artificial intelligence, machine learning, and blockchain: convergence and implications for cybersecurity. IEEE Syst J 16(3):1668–1680. https://doi.org/10.1109/JSYST.2022.3189369
65. Kumar V, Kaur I (2022) Convergence of artificial intelligence, machine learning, and blockchain for cybersecurity: a systematic literature review. Inf Sci 565:1314–1336. https://doi.org/10.1016/j.ins.2022.01.068
66. Mishra S, Saini A, Singh S (2023) Convergence of artificial intelligence, machine learning, and blockchain for cybersecurity: a systematic review. J Netw Comput Appl 189:103173. https://doi.org/10.1016/j.jnca.2022.103173
67. Rathee A, Anoop R (2022) The convergence of artificial intelligence, machine learning, and blockchain: a survey of security challenges and solutions. Comput Secur 115:102502. https://doi.org/10.1016/j.cose.2022.102502
68. Sharma AK, Mishra S, Singh S (2022) Convergence of artificial intelligence, machine learning, and blockchain: a systematic review of security and privacy challenges. Secur Priv 10(1):1–24. https://doi.org/10.1002/sp.2171
69. Singh M, Kaur J (2022) Convergence of artificial intelligence, machine learning, and blockchain: a survey of security and privacy challenges. Secur J 35(2):299–319. https://doi.org/10.1057/s41284-022-00313-1
70. Alotaibi B, Alghamdi A, Yaseen M (2022) Security challenges and future directions in the internet of things. Comput Secur 116:102548. https://doi.org/10.1016/j.cose.2022.102548
71. Anoop R, Rathee A (2023) Challenges and future directions in blockchain security. Int J Inf Secur 22(2):20–41. https://doi.org/10.1007/s10207-022-00536-2
72. Chang H, Lee J (2022) Challenges and future directions in artificial intelligence-based cybersecurity. IEEE Access 10:78145–78160. https://doi.org/10.1109/ACCESS.2022.3152334
73. Dwivedi A, Jain R (2023) Challenges and future directions in cybersecurity for cloud computing. Comput Secur 116:102549. https://doi.org/10.1016/j.cose.2022.102549
74. Gupta R, Kumar S (2022) Challenges and future directions in cybersecurity for internet of things. IEEE Syst J 16(3):1681–1692. https://doi.org/10.1109/JSYST.2022.3189370
75. Kumar V, Kaur I (2022) Challenges and future directions in cybersecurity for artificial intelligence. Inf Sci 565:1337–1359. https://doi.org/10.1016/j.ins.2022.01.069
76. Mishra S, Singh S (2023) Challenges and future directions in cybersecurity for blockchain. J Netw Comput Appl 189:103174. https://doi.org/10.1016/j.jnca.2022.103174
77. Rathee A, Anoop R (2022) Challenges and future directions in cybersecurity for artificial intelligence, machine learning, and blockchain. Comput Secur 115:102501. https://doi.org/10.1016/j.cose.2022.102501
78. Sharma AK, Mishra S, Singh S (2022) Challenges and future directions in cybersecurity for artificial intelligence, machine learning, and blockchain: a systematic review. Secur Priv 10(1):25–50. https://doi.org/10.1002/sp.2170
79. Singh M, Kaur J (2022) Challenges and future directions in cybersecurity for artificial intelligence, machine learning, and blockchain: a survey and research directions. ACM Comput Surv (CSUR) 55(4):1–41. https://doi.org/10.1145/3507495

Opportunities and Challenges in New Generation Cyber Security Applications Using Artificial Intelligence, Machine Learning and Block Chain

Sarthak Malik, Praveen Kumar Malik, and Arshi Naim

Abstract This chapter explores the dynamic landscape of new-generation cyber security applications, delving into the integration of Artificial Intelligence (AI), Machine Learning (ML), and Block chain technologies. The opportunities presented by this amalgamation are significant, with AI and ML offering unparalleled capabilities in adaptive threat detection, predictive analysis, and automated response mechanisms. The decentralized and tamper-resistant nature of Block chain adds transparency and integrity to data security and transaction verification. However, these opportunities are accompanied by challenges, including the imperative for continuous innovation to combat evolving cyber threats, ethical considerations in technology utilization, and the need to strike a balance between security and privacy. This abstract underscores the transformative potential of AI, ML, and Block chain in reshaping cyber security paradigms, emphasizing the ongoing necessity for the cyber security community to address challenges and harness the synergies of these technologies responsibly and effectively.

Keywords Artificial learning · Challenges · Cyber security · Machine learning · Next generation

S. Malik
School of Computer Science Engineering, Lovely Professional University, Phagwara, Punjab, India
e-mail: sarthakmalikmeerut@gmail.com

P. K. Malik (✉)
School of Electronics and Electrical Engineering, Lovely Professional University, Jalandhar, Punjab, India
e-mail: pkmalikmeerut@gmail.com

A. Naim
Business Management, King Khalid University, AlSamer, University Campus, Aseer, Abha, KSA, Saudi Arabia
e-mail: naimarshi123@gmail.com

K. Kaushik and I. Sharma (eds.), *Next-Generation Cybersecurity*, Blockchain Technologies, https://doi.org/10.1007/978-981-97-1249-6_2

1 Introduction

The landscape of cybersecurity is undergoing a profound transformation with the integration of cutting-edge technologies such as Artificial Intelligence (AI), Machine Learning (ML), and Blockchain. This introduction explores the opportunities and challenges that arise from the synergies of these new-generation cybersecurity applications [1]. The advent of AI and ML introduces adaptive capabilities for threat detection, predictive analysis, and automated responses, promising a paradigm shift in cybersecurity resilience. Concurrently, the decentralized and tamper-resistant attributes of Blockchain offer unprecedented transparency and integrity to data security and transaction verification. However, with these opportunities come multifaceted challenges, encompassing the imperative for continuous innovation to counter ever-evolving cyber threats, ethical considerations in the application of advanced technologies, and the delicate balance between security imperatives and privacy concerns. As we embark on this exploration, it becomes evident that the convergence of AI, ML, and Blockchain stands at the forefront of reshaping cybersecurity, requiring a nuanced understanding of the opportunities and challenges that accompany this transformative journey. In an era where the digital landscape is evolving at an unprecedented pace, the integration of AI, ML, and Blockchain into cybersecurity represents a frontier of innovation and adaptation. AI and ML technologies empower cybersecurity systems to dynamically learn and respond to emerging threats, offering a proactive defense mechanism that goes beyond traditional reactive approaches. Simultaneously, Blockchain's decentralized architecture addresses the vulnerabilities associated with centralized systems, ensuring data integrity and authentication in an increasingly interconnected world. As organizations strive to harness the advantages of these technologies, they are confronted with challenges ranging from the need for skilled expertise in managing sophisticated systems to ethical dilemmas surrounding privacy and security trade-offs. This exploration delves into the intricate interplay between opportunities and challenges, shedding light on the transformative potential of AI, ML, and Blockchain in shaping the future of cybersecurity. Figure 1 illustrates various resources that can be employed for cyber-security within the AI-powered metaverse.

It is possible to use block chain to move data between parties in a secure manner, which can help reduce the likelihood of data theft. In addition, these technologies can be utilised to automate a large number of the processes and procedures that are presently controlled manually [3–5]. As a result, the workload of security teams can be lightened, and resources can be reallocated to other projects. Artificial intelligence (AI) and machine learning (ML) systems are only as successful as the data they are trained on, which means that they can be deceived by malicious actors. However, the adoption of these technologies also brings with it a variety of obstacles. It is also possible for hostile actors to get access to the ledger and read or edit data using block chain technology; hence the system is not entirely safe. In addition, these technologies are still in their infancy, so their efficacy is still being assessed. Because of this, businesses need to be ready to adjust as the technology advances

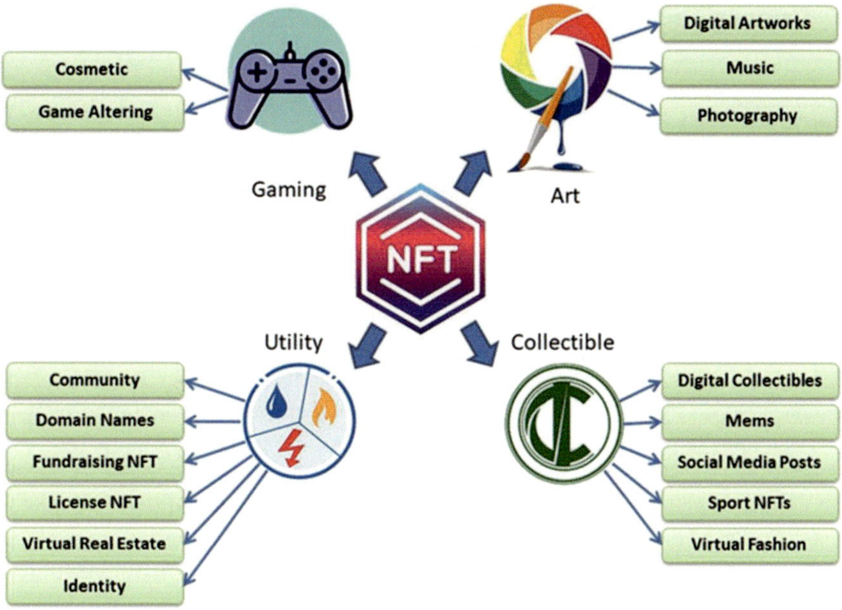

Fig. 1 Assets which can be used in cyber-security in the AI-based metaverse [2]

[6, 7]. The implementation of next-generation applications for cyber security, such as those that make use of artificial intelligence, machine learning, and block chain, presents organisations with opportunities as well as obstacles which can also be seen in Table 1.

It is possible to use these technologies in order to increase cyber security; but, in order to ensure that they are effective; they need to be properly installed and monitored. It is important for organisations to take the time to carefully consider the potential drawbacks and advantages of using these technologies before deciding whether or not they should use them [8, 9].

Table 1 Relation between different entities of cyber security applications using artificial intelligence, machine learning and block chain

Cyber security apps	AI	ML	Block chain
Antivirus software	✓	✓	
Firewalls	✓	✓	
Intrusion detection	✓	✓	
Data encryption	✓		✓
Password management	✓	✓	
DDoS protection	✓	✓	
Vulnerability scan	✓	✓	
Security audits	✓	✓	

2 Understanding the Basics of Cyber Security

Protecting computer networks, computer systems, computer programmes, and data from being attacked digitally is the practise known as cyber security. Understanding the basics of cybersecurity is paramount in navigating the increasingly complex digital landscape. At its core, cybersecurity is a discipline dedicated to safeguarding digital systems, networks, and data from unauthorized access, attacks, and damage. It involves the implementation of robust measures to ensure confidentiality, integrity, and availability of information. Key components of cybersecurity include firewalls, antivirus software, encryption, and secure authentication methods. A fundamental aspect is user awareness, emphasizing the importance of recognizing and mitigating potential threats such as phishing, malware, and social engineering. As technology advances, so do cyber threats, making continuous learning and adaptation integral to effective cybersecurity practices. Ultimately, a strong foundation in the basics of cybersecurity is crucial for individuals and organizations alike, forming the front-line defense against an ever-evolving array of digital risks. In addition to the technical aspects, a comprehensive understanding of cybersecurity involves recognizing the human element as a crucial factor in maintaining a secure environment. This includes promoting a culture of cybersecurity awareness and education within organizations, ensuring that employees are informed about best practices and potential risks. Regular updates and patches for software and systems play a pivotal role in closing vulnerabilities that cyber attackers may exploit. Additionally, cybersecurity measures extend beyond the confines of personal computers to encompass mobile devices, IoT (Internet of Things) devices, and cloud-based services. Collaborative efforts between individuals, businesses, and governments are essential for creating a resilient cybersecurity ecosystem. By grasping the basics of cybersecurity, individuals can contribute to a collective defense against cyber threats and play an active role in fostering a secure digital environment for themselves and the broader community [10]. It is a process, not a product, and in order for it to be effective, it requires a combination of different technologies, practises, and other procedures. Organisations are able to guarantee the safety of their data and systems if they have a fundamental awareness of, and adherence to, cyber security best practises. Following are some of the important aspects of cyber securities (Fig. 2).

2.1 Understand the Importance of Cyber Security

Protecting computer networks, computer systems, and computer programmes from being attacked digitally is the practise known as cyber security. The goal of these attacks is to acquire access to sensitive information, disrupt or destroy that information, extort money from an organisation or individual, or cause harm in some other way. As both the number and the level of sophistication of online attacks continue to rise, it is essential for businesses to incorporate cyber security into their overall risk

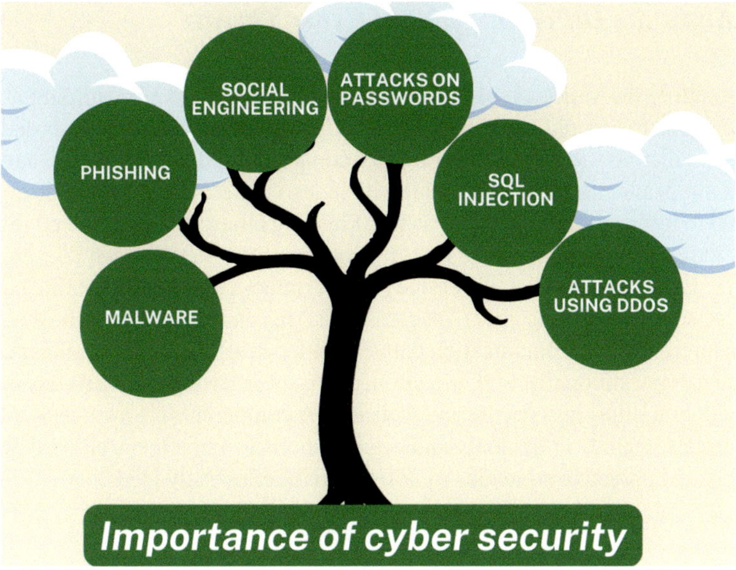

Fig. 2 Importance and understanding of cyber-security

management plan. It is imperative for businesses to take a preventative approach to controlling their cyber security risk by putting in place the necessary safeguards to secure their data, systems, and networks. This involves doing things like encrypting data and routinely patching and updating software, as well as implementing robust authentication techniques. In addition, companies should devise acceptable use standards for their computer networks and systems, make sure staffs receives frequent training on the latest cyber security best practises, and make sure the policies are enforced [11].

In addition, companies and other organisations must to have a thorough incident response plan in place, which will assist them in swiftly reacting to any cyber security events that may arise. This includes the establishment of distinct roles and duties for responding to incidents, identifying and analysing risks, and designing and putting into action strategies for remediation. In the modern digital world, when cyber dangers are becoming more frequent, having adequate cyber protection is absolutely necessary. Organisations are obligated to protect their networks, systems, and data by taking preventative measures and developing detailed incident response plans that can be implemented in the event that an incident occurs [12]. By doing so, organisations will be able to reduce their exposure to risk, safeguard their assets, and ensure the continuation of their operations.

2.2 Know the Different Types of Cyber Threats

Understanding the various types of cyber threats is imperative in building a robust defense against potential security breaches. Cyber threats encompass a broad spectrum of malicious activities designed to compromise digital systems and data. Common types include malware, which includes viruses, worms, and ransomware, designed to infiltrate and damage systems. Phishing attacks involve deceptive tactics to trick individuals into revealing sensitive information, often through fake emails or websites. Distributed Denial of Service (DDoS) attacks aim to overwhelm a target's online services, rendering them inaccessible. Man-in-the-Middle (MitM) attacks involve intercepting communication between two parties, posing a significant risk to data integrity. Additionally, SQL injection, cross-site scripting, and zero-day exploits target vulnerabilities in software and websites. A comprehensive awareness of these cyber threats enables individuals and organizations to implement tailored security measures and proactive strategies to mitigate risks effectively [13].

2.2.1 Malware

It is a type of harmful software that is meant to cause damage to or collect information from a computer system. Malware also goes by the name "spyware." Email attachments, software download, and malicious websites are all potential vectors for the propagation of malware. Malware can take many different forms, some of which are viruses, worms, Trojan horses, rootkits, ransom ware, and spyware.

2.2.2 Phishing

Phishing is a type of cyber-attack in which an attacker tries to deceive a victim into disclosing sensitive information such as passwords, credit card details, or bank account numbers. These assaults are frequently carried out through the use of malicious websites or emails.

2.2.3 Social Engineering

Creating a fake identity or using social media to win someone's trust in order to obtain sensitive information is a common tactic utilised in these types of attacks.

2.2.4 Attacks Using DDoS

DDoS is an abbreviation that stands for distributed denial of service. This kind of assault inundates a server with requests, which eventually leads to the server becoming overloaded and crashing.

2.3 Understand the Roles and Responsibilities of a Cyber-Security Professional

Comprehending the roles and responsibilities of a cybersecurity professional is crucial in addressing the escalating challenges of the digital landscape. These professionals are tasked with safeguarding sensitive information, networks, and systems from a diverse array of cyber threats. Their responsibilities often include implementing security measures such as firewalls, encryption, and intrusion detection systems to fortify digital infrastructure. Regularly assessing vulnerabilities, conducting penetration testing, and staying abreast of the latest cyber threats are integral components of their duties. Cybersecurity professionals also play a pivotal role in educating end-users and colleagues about best practices and potential risks. Moreover, they are at the forefront of incident response, swiftly identifying and mitigating security breaches when they occur. The evolving nature of cyber threats demands continuous learning, adaptability, and a commitment to staying ahead of emerging risks. In essence, the role of a cybersecurity professional is multifaceted, requiring a holistic approach to protect digital assets and ensure the integrity of information in an ever-changing technological landscape. They have to be able to strike a balance between the many privacy laws and regulations and the security requirements. A cyber security professional's responsibilities may extend to advising staff members and end users on the most effective ways to safeguard their organization's information systems and data in some businesses. This involves the provision of training on subjects such as the management of passwords and the secure use of the internet. In the modern digital world, the function of a cyber-security professional is an extremely important one. Because of the vast amounts of data and information that are kept in digital form, organisations need to make sure that they have the appropriate expertise on staff in order to keep their networks and systems secure [14].

2.4 Understand the Different Cyber Security Tools Available

Software applications that assist safeguard networks, systems, and data against cyber-attacks, cyber criminals, and other types of cyber dangers are known as cyber security tools. These instruments can perform a wide range of security functions, such as malware identification, vulnerability scanning, patch update administration, network monitoring, intrusion prevention, and data protection. A firewall is one of the most important parts of a strong cyber security system. By restricting the types of data traffic that are permitted to pass through the firewall, firewalls serve as a barrier between an internal network and the outside world. Firewalls can be configured to block malicious traffic, including that generated by viruses, malware, and phishing scams. Antivirus software can assist in the detection, prevention, and removal of malicious software, also referred to as malware, from a system.

It can also detect unusual activities, such as attempts to access restricted portions of the system, which may indicate that the user is trying to gain unauthorised access. Intrusion detection and prevention systems, often known as IDS/IPS, are able to identify any attempts made to obtain unauthorised access to a system or network. IDS and IPS systems have the ability to identify potentially malicious behaviour, such as port scans and attempts to exploit vulnerabilities, and then take preventative measures to thwart an attack [15].

Spotting vulnerabilities in a network or computer system is the primary purpose of vulnerability scanners. These scanners are able to identify software flaws such as vulnerabilities, misconfigurations, and missing patches, as well as provide recommendations on how to fix the problems they find. Tools for encrypting data are used to protect the information that is saved on computers and servers. Encryption algorithms transform data into a form that can only be read and understood by those who have been granted permission to do so. Users are assisted in the creation of, storage of, and administration of secure passwords via tools for password management. These solutions have the ability to generate robust passwords, store them in a secure location, and control access permissions. These are only a few of the many cyber security products that are currently accessible. Because each tool possesses a distinct set of features and capabilities, it is essential to select the tool that is most suited to meet the requirements for security imposed by your organisation [16].

2.5 Understand Data Privacy and Protection Regulations

Understanding data privacy and protection regulations is paramount in the contemporary digital landscape, where the handling of personal information is ubiquitous. These regulations, such as the General Data Protection Regulation (GDPR) in Europe or the California Consumer Privacy Act (CCPA) in the United States, are designed to safeguard individuals' privacy rights and impose responsibilities on organizations that collect, process, and store personal data. The regulations typically outline strict guidelines on consent mechanisms, data breach notifications, and the rights of individuals to access and control their data. Compliance with these regulations is not only a legal requirement but also a fundamental ethical obligation. It involves implementing robust data protection measures, including encryption, access controls, and secure storage practices. Understanding and adhering to data privacy regulations not only mitigate legal risks for organizations but also foster trust with users by demonstrating a commitment to respecting their privacy and safeguarding their sensitive information. Regulations designed to protect individuals' privacy, as well as the data they voluntarily contribute to businesses, are critically vital. In order to guarantee that they are processing data in a responsible and secure manner, businesses are required to comply with certain requirements [17, 18].

2.6 Recognize the Signs of a Cyber-attack

Attacks in cyberspace can take many different forms, and it can be challenging to spot the tell-tale symptoms of an impending attack. On the other hand, there are specific markers that can assist in determining whether or not an attack is taking place.

1. If you suddenly notice that your network is performing more slowly than usual, this could be an indication that an attack is taking place. Attackers in the digital realm frequently deploy malicious malware to bog down networks and make them run more slowly.
2. Unusual emails or website visits: You should be on the lookout for symptoms of a cyber-attack if you start receiving emails from senders you aren't acquainted with or if you start visiting websites that you wouldn't typically go to.
3. Account access without authorization: If you find that someone has accessed your account without your authorization, this may be a sign of a cyberattack.
4. Unexpected file modifications: If you discover that files have been altered without your knowledge, this could be a sign that your system is under attack.
5. Unusual activity in your log files: If you observe unusual activity in your log files, it may indicate that you are the victim of a cyberattack.

If you can identify the telltale signs of a cyberattack and take appropriate action, you will be able to protect your data and systems from malicious activity. To ensure the security of your computer systems, you must remain vigilant and take the necessary precautions [19, 20].

2.7 Understand the Basics of Cryptography

The most prevalent form of cryptography is called symmetric cryptography, and it entails using a single cryptographic key to both encrypt and decrypt data. Symmetric cryptography is also the most secure form of cryptography. In asymmetric cryptography, each step—encryption, decryption, and verification—requires the use of a distinct key. In addition, cryptography encompasses the utilisation of digital signatures, which are used to validate the genuineness of data or messages. The authenticity of a document can be validated with the help of something called a digital signature, which is a form of cryptographic hash. A private key, which is used to sign the document, and a public key, which is used to verify the signature, are both incorporated into its generation. The private key is used to sign the document. In today's digital age, one of the most important tools for protecting one's data and one's communications is cryptography. Its purpose is to protect the privacy of data and communications while also guaranteeing their integrity and accessibility.

2.8 Learn How to Create an Incident Response Plan

Every strong cyber-security programme needs to have a solid incident response plan that can handle any situation that may arise. It specifies the procedures that need to be taken in order to respond to a security incident in a manner that guarantees the response will be effective, uniform, and delivered in a timely manner. The first thing that has to be done when building a plan for responding to incidents is to determine the different kinds of occurrences that could arise. This should cover both harmful and non-malicious incidents, such as data breaches, system failures, and unauthorised access, among other things. Following the identification of these incidents, the response plan should detail the measures to take when responding to each distinct type of incident. Additionally, the plan ought to establish who is accountable for providing a response to each incident. This should include both internal staff and external resources, such as law enforcement and cyber-security experts, among other types of professionals. In addition to this, the plan should detail the roles and duties of each member of the team, in addition to providing a timeframe for responding to each incident.

3 The Benefits of Artificial Intelligence in Cyber Security

Artificial Intelligence (AI) has emerged as a crucial ally in the realm of cybersecurity, offering a myriad of benefits that significantly enhance our ability to safeguard digital assets. One of the key advantages lies in AI's capacity to swiftly analyze vast amounts of data in real-time, enabling the rapid detection of anomalies and potential security threats. Machine learning algorithms, a subset of AI, excel at recognizing patterns and adapting to evolving cyber threats, thereby fortifying defense mechanisms against sophisticated attacks. Additionally, AI facilitates predictive analysis, allowing cyber-security systems to anticipate potential vulnerabilities and proactively address them before they can be exploited. Moreover, AI-powered automation streamlines routine security tasks, reducing the burden on human operators and minimizing the risk of human error. This not only enhances the overall efficiency of cybersecurity protocols but also ensures a quicker response to emerging threats. Ultimately, the integration of AI in cybersecurity not only fortifies our digital defenses but also empowers

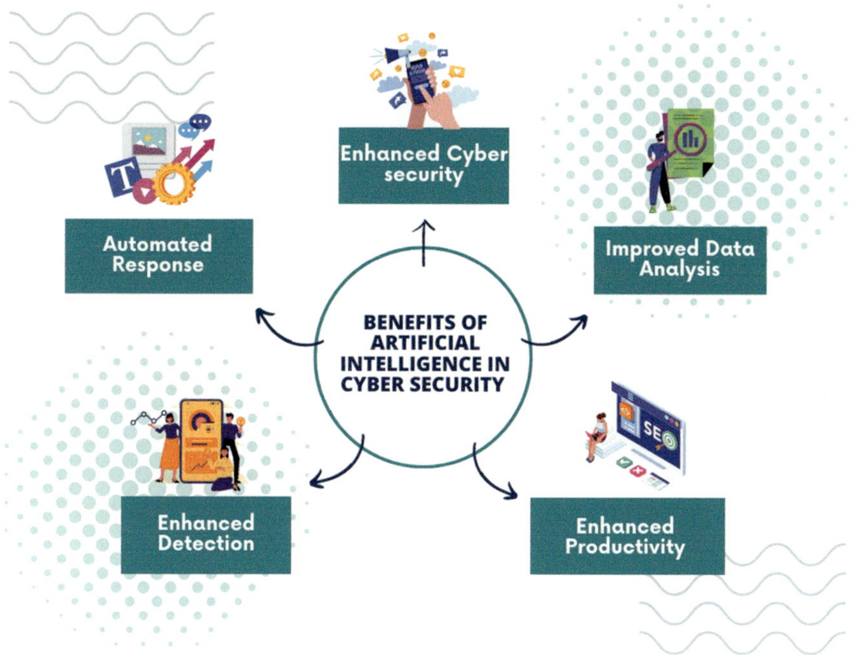

Fig. 3 Artificial intelligence in cyber security, benefits and advantages

organizations to stay one step ahead in an increasingly complex and dynamic threat landscape.

Following are some of the benefits of AI in cyber security as shown in Fig. 3 also:

- **Enhanced Detection**: Compared to more conventional approaches, artificial intelligence can detect potential dangers both more quickly and more correctly.
- **Automated Response**: Artificial intelligence has the ability to automatically respond to cyber-attacks, which can significantly cut down on the amount of time and effort spent on manual responses.
- **Enhanced Cyber security**: AI may be used to detect and prevent unwanted activity, such as malicious network invasions, hacking attempts, and phishing. This can be a significant benefit to organisations concerned about their online safety.
- **Improved Data Analysis**: AI may be used to analyse vast volumes of data in order to spot patterns and abnormalities that may point to a possible breach in security. This can be done in order to reduce the likelihood of a breach occurring.
- **Enhanced Productivity**: Artificial Intelligence has the potential to cut down on the amount of time and resources needed to keep networks under surveillance and protect them.

4 Machine Learning and Cyber Security

Machine Learning (ML) has become a linchpin in the field of cybersecurity, revolutionizing the way we detect, prevent, and respond to evolving cyber threats. At its core, machine learning equips cybersecurity systems with the ability to learn from data patterns and adapt in real-time, offering a dynamic and proactive defense mechanism. ML algorithms excel in identifying anomalies and unusual patterns within vast datasets, enabling the swift detection of potential security breaches that may go unnoticed by traditional methods. By continuously analyzing and learning from new data, machine learning models can evolve to recognize emerging threats, providing a robust defense against sophisticated and constantly evolving cyber attacks. The predictive capabilities of ML contribute significantly to cybersecurity by allowing for the identification of vulnerabilities before they are exploited. This proactive approach enhances overall resilience and responsiveness, as security measures can be adjusted in anticipation of potential threats. In essence, the integration of machine learning into cybersecurity not only fortifies digital defenses but also represents a critical paradigm shift in adapting to the ever-changing landscape of cyber threats.

It is possible for businesses to increase the efficiency of their security operations and cut down on the amount of time and resources needed to detect and respond to cyber threats by making use of the power of machine learning (ML). Following are the main connection between machine learning and cyber security as shown in Fig. 4 also:

- Identifying malicious activity can be accomplished with the use of machine learning by analysing user behaviour, system logs, and network traffic for recurring trends.

Fig. 4 Cyber security and machine learning connections

- Classifying malicious files and identifying malicious code patterns can both be accomplished through the application of machine learning.
- Vulnerabilities in software, networks, and other systems can be discovered with the help of machine learning.
- Zero-day vulnerabilities can be identified and protected against with the use of machine learning.
- Monitoring and analysing huge datasets for suspicious behaviour can be accomplished with the help of machine learning.
- Detection and warning of potential harmful risks can be achieved through the application of machine learning.
- Identifying malicious actors and their strategies can be accomplished through the use of machine learning.
- Dangerous websites and networks can be identified and blocked using machine learning, which can also be used to identify dangerous websites.
- Phishing attempts and malicious email campaigns can be uncovered with the use of machine learning techniques.
- Both detecting and preventing data breaches can be accomplished with the help of machine learning.

5 The Potential of Block Chain in Cyber Security

The world is getting more digitalized at a rapid pace, and as a result, the demand for a safe and secure online environment is higher than it has ever been. Blockchain technology holds immense potential in reshaping the landscape of cybersecurity, offering innovative solutions to address longstanding challenges in securing digital transactions and data. One of the key strengths of blockchain lies in its decentralized and tamper-resistant nature. By utilizing a distributed ledger, blockchain ensures that once data is recorded, it becomes virtually immutable, reducing the risk of unauthorized alterations or malicious tampering. This characteristic enhances the integrity and transparency of digital records, providing a robust foundation for secure and verifiable transactions. Additionally, blockchain's consensus mechanisms and smart contracts contribute to the automation of security processes, reducing the reliance on centralized authorities. This decentralized approach not only mitigates the risk of a single point of failure but also enhances the resilience of cyber security infrastructure against potential attacks. As block chain continues to mature, its potential applications in securing identity management, protecting sensitive information, and enabling secure peer-to-peer transactions underscore its pivotal role in fortifying the foundations of cyber security for the digital age.

6 Conclusion

In conclusion, the fusion of Artificial Intelligence (AI), Machine Learning (ML), and Block chain in the realm of cyber security presents a landscape rich with opportunities and challenges. The integration of AI and ML brings forth the promise of unparalleled threat detection capabilities, enabling systems to adapt and evolve in real-time. The predictive analysis and automated response mechanisms enhance overall cyber security resilience. Meanwhile, Block chain's decentralized and tamper-resistant nature adds a layer of transparency and integrity, revolutionizing data security and transaction verification. However, with these opportunities come challenges, such as the need for continuous innovation to keep pace with evolving cyber threats and the potential ethical considerations surrounding the use of advanced technologies. Striking the right balance between security and privacy, ensuring regulatory compliance, and addressing the dynamic nature of cyber threats are ongoing challenges that must be navigated. As we navigate this complex terrain, it is evident that the synergies between AI, ML, and Block chain hold the key to shaping the future of cyber security, requiring a concerted effort from the cyber security community to unlock their full potential while mitigating associated risks.

References

1. Lee S, Kim S (2022) Blockchain as a cyber defense: opportunities, applications, and challenges. IEEE Access 10:2602–2618. https://doi.org/10.1109/ACCESS.2021.3136328
2. Pooyandeh M, Han K-J, Sohn I (2022) Cyber security in the AI-based metaverse: a survey. Appl Sci 12:12993. https://doi.org/10.3390/app122412993
3. Zhang H, Li P, Du Z, Dou W (2020) Risk entropy modeling of surveillance camera for public security application. IEEE Access 8:45343–45355. https://doi.org/10.1109/ACCESS.2020.2978247
4. Akram SV, Alshamrani SS, Singh R, Rashid M, Gehlot A, AlGhamdi AS, Prashar D (2021) Blockchain enabled automatic reward system in solid waste management. Secur Commun Netw
5. Sun N et al (2022) Defining security requirements with the common criteria: applications, adoptions, and challenges. IEEE Access 10:44756–44777. https://doi.org/10.1109/ACCESS.2022.3168716
6. Zhang Z, Hamadi HA, Damiani E, Yeun CY, Taher F (2022) Explainable artificial intelligence applications in cyber security: state-of-the-art in research. IEEE Access 10:93104–93139. https://doi.org/10.1109/ACCESS.2022.3204051
7. Bhatt H, Bahuguna R, Singh R, Gehlot A, Akram SV, Priyadarshi N, Twala B (2022) Artificial intelligence and robotics led technological tremors: a seismic shift towards digitizing the legal ecosystem. https://doi.org/10.3390/app122211687
8. Choudhury S, Singh R, Gehlot A, Kuchhal P, Akram SV, Priyadarshi N, Khan B (2022) Agriculture field automation and digitization using internet of things and machine learning. https://doi.org/10.1155/2022/9042382
9. Elsisi M, Tran M-Q, Mahmoud K, Mansour D-EA, Lehtonen M, Darwish MMF (2021) Towards secured online monitoring for digitalized GIS against cyber-attacks based on IoT and machine learning. IEEE Access 9:78415–78427. https://doi.org/10.1109/ACCESS.2021.3083499

10. Shaukat K, Luo S, Varadharajan V, Hameed IA, Xu M (2020) A survey on machine learning techniques for cyber security in the last decade. IEEE Access 8:222310–222354. https://doi.org/10.1109/ACCESS.2020.3041951
11. Dua S, Kumar SS, Albagory Y, Ramalingam R, Dumka A, Singh R, Rashid M, Gehlot A, Alshamrani SS, Alghamdi AS (2022) Developing a speech recognition system for recognizing tonal speech signals using a convolutional neural network. Appl Sci (Switzerland) 12(12). https://doi.org/10.3390/app12126223
12. Gehlot A, Malik PK, Singh R, Akram SV, Alsuwian T (2022) Dairy 4.0: intelligent communication ecosystem for the cattle animal welfare with blockchain and IoT enabled technologies. Appl Sci (Switzerland) 12(14). https://doi.org/10.3390/app12147316
13. Ferrag MA, Friha O, Hamouda D, Maglaras L, Janicke H (2022) Edge-IIoTset: a new comprehensive realistic cyber security dataset of IoT and IIoT applications for centralized and federated learning. IEEE Access 10:40281–40306. https://doi.org/10.1109/ACCESS.2022.3165809
14. Kumar SC, Thakur AK, Aseer JR, Natarajan SK, Singh R, Priyadarshi N, Twala B (2022) An experimental analysis and ANN based parameter optimization of the influence of microalgae spirulina blends on CI engine attributes. Energies 15(17). https://doi.org/10.3390/en15176158
15. Ferrag MA, Friha O, Maglaras L, Janicke H, Shu L (2021) Federated deep learning for cyber security in the internet of things: concepts, applications, and experimental analysis. IEEE Access 9:138509–138542. https://doi.org/10.1109/ACCESS.2021.3118642
16. Madan P, Singh V, Chaudhari V, Albagory Y, Dumka A, Singh R, Gehlot A, Rashid M, Alshamrani SS, Alghamdi AS (2022) An optimization-based diabetes prediction model using CNN and Bi-directional LSTM in real-time environment. Appl Sci (Switzerland) 12(8). https://doi.org/10.3390/app12083989
17. Liang F, Hatcher WG, Liao W, Gao W, Yu W (2019) Machine learning for security and the internet of things: the good, the bad, and the ugly. IEEE Access 7:158126–158147. https://doi.org/10.1109/ACCESS.2019.2948912
18. Malik P, Gehlot A, Singh R, Gupta LR, Thakur AK (2022) A review on ANN based model for solar radiation and wind speed prediction with real-time data. Arch Comput Methods Eng. https://doi.org/10.1007/s11831-021-09687-3
19. Hussain F, Hussain R, Hassan SA, Hossain E (2020) Machine learning in IoT security: current solutions and future challenges. IEEE Commun Surv & Tutor 22(3):1686–1721. thirdquarter 2020. https://doi.org/10.1109/COMST.2020.2986444
20. Verma P, Dumka A, Singh R, Ashok A, Gehlot A, Malik PK, Gaba GS, Hedabou M (2021) A novel intrusion detection approach using machine learning ensemble for iot environments. Appl Sci (Switzerland) 11(21). https://doi.org/10.3390/app112110268

AI and Blockchain for Secure Data Analytics

Shweta Mayor Sabharwal, Sonia Chhabra, and Manpreet Kaur Aiden

Abstract In the age of big data and sophisticated analytics, organizations experience major barriers to ensuring the privacy, security, and dependability associated with their data-driven projects. This chapter examines how integrated blockchain and machine learning might effectively handle these issues and improve the reliability of data analytics procedures. In addition to providing users with tokenized benefits, the artificial intelligence and blockchain integration offer distinctive characteristics to protect the confidentiality of data, assure data integrity, and allow the safe exchange of information and partnership. This chapter emphasizes the crucial relevance of the security and confidentiality of data. It examines how blockchain's decentralized and irreversible characteristics can offer a solid platform for safeguarding private information, preventing unauthorized access, and reducing the danger of security breaches. In order to demonstrate how the privacy of information can be protected while utilizing intelligence that is distributed, the idea of federated learning, where artificial intelligence models undergo training on decentralized data sources, is investigated in regards to the blockchain. This chapter also discusses how the safe sharing of data and teamwork could be enabled by blockchain technology and blockchain-based smart contract systems, and access restrictions allow businesses to specify data usage commitments, guaranteeing that data is distributed and retrieved following established guidelines and privileges. Lastly, this chapter discusses the technological issues, like scalability, interoperability, and processing cost, that emerge when integrating AI and blockchain. It additionally emphasizes how crucial it is to take legal and regulatory considerations into account, especially when it comes to safeguarding information and adhering to standards, to achieve a strong and ethical execution. In general, this chapter provides a thorough review of the connections among blockchain

---

S. M. Sabharwal (✉)
Galgotia's University, Greater Noida, India
e-mail: Shweta.mayor@gmail.com

S. Chhabra
Sharda University, Greater Noida, India

M. K. Aiden
Maharishi University, Noida, India

K. Kaushik and I. Sharma (eds.), *Next-Generation Cybersecurity*, Blockchain Technologies, https://doi.org/10.1007/978-981-97-1249-6_3

and AI for reliable data analytics. This integration helps in increasing the confidentiality of data, integrity, and partnership while highlighting the necessity of careful planning and execution to reap its full rewards.

Keywords Artificial Intelligence · Blockchain · Security · Data analytics · Federated learning · Privacy

1 Introduction

Organizations confront considerable challenges in maintaining the confidentiality, safety, and reliability connected to their initiatives based on data in the era of big data and smart analytics. The significance of protecting private information is increasing as a result of a sharp rise in the amount of data and developments in analytics and machine learning [1]. These issues are frequently not adequately addressed by conventional centralized storage and retrieval methods, which leaves potential for data breaches, unauthorized use, and damaged integrity. The combination of blockchain with artificial intelligence appears as a potent approach to increase the accuracy of the data analytics process as a response to these problems [2, 3].

Organizations have become increasingly reliant on decisions based on data, which has forced them to gather, store, and analyse huge volumes of data. While this development opens up a lot of doors for new ideas and innovations, it also raises serious privacy and security issues. Alarm bells have been rung by prominent data theft and misuse instances, forcing organizations to review their information security plans [4].

Due to its decentralized and irreversible characteristics, blockchain technology—the foundation of cryptocurrency like Bitcoin—has attracted a lot of attention. Every block in a blockchain, which forms a series of data that is linked, includes a secure hash of the block before it, making the record of the transaction or information unbreakable by design [5, 6]. Data security and reliability are solidly supported by this innovative design, which makes sure that once information is stored on the blockchain, information cannot be accessed or changed.

In contrast, algorithms for machine learning and AI have proven to have extraordinary talents in gleaning worthwhile insights from huge datasets. Privacy issues arise, especially when confidential user data is involved because these algorithms frequently need access to large volumes of data for proper training [7].

A potential answer to these problems is the combination of blockchain technology and artificial intelligence. Blockchain's decentralized structure improves privacy and security of information, whereas AI expands the potential of data analytics [8, 9]. A more dependable and inclusive data analytics environment is facilitated by this synergy, which enables reliable information sharing, teamwork, and the creation of tokenized incentive.

With particular objectives in consideration, this chapter intends to provide a thorough analysis of the use of artificial intelligence and blockchain in reliable analytics.

It explores the difficulties businesses encounter when it comes to maintaining the accuracy, protection, and confidentiality of their data. The chapter discusses how blockchain technology relates to data analytics by examining its core ideas and characteristics [10]. To improve the security and confidentiality of information, the advantages and implications of combining machine learning and artificial intelligence with blockchain are examined. Additionally, the idea of federated learning is examined to see if it may protect the confidentiality of data in widespread training for artificial intelligence environments. The chapter explores how the blockchain relies on smart contract platforms to enable safe data exchange and teamwork while assuring compliance with privacy contracts [11, 12].

It also investigates the use of tokenized rewards and the use of tokens in rewarding data producers, intelligence designers, and validators. The chapter emphasizes the significance of constitutional and regulatory factors, particularly with regard to data protection and social execution, while discussing technical difficulties including flexibility, connectivity, and data processing costs [13, 14]. This chapter aims to provide an in-depth knowledge of how blockchain technology and artificial intelligence collaboration might revolutionize data analytics, ensuring increased security, privacy, and reliability in the age of data. It does this by offering suggestions for potential futures and new developments.

2 The Era of Big Data

Authorities, businesses, and advertisers all have access to a growing amount of data about areas of our daily lives that, in earlier times, we would have assumed to be relatively confidential. This is the Big Data age. While the processing capacity needed to manage such information is always growing, technology that extracts, gathers, keeps, and analyses data is getting less expensive and quicker. The 'datafication' of the community, which impacts all spheres of life, has been made feasible by electronic devices [15, 16]. It is undeniable that data is becoming more and more important for business and the community, and this trend will continue.

So what does the term "Big Data" indicate? Despite being often used, the phrase has no established meaning. It is typically related to complicated and enormous datasets, whereby specialized techniques and instruments are used to carry out operations to extract valuable data to make better decisions. Big Data isn't merely about the amount of data that is available, though; it also includes fresh approaches to data analysis and the creation of novel information [17]. The phrase is frequently used in the media to describe the growing abundance of data, the scale of information sets, the expansion of digital data, as well as emerging or different data sources.

2.1 Features of Big Data

Big Data includes five critical features considering a more specialized technological standpoint.

2.1.1 Volume

The amount of data produced and preserved, especially in terms of its size. The worth and possibility of understanding the data are determined by its amount [18]. Big Data requires an enormous amount to be able to exist.

2.1.2 Variety

The variety and type of information, in addition to how it was organized. Big Data can incorporate text, images, audio, and video clips, and data synthesis can fill in the gaps. It can also be organized, semi-organized, or unorganized. Data can be gathered from a wide range of sources, and its relevance changes based on the type of analysis being done, from online communities to internal devices to cell phone navigational systems [19]. The multiple levels and forms of big data might also vary.

2.1.3 Velocity

The duration required to produce and deal with data. In a corporate setting, rapid transmission can undoubtedly provide an edge over others, thus data must flow fast and as close to real-time as feasible [20].

2.1.4 Veracity

Data accuracy and dependability; it's crucial to have methods for spotting and fixing any fake, erroneous, or inaccurate information.

2.1.5 Value

The evaluation of trustworthy data enhances knowledge between and within specialties and contexts. The creation of helpful data results in wealth.

2.2 *Ethical Challenges of Big Data*

Big Data is becoming more widely acknowledged as a catalyst for change in the private and public sectors today. With advantages spanning from financial to medical, meteorological to genome research, medical or ecological study to analytics and commerce, databases are enabling broad changes in society that are progressively becoming an integral component of our daily lives [21].

The techniques of manufacturing, utilization, and general living will be revolutionized by data. Its benefits will penetrate every aspect of our living and include more knowledge regarding energy consumption, greater product, substances, and agricultural accountability, as well as better medical care and wellness. Data serves as the building block for novel goods and services, fostering improved output and resource effectiveness throughout every industry. Data is the essence of commercial advancement. It enables more personalized products, improved legislation, and improved governmental services [22].

Due to the quick transition of goods and services through recognizing patterns and insight creation to more complex forecasting approaches and, consequently, enhanced decision-making, the accessibility of data is crucial when developing machine learning systems. To address social, environmental, and climate concerns and promote happier, wealthier, as well as more equitable societies, additional information must be made accessible, and data utilization must be improved. To accomplish the goals of the European Green Deal, for instance, it will result in improved legislation [23, 24].

Making choices, client interaction, demand-side prediction, brand and marketplace growth, and operational effectiveness are just a few of the areas where big data may be used to great effect. Automotive dominates all other industries in keeping records, according to a McKinsey & Company analysis, rendering Big Data an essential part of the latest industrial revolution, sometimes known as "Industry 4.0." With improved handling of supply chains and improved mitigation of risks mechanisms, led by smart choices, this renaissance has the potential to increase efficiency. Additionally, Industry 4.0 emphasizes developing sophisticated goods that can collect and send massive volumes of data across the course of their manufacture and consumption lifetimes. In order to identify consumer needs and influence next items, such information must be acquired and evaluated in real-time. The enormous adoption of transformational practices, like the usage of computerized doubles in production, is also anticipated to be fuelled by data [25, 26].

As previously indicated, Big Data also adds value in a wide range of other industries, such as government functions, learning, and the medical field. The implementation of democratic government and openness principles is anticipated to improve many facets of citizens' life. Along with perhaps more obvious applications, like better preventive medicine in the medical field or monitoring their own in the educational sector, this may end up resulting in the creation of increasingly fair and participatory societies [27]. These advantageous outcomes must be balanced off against intricate and multifaceted problems, though. Some worries in the medical sector, which stand to gain greatly from Big Data remedies, pertain, for example, to the challenge of upholding rules of ethics regarding confidential information, where the huge amount of data may make it difficult to obtain the updated and particular consent necessary before every instance of handling it requires place. A further instance comes from the danger that learners in educational institutions feel constantly watched due to the ongoing gathering and analysis of their information, which could reduce their ability to innovate and/or increase their anxiety levels [28].

The discussion of Big Data must emphasize the various potentially moral and social factors that may arise and examine the related legal, sociological, and moral difficulties. Here, it is necessary to develop an ecological and moral structure for defending rights for individuals, reducing risks, and making sure that ethical principles and actions are consistently in line [29]. A structure like this ought to be able to boost public and corporate trust in big data as well as the data revolution. Enormous data involves enormous responsibility, so as the European Data Protection Supervisor (EDPS) stated, "suitable data security measures need to be in existence."

Modern ethical conversations have centred on issues relating to trust beyond all else, confidentiality, anonymity, encoding, and monitoring. As innovation improves, the discussion is shifting more and more towards self-driving technologies and machine learning. Future hazards of novel kinds may also be recognized and discussed when technology develops even more, it is probable [30].

2.3 Research Ethics

Considering the advancements in heavily data-driven study, it is important to consider how well ethical standards created in the fields of psychology and sociology using fundamental ideas like informed consent, threat, harm, possession, etc. can be instantly implemented in the field of data science or whether they need to be supplemented with other ideas aimed at safeguarding "living subjects" in heavy data investigations. When data may easily be linked to a person who faces the possibility of harm in daily life, personal objects' safety is typically applied. Big Data, however, may put a significant barrier between how sensitive information is used in daily life. Because of the numerous, intricate justifications for collaborating, utilizing, and disseminating findings from studies, it is difficult to forecast the kinds of potential consequences to people if technological safeguards are insufficient and fail to avoid the reuse of confidential information across multiple repositories [31].

We must re-evaluate the conventional perspectives if these obstacles are too great to be overcome within the current frameworks of research morality. In this section, an innovative model of scientific ethics designed for data science might be developed to better place the "individual" at the core of the discussion. The growing body of research on confidentiality as well as other rights of citizens supports the notion that the moral consequences of big data are taking centre stage in the European Union (EU) discussion, and that everyone is working towards finding workable solutions that strike a balance between maximizing the worth of big data despite threatening basic human rights [32].

2.4 *Big Data and Its Impact on Privacy*

Big Data's widespread gathering, preservation, and evaluation of enormous amounts of personal data have a significant influence on privacy. As additional information is collected and acquired from many sources, worries about privacy violations and potential exploitation of confidential information have grown significantly. A few significant ways that Big Data impacts privacy are listed below.

2.4.1 'Creep Factor' of Big Data

A crucial development for the European economy is the utilization of Big Data, contemporary surveillance technologies, as well as data collection methods. In virtue of the updated legislative structure, it also presents substantial legal issues from a data safety standpoint. While the information is frequently utilized and repurposed in ways that were practically unthinkable when the data were obtained, conventional processes and conceptions of confidentiality protections (such as informed permission approaches) may be insufficient in some cases under the Big Data model [33].

As noted through the EDPS, the regard for a person's right to confidentiality and the freedom to the safeguarding of private information is inextricably linked with the regard for the worth of others. The European Charter of Fundamental Rights recognizes that one's right to respect for humanity is unalienable. This fundamental right may be violated by acts like dehumanization, which take place when a person is considered as an object used for another person's benefit (European Data Protection Supervisor, Opinion 4/2015).

Big Data innovations' effects on confidentiality span from data inequality and automated choice to category confidentiality and sophisticated segmentation. It is much more important if individuals share sensitive data online while experiencing the major stages of their lives and do so with varying degrees of knowledge. Individuals can frequently almost disappear from view here for data mining applications that utilize publicly available information from social media platforms and other information connected to an Internet Protocol (IP) address for analytical reasons [34].

Because of unethical and intentional practices, Big Data has a "creep factor" that defeats the purpose of confidentiality laws. Such practices, which frequently have the end goal of addressing and categorizing clients, are made possible by developments in the analysis and use of Big Data.

The potential for retrieval of the information's source following the anonymization procedure is a further concern in regard to big data. The growing processing capacity of contemporary personal computers has made de-anonymization technologies accessible, allowing a track directly to the initial private information. Traditional anonymization methods, which make every data access non-identifiable by eliminating distinctively recognizable details, do have limitations. For example, even when personal data submitted by users is substituted in a dataset, de-anonymization can be accomplished in a relatively short amount of time through straightforward connections among such unidentified databases, other databases, and personal data. Reconstruction of the information source may result from the potent insights gained through the joining of various, particular datasets from various sources. This could make it possible for those with an interest to definitively pin down certain anatomical individuals or a restricted number of individuals [35].

Data and new identification raise major issues regarding privacy since once anonymized (or pseudo-anonymized), the data can be treated outside the subject's permission at any time before the individual is re-identified. Lack of proper openness about the utilization of Big Data makes the circumstance worse since it impacts a data the subject's capacity to manage the exposure of their personal details to third parties along with access to those data, which also has an effect on privacy rights.

It is advised that businesses interested in using big data create open processes and make sure the general public can comprehend them. In this approach, development and boundary-setting could be genuinely driven by ethical considerations, effectively taking into consideration the requirement for confidentiality and freedom of choice on the part of an individual.

2.4.2 Stigmatization and Manipulation

There is a change in the confidentiality right. Primarily developing as a privilege to privacy and exclusivity, it has evolved through time into the right to regulate personal data and is currently heading even farther in the path of enhanced control. The present movement is approaching the right to control identification and the statistical picture developed by outside parties that choose the pertinent trends to be taken into account in metadata. The third stage focuses more on knowledge that can be used to identify specific behavioural characteristics than it does on data that can be used to identify specific physical individuals, such as health information, purchasing choices, medical history, sleep schedules, patterns of movement, access to the internet, connections, etc. Organizations have grown more open regardless of the fact that the data is unidentified; in fact, removing all characteristics of group feeling connected from that data would obstruct the gathering of its substance and, thus, render it useless [35].

Big Data information can be utilized to specifically influence how individuals conduct or what they buy. Targeted advertising is one instance, but additional strategies that depend on Big Data's capacity to uncover concealed relationships and on the assumed choices and circumstances of a particular group, might be embraced to promote or dissuade a particular behaviour with rewards whose objectives are less open to scrutiny.

Stigmatization may also take on new forms, such as when it comes to a group's purchasing habits or other private information. Aspects that discriminate are probably going to occur, particularly as the categories get narrower. Big Data technology may supersede steadfast fundamental rights safeguards in this regard [36].

The associated massive Big Data gathering and accumulation, as well as the qualitative evaluation and organized information produced for this purpose but not subjected to the implementation of the present privacy regulations, constitute what raises ethical concerns. Therefore, new and creative approaches to considering the safety of citizens are required, capable of providing sufficient and complete protection.

3 The Convergence of AI and Blockchain

Before delving into the convergence of blockchain and AI, it is essential to establish a basic comprehension of both technologies and their extensive array of potential applications.

3.1 Blockchain Technology

As was already said, the blockchain is regarded as a cutting-edge technology that has the potential to revolutionize human transactions. The blockchain is a configuration of distributed ledger technology (DLT) that transmits digital data securely to all nodes linked in a peer-to-peer (P2P) network before storing it. Blockchain-related technologies that increase trust, data security, and transparency include shared ledgers and cryptography. The records of open transactions that are carried out upon authentication are stored as blocks in the blockchain. The linked user or group of users must agree to any changes made to the records that have been recorded in this manner [37].

3.1.1 Blockchain Components

Blockchain technology comprises several essential elements, including below.

Blocks

These are the fundamental units of data that contain a set of transactions. Each block is linked to the previous one, forming a chain of blocks, hence the term "blockchain."

Transactions

Transactions represent the exchange of data or assets between participants on the blockchain network. These can include financial transactions, contracts, or any other data that needs to be recorded and verified [33].

Ledgers

The blockchain operates as a decentralized and distributed ledger, meaning that all participants on the network maintain a copy of the entire transaction history. This shared ledger ensures transparency and immutability of the recorded data.

Asymmetric or Public-Key Cryptography

This cryptographic method uses a pair of keys, a public key, and a private key, to secure transactions and ensure that only the intended recipient can access the data. The public key is used for encryption, while the private key is used for decryption [34].

Cryptographic Hash Functions

These functions play a crucial role in securing the integrity of data within each block. They convert data of variable sizes into fixed-length hash codes, making it challenging for any malicious party to alter the content of a block without detection.

3.1.2 Blockchain Categories

According to Fig. 1, various categories of blockchain exist. Based on data accessibility, blockchain can be classified into the following four types.

Public Blockchain

In this type, anyone can access and submit transactions to the blockchain network.

Private Blockchain

Only a designated group of individuals or entities has the authority to access and submit transactions on this type of blockchain.

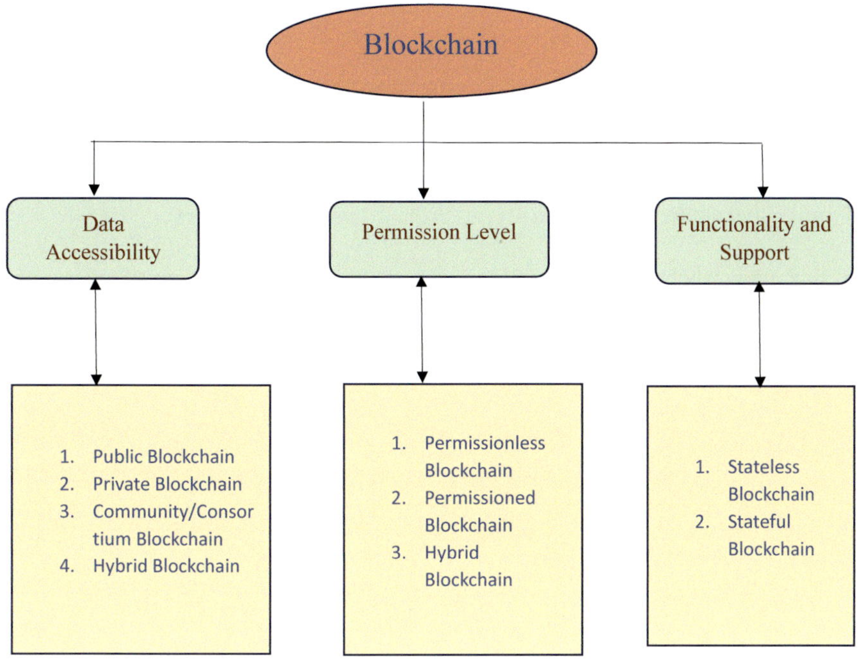

Fig. 1 Categories of blockchain

Community/Consortium Blockchain

Multiple groups are allowed to access and submit transactions on this blockchain category.

Hybrid Blockchain

This type combines different blockchain models to facilitate transactions.

Additionally, blockchain can be categorized based on permission levels.

Permissionless Blockchain

No advance permission is required for participants to access and engage in transactions on this type of blockchain.

Permissioned Blockchain

Advance permission is necessary for users to access and participate in transactions within this blockchain network.

Hybrid Blockchain

In this case, there is a possibility of switching between permissionless and permissioned blockchains as needed.

These categorizations help define the structure and access controls of blockchain networks, offering various options for specific use cases and requirements [35].

3.1.3 Blockchain Characteristics

The blockchain technology encompasses several distinct characteristics, including below.

Decentralization

The data and control in a blockchain network are distributed across multiple nodes, ensuring that no single entity has complete control, enhancing security and resilience.

Continuous Data Flow

Transactions are continuously added to the blockchain, creating an immutable and chronological record of all activities on the network.

Validated Transactions

Transactions on the blockchain are verified and validated by consensus mechanisms, ensuring the integrity and accuracy of the recorded data [35].

Transparency

The blockchain's public nature allows all participants to view and audit the entire transaction history, promoting trust and accountability.

Anonymity

While transactions are transparent, participants' identities are often represented by cryptographic keys, providing a certain level of anonymity.

Autonomy

The blockchain operates based on pre-defined rules and smart contracts, enabling automatic execution without the need for intermediaries.

3.2 *Power of AI*

Indeed, decentralization, continuous data flow, validated transactions, transparency, privacy, and autonomy are key characteristics of blockchain technology. Figure 2 illustrates a resource management framework leveraging blockchain technology. This framework comprises various elements and relies on three primary resources: energy and infrastructure, communication, and computing [36]. The blockchain functions as a central hub for data exchange and also serves as a regionally distributed system for resource management.

By implementing such a system, an open and transparent market is established, facilitating seamless transactions between different entities. The use of blockchain transactions further enhances the efficient utilization of available resources, optimizing their allocation and usage. This combination of characteristics and the integration of blockchain technology offers promising solutions for resource management and transactional efficiency.

3.2.1 Blockchain and AI Applications

The vast potential of intelligent computers equipped with machine learning capabilities has sparked a remarkable surge in the scope of AI applications, affecting corporations, governments, and society as a whole [37]. As the manuscript proceeds, it offers further background information on both blockchain and AI, with a primary focus on the following aspects as shown in Fig. 2.

Supply Chain Management

With the help of blockchain, every transaction and operation in a supply chain can be recorded transparently and irrevocably. The procedure for supply chains may be enhanced by using AI to analyse this data and provide insights into inconsistencies, identifying fraud, and statistical analysis [38].

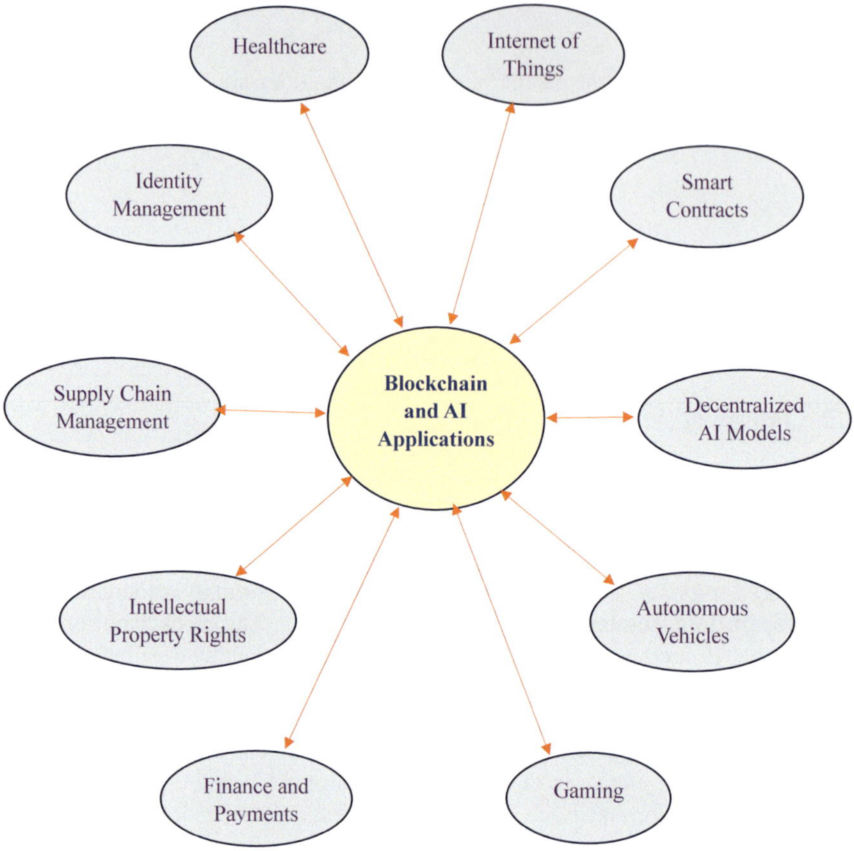

Fig. 2 Blockchain and AI applications

Identity Management

Blockchain can be used to verify identities in a decentralised and safe manner. The preciseness and dependability of identity verification processes can be improved by using AI-powered identification of faces and biometric identification.

Healthcare

Blockchain can enable safe and accessible medical records, enabling smooth access to data and sharing for both patients and healthcare professionals. Huge volumes of health information may be analysed by AI to help with diagnosis, therapy suggestions, and medication discovery [39].

Internet of Things (IoT)

By integrating AI and blockchain, IoT systems and networks may be made more secure and private. While AI can evaluate data produced by IoT devices to arrive at wise judgements, blockchain can offer a decentralised framework for device connectivity and verification of transactions.

Smart Contracts

These self-executing agreements have specific terms and constraints incorporated directly into the programme code. Smart contracts can incorporate AI to autonomously initiate particular operations based on predetermined conditions without the involvement of a human.

Decentralised AI Models

Blockchain can make it easier for AI models to be created and shared decentralized. Smart contracts enable data owners to monetize the use of their data while maintaining control over access, promoting fair reimbursement and security [40].

Autonomous Vehicles

Blockchain is capable of being used to safely store and distribute information about the operation, upkeep, and accidents of self-driving automobiles. This data can be processed by AI to enhance the driving skills and general safety of the cars.

Intellectual Property Rights

Blockchain can create a verifiable and unchangeable record of rights to intellectual property. By examining a large amount of online content, AI can help with safeguarding intellectual property and infringement of copyright identification.

Finance and Payments

Blockchain-based digital currencies and smart contracts have the potential to revolutionise financial interactions and make international payments easier and safer. In order to identify fraud, evaluate credit risk, and improve strategies for investment, AI can analyse financial information [41].

Gaming

Blockchain can be used in gaming to support asset ownership, allowing gamers to truly own the characters and objects they use in-game. AI can improve game experiences by creating dynamic and customised content.

By addressing a variety of issues with confidence, openness, and efficiency across a variety of disciplines, integrating artificial intelligence with blockchain has the ability to open up new possibilities and upend established business models. It's crucial to remember that implementing these technologies comes with its own set of difficulties and concerns, including flexibility, data protection, and compliance with laws and regulations.

3.3 The Convergence of AI and Blockchain for Data Analytics

The fusion of blockchain with AI has shown to be a formidable and revolutionary combination that has the ability to change data analytics. The combination of each of these technologies offers special capabilities to improve the safety, confidentiality, and dependability of data-driven operations as organizations progressively depend on data to gather insightful knowledge and make knowledgeable decisions [42].

AI has made strides in predictive modelling, machine learning, machine vision, as well as other applications of data analytics by handling and analysing enormous quantities of data. However, as AI algorithms frequently need access to large volumes of data, issues about security and confidentiality of information are raised.

Such problems are addressed by the use of blockchain technology, which offers a safe and open framework for storage of information and administration. Blockchain technology is recognized for its decentralized and irreversible ledger. Data integrity is improved and unauthorized access is prevented through blockchain, which guarantees that information cannot be changed or tampered therewith once it has been recorded [43].

3.3.1 Advantages

The convergence of AI and blockchain brings several advantages to data analytics.

Secure Data Sharing

Organizations have the ability to safely share data with a variety of partners using the decentralized architecture of the blockchain while maintaining authority over its accessibility and utilization. Smart contracts make it possible to automate and permissioned share data, protecting the safety and confidentiality of such data.

Data Lineage and Auditing

The immutability of blockchains makes it possible to track the origins of data in an open and accountable manner. This characteristic increases the credibility of AI-generated findings by enabling partners to confirm the reliability of information sources and the analytical techniques [44].

Analytics that Protect Privacy

By using artificial intelligence algorithms to secure information that is kept on the blockchain, businesses may run data analytics without disclosing the unprocessed information. Data privacy and security are ensured by this method.

Tokenized Incentives

The capacity of blockchain to produce electronic currencies paves the way for incentive structures for data producers, artificial intelligence (AI) model creators, and validators [45]. For the contributions they make, people or organizations might receive tokens, encouraging a cooperative data analytics ecosystem.

Blockchain may facilitate federated learning, in which artificial intelligence (AI) models undergo training using global sources of data instead of centralized data storage. This strategy protects confidentiality of information and promotes data exchange for better analytics results.

Although the fusion of blockchain with AI offers exciting possibilities for analytics of data, it also comes with a number of distinct difficulties, such as flexibility, interoperability, and processing overhead. In order to guarantee ethical deployment, legislative and regulatory considerations like confidentiality of information compliance and rights to intellectual property have to be properly considered [46].

In the end, the fusion of AI with blockchain results in an effective blend that strengthens safety, confidentiality, and confidence in data-driven operations, improving data analytics [47]. Organizations may open up new avenues for safe and dependable data analytics, resulting in more knowledgeable choices and game-changing insights, by combining the benefits of the two platforms.

4 Enhancing Data Privacy with Blockchain

One of the key advantages of combining these technologies is the enhancement of data privacy through blockchain. Blockchain's decentralized and impenetrable design offers a solid framework for protecting sensitive data and guaranteeing user privacy. The following are some ways that blockchain improves data privacy.

4.1 Decentralization and Immutability

The two core characteristics of blockchain technology—decentralization and immutability—play a key role in increasing data privacy.

4.1.1 Decentralization

Blockchain runs on a dispersed system of nodes, as opposed to conventional centralized systems, where information is held and managed by an individual body. Every member of the group has an exact replica of the complete blockchain ledger. By doing away with the requirement for a centralized middleman, this decentralized architecture lowers the possibility of one single point of breakdown or unauthorized use of data [48].

Information is not entirely managed by any one organization in a decentralized blockchain network, which renders it more resistant to breaches and ensures no one organization has complete authority over the information. As individuals are able to independently verify the veracity and reliability of data without depending on a centralized authority, decentralization promotes confidence among users.

4.1.2 Immutability

The failure to change or remove data once it has been stored on a blockchain is referred to as immutability. The blockchain is a network of data links where each block is an encrypted version of the one before it. Any effort to change data would necessitate altering the information in following blocks, which is practically impossible because of the consensus technique employed by blockchain systems [49].

When data is stored on the distributed ledger, its immutability guarantees that it will remain accurate and open during its entire existence. An accountable and open record of data operations and modifications is provided by this attribute, which improves the integrity of data and minimizes unauthorized updates.

Blockchain technology establishes a framework for private data that is safe and reliable by fusing decentralization and immutability. Stakeholders can confidently communicate confidential information because they understand that the data is transferred reliably via a network along with any modifications to the information have been openly documented and verified. These characteristics are especially useful in applications in which security and confidentiality of information are essential, such as healthcare, financial services, distribution chains, and identity administration, enabling organizations and individuals to preserve authority over personal data and protecting data confidentiality in a quickly changing environment.

4.2 Preventing Unauthorized Access

4.2.1 Blockchain networks use access controls and encryption techniques to handle issues with data security. Operations on open blockchains like Bitcoins are fictitious, which means that user identities are hidden behind encrypted addresses. Whereas secure blockchains can add more severe access restrictions to limit visibility of information to authorized parties, public ones prioritize openness [50].

4.2.2 Data access control at the granular level is made possible by the use of smart contracts, executable code on the distributed ledger. The smart contract can set access rights that permit particular individuals or organizations to see or alter data in accordance with established regulations. This strategy minimizes the risk of unauthorized data breaches by ensuring that only authorized stakeholders can access confidential data [51].

4.3 Mitigating Security Breach Risks

In data-driven initiatives, reducing the likelihood of privacy thefts is of the utmost significance. Blockchain technology offers a number of techniques to improve security of information and reduce the chance of breaches [52]. The following are important tactics to use blockchain to reduce the risk of privacy theft [53].

4.3.1 Single Point of Failure

Blockchain is based on a distributed system of nodes, which eliminates single points of failures. Such attacks are less damaging when a single node has a security breach because the whole network is not compromised [54].

4.3.2 Consensus Procedure

Blockchain networks use consensus procedures to verify and concur on the accuracy of data. By preventing hostile actors from changing data without agreement, the blockchain's entries are very difficult to tamper with [55].

4.3.3 Immutability

Data that has been stored on the distributed ledger is no longer able to be changed or erased. By offering an impermeable to record of auditing that discourages possible attackers, this function protects the reliability and validity of data [56].

4.3.4 Encryption

Blockchain networks are capable of safeguarding data throughout its transfer and preservation by utilizing cutting-edge encryption methods. By encrypting data, you can make certain that even if unauthorized individuals access the distributed ledger, the data will still be encrypted and impossible to decrypt without the right keys for decryption [56].

4.3.5 Identity Management

Cryptographic keys are used in blockchain to identify users, ensuring safe authentication. The risk of unauthorized utilization of a network or confidential information is decreased by effective management of identities [57].

4.3.6 Access Controls

The blockchain's smart contracts allow for the definition of granular controls on access for data. The danger of hacking into data can be reduced by including access authorizations in the smart contract, which will ensure that only authorized individuals can access particular data [58].

4.3.7 Regular Monitoring and Periodic Inspections

Regular monitoring and periodic inspections of the distributed ledger system help quickly identify unusual activity and possible breaches of confidentiality. Early identification enables rapid responses and mitigation measures.

4.3.8 Bug-Free Smart Contracts

To prevent weaknesses, smart contracts must be carefully developed and audited. Prioritizing secure coding practices is crucial since smart contract flaws or bugs might result in data loss or security violations [59].

4.3.9 Blockchains with Access Restrictions

Using a blockchain with restricted access is recommended for scenarios where confidentiality of information is of utmost importance. As a result, there is less chance of confidential data being accessed by unauthorized people [60].

4.3.10 Data Backups and Redundancy

Keeping blockchain's private data redundant and regularly backing it up reduces the consequences of possible breaches of security by guaranteeing accessibility in the event of infrastructure failures or crimes [61].

Organizations may considerably improve security, lower the danger of data theft, and guarantee the accuracy of data in initiatives that are data-driven by putting these techniques into practice. It is vital to understand that no system is completely impervious to privacy hazards, and that constant enhancement and continued diligence are required to keep on top of emerging threats.

5 Secure Data Sharing and Collaboration

The exchange and cooperation of protected data are essential components of data-driven initiatives, and the use of blockchain technology is vital in strengthening the safety and confidentiality of these procedures. Historically, security breaches, unauthorized use, and a lack of user confidence have presented problems for sharing information and engagement [62]. The issues can be addressed through the utilization of decentralized, immutable, and transparent blockchain technology.

5.1 Securing Remedies

To guarantee that people and organizations have the opportunity to seek fair and reasonable solutions whenever challenged with legal conflicts or unfair practices, it is essential to secure remedies [63]. Remedies are used to remedy harm, right injustices, and defend the rule of law, whether in terms of legal systems, economic dealings, or societal challenges.

5.1.1 Decentralization and Immutability

Blockchain's decentralized structure of nodes and immutability make it possible to function without a governing body. Data is spread across numerous nodes in the network, increasing its resistance to single points of failures and lowering the possibility of unauthorized access [64]. Data that has been stored on the blockchain remains immutable, which means that it cannot be changed or removed without the entire network's approval. These characteristics foster trust between stakeholders by ensuring the validity and integrity of provided data.

5.1.2 Consensus Mechanisms

Consensus techniques are used by blockchain systems to verify and consent to the legitimacy of data contributed to the blockchain. These procedures, such as Proof of Work (Pow) or Proof of Stake (PoS), guarantee that data is validated and approved by a significant number of network members prior to being added to the network. By preventing malevolent actors from altering or interfering with data, this consensus method improves data security [65].

5.1.3 Encrypted Data Sharing

Blockchain can help with confidential information sharing, allowing for the safe transmission and storage of confidential information. Information that is shared is further protected by encryption, which guarantees that only authorized individuals with the necessary decryption credentials are allowed to view the data [66].

5.1.4 Smart Contracts

Self-executing code known as "smart contracts" is kept on the distributed ledger and automatically executes specified actions when certain criteria are satisfied. Communication contracts can be implemented via smart contracts, which guarantee that information can only be used and utilized in line with predetermined rules and rights. As a result, there are no longer any intermediaries required in the processes for sharing data, which lowers costs and minimizes security risks [67].

5.1.5 Tokenized Incentives

Blockchain enables the development of digital currencies that can be utilized for monetary engagement and the exchange of data. Tokens may be awarded to users for exchanging data, assisting with analytics procedures, or offering insightful commentary. Such tokenized incentives promote information exchange and build a cooperative ecology [68].

Reliable collaboration and the sharing of data are made possible by utilizing blockchain's characteristics without sacrificing security or confidentiality of information. The use of distributed ledgers in data-driven projects encourages member accountability, openness, and confidence, which improves data sharing procedures across a range of businesses.

Blockchain-based smart contract systems are a foundational element in ensuring secure data sharing and collaboration. Smart contracts are self-executing contracts with predefined rules and conditions written in code. They reside on the blockchain and are automatically executed when specific conditions are met, without the need for intermediaries. These contracts enable trust less interactions, as all parties involved can rely on the transparent and immutable nature of the blockchain to enforce the agreed-upon terms.

5.2 Advantages

In the context of data sharing and collaboration, blockchain-based smart contract systems offer several key advantages [69].

5.2.1 Transparency and Trust

The blockchain makes smart contracts accessible to all users, fostering faith in the sharing information process. There is not a requirement to depend on a single source for validation because anyone may examine the agreement's logic and check that it is being executed.

5.2.2 Autonomy and Decentralization

Once implemented, smart contracts run independently without outside assistance. This decentralized structure reduces the possibility of human mistake or manipulation and eliminates the likelihood of single points of failure.

5.2.3 Data Security and Privacy

Smart contracts provide the ability to impose access controls and define who has access to and is allowed to use the shared data. Security and confidentiality of information can be ensured by storing encrypted data on a distributed ledger and limiting access to critical information using secret keys [70].

5.2.4 Immutable Records

On the distributed ledger, smart contracts keep track of all communications and data transfers, establishing a lasting record of every transaction. By offering an auditable track, this feature improves data reliability and transparency.

5.2.5 Automatic Execution

Smart contracts speed up data-sharing procedures by controlling the carrying out of agreement clauses. This effectiveness lowers expenses while accelerating member engagement.

Blockchain coins can be used by smart contracts as rewards for data exchange and cooperation. Tokens can be given to members in exchange for their data, knowledge, or computing power, promoting a cooperative and financially motivated ecosystem.

However, as defects or weaknesses in the code might have serious repercussions, it is crucial to carefully create smart contracts and properly analyse them for possible weaknesses. To guarantee adherence to the relevant rules and laws, legislative and governmental considerations must also be made while creating agreements [71].

In general, by offering a safe, open, and self-governing framework for data-driven conversations based on blockchain technology, smart contract networks play a critical part in improving data sharing and cooperation. Smart contract incorporation in data-driven initiatives enables members to participate in secure and reliable data transfers, fostering innovation and boosting productivity in a variety of industries.

5.3 Data Usage Agreements

A key step in guaranteeing secure data exchange and teamwork within the framework of the blockchain system is the definition of use of information contracts. These contracts specify the circumstances under which parties can obtain, utilize, and exchange data [72]. Organizations can increase confidence and dependability across providers of data, customers, and various other partners by implementing simple and open use of data contracts.

5.3.1 Data Usage Agreement Element

Key elements in defining data usage agreements on the blockchain include below.

Data Use Agreements

These agreements define who is granted access to the data that is shared and what rights they have. Varying roles or entities associated with the partnership may have varying access privileges. For example, although some individuals may just have permission to peruse the data, individuals might be given the authority to edit or analyse it [73].

Restrictions on Data Usage

Limitations on the use of data may be outlined in Data Usage Agreements. This may include limitations on the utilization of the data for business reasons, beyond the parameters of the partnership, or even dissemination. These limitations assist in making certain that the information is utilized sensibly and for the purpose it was created.

Intellectual Property Rights and Data Ownership

The contract should specify the ownership of data responsibilities and handle any issues with intellectual property. Either data producers keep control over their data or provide that ownership to the collaborative organization should be made clear [74].

Safety and Protection

Clauses addressing the safety and privacy of data should be included in data usage contracts. This involves verifying adherence to pertinent data protection laws and defining safeguards to guard against unauthorized access, manipulation, or violations of data security.

Lifespan and Termination

The contract should outline the collaboration's lifespan as well as the circumstances under which it may be terminated. It should specify how information will be managed after the cooperation is finished, like by removal or return to the original data providers [75].

Consent and Ethical Considerations

Data usage contracts should include provisions addressing consent specifications, particularly for critical or private information. User consent for sharing information must be obtained in writing, and participants must be made aware of how their information will be utilized.

Dispute Resolution

In the event of conflicts or contract violations, the contract may include procedures for resolving disputes. By establishing an accountable record of conversations, the openness and inviolability of blockchain technology can help resolve conflicts.

Participants can gain from an automatic and impenetrable system that upholds the agreed-upon rules by putting such information usage contracts on the distributed ledger using smart contracts. The conditions are open, accountable, and implemented independently because of the implementation of blockchain technology, which reduces the need for faith in a centralized authority and improves security.

5.4 Safe Data Exchange and Collaboration

Data-driven initiatives must enable secure data interchange and cooperation, and blockchain technology provides helpful solutions to improve trust and safety in these procedures [76]. Several major advantages result from incorporating blockchain technology into data collaboration and exchange.

5.4.1 Decentralization and Security

The decentralized nature of blockchain eliminates the requirement for a centralized organization to promote interaction and cooperation. Because each member of the system has an exact replica of the blockchain, redundancy of information is ensured and the likelihood of a single point of failures is decreased [77]. By reducing dependence on a single body for handling data, its decentralized nature improves data security.

5.4.2 Immutable Data Files

Information transmitted and stored on the blockchain is inviolable, which means that once it has been done so, it cannot be changed or removed without the entire network's approval. Through the provision of an accountable record of all information exchanges and activities operations, this feature protects the reliability and legitimacy of shared data.

5.4.3 Encrypted Security

To protect data, blockchain uses cutting-edge encryption techniques. Because transactions on the distributed ledger are by encryption hashed and connected, it is very challenging for attackers to covertly change data. Using both public and private key encryption, only authorized parties may safely access and share data.

5.4.4 Using Smart Contracts to Facilitate Automated Collaboration

The blockchain's smart contract structures, which are pieces of executable code, allow for autonomous and automatic collaboration. These agreements ensure that collaboration and sharing of data take place in accordance with predetermined terms by automatically enforcing specified norms and restrictions [78]. This decreases the possibility of human error and does away with the necessity for middlemen.

5.4.5 Access to Permissioned Data

Blockchain networks may be created as permissioned systems, in which users must obtain permission before they can participate in the system to utilize its shared data. This makes it possible to have exact control about data access and guarantees that only parties with permission can work together to contribute to the system [79].

Blockchain currencies can be utilized as rewards for secure data sharing and teamwork. Tokens may be given to members in exchange for their data, knowledge, or computational assets. The collaborative and rewarded sharing information ecology is fostered by these tokenized incentives, which promote active engagement.

5.4.6 Superior Data Privacy

Sensitive data is safely exchanged and safeguarded on the blockchain, facilitating secure information sharing. An additional degree of security and confidentiality is provided by encryption, which makes sure that only people with the necessary permissions and keys for decryption can access the data.

Organizations can establish a safe, open, and cooperative platform for sharing information by utilizing blockchain technology. By incorporating blockchain technology into data-driven initiatives, participants are more likely to feel trusted, data security is improved, and the collaboration process is streamlined, resulting in insights based on data and developments that are more dependable and effective.

6 Federated Learning and Data Privacy

Organizations can leverage federated learning to fully realize the benefits of cooperative analysis of data and artificial intelligence while maintaining the confidentiality of data. This strategy is particularly pertinent to sectors like medical care, banking, and communications that have strict regulations on the confidentiality of information [80]. Federated learning is constantly changing as advancements in technology occur, and continuing research intends to further enhance its secure characteristics, making it an important tool for safe and considerate data analytics.

6.1 The Concept of Federated Learning

A decentralized machine learning method called federated learning enables different devices or organizations to cooperatively train a single network without exchanging their unprocessed information with a centralized server. When using confidential information, security and confidentiality problems may arise since in conventional artificial intelligence, data is gathered and centralized in one place for model training [81].

Federated learning allays these worries by limiting updates to models to being sent to a centralized server while maintaining data on personal servers or endpoints [82]. The following steps are often included in the federated learning procedure, as shown in Fig. 3.

Organizations can leverage federated learning to fully realize the benefits of cooperative analysis of data and artificial intelligence while maintaining the confidentiality of data. This strategy is particularly pertinent to sectors like medical care, banking, and communications that have strict regulations on the confidentiality of information [80]. Federated learning is constantly changing as advancements in technology occur, and continuing research intends to further enhance its secure characteristics, making it an important tool for safe and considerate data analytics.

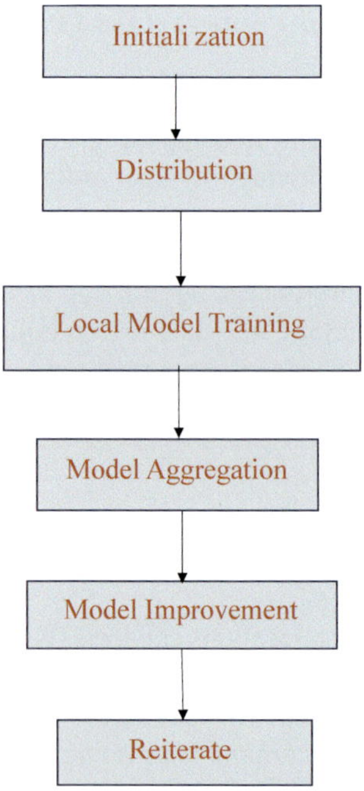

Fig. 3 Federated learning

6.1.1 Initialization

At the central server's level, the worldwide model is constructed and initialized.

6.1.2 Distribution

Each device or network holding local information receives the worldwide model.

6.1.3 Local Model Training

Despite exchanging the unprocessed information with the centralized server, every device independently develops the worldwide model with its own individual data [83]. The efficiency of the model may be enhanced through this training procedure over several cycles.

6.1.4 Model Aggregation

Every equipment transmits its modified model parameters—not unprocessed data—back to the centralized server after regional training.

6.1.5 Aggregated Model Improvement

Using the information gleaned from multiple data sources, the central server collects the model upgrades across every device.

6.1.6 Reiterate

To further enhance the worldwide model, the repeated localized training, model accumulation, and model updating processes are carried out.

6.2 Advantages

Federated learning offers several advantages.

6.2.1 Data Security

By keeping data decentralized, federated learning lowers the possibility of data disclosure, compromises, or unauthorized access. To improve the security and confidentiality of data, confidential information is kept on devices that are local, and just latest models are transferred [84].

6.2.2 Efficiency

For huge datasets, centralized training of models may not be feasible because it involves data transport and can be computationally challenging. Federated learning divides up the training procedure, minimizing the demand for substantial transfer of information and conserving computer resources.

6.2.3 Flexibility

Federated learning is perfect for situations with a variety of data distributions since it enables diverse devices and information sources to take a role in the process of training models.

6.2.4 Edge Computing

Federated learning is particularly appropriate for edge computing situations, wherein endpoints have restricted processing capacity or connectivity, as information is handled and taught directly on devices [85].

In many fields where confidentiality of information, safety, and communication are crucial, such as healthcare, banking, mobile technology, the Internet of Things (IoT), and federated analysis, federated learning is used. While federated learning overcomes issues with data privacy, it also raises issues with model integration, communication effectiveness, and managing uneven distributions of data among devices. These issues will be addressed, and the entire potential of federated learning for cooperative and private machine learning will be realized, thanks to continuous studies and developments in the field.

6.3 Preserving Data Privacy with AI and Blockchain Integration

An effective way to protect confidentiality of information in projects based on data is through the combination of artificial intelligence and blockchain. Businesses may improve security and confidentiality of information while allowing sophisticated machine learning and analytics capabilities by integrating the best features of the two technologies [86]. The combination protects the privacy of data in the following significant ways.

6.3.1 Decentralized Data Preservation

The decentralized nature of the blockchain reduces the danger of one point of failures and unauthorized access by allowing data to be disseminated throughout the network. Highly confidential information can be less exposed since artificial intelligence algorithms can work instantly on data that is kept in the blockchain rather than having to send it to a centralized server [87].

6.3.2 Protected Information Sharing

Blockchain permits the storing and transfer of encrypted data, guaranteeing that data is kept private and secure throughout sharing and interaction. Further preserving data privacy, artificial intelligence algorithms are capable of handling data that is encrypted without entirely decrypting it.

6.3.3 Smart Contracts with Data Access Management

Smart contracts on the blockchain let you control data access in very specific ways. Data confidentiality agreements can be rigorously enforced by organizations by defining who has permission to see what information and under what circumstances.

6.3.4 Tokenized Data Accessibility

With blockchain currencies, information use may be restricted as well as information sharing can be rewarded. The owners of data can create tokens that allow permission to their information, and currencies can be awarded to artificial intelligence algorithms for analysing information without actually gaining access to it [88].

6.3.5 Auditable Data Provenance

The transparency and immutability of blockchain technology create a verifiable record of data exchanges and exchanges. Data owners can monitor how their information is used, increasing responsibility as well as openness.

6.3.6 Data Privacy and De-identification

When analysing and interpreting data, artificial intelligence approaches like different levels of privacy and the data anonymity abilities of blockchains can be utilized to secure people's identities as well as confidential information [89].

6.3.7 Regulation Adherence

Blockchain's secure nature and AI's capacity to implement data privacy standards ensure adherence to data protection laws like GDPR, HIPAA, and more.

6.3.8 User-Controlled Data Exchange

By combining blockchain and artificial intelligence, consumers can take more ownership over their information, choose who has permission to use it and how, and maintain their personal information security choices.

Organizations may create a safe and confidentiality data environment by utilizing the convergence of artificial intelligence and blockchain. Researchers can maintain authority over what they share, and artificial intelligence algorithms can learn from a large body of data without jeopardizing the privacy of individual users. Through this combination, people can collaborate and innovate in a protection-conscious way while also improving the security of their data.

7 Tokenized Incentives for Users

The process for launching tokenized rewards for individuals includes giving users digital tokens as prizes or rewards for their involvement, engagement, or certain acts inside an environment or ecosystem. These tokens have value and may be traded, redeemed, or applied in a variety of ways throughout the system [90]. In blockchain-based initiatives and decentralized platforms, tokenized rewards have become popular as a means of promoting user engagement, adherence, and contributions. By making tokens available, companies can build a community in which users are encouraged to take an active role in the development and accomplishments of the platform.

7.1 Introducing Token Economies

A decentralized system based on the blockchain system called a "token economy" makes use of tokens as virtual assets that symbolize value and reward different acts or behaviours inside a network. Smart contracts, which serve as autonomous agreements on the distributed ledger, are used for manufacturing and distributing tokens. These tokens of value have uses beyond typical cryptocurrency and can be utilized as a sort of virtual currency inside the overall system [91].

Token economies are intended to increase participation among users, encourage desirable behaviours, and build a network that can support itself. In accordance with the guidelines and norms outlined in the smart contracts, network users can acquire, trade, and use tokens for particular tasks, products, or benefits [92].

Key elements and characteristics of token-based economies are as follows.

7.1.1 Token Creation

Tokens frequently arise through token generation events (TGEs) or initial coin offers, in which a particular number of tokens are distributed and sold to the attendees or shareholders. The token economy can be developed and expanded using the money earned from such occurrences.

7.1.2 Token Utility

In the economy of tokens, tokens may serve a variety of purposes, including granting users' access to products and services, the right to vote, or giving them incentives for making contributions. The value of the token's layout plays a crucial role in encouraging desirable behaviours and increasing engagement among users.

7.1.3 Incentives and Rewards

For certain behaviours or services to the ecosystem, users are rewarded and incentivized with tokens. For instance, users might be compensated with cryptocurrencies for data communication, transaction validation, platform creation work, or customer referrals [93].

7.1.4 Transparency and Decentralization

Token economies run on decentralized blockchain systems, which have both of these qualities. On the blockchain, every transaction and token transfer are tracked, promoting openness and accountability.

7.1.5 Interoperability

Users may exchange or transfer tokens for other digital currencies or fiat money on a variety of cryptocurrency trading platforms. The flexibility and increased value of the token result from this interconnectivity.

7.1.6 Governance

Decentralized governance methods may be implemented in token-based economies, allowing users of tokens to take part in making decisions and cast votes on crucial matters affecting the growth and functioning of the ecosystem.

7.1.7 Trust and Security

The immutability and safety of token exchanges are guaranteed by the use of blockchain technology, lowering the possibility of theft or tampering. Members can rely on the token economic system's regulations and procedures to be followed as specified in smart contract agreements.

Numerous industries, like banking, gaming, supply chain management, and decentralized applications (DApps), have adopted token economies [94]. They provide a new framework for exchanges of value and are driven by communities' ecosystems, as well as enabling creative business strategies. However, the success of a token economic system depends on careful planning, a large member base, and ongoing governance to guarantee justice and stability.

7.2 Rewarding Data Contributors and Validators

Token economies can compensate Data Contributors and Validators for their assistance to the ecosystem of actors in initiatives that are driven by data.

7.2.1 Data Contributors

As a reward for providing their data, people or organizations that contribute relevant information to the network may be given tokens. These tokens serve to acknowledge the value of the data and to motivate its owners to actively engage with the ecosystem. Token economies encourage collaboration and data sharing by compensating data contributors [95].

7.2.2 Validators

Validators are essential to maintaining the system's data reliability and accuracy. They check and confirm the accuracy of data changes and update the model. Validators in a federated learning environment could be in charge of collecting model updates from various devices [96].

Smart contracts that automatically carry out token transfers depending on specified requirements, such as the quantity and quality of information given or the accuracy of validation of models, can be used to disperse token incentives following predefined regulations. The system of rewards is made more transparent and equitable thanks to this automated procedure, which also lessens the demand for centralized administration [97].

Data-driven initiatives can encourage active involvement, enhance the accuracy of data, and foster innovation inside the ecosystem by offering tokenized rewards for users. By bringing members' interests into line with the network's achievement, token-based economies foster engagement and value generation through fostering a win–win situation.

8 Technical Challenges of AI and Blockchain Integration

To fully realize the promise of this strong combination, a number of technological obstacles relating to the combination of artificial intelligence and blockchain technology must be resolved [98]. Among the principal difficulties are below.

8.1 Scalability Issues

Due to the fundamental structure and properties of both methods, combining artificial intelligence with blockchain poses serious scalability issues. Due to the requirement for every network node to handle and maintain each transaction and piece of information on the chain, blockchain technology faces a scalability problem. The blockchain's size continuously expands with the number of users and operations, requiring more storage space and increasing the duration of processing [99].

The resource-intensive characteristics of sophisticated machine learning algorithms present scaling issues for artificial intelligence, particularly when handling enormous quantities of data. Large artificial intelligence models demand a lot of compute and memory, which can be taxing on blockchain nodes' resources and slow down the whole network.

The difficulties with scaling are exacerbated when artificial intelligence and blockchain are combined because both technologies demand a lot of resources. These problems are currently being worked on, and efforts include investigating outside-of-the-chain remedies, sharding strategies, and resource-saving AI algorithm optimization [100]. To increase scalability without sacrificing the safety and decentralization features of the blockchain, layer-two alternatives like secondary chains and state networks are also being investigated.

8.2 *Interoperability Challenges*

Interoperability is the efficient communication and interaction between various blockchain-based networks and artificial intelligence systems. Data exchange, the implementation of smart contracts, or transactions across chains may all be a part of the artificial intelligence and blockchain incorporation process. Interoperability is difficult to establish since different blockchain systems have different protocols, consensus procedures, and data structures [101].

When combining artificial intelligence models developed on several blockchain systems or exchanging data among different blockchain platforms, interoperability issues may appear. Standardization of data standards and methods of communication may also be necessary to ensure effective interaction between applications based on artificial intelligence and smart contracts.

The creation of bridging methods, communication across chain regulations, and compatible protocols are necessary to tackle interoperability difficulties. The blockchain ecosystem is working to develop interoperability options that enable the seamless fusion of blockchain with artificial intelligence systems.

8.3 *Processing Costs and Efficiency*

Blockchain technology and artificial intelligence both entail relatively expensive procedures that might result in high expenses for processing and decreased efficiency as a whole. As a result, whenever AI-related operations are done on the blockchain, fees for transactions may increase. Models based on deep learning, in particular, require a lot of processing power.

Additionally, the Proof of Work (Pow) and Proof of Stake (PoS) consensus techniques used by many blockchains consume a significant amount of energy, resulting in computational inefficiencies and problems with the environment [102].

Research focuses on increasing the computing efficiency of AI systems to increase processing effectiveness and decrease expenses. In order to save energy and boost transaction performance, blockchain-based systems are also investigating different consensus procedures, including Proof of Authority (PoA) and Delegated Proof of Stake (DPoS).

The combination of AI with blockchain can open up new opportunities and applications in a variety of sectors, including healthcare, banking, supply chain, and IoT, by overcoming these technical obstacles. To get over these challenges and fully realize the endless possibilities of this revolutionary integration, further study and development are essential.

9 Legal and Regulatory Considerations

To maintain compliance and promote appropriate use of these technologies, it is important to properly handle the legal and regulatory issues that arise when combining blockchain and artificial intelligence technologies. Among the most important legal and regulatory factors are below.

9.1 Data Privacy Compliance

Organizations must solve complicated data privacy issues to meet legal and regulatory obligations when integrating AI and blockchain. Organizations must abide by pertinent data protection rules, such as the General Data Protection Regulation (GDPR) of the European Union or other local data privacy legislation because these technologies entail the handling and retention of confidential information.

Data minimization techniques, the right to be forgotten (data erasure), explicit user authorization for the processing of data, and confidentiality and security of data are all part of data privacy compliance. Moreover, companies must notify customers in an open way about how they use, store, and share their data [103].

The decentralized characteristics of the blockchain system can make it difficult to comply with data privacy laws because data stored there is frequently immutable. To guarantee that private and sensitive information is properly safeguarded, and that the privacy of data subjects may be upheld, organizations must carefully build their blockchain applications.

Conducting privacy impact analyses, putting in place strict access restrictions, and adhering to best practices for encryption of data and pseudonymization are all things that organizations should do to ensure data security compliance in the combination of artificial intelligence and blockchain [104]. To manage the complexity of data protection rules and ensure that the combination complies with regulatory standards, interactions with legal specialists are also crucial.

9.2 Ensuring Ethical Implementation

To preserve public faith and confidence, ethical issues related to the combination of artificial intelligence and blockchain technology must be addressed. Providing that the algorithms used by AI are impartial, fair, and open, as well as that they don't support or infringe against human rights, requires ethical execution [105].

Organizations should be open and honest about how artificial intelligence (AI) algorithms are developed, the data utilized for education, and any potential repercussions of the applications powered by AI when combining AI with blockchain. Understanding how decisions are made and seeing possible biases can be accomplished by incorporating explainability and interpretability methods into artificial intelligence models [106].

Adhering to principles like informed consent, user freedom, and responsibility is part of ethical execution, along with taking technological factors into account. Data should be under people's authority, and they should be aware of how their information is used in the blockchain and AI ecosystems.

Additionally, organizations should think about how integrating AI and blockchain would affect society as a whole. Evaluating potential risks, unforeseen effects, and making sure the integration is consistent with moral principles and social conventions are all included in this.

Collaboration between ethicists, sociologists, and all other appropriate experts is crucial for promoting ethical execution. Businesses may address moral issues and make sure that blockchain and artificial intelligence deployment are done responsibly by engaging in a candid conversation with stakeholders, such as users and groups affected by the connection.

Finally, it should be noted that merging blockchain with AI requires careful attention to legal, regulatory, and ethical issues. Organizations may foster trust, safeguard user confidentiality, and ensure the moral and responsible use of these innovative innovations by actively dealing with these issues.

10 Conclusion

The combination of artificial intelligence with blockchain enabling data analytics presents a viable approach to the problems of data confidentiality, safety, and cooperation. This chapter has examined the difficulties organizations confront in protecting the safety and confidentiality of data when working on data-driven initiatives. We have examined the foundational ideas, characteristics, and applicability of the technology of blockchain to data analytics. Additionally, we have emphasized how combining blockchain with AI and machine intelligence may improve data confidentiality and security. In distributed AI training environments, the idea of federated training has been touted as a potent method to protect data privacy. Furthermore, we have looked at how the blockchain relies on smart contract solutions that allow for safe data exchange and teamwork while enforcing access restrictions as well as data usage contracts. Additionally, the use of token-based economies as a way to motivate and recognize data providers and validators inside the blockchain ecosystems has been discussed. Nevertheless, this integration has several technological difficulties that must be successfully controlled, such as scalability problems, difficulties with interoperability, and processing expenses. The significance of legal and governmental issues, such as information security compliance and moral execution, has also been emphasized.

The combined use of artificial intelligence and blockchain for data analytics has bright future possibilities and many facets. We may anticipate advances in decentralized federated learning, increased data privacy, scalability remedies, token-based innovation, ethical artificial intelligence governance, acceptance in a variety of sectors and laws and regulations as these innovations continue to advance. The combination of artificial intelligence and blockchain technology will provide effective privacy-preserving methods that will ensure confidential sharing of information while retaining security. Artificial intelligence models may be trained effectively on dispersed devices thanks to decentralized federated learning. Enhancing interoperability will enable easy data sharing between multiple blockchain networks. In order to support massive data analysis applications, solutions for scaling will solve processing constraints. A new token economy will reward user engagement and promote the expansion of blockchain ecosystems. Openness, equity, and responsibility will be given top priority in ethical AI governance. Blockchain technology and artificial intelligence (AI) will be used in a variety of businesses, revolutionizing data analysis and decision-making. Regulations will strike a balance between user safety and creativity.

In the end, the fusion of blockchain with AI has tremendous potential to revolutionize data analytics by improving security, confidentiality, and teamwork. These innovations will influence the development of data-driven initiatives as studies and innovations proceed, opening up new opportunities for both businesses and people. The ethical, moral, and technical facets of the data analytics environment must be carefully taken into account in order for artificial intelligence and blockchain technology to be integrated responsibly and sustainably.

References

1. Xu L, Wang C, Chen X, Shen X (2020) A survey on blockchain-based data security for industrial Internet of Things. IEEE Commun Surv Tutor
2. Bernardino M, Ferreto TC et al (2019) Blockchain for AI: review and open research challenges. IEEE Access
3. Shakir MZ, Saad W et al (2020) Blockchain and AI for federated learning in 6G networks: opportunities, challenges, and solutions. IEEE Open J Commun Soc
4. Puthal D et al (2019) Secure data sharing in cloud computing using blockchain and attribute-based encryption. Future Gener Comput Syst
5. Li R, Li S, Jiang T (2020) Secure data analytics in federated cloud-edge computing with blockchain. IEEE Trans Ind Inform
6. Gondal MW et al (2019) AI-driven blockchain for secure and privacy-preserving decentralized applications. Sensors
7. Khan SU et al (2019) Securing data sharing with blockchain and attribute-based encryption for IoT-enabled smart grids. J Netw Comput Appl
8. Anwar F et al (2020) Blockchain and AI-based secure and privacy-preserving data sharing framework for industrial Internet of Things. Comput Electr Eng

9. Zhang Y, Liu L, Liu Y (2019) Blockchain-based secure data sharing scheme with fine-grained access control in cloud storage. Comput Electr Eng 76:66–77
10. Li J, Wu Y, Fan P (2020) A secure data storage scheme using blockchain and attribute-based encryption for industrial Internet of Things. Comput Electr Eng 85:106658
11. Zhang L, Zhao Y, Liu J (2019) Blockchain-based secure and privacy-preserving content sharing scheme for industrial Internet of Things. IEEE Trans Industr Inf 15(3):1798–1807
12. Gao F, Qu M, Sun X (2020) Secure data sharing for smart city using blockchain and attribute-based encryption. Comput Electr Eng 84:106616
13. Yu Y, Zhang T, Xu W (2019) Blockchain-based secure data sharing scheme with user revocation in cloud storage. Clust Comput 22(S1):1465–1476
14. Cao Y, Wang X (2020) A secure data sharing scheme based on blockchain and smart contracts for industrial Internet of Things. Secur Commun Netw 2020:1–13
15. Sobrinho OG, Bernucci LLM, Corrêa PLP, Motta RS, Machicao J, Junqueira AS, Lopes F, Cassaro L, Oliveira L (2021) Big data analytics in support of the under-rail maintenance management at Vitória—Minas Railway. In: Proceedings of the 2021 IEEE international conference on big data (big data), pp 126–133
16. Cuzzocrea A, Leung CK, Hajian M, Jackson MD (2022) Effectively and efficiently supporting predictive big data analytics over open big data in the transportation sector: a Bayesian network framework. In: Proceedings of the 2022 IEEE international conference on dependable, autonomic and secure computing, international conference on pervasive intelligence and computing, international conference on cloud and big data computing, international conference on cyber science and technology congress (DASC/PiCom/CBDCom/CyberSciTech)
17. Rawat DB, Doku R, Garuba M (2021) Cybersecurity in big data era: from securing big data to data-driven security. IEEE Trans Serv Comput 14(6):2055–2072
18. Wang F, Wang H, Chen X (2023) Research on access control technology of big data cloud computing. In: Proceedings of the 2023 IEEE 3rd international conference on information technology, big data and artificial intelligence (ICIBA), p 10165326. https://doi.org/10.1109/ICIBA56860.2023
19. Mayer-Schönberger V, Cukier K (2013) Big data: a revolution that will transform how we live, work, and think
20. Mayer-Schönberger V, Cukier K (2013) Big data: a twenty-first century arms race. Foreign Affairs
21. Smolan R, Erwitt J (2012) The human face of big data
22. Marz N, Warren J (2015) Big data: principles and best practices of scalable real-time data systems
23. Chen M, Mao S, Liu Y (2014) Big data analytics: a survey. Mob Netw Appl
24. Heeks R (2017) Big data for development: challenges and opportunities. Inf Technol Dev
25. David RC, Fajar JK, Siahaan ML (2019) Big data: new oil or just semantics? A systematic literature review of big data concepts in marketing. Commun Comput Inf Sci
26. Bollier D, Bollier K (2010) The promise and peril of big data. The Aspen Institute
27. Zheng Z, Zhao Y, Wei J (2013) Big data and its technical challenges. J Softw Eng Appl
28. Lin J, Dyer C (2010) Data-intensive text processing with MapReduce
29. Andrejevic M (2014) The big data divide. Int J Commun 8:1673–1689
30. COM (2019) 168 final "Building trust in human-centric artificial intelligence". https://eur-lex.europa.eu/legal-content/EN/ALL/?uri=CELEX:52019DC0168. Accessed 26 July 2021
31. COM (2020) 65 final "White paper on artificial intelligence—a European approach to excellence and trust". https://ec.europa.eu/info/sites/default/files/commission-white-paper-artificial-intelligence-feb2020_en.pdf. Accessed 26 July 2021
32. COM (2020) 66 final "A European strategy for data". https://eur-lex.europa.eu/legal-content/EN/TXT/?uri=CELEX%3A52020DC0066. Accessed 26 July 2021
33. COM (2020c) 767 final "Proposal for a regulation of the European Parliament and of the Council on European data governance (Data Governance Act)". https://eur-lex.europa.eu/legal-content/EN/TXT/?uri=CELEX%3A52020PC0767. Accessed 26 July 2021

34. ENISA (2015) Privacy by design in big data. An overview of privacy enhancing technologies in the era of big data analytics. www.enisa.europa.eu. Accessed 26 July 2021
35. Favaretto M, De Clercq E, Elger BS (2019) Big data and discrimination: perils, promises and solutions. A systematic review. J Big Data 6(1):12
36. van Brakel R (2016) Pre-emptive big data surveillance and its (dis)empowering consequences: the case of predictive policing. In: van der Sloot B, Broeders D, Schrijvers E (eds) Exploring the boundaries of big data. Amsterdam University, Amsterdam, pp 117–141
37. Gupta M, Dey S, Sharma SK, Singh J, Chang V (2020) AI and blockchain-enabled privacy-preserving big data analytics for smart healthcare. J Big Data 7(1):1–21
38. Cai Q, Zhu L, Peng W (2020) Blockchain and AI in edge computing: opportunities and challenges. IEEE Netw 34(5):126–133
39. Nguyen DT, Kim YH, Kim S (2019) AI-based blockchain technology for secure IoT communications. IEEE Access 7:101553–101562
40. Lemieux VL (2020) Democratizing AI and blockchain: are we headed towards a distributed utopia or digital feudalism? Proc IEEE 108(3):440–451
41. Suo H, Wan J, Zhang C, Liu J (2020) A survey on blockchain and AI technologies for healthcare. IEEE Access 8:121034–121054
42. Yazdavar AH, Li S, Fathy M (2019) Blockchain for AI-enabled manufacturing. Procedia CIRP 84:830–835
43. Park S, Yang JY, Lee J (2021) Blockchain meets AI: challenges and opportunities. J Commun Netw 23(1):1–11
44. Talwar R, Chatterjee S, Rana S, Raghavendra R (2020) Towards an AI and blockchain enabled integrated edge-cloud computing framework. Futur Gener Comput Syst 111:1045–1059
45. Sang J, Wan J, Chen C, Wang S (2020) A review of blockchain technologies for different IoT applications. J Netw Comput Appl 168:102675
46. Lam WWY, Lam JSL (2020) Blockchain and AI-enabled healthcare applications: a systematic review. Healthcare 8(3):232
47. Maheshwari H, Chandra U, Yadav D, Gupta A, Kaur R (2023) Machine learning and blockchain: a promising future. In: Proceedings of the 2023 4th international conference on intelligent engineering and management (ICIEM)
48. Nakamoto S (2008) Bitcoin: a peer-to-peer electronic cash system. https://bitcoin.org/bitcoin.pdf
49. Swan M (2015) Blockchain: blueprint for a new economy. O'Reilly Media
50. Buterin V (2013) Ethereum white paper: a next-generation smart contract and decentralized application platform. https://github.com/ethereum/wiki/wiki/White-Paper
51. Narayanan A, Bonneau J, Felten E, Miller A, Goldfeder S (2016) Bitcoin and cryptocurrency technologies: a comprehensive introduction. Princeton University Press
52. Yli-Huumo J, Ko D, Choi S, Park S, Smolander K (2016) Where is current research on blockchain technology? A systematic review. PLoS ONE 11(10):e0163477
53. Bechini A, Cioni A, Gori A (2019) Blockchain as a service for data privacy management in IoT scenarios. IEEE Internet Things J 6(2):2625–2637
54. Qin J, Strinati EC (2020) Toward user-centric identity management systems with blockchain. IEEE Trans Dependable Secure Comput 17(6):1339–1353
55. Sharma PK, Zhang H, Li Z (2019) A privacy-preserving framework for EHR using blockchain and IPFS. J Med Syst 43(5):133
56. Pawani H, Desai A, Wang S (2019) A blockchain based data privacy and integrity preservation framework for smart grid communication networks. J Netw Comput Appl 126:85–97
57. Garg D, Chaudhary P, Sharma P (2020) Blockchain based privacy-preserving healthcare system. Sustain Cities Soc 54:101945
58. Zhang Q, Wen Y, Sun X (2019) Blockchain-based privacy-preserving data sharing scheme for internet of vehicles. IEEE Internet Things J 7(2):1194–1204
59. Fang Y, Kanellopoulos D, Xu Y, Ramachandran M (2019) A blockchain-based approach for enhancing data privacy in vehicle-to-grid networks. IEEE Trans Intell Transp Syst 21(9):3720–3732

60. Dutta R, Chatterjee S, Nigam N (2020) Secure data sharing using blockchain for privacy-preserving in social IoT. J Netw Comput Appl 149:102474
61. Fernandes CA, Ferreira R (2020) Enhancing privacy for IoT and mobile devices through blockchain. J Netw Comput Appl 168:102711
62. Liu Y, Huang X (2020) Blockchain-based efficient and privacy-preserving user revocation scheme for fog computing. IEEE Trans Industr Inf 17(5):3488–3498
63. Rashidi B, Gharibzadeh S (2020) A blockchain-based secure data sharing framework for smart city applications. Futur Gener Comput Syst 111:843–855
64. Elgamal MA, Elmahdy N (2020) A secure and decentralized data sharing model for cloud-based healthcare systems using blockchain technology. Int J Inf Manag 50:416–427
65. Zhang Q, Zhang Y, Shi W (2019) A survey of blockchain-based secure sharing of big data. IEEE Access 7:151485–151497
66. Ghosh A, Banerjee T, Ruj S (2019) Secure data sharing in a consortium blockchain network. Futur Gener Comput Syst 98:542–555
67. Keshavarz-Haddad A, Mousavi SM (2019) A secure and dynamic data sharing mechanism for cloud storage using blockchain. J Netw Comput Appl 147:52–62
68. Li Z, Yuan Y (2019) PFS: A privacy-aware data sharing scheme based on blockchain for IoT environments. IEEE Internet Things J 6(4):6548–6558
69. Ruj S, Mukherjee S (2020) Secure data sharing in consortium blockchains: a survey. J Parallel Distrib Comput 140:35–53
70. Song C, Lee J (2019) Trust-free collaborative data sharing via blockchain in industrial Internet of Things. IEEE Trans Industr Inf 16(6):4102–4110
71. Tang Y, Wang R, Xie D (2019) A decentralized data sharing scheme based on consortium blockchain for IoT. IEEE Internet Things J 6(3):4137–4145
72. Yue X, Wang H, Jin D (2019) FSSDS: a flexible and secure searchable data sharing scheme over cloud storage. Futur Gener Comput Syst 91:341–351
73. Javaid U, Li Z, Aman MN, Shao D, Yee K, Sikdar B (2022) Blockchain based secure group data collaboration in cloud with differentially private synthetic data and trusted execution environment. In: Proceedings of the 2022 IEEE international conference on big data (big data)
74. Dong L, Zhao J, Chen T, Yu Y, Duan Z, Zhu J (2022) The secure data sharing and interchange model based on blockchain for single window in trade facilitation. In: Proceedings of the 2022 international conference on blockchain technology and information security (ICBCTIS)
75. Cheng X, Zhang X, Zhang H (2019) Blockchain-based secure data sharing scheme for IoT with fine-grained access control. J Netw Comput Appl 136:1–12
76. Qu Y, Zhang H, Li L (2020) Secure and privacy-preserving data sharing in industrial Internet of Things based on consortium blockchain. Futur Gener Comput Syst 103:532–541
77. Liu L, Yu S, Wang L (2019) Secure data sharing scheme based on consortium blockchain for industrial Internet of Things. J Netw Comput Appl 133:13–21
78. Dinh TTA, Liu D, Zhang M (2020) A blockchain-based framework for secure and efficient data sharing in industrial IoT. IEEE Trans Industr Inf 16(2):1103–1111
79. Hengameh M, Hosseinzadeh M, Bozorgi M (2019) A secure and efficient data sharing scheme for IoT using blockchain and smart contracts. Futur Gener Comput Syst 98:489–500
80. Konečný J, McMahan HB, Yu FX, Richtárik P, Suresh AT, Bacon D (2016) Federated learning: strategies for improving communication efficiency. arXiv:1610.05492
81. McMahan HB, Moore E, Ramage D, Hampson S, Arcas BA (2017) Communication-efficient learning of deep networks from decentralized data. In: Artificial intelligence and statistics (AISTATS)
82. Smith V, Chiang K, Sanjabi M, Recht B (2017) Federated multi-task learning. In: Advances in neural information processing systems (NeurIPS)
83. Kairouz P, McMahan HB, Avent B, Bellet A, Bennis M, Bhagoji AN, Singh V (2019) Advances and open problems in federated learning. arXiv:1912.04977
84. Hard A, Rao V, Mathews A, Ramachandran B, Beutel A, Smelyanskiy M, Agrawal S (2018) Federated learning for mobile keyboard prediction. arXiv:1811.03604

85. Yang Q, Liu Y, Chen T, Tong Y (2019) Federated machine learning: Concept and applications. ACM Trans Intell Syst Technol (TIST) 10(2):1–19
86. Bonawitz K, Eichner H, Grieskamp W, Huba D, Ingerman A, Ivanov V, Velez F (2019) Towards federated learning at scale: system design. arXiv:1902.01046
87. Smith AJ, Chiang K, Sanjabi M, Recht B (2017) Federated multi-task learning. In: Advances in neural information processing systems (NeurIPS)
88. Wang Z, Liu Z, Zhu L, Ma Y (2018) Secure federated transfer learning. In: Proceedings of the 24th ACM SIGKDD international conference on knowledge discovery and data mining
89. McMahan B, Moore E, Ramage D, Hampson S, Agüera y Arcas B (2017) Communication-efficient learning of deep networks from decentralized data. In: Artificial intelligence and statistics
90. Castellanos JAF, Coll-Mayor D, Notholt JA (2017) Cryptocurrency as guarantees of origin: simulating a green certificate market with the Ethereum blockchain. In: Proceedings of the 2017 IEEE international conference on smart energy grid engineering (SEGE)
91. Abognah A, Basir O (2022) Distributed spectrum sharing using blockchain: a hyperledger fabric implementation. In: Proceedings of the 2022 IEEE 1st global emerging technology blockchain forum: blockchain and beyond (iGETblockchain)
92. Cheung J (2021) Real estate politik: democracy and the financialization of social networks. J Soc Comput
93. Zhang H, Xu C, Huang C (2020) Tokenomics in blockchain: a comprehensive survey. IEEE Access 8:188689–188706
94. Han T, Kim Y (2019) A survey on token economics: design principles and blockchain applications. Sustainability 11(10):2932
95. Li Z, Ning H (2020) Token economics in blockchain networks: a survey. IEEE Commun Surv Tutor 22(4):2877–2895
96. Kumar R, Dhiman G (2020) Incentive mechanisms for token-based blockchain networks: a survey. In: Blockchain and Internet of Things: proceedings of IIoTS 2019. Springer, Singapore, pp 105–115
97. Zhou M, Lu R (2021) A survey on token economy in blockchain-based crowdfunding. IEEE Access 9:87175–87192
98. Xu Y, Huang X, Liu L, Zhang J (2019) Blockchain-based decentralized AI: a survey. arXiv: 1908.00700
99. Zhang H, Li Y, Zhang Y (2020) Integrating Blockchain and Artificial Intelligence: challenges and Opportunities. IEEE Intell Syst 35(4):92–96
100. Kuo TT, Kim HE, Ohno-Machado L (2017) Blockchain distributed ledger technologies for biomedical and health care applications. J Am Med Inform Assoc 24(6):1211–1220
101. Bouhlel MS, Ben Amor N, Ben Ahmed M (2018) Challenges and opportunities of blockchain-based artificial intelligence in healthcare. In: 2018 3rd international conference on image, vision and computing (ICIVC). IEEE, pp 1–5
102. Tsai CW, Lai CF (2019) Applying blockchain in securing Internet of Things. IEEE Internet Things J 7(5):4502–4510
103. Tsang DH, Gilder A (2021) Blockchain and the general data protection regulation (GDPR). Comput Law Secur Rev 40:105386
104. De Filippi P, Wright A (2018) Blockchain and the law: the rule of code. Harvard University Press
105. Choudhury OR, Sarker MH (2021) Legal and regulatory issues in blockchain technology: a comprehensive review. In: Handbook of blockchain, digital finance, and inclusion, vol 2. Springer, Cham, pp 129–149
106. Henderson MA, King DL (2018) The potential and limits of blockchain for state taxation. Natl Tax J 71(4):725–752

Synergizing Artificial Intelligence and Blockchain

Priyanka Tyagi, Neelam Shrivastava, Sakshi, and Vishal Jain

Abstract This chapter provides a comprehensive examination of the confluence of Artificial Intelligence (AI) and Blockchain, which are two of the most notable technical advancements in our day. This explores the distinct methods in which the powerful data processing skills of AI and the solid security features of Blockchain might be merged to transform different sectors. The combination of AI and Blockchain is examined through real-world implementations in healthcare to enhance patient care and safeguard data, in supply chain management to enhance transparency and efficiency, in finance to detect fraud and ensure secure transactions, and in legal systems to automate and secure contract management. The chapter delves into the obstacles and potential future developments of integrating AI with Blockchain. It highlights the possibility of AI to enhance the intelligence and autonomy of Blockchain systems, while highlighting the role of Blockchain in guaranteeing the trustworthiness and consistency of AI applications. The integration of these methods has the potential to develop advanced, protected, and streamlined systems, hence enabling the emergence of inventive solutions in data administration and industrial processes.

P. Tyagi · V. Jain
Department of Computer Science and Engineering, Sharda University, Greater Noida, Uttar Pradesh, India

N. Shrivastava
Department of Computer Science and Engineering, Hi-Tech Institute of Engineering and Technology, Ghaziabad, Uttar Pradesh, India

Sakshi (✉)
Department of Computer Science and Applications, Sharda University, Greater Noida, Uttar Pradesh, India
e-mail: Asakshi541@gmail.com

K. Kaushik and I. Sharma (eds.), *Next-Generation Cybersecurity*, Blockchain Technologies, https://doi.org/10.1007/978-981-97-1249-6_4

1 Introduction to AI and Blockchain

Artificial Intelligence (AI) and Blockchain are two revolutionary technologies that are transforming industries and reshaping the way we interact with the digital world. AI encompasses the emulation of human intelligence in machines, enabling them to learn, reason, and make decisions. On the other hand, Blockchain is a decentralized and tamper-resistant digital ledger that ensures transparency and security in transactions. While they serve distinct purposes, the convergence of AI and Blockchain holds the potential to unlock new frontiers of innovation [1]. AI, with its subsets like Machine Learning and Deep Learning, has shown its prowess in tasks ranging from image recognition to natural language processing. By analyzing massive datasets and learning patterns, AI systems can make predictions and decisions that were once reserved for human experts. Additionally, AI-powered algorithms are driving automation, efficiency, and personalization across various industries.

Blockchain, known for its secure and transparent nature, is disrupting industries like finance, supply chain, and healthcare. It eliminates the need for intermediaries, reduces fraud, and ensures data integrity through its decentralized architecture. Transactions recorded on the Blockchain are immutable and traceable, fostering trust in digital interactions. The synergy between AI and Blockchain brings forth exciting possibilities. Blockchain can enhance the security and privacy of AI applications by safeguarding data and ensuring transparent data usage. Simultaneously, AI can empower Blockchain networks by optimizing processes, analyzing data, and enabling predictive insights. In sectors like healthcare, the combination of AI and Blockchain can facilitate secure sharing of patient data for research while maintaining privacy [2]. Supply chains can leverage AI's predictive capabilities to optimize logistics, while Blockchain ensures the authenticity and traceability of products. The intersection of AI and Blockchain also paves the way for decentralized AI marketplaces, where individuals can access and contribute AI services

without intermediaries. This democratization of AI could fuel innovation and empower a broader range of stakeholders. As AI and Blockchain continue to evolve, their convergence holds the promise of creating more robust, transparent, and intelligent systems. This introduction scratches the surface of these complex technologies, which individually have already reshaped industries, and together have the potential to reshape our digital landscape even further.

2 Foundations of AI and Blockchain

AI Techniques and Their Relevance in Data Analysis:

Artificial Intelligence (AI) encompasses several techniques that enable machines to learn from data and perform tasks that mimic human cognitive abilities. Three prominent techniques within AI—Machine Learning, Deep Learning, and Natural

Language Processing (NLP)—are integral to data analysis, as they extract insights from vast and complex datasets, enabling informed decision-making.

Machine Learning (ML): Machine Learning encompasses the utilization of algorithms enabling computers to glean knowledge from data sans explicit programming. It encompasses supervised learning, unsupervised learning, and reinforcement learning. Significance in Data Analysis: ML algorithms are instrumental in categorizing data, making projections, and uncovering patterns within datasets. They have the capacity to discern trends, segment data, and prognosticate results, rendering them indispensable for tasks such as customer categorization, fraud identification, and anticipatory upkeep [3].

Deep Learning (DL): Deep Learning constitutes a subset of Machine Learning that employs artificial neural networks with numerous layers (deep neural networks). It excels in processing extensive volumes of unstructured data, including images and textual content. Significance in Data Analysis: DL is especially adept at tasks such as image recognition, speech understanding, and natural language processing. It autonomously acquires intricate data features, thereby facilitating precise recognition of objects, sentiments, or contextual information.

Natural Language Processing (NLP): NLP entails the interaction between computers and human language. It empowers machines to comprehend, construe, and generate human language, thereby fostering seamless communication between humans and computers. Significance in Data Analysis: NLP is pivotal in the scrutiny of unstructured textual data, encompassing social media posts, customer feedback, and documents. Sentiment evaluation, identification of named entities, and topic modeling represent a subset of applications that extract meaningful insights from textual data, aiding in the comprehension of customer sentiments and trends.

2.1 Relevance of AI Techniques in Data Analysis

1. **Data Pattern Recognition**: Machine Learning and Deep Learning can identify complex patterns in large datasets. This helps in making accurate predictions and uncovering hidden insights, such as customer preferences, market trends, and potential risks.
2. **Image and Video Analysis**: Deep Learning excels in image and video analysis tasks. It enables object detection, facial recognition, and even autonomous driving by processing visual data (Fig. 1).
3. **Text Analysis**: NLP techniques allow machines to understand and analyze text data. Sentiment analysis helps gauge customer opinions, while text summarization condenses lengthy content for quick insights.
4. **Personalization and Recommendation**: Machine Learning algorithms power recommendation systems that suggest products, services, or content based on user behavior and preferences, enhancing user experiences.

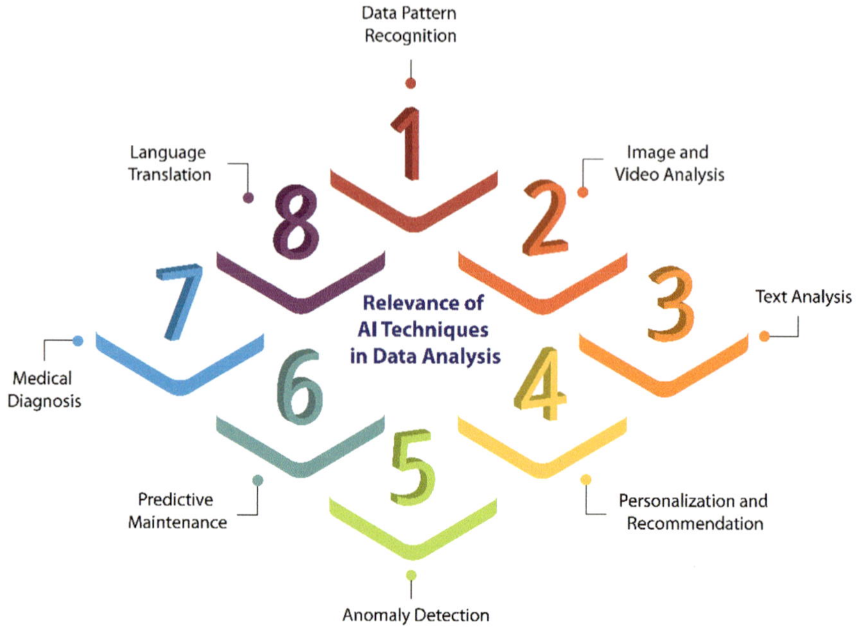

Fig. 1 Relevance of AI techniques in data analysis

5. **Anomaly Detection**: Machine Learning models can identify anomalies in data, aiding in fraud detection, network security, and quality control.
6. **Predictive Maintenance**: Machine Learning algorithms predict equipment failures by analyzing sensor data, minimizing downtime and maintenance costs.
7. **Medical Diagnosis**: Machine Learning and Deep Learning are used to analyze medical images, assisting doctors in diagnosing diseases like cancer from scans.
8. **Language Translation**: NLP techniques enable real-time language translation, breaking down language barriers and facilitating global communication.

2.2 Blockchain: Components and Principles of Decentralized Consensus

Blockchain is a transformative technology that underpins various digital systems with its decentralized, secure, and transparent nature. It is a distributed digital ledger that records transactions in an immutable and chronological manner. Below is an outline of its components and the principles of decentralized consensus:

Blockchain Components:

1. **Blocks**: Transactions are organized into blocks, which contain information like transaction details, timestamps, and a reference to the previous block.

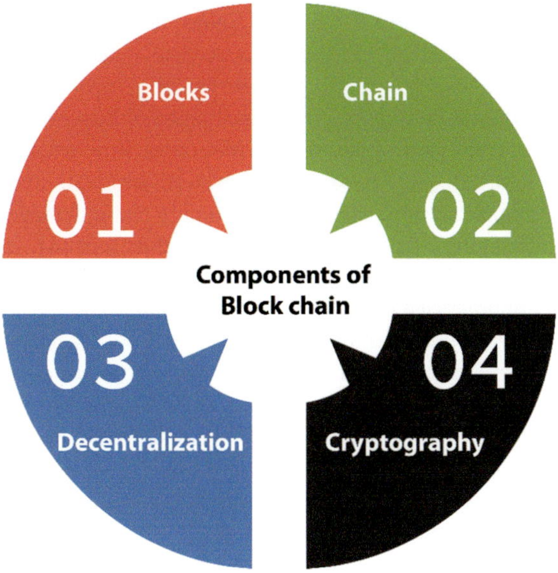

Fig. 2 Components of block chain

2. **Chain**: Blocks are interconnected in chronological order, creating a chain. The cryptographic hash of each block is used to connect it securely to the preceding block, ensuring resistance to tampering.
3. **Decentralization**: Instead of relying on a central authority, blockchain operates on a network of nodes (computers) that collaboratively validate transactions and uphold the integrity of the ledger (Fig. 2).
4. **Cryptography**: Transactions and blocks are secured using cryptographic techniques. Each block's hash depends on the data within the block and the previous block's hash, making it computationally infeasible to alter a block without affecting subsequent blocks.

2.3 Principles of Decentralized Consensus

Principles of Decentralized Consensus form the bedrock of blockchain technology, reshaping the way trust and agreement are achieved in digital systems. At the core of decentralized consensus lies the principle of eliminating reliance on central authorities by distributing decision-making across a network of nodes. These nodes collaborate to validate transactions, ensuring that all participants agree on the state of the shared ledger without the need for intermediaries. This consensus is achieved through mechanisms like Proof of Work (PoW) or Proof of Stake (PoS), where participants compete or are chosen based on their computational effort or stake in the network. This approach not only enhances security by removing single points of failure but also fosters transparency through the verification of transactions by multiple parties. By leveraging cryptographic techniques, decentralization, and the immutable nature of the blockchain, decentralized consensus establishes a trustless system where interactions can occur securely and transparently, revolutionizing industries and challenging traditional models of governance and authority. In essence, blockchain's decentralized consensus principles revolutionize how trust is established, transactions are conducted, and data is secured. Its transparent, tamper-resistant, and trustless nature has applications across finance, supply chain management, healthcare, and beyond, offering new paradigms for digital interactions [4].

Principles of Decentralized Consensus are at the heart of blockchain technology, redefining trust and cooperation in digital ecosystems. Instead of relying on a central authority, these principles establish a network-wide agreement mechanism, ensuring the integrity of transactions and data. Through consensus mechanisms like Proof of Work (PoW) or Proof of Stake (PoS), nodes within the decentralized network collaborate to validate transactions and create new blocks. This decentralized validation process not only enhances security by eliminating single points of failure but also fosters transparency, as each participant can independently verify the correctness of transactions. Immutability, another key principle, ensures that once data is added to the blockchain, it becomes tamper-resistant and practically immutable due to cryptographic linkage and computational difficulty [5]. These principles empower blockchain networks to operate trustlessly, allowing participants to interact and transact without requiring explicit trust in one another, thereby ushering in a new era of secure, transparent, and autonomous digital collaboration.

3 Challenges in Secure Data Analytics

Challenges in Secure Data Analytics encompass a spectrum of issues that impede the effective utilization of data for valuable insights while maintaining robust security measures. These challenges highlight the complexities involved in reconciling data analysis and privacy in today's digital landscape. Here are the outlined challenges:

- **Data Privacy Concerns and Regulatory Issues**:

As data becomes a valuable commodity, concerns over the privacy of personal and sensitive information have escalated. Striking a balance between data utility and individual privacy is crucial. Regulations like the General Data Protection Regulation (GDPR) and the California Consumer Privacy Act (CCPA) impose strict requirements on how data is collected, stored, and used. Meeting these regulatory demands without compromising analytical potential is a significant challenge [6]. Data privacy concerns and regulatory issues have become paramount as the digital landscape evolves, requiring organizations to navigate intricate legal frameworks while safeguarding individual privacy. Here are key notes on the challenges and considerations within this domain (Fig. 3):

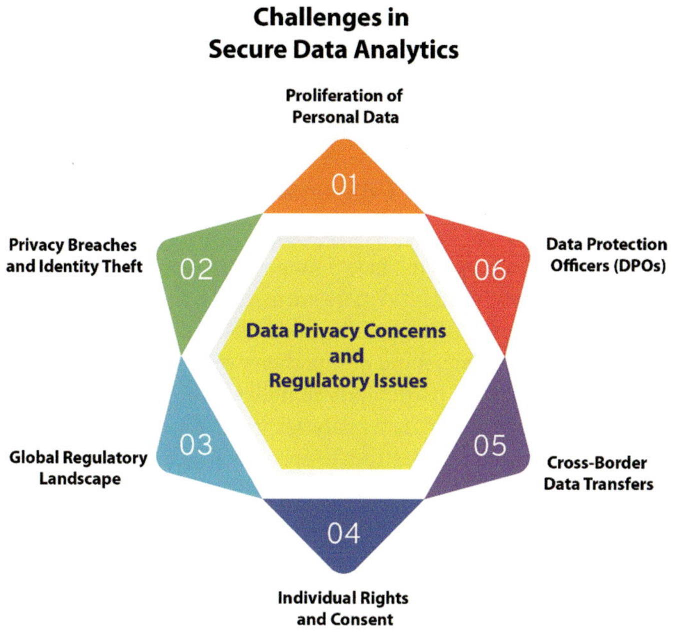

Fig. 3 Challenges in secure data analytics

1. **Proliferation of Personal Data**: The exponential growth of data collection, driven by IoT devices, social media, and digital transactions, has led to vast repositories of personal information. This includes sensitive data like financial details, medical records, and biometric information.
2. **Privacy Breaches and Identity Theft**: High-profile data breaches have exposed millions of individuals to identity theft, financial fraud, and reputational damage. The mishandling of personal data can lead to severe consequences for both individuals and organizations.
3. **Global Regulatory Landscape**: Various regions have introduced stringent data protection regulations to address these concerns. Notably, the General Data Protection Regulation (GDPR) in the EU and the California Consumer Privacy Act (CCPA) in the US have set a precedent for data privacy rights and transparency [7].
4. **Individual Rights and Consent**: Regulations emphasize individual rights, including the right to access, rectify, and erase personal data. Organizations must obtain explicit and informed consent from individuals before collecting and processing their data.
5. **Cross-Border Data Transfers**: Regulations restrict the transfer of personal data to countries with inadequate data protection standards. Adequacy decisions, standard contractual clauses, and binding corporate rules govern cross-border data flows.

6. **Data Protection Officers (DPOs)**: Some regulations mandate appointing DPOs within organizations to oversee data protection efforts, facilitate compliance, and act as a point of contact for data protection authorities.

3.1 Security Weaknesses in Conventional Data Analytics

Conventional data analytics systems face a range of cybersecurity threats, including data breaches, unauthorized entry, and malware assaults. As data analytics workflows encompass multiple phases of data handling and transmission, each juncture presents a potential vulnerability. Establishing comprehensive end-to-end security for these workflows is a formidable task, particularly in light of the ever-evolving landscape of cyber threats. While conventional data analytics methods are valuable for extracting insights, they are vulnerable to a spectrum of security weaknesses that could compromise data integrity and confidentiality. Recognizing these vulnerabilities is imperative to implement suitable safeguards for data protection and to uphold the confidence of stakeholders. The following are key points regarding security vulnerabilities in conventional data analytics:

Data Breaches: Traditional data analytics workflows encompass various stages of data gathering, storage, and transmission. Weaknesses at any of these stages can result in data breaches, where unauthorized parties gain access to sensitive information.

Unauthorized Access: Inadequate access controls and feeble authentication mechanisms may permit unauthorized individuals to access data and analytics systems, potentially resulting in data theft or manipulation.

Malware and Cyberattacks: Traditional data analytics systems may be vulnerable to malware attacks such as viruses, worms, and ransomware, which can disrupt operations and compromise data integrity.

1. **Insider Threats**: Malicious actions by employees or individuals within an organization, known as insider threats, can expose sensitive data to unauthorized parties (Fig. 4).
2. **Injection Attacks**: SQL injection, a common vulnerability, occurs when attackers manipulate input fields to execute malicious SQL commands in databases, potentially leading to data leakage or unauthorized access.

3.2 Trust and Transparency Gaps in Data Sharing

Collaborative data sharing is pivotal for holistic insights. However, establishing trust among different entities to share their data is challenging. Concerns about data misuse, ownership, and the lack of transparency in how data will be handled hinder seamless data sharing. Overcoming these trust issues while maintaining transparency and safeguarding the interests of all parties involved is a complex endeavor [4, 4].

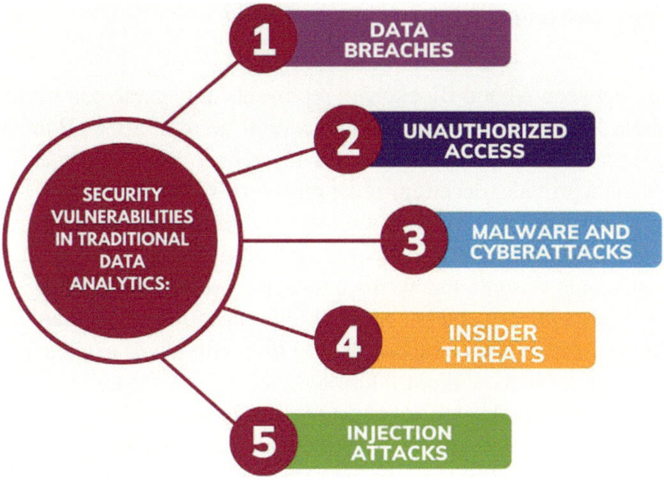

Fig. 4 Security vulnerabilities in traditional data analytics

Trust and transparency are essential cornerstones of effective data sharing in the digital age. However, several challenges and gaps in these areas can hinder the successful exchange of data among organizations and individuals. Understanding these issues is crucial for building a more trustworthy and transparent data-sharing ecosystem. Here are key insights into the trust and transparency gaps in data sharing:

1. **Data Ownership and Control**: The ambiguity surrounding data ownership and control often leads to mistrust. Organizations and individuals are concerned about losing control over their data once it's shared, fearing potential misuse or unauthorized access.
2. **Data Misuse and Exploitation**: Data owners worry that shared data might be exploited for unintended purposes, leading to adverse consequences. Lack of transparency about how shared data will be utilized exacerbates these concerns.
3. **Lack of Accountability**: In some cases, there's a lack of clarity about who is responsible for the security and appropriate use of shared data. This absence of accountability undermines trust.
4. **Opaque Data Handling**: When data recipients don't provide insights into how shared data will be processed, concerns about data integrity, anonymization, and adherence to privacy regulations arise.
5. **Data Quality and Accuracy**: The accuracy and reliability of shared data can be questionable, eroding trust. Without transparency in data collection and validation processes, recipients may doubt the quality of the shared information.

4 Synergy Between AI and Blockchain

The synergy between AI and Blockchain represents a dynamic convergence of two transformative technologies, fostering innovation across various domains. Artificial Intelligence (AI) empowers machines to learn, reason, and make decisions, while Blockchain provides decentralized, tamper-resistant data storage [8]. Together, they create a powerful combination with the potential to revolutionize industries. AI enhances data analysis on the Blockchain, extracting insights from distributed ledgers. Blockchain ensures the security and transparency of AI data, addressing concerns about data privacy and integrity [1]. The decentralized nature of Blockchain can support AI models' democratization, enabling individuals to contribute, access, and monetize AI services without intermediaries. This synergy also bolsters trust in AI decision-making through transparent and auditable records. The marriage of AI and Blockchain paves the way for secure, intelligent systems that embrace transparency, privacy, and empowerment, ushering in a new era of technological advancement [2]. This synergy is driven by the following key aspects.

Enhancing Data Privacy Through Encryption and Decentralized Control:

Blockchain's decentralized architecture offers a secure and transparent environment for data storage and sharing. AI's encryption capabilities further bolster data privacy by encoding sensitive information. Together, they provide a robust framework where data remains confidential, and access is controlled through cryptography and decentralized consensus mechanisms. The collaboration between encryption and decentralized control within the realms of data privacy signifies a robust approach to safeguarding sensitive information in today's digital landscape [3]. Encryption, a technique that transforms data into unreadable code, acts as a shield against unauthorized access. When coupled with decentralized control, as facilitated by blockchain technology, a powerful framework emerges to ensure data privacy.

Encryption serves as a formidable barrier against data breaches and unauthorized viewing. By converting data into an encrypted format, even if unauthorized entities gain access, the information remains incomprehensible without the decryption key [12]. This approach thwarts potential hackers and safeguards confidential data. In tandem, decentralized control introduced by blockchain technology addresses the vulnerabilities associated with centralized data storage. Instead of relying on a single point of control, data is distributed across a network of nodes, each holding a copy of the encrypted information. Changes to the data require consensus among multiple nodes, ensuring tamper resistance and reducing the risk of unauthorized alterations.

The synergy between encryption and decentralized control empowers individuals and organizations to retain ownership and control over their data while enabling secure sharing. This collaborative approach assures data subjects that their information remains confidential, accessible only to authorized parties [5]. It reinforces privacy compliance and engenders a digital environment where data protection is paramount, thus fostering trust and accountability in the realm of data privacy.

Using AI for Anomaly Detection, Pattern Recognition, and Predictive Analysis: AI's cognitive abilities, particularly in anomaly detection, pattern recognition, and predictive analysis, complement Blockchain's data integrity. AI algorithms can monitor Blockchain transactions for unusual patterns, potentially indicating fraudulent activities. Moreover, AI-driven predictive analysis can unveil trends and insights from the Blockchain's decentralized data, empowering informed decision-making.

Smart Contracts and Self-Executing Agreements to Automate Data-Sharing Protocols: Blockchain's smart contracts enable self-executing agreements with predefined conditions. This automation streamlines data-sharing protocols, ensuring compliance and reducing the need for intermediaries. AI can enhance these contracts by autonomously monitoring and triggering actions based on data inputs, optimizing processes and reducing human intervention.

5 Applications of AI-Blockchain Integration

The fusion of artificial intelligence (AI) and blockchain technology has ushered in a wave of transformative applications across diverse sectors. This integration leverages AI's robust data analysis capabilities and pattern recognition alongside blockchain's inherent security and transparency [6, 7]. One prominent application lies in supply chain management, where AI-driven sensors capture real-time data on products, securely documented on the blockchain. This synergy ensures end-to-end traceability, minimizing fraud and enhancing provenance. In the healthcare realm, blockchain's secure data storage complements AI's diagnostic prowess, enabling the analysis of medical information while preserving patient privacy. Moreover, the financial landscape benefits from AI-powered market predictions executed as smart contracts on blockchain, resulting in heightened transparency and automation within cryptocurrency transactions. The legal industry embraces this synergy by utilizing AI to simplify the creation

and execution of complex smart contracts, securely recorded on the blockchain. Finally, the establishment of tamper-proof digital identities through AI-based verification combined with blockchain immutability holds promise for secure authentication and streamlined Know Your Customer (KYC) processes. These applications collectively underscore the potential of AI- blockchain integration to reshape industries by fortifying security, driving efficiency, and ushering in a new era of transparency.

Supply Chain Management:

AI Sensors and Real-Time Tracking: Artificial intelligence integrated with sensors provides real-time tracking of products throughout the supply chain. Blockchain Security, Data collected by AI sensors is securely recorded on a blockchain, ensuring the integrity and immutability of the information [18]. Traceability and Authenticity, this integration guarantees the traceability of products from origin to destination, enhancing transparency in supply chain operations. Fraud Reduction, by recording each step on the blockchain, fraudulent activities such as counterfeiting can be detected and minimized.

Healthcare:

Blockchain for Data Security, Blockchain technology securely stores sensitive medical data, making it tamper-proof and easily auditable. AI-Driven Diagnosis, AI analyzes medical data, such as diagnostic images and patient records, to provide accurate and timely diagnoses. Patient Privacy, Blockchain's decentralized and encrypted nature ensures patient data privacy while allowing authorized parties to access necessary information [10]. Interoperable Data Sharing: Different healthcare providers can securely share patient data, fostering better collaboration and improving patient care.

Finance and Cryptocurrency:

AI Market Predictions: Artificial intelligence algorithms analyze market trends and predict future outcomes, guiding investment decisions. Smart Contracts in Finance, Smart contracts automatically execute financial agreements once predefined conditions are met, minimizing manual intervention and the potential for errors [8]. Blockchain for Cryptocurrency, Blockchain ensures transparency and security in cryptocurrency transactions, preventing double-spending and unauthorized access. Automated Transactions, Integration of AI with blockchain facilitates autonomous execution of transactions based on AI predictions.

Smart Contracts and Legal Industry:

AI-Assisted Contract Creation, AI simplifies the creation of complex contracts by identifying relevant clauses and legal terminology. Secure Record Keeping, Smart contracts are recorded on a blockchain, ensuring an immutable and tamper-proof record of contractual agreements.

Reducing Intermediaries, Smart contracts eliminate the need for intermediaries, reducing costs and increasing the efficiency of contract execution. Enhanced Contract Efficiency, The combination of AI and blockchain streamlines contract management and enforcement, minimizing disputes and delays [9].

Identity Verification:

AI-Driven Identity Verification, AI analyzes biometric and personal data for identity verification, enhancing accuracy and fraud detection. Blockchain Immutability, Verified identities are stored on a blockchain, making them tamper-proof and enabling secure authentication. Applications in KYC and Voting [10]. This technology has applications in Know Your Customer (KYC) processes and digital voting, ensuring secure and transparent verification.

6 Case Studies

Health Data Analysis Using AI and Blockchain

Background A: healthcare organization aimed to improve patient outcomes through advanced data analysis while maintaining patient data privacy and security.

Solution:

- **Data Collection**: AI-powered devices collected patient health data such as vital signs, activity levels, and medical history.
- **Data Analysis**: AI algorithms processed the collected data to identify patterns, correlations, and potential health risks.
- **Blockchain Integration**: Analyzed data was securely stored on a blockchain, ensuring tamper-proof records.
- **Patient Control**: Patients were given control over their data and could grant permission for research purposes through smart contracts.
- **Research Collaboration**: Researchers accessed anonymized and authorized data through the blockchain, accelerating medical research.

Outcome:

- Improved patient care through personalized health insights and early risk detection.
- Enhanced data security and privacy due to blockchain's encryption and decentralized nature.
- Efficient research collaboration while respecting patient data ownership.

7 Supply Chain Traceability and Verification

In the complex landscape of supply chains, ensuring the authenticity and traceability of products has long been a challenge, often resulting in fraudulent activities and operational inefficiencies. However, a consortium of supply chain stakeholders embraced a transformative solution by integrating artificial intelligence (AI) and blockchain technology. By combining these two cutting-edge technologies, they revolutionized the way products are tracked and verified throughout the entire supply chain [11].

In this case, AI-driven sensors were strategically integrated into products, enabling real-time data collection as these items traversed the supply chain journey. The collected data encompassed crucial information such as origin points, transportation conditions, and storage environments. This data was then securely recorded on a blockchain, creating an immutable and tamper-proof digital ledger.

The integration's implications extended far and wide. Stakeholders were afforded an unprecedented level of visibility into the supply chain, enabling them to validate the authenticity and integrity of each product at any point in its journey [12]. This transparency not only thwarted the infiltration of counterfeit goods but also cultivated a heightened level of trust among all participants. Additionally, the automated process of collecting and verifying data streamlined supply chain operations, significantly reducing the delays and errors that had traditionally plagued the process.

From a consumer perspective, the benefits were evident. The capacity to verify the origin and authenticity of products through a secure and transparent blockchain-based system significantly bolstered consumer confidence. Consequently, trust in the supply chain increased, ensuring that consumers received authentic, high-quality products [13]. This case study underscores the transformative potential of integrating AI and blockchain in addressing long-standing challenges within supply chain management. By offering real-time tracking, irrefutable verification, and heightened transparency, this approach serves as an exemplar of how technology can reshape industries, enhancing both operational efficiency and consumer trust.

8 Conclusion

The fusion of Artificial Intelligence (AI) and Blockchain technology has initiated a new era of data analytics security, effectively addressing enduring challenges related to data privacy, security, transparency, and trust. As the digital landscape continues to evolve, this collaboration presents a potent framework that not only unlocks valuable insights but also guarantees the integrity and confidentiality of the underlying data that informs these insights. In this chapter, we embarked on a journey through the domains of AI, Blockchain, and their seamless integration. We delved into the numerous advantages that arise from merging these technologies and observed how their individual strengths lay a solid foundation for secure data analytics. From the precision of AI algorithms to the immutability of Blockchain, the resulting combination offers a comprehensive solution that resonates across various industries. The significance of harnessing the potential of AI and Blockchain for secure data analytics cannot be overstated. The insights derived from AI-driven analytics hold the potential to steer strategic decision-making processes, to optimize processes and elevate customer experiences has been a central goal of the data-driven revolution. However, the increasing frequency of data breaches and mounting privacy concerns have cast a cloud over these aspirations. Enter Blockchain, with its decentralized structure and robust cryptographic protections, which guarantee the immutability and transparency of data at every stage of its existence. By exploring real-world applications across various sectors, we have witnessed the concrete benefits of this fusion. Whether it's ensuring supply chain authenticity or enabling secure healthcare data sharing, the synergy between AI and Blockchain is reshaping the way industries engage with data analytics. The confidence instilled by Blockchain's trustless nature and AI's analytical prowess serves as a cornerstone for innovation and collaboration in a data-driven world.

Yet, this journey is not without its challenges. The intricate dance between AI and Blockchain brings about computational complexities, scalability concerns, and regulatory considerations. These hurdles are not insurmountable but beckon for continued research and innovation. As we gaze into the horizon, we glimpse the emergence of Federated Learning and privacy-preserving algorithms, hinting at a future where data can be analyzed without leaving its secure enclave.

In conclusion, the union of AI and Blockchain holds immense promise for the future of secure data analytics. It heralds a paradigm shift wherein data-driven insights need not come at the cost of privacy and security. With each passing day, advancements in both fields pave the way for a more seamless and trustworthy data analytics ecosystem. This chapter serves as a testament to this exciting journey, inviting researchers, practitioners, and enthusiasts to explore, collaborate, and propel us towards a more secure and insightful digital future.

References

1. Wang Q, Su M (2020) Integrating blockchain technology into the energy sector—from theory of blockchain to research and application of energy blockchain. Comput Sci Rev 37:100275.6
2. Sengupta U, Kim H (2021) Meeting changing customer requirements in foodand agriculture through the application of blockchain technology. Frontiers in Blockchain
3. Düdder B, Fomin V, Gürpinar T, Henke M, Iqbal M, Janavičienė V, Matulevičius R, Straub N, Wu H (2021) Interdisciplinary blockchain education: utilizing blockchain technology from various perspectives. Frontiers in Blockchain
4. Zhang Z, Song X, Liu L, Yin J, Wang Y, Lan D (2021) Recent advances in blockchain and artificial intelligence integration: feasibility analysis, research issues, applications, challenges, and future work. Secur Commun Netw, 15. Article ID 9991535. https://doi.org/10.1155/2021/9991535
5. Xing B, Marwala T (2018) The synergy of blockchain and artificial intelligence. Available at SSRN: https://ssrn.com/abstract=3225357, https://doi.org/10.2139/ssrn.3225357
6. Taherdoost H (2022) Blockchain technology and artificial intelligence together: a critical review on applications. Appl Sci 12:12948. https://doi.org/10.3390/app122412948
7. Chamola V, Goyal A, Sharma P, Hassija V, Binh HTT, Saxena V (2022) Artificial intelligence-assisted blockchain-based framework for smart and secure EMR management. Neural Comput Appl:1–11. https://doi.org/10.1007/s00521-022-07087-7. Epub ahead of print. PMID: 35310553; PMCID: PMC8918902
8. Turner SW, Karakus M, Guler E, Uludag S (2023) A promising integration of SDN and blockchain for IoT networks: a survey. IEEE Access 11:29800–29822.https://doi.org/10.1109/ACCESS.2023.3260777
9. Lin S-Y, Zhang L, Li J, Ji L-l, Sun Y (2022) A survey of application research based on blockchain smart contract Wireless Netw 28(2):635–690. https://doi.org/10.1007/s11276-021-02874-x
10. Jie X, Zhou W, Zhang S, Jinhua F (2021) A review of the technology and application of deposit and traceability based on blockchain. J High Speed Netw 27(4):335–359. https://doi.org/10.3233/JHS-210671
11. Wazid M, Das AK, Park Y (2021) Blockchain-enabled secure communication mechanism for IoT-driven personal health records. Trans Emerg Telecommun Technol 33:4.https://doi.org/10.1002/ett.4421
12. Sharma C, Sharma S, Sakshi (2022) Latent DIRICHLET allocation (LDA) based information modelling on BLOCKCHAIN technology: a review of trends and research patterns used in integration. Multimedia Tools and Applications 81(25):36805–36831. https://link.springer.com/article/10.1007/s11042-022-13500
13. Sakshi, Kukreja V (2023) Machine learning and non-machine learning methods in mathematical recognition systems: Two decades' systematic literature review. Multimedia Tools and Applications 1–70. https://link.springer.com/article/10.1007/s11042-023-16356-z

Blockchain-Based Smart Contracts: Technical and Usage Aspects

Gulbir Singh

Abstract Blockchains are decentralized, distributed, and immutable ledgers that can execute contracts on the blockchain, called smart contracts. Blockchain-based smart contracts have different challenges, opportunities, and applications in various fields like cryptocurrency systems, real estate, agriculture, logistics, and fire industry. The codes that can be executed and which operate on top of a blockchain to facilitate, implement, and enforce agreements between numerous parties in the absence of a trusted third party are called smart contracts. This chapter uses a taxonomy to classify different processes, such as comparing and contrasting the strengths and weaknesses of various methods and identifying research gaps and unanswered questions. The chapter also outlines some promising research guidelines for blockchain-based smart contracts and discusses the benefits and challenges. Researchers and practitioners in the field of blockchain-based smart contracts may find this chapter useful because it offers an overview of current and emerging developments in the field.

Keywords Blockchain · Smart contracts · Taxonomy · Future directions

1 Introduction

The purpose of this study is to give a complete assessment of the existing literature on blockchain-based smart contracts from both a technical and an application standpoint. We make use of a taxonomy to classify the papers in accordance with several criteria, such as blockchain platforms, smart contract languages, security issues, scalability solutions, privacy upgrades, and programmability factors, among others [1–3]. We evaluate and contrast the positives and negatives of a variety of approaches, as well as uncover research gaps and outstanding issues. We also examine the possible benefits

G. Singh (✉)
Assistant Professor, Department of Computer Science and Engineering, Graphic Era Hill University, Haldwani Campus, Haldwani, Uttarakhand, India
e-mail: gulbir.rkgit@gmail.com

K. Kaushik and I. Sharma (eds.), *Next-Generation Cybersecurity*, Blockchain Technologies, https://doi.org/10.1007/978-981-97-1249-6_5

and obstacles of pursuing these interesting future research avenues for blockchain-based smart contracts and offer some future research directions for blockchain-based smart contracts.

A blockchain is a distributed, decentralized, and unchangeable ledger that is built of a chain of record collection that is linked cryptographically. Blockchains are used in cryptocurrencies like Bitcoin and Ethereum. Transactions and events are terms that are used to refer to the individual records themselves, while blocks are terms that are used to refer to groups of data. Each individual who is a part of the blockchain network has a copy of the distributed ledger stored in the account that they use to access the network. These transactions are added to the distributed ledger only after passing through a process of verification and agreement with the other parties involved in the blockchain. Blockchain technology is distinguished by a number of basic qualities, the most important of which are immutability, decentralization, and a cryptographic link.

Immutability: The fact that the records of transactions in the ledger, which continue to be spread across the nodes, are permanent and cannot be modified is what is meant by the term "immutability." The immutability of the data stored on a blockchain is a property that distinguishes it from centralized database systems and raises to a higher level the level of data integrity that may be maintained on a ledger.

As long as the cryptographic links are in place, it will be difficult to manipulate the records via computational means.

Decentralization: As a result of blockchain's decentralization, the network's participants are given equal authority over the distributed ledger. In contrast to centralized systems that are managed by a reliable third party, the blockchain has a feature that ensures redundancy, and this sets it apart from other types of systems. Through decentralization, the service availability can be guaranteed, the risk of failure can be reduced, and ultimately, the trust in the service may be improved [4–6].

Cryptographic Link: The block is what establishes the chain of integrity throughout the remainder of the blockchain. It is a link between each record ordered in chronological order. A blockchain is a decentralized database that logs transactions. During the validation of each record's digital signature, hashing techniques and asymmetric key cryptography are employed. This ensures the record's authenticity and prevents manipulation. Any modification to a block or transaction record constitutes a violation of integrity [7], resulting in the invalidation of both the record and the block.

In addition, cryptocurrencies enable seamless peer-to-peer payments and disseminate the financial context while emphasizing that they are the first public blockchain generation. Moreover, cryptocurrencies enable payments to be made between parties who are not acquainted with one another. Cryptocurrencies are a type of digital currency that can only be used on the blockchain, which is an online-only distributed ledger. Bitcoin [8] is the world's first significant cryptocurrency that facilitates peer-to-peer financial transactions without the need for trusted third parties such as international payment channels. Previously, these types of transactions were only possible through international payment channels. Prior to this juncture, this type of transaction could only be completed using foreign payment methods.

A third party is not required for the operation of the system, and the verification of transactions that are committed to the network is taken care of by specialized nodes known as miners who employ cryptographic techniques.

In centralized systems, the use of a trustworthy third party, like a bank, is required so that participants who do not trust one another can connect with one another and send and receive financial transactions with one another. When you rely on a trustworthy third party, on the other hand, you run the risk of encountering issues with both your privacy and your data's security [9], in addition to experiencing higher transactional expenses. Blockchain technology was created with the intention of resolving this issue by allowing parties who do not trust one another to reach an accord over their transactions and communications without a trusted third party. The blockchain can be viewed as a decentralized database that documents and maintains the history of all blockchain network transactions [10, 11]. This is one approach to conceptualize the blockchain. Bitcoin, the first decentralized digital money, was constructed on top of a technological advancement known as the blockchain, which acted as the underlying infrastructure for the currency. In addition to its use in the financial sector, blockchain technology has progressed to the point that it can now support a wide range of decentralized applications. A number of these applications depend on the deployment and execution of smart contracts on the blockchain in order to function properly.

A smart contract [12] is a computer program that encodes the agreement between parties that do not trust one another and that is carried out in accordance with certain pre-defined circumstances. This agreement is carried out in accordance with some pre-defined conditions. As a component of a blockchain transaction, smart contracts are gaining in popularity, and one or more blockchain systems may be used to construct or carry out a smart contract [8]. A smart contract may also be referred to as an intelligent contract. Miners are a special category of participants in the blockchain network that are responsible for the deployment of new contracts as well as the execution of contracts that already exist in the system. Miners are paid for their efforts based on the amount of computational work that is required to fulfill the contracts. This determines how much compensation miners earn. Currently, Ethereum and Hyperledger Fabric are the two platforms that have the largest market share when it comes to providing support for the deployment and execution of smart contracts.

The usage of smart contracts enables the regulations of the company to be rewritten as computer programs. As a result of the diverse requirements imposed by various industries, a number of distinct platforms for smart contracts have recently come into existence [13].

Every single one of the several platforms for smart contracts comes with its own one-of-a-kind collection of features that are geared specifically for the use case in question. To give just one illustration, the primary concentration of effort that has been put into the development of Ethereum [14] has been on applications that require tokenization. Almost every platform includes the fundamental components of a smart contract system, including immutable computer code, a decentralized ledger, and a consensus layer. These components are necessary for the system to implement smart contracts. The consensus layer is one of these constituents [15, 16]. Figure 1 depicts the most prominent smart contract systems that are presently available.

In order to achieve its objectives, this study will first identify and categorize all previous research conducted on the topic of smart contract technology, and then have other academicians evaluate the relevant prior research. It is essential to note that we do not endeavor to cover all of the most recent breakthroughs in technical, financial,

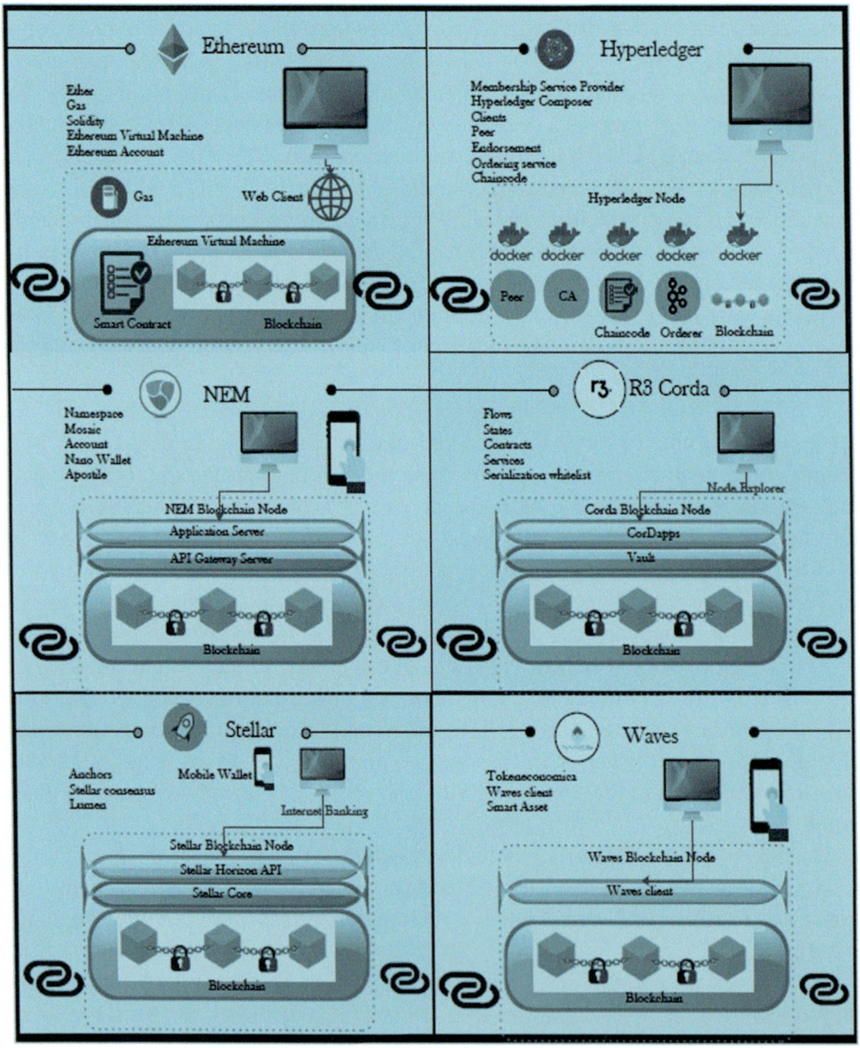

Fig. 1 Different smart contract platforms

or political matters that have been disseminated via other channels, primarily the Internet [17, 18]. This is something we believe should be clarified. The longitudinal (year after year) characteristics of academic contributions to smart contracts are of particular interest to us because they will allow us to not only document and evaluate the growth in research outputs, but also compare the emphasis placed on various issues across the board. In addition, the purpose of this study is to identify open issues in smart contracts that, once resolved, will necessitate further research. We came to the conclusion that the most efficient approach to proceed would be to use the method of systematic mapping study described in [1] to construct a classification map and search for relevant papers in the most important scientific databases. We reached this conclusion after determining that this method would be the most effective means of achieving our goal. The generated map can be used to obtain a deeper understanding of both the issues of interest and the gaps that must be filled in terms of future work [19, 20].

This study makes a contribution to the existing body of research by offering a comprehensive, up-to-date, and methodical review of blockchain-based smart contracts. This overview includes both theoretical and practical elements. Additionally, it provides some observations and suggestions for future work that can assist in advancing the research and development of smart contracts that are based on blockchain technology.

The remaining parts of this essay are structured as follows: In Sect. 2, we cover some of the fundamentals of blockchain technology and smart contracts. The methods that we used to conduct our literature review are discussed in Sect. 3. In Sect. 4, we will discuss the findings that emerged from our analysis of the relevant literature in light of our taxonomy. In the fifth section, we address potential future research directions for smart contracts that use blockchain technology. The final section of the paper is Sect. 6.

2 Background

We will provide some background knowledge on blockchain technology and smart contracts in the next section. We discuss the fundamental ideas, characteristics, and issues associated with these technologies, as well as how they are related to one another.

2.1 Blockchain Technology

The unchangeable, decentralized, and distributed ledger that blockchain technology uses to record transactions that take place between several parties in a peer-to-peer network is known as the blockchain. A blockchain is a distributed ledger that is made up of a chain of blocks that are connected by cryptographic hashes [19, 21]. A consensus process used by the nodes in the network verifies the validity of a set of transactions that are contained within each block. The consensus mechanism ensures that all nodes agree on the validity and sequence of transactions, preventing double spending as well as malicious attacks. It also ensures that all nodes agree on the validity and order of transactions. It is impossible to modify or delete a block once it has been added to the blockchain since doing so would break the chain of hashes, which renders the blockchain both tamper-proof and transparent [22, 23].

Blockchain technology has several features that make it suitable for various applications, such as:

- **Decentralization**: The blockchain network is not governed or controlled by a centralized authority or an intermediary because there isn't one. Each node in the network is capable of independently verifying transactions and taking part in the network. This lowers the likelihood of there being a single point of failure, as well as the chances of censorship and corruption [24, 25].
- **Immutability**: Transactions that are recorded on a blockchain cannot be undone and are therefore permanent. Because of this, the data that is recorded on the blockchain is guaranteed to be accurate and trustworthy, and it also provides auditability and accountability [26, 27].
- **Cryptographic link**: Cryptographic methods, like as digital signatures and encryption, are utilized to ensure the safety of all transactions carried out on the blockchain. This not only protects the transactions from illegal access and modification, but also guarantees the transactions' legitimacy and confidentiality [28].
- **Smart contracts**: It is possible to pre-program transactions on the blockchain so that they carry out automatically when certain circumstances are satisfied. These exchanges are what are known as "smart contracts," and they are computer protocols that facilitate, verify, and enforce agreements between many parties.

However, blockchain technology also faces some challenges that limit its scalability, efficiency, and usability, such as:

- **High resource consumption**: The majority of blockchains use a consensus mechanism, which requires a large amount of both processing power and energy. This is done to validate transactions and assure the continuous safety of the blockchain network. As a consequence of this, the operation and upkeep of a blockchain network incur enormous expenses and have an adverse impact on the natural environment [29].
- **Low throughput**: The amount of transactions that may be handled in a single second by a blockchain network is constrained by the block size as well as the block time. This leads to low throughput and excessive latency of transactions, which affects the performance and user experience of blockchain apps.
- **Privacy issues**: All nodes in the network are able to view all transactions that take place on the blockchain publicly. Concerns about users' ability to maintain the confidentiality of their identities and data are prompted by this development.

In addition, certain transactions might include private or confidential information that shouldn't be seen by the public on the blockchain.
- **Interoperability issues**: Architectures, protocols, standards, and language implementations might vary widely among different blockchains. Because of this, it is difficult for them to communicate and share data with other internal systems as well as with systems that are external to their organization. This makes it more difficult to integrate blockchain apps in a variety of contexts and ensure that they are compatible with one another.

2.2 *Smart Contracts*

- Smart contracts are executable codes that operate atop a blockchain to facilitate, implement, and enforce agreements between multiple parties in the absence of a trusted third party. Other programming languages, such as Solidity for Ethereum, Java for Hyperledger Fabric, and JavaScript for NXT, can also be used to construct smart contracts. The ability of smart contracts to communicate with one another and with other data sources, commonly referred to as oracles, enables the execution of complex logic and actions. The use of smart contracts rather than regular contracts has a number of benefits, including the following:
- **Automation**: Smart contracts can execute automatically when pre-defined conditions are met. This reduces the need for manual intervention and human errors [30].
- **Efficiency**: Smart contracts can process transactions faster and cheaper than traditional contracts. This saves time and money for the parties involved.
- **Security**: Smart contracts are secured by cryptography and the immutability of the blockchain. This prevents fraud, manipulation, and breach of contract.
- **Transparency**: Smart contracts are visible and verifiable by all parties on the blockchain. This ensures trust and accountability among the parties.

However, smart contracts also have some limitations and challenges that need to be addressed, such as:

- **Design flaws**: Smart contracts are prone to bugs, errors, or vulnerabilities in their code or logic. These flaws can lead to unexpected outcomes or malicious attacks that can compromise the functionality or security of smart contracts [31].
- **Legal issues**: In the majority of nations, smart contracts are not recognized by the law and are not enforceable. The use of smart contracts is not governed by any specific legal framework or law. In addition, there is no process for the resolution of disputes or any other kind of remedy for smart contract disagreements [32].
- **Usability issues**: In order to design, deploy, and use smart contracts, one must first acquire the necessary technical skills and knowledge. For the creation of smart contracts and their subsequent interactions, there is a dearth of user-friendly tools and interfaces. In addition, there is a lack of standardization as

Table 1 Summarizes the main application contexts and related works

Platforms	Major application contexts	Related works
Ethereum	• Financial • Asset trading	• DAI [21, 35] • Cryptokitties [36]
Corda	• Insurance	• Energy block exchange [37] • MonetaGo [20]
Hyperledger fabric	• Supply chain management • Stock trading • Trade finance	• Everledger diamond blockchain [38] • IBM food trust [39]
Stellar	• Remittance	• StellarX [40] • Tempo [41] • TillBilly [42]

well as interoperability between the various platforms for smart contracts [33, 34] (Table 1).

3 Methodology

In this section, we describe the methodology of our literature review. We explain how we searched, selected, categorized, and analyzed the relevant papers on blockchain-based smart contracts.

3.1 Search Strategy

To conduct a comprehensive and systematic literature search, we followed the PRISMA guidelines. We used the following keywords and their combinations to search for papers on blockchain-based smart contracts: blockchain, smart contract, distributed ledger, decentralized application, DApp, Ethereum, Hyperledger Fabric, NXT, Solidity, security, scalability, privacy, programmability. We searched the following databases: Google Scholar, Scopus, Web of Science, IEEE Xplore, and ACM Digital Library. We also searched the references of the retrieved papers to identify any additional relevant papers.

We applied the following inclusion and exclusion criteria to filter the papers:

- **Inclusion criteria**: The paper must be written in English; it must be published in a journal or conference that is subject to peer review; the study's primary focus must be on blockchain-based smart contracts, either from a technical or an application standpoint; and the paper's publication date must fall between 2015 and 2023.
- **Exclusion criteria**: The paper is not written in English; the article is not published in a journal or conference that is subject to peer review; the paper does not focus on blockchain-based smart contracts or only briefly mentions them; the paper was

published before 2015 or after April 2023; and the paper was published before or after those dates.

In April of 2023, we carried out a search of the relevant literature and came up with a total of 3356 papers. Following the application of the criteria for inclusion and exclusion, we settled on 539 papers for further investigation.

3.2 Categorization and Analysis

A taxonomy that we built ourselves based on the previous research was utilized by us in order to classify and evaluate the papers that were chosen. The taxonomy is comprised of six different dimensions: a blockchain platform, a smart contract language, a security issue, a solution to scalability, an enhancement to privacy, and an aspect of programmability. Each dimension is broken down into multiple subcategories that each indicate a particular strategy or method that was utilized by the papers. As an illustration, some of the platforms that fall under the umbrella of blockchain technology are Ethereum, Hyperledger Fabric, NXT, and others. Vulnerabilities can be broken down into its component parts, such as code vulnerability, transaction vulnerability, network vulnerability, and others [43, 44].

We classified each publication by assigning it to one or more subcategories for each dimension according to the primary contribution or focus that it presented in the paper. In addition, we retrieved some descriptive information from each of the papers, such as the title, authors, year of publication, kind of publication (journal or conference), type of research (empirical or theoretical), type of technique (qualitative or quantitative), and the primary findings or implications.

In order to provide a concise summary of the distribution and frequency of the papers across a variety of dimensions and subcategories, we made use of descriptive statistics. In addition, we utilized thematic analysis in order to recognize the predominant patterns and ideas that surfaced across the papers. We analyzed the various approaches and methods described in the papers and compared and contrasted their positive and negative characteristics. In addition to this, we uncovered the study voids and unanswered questions that must be answered by subsequent investigations.

4 Results

In this section, we will describe the findings of our literature review in accordance with the taxonomy that we developed. We describe the distribution of the papers throughout the various dimensions and subcategories, as well as the frequency of their appearance. In addition, we analyze the papers to determine the most prominent recurring themes and trends.

4.1 Distribution and Frequency of Papers

The breakdown of the papers into their respective publication years is presented in Fig. 2. Since 2015, we can see that the number of papers on blockchain-based smart contracts has dramatically increased, hitting a peak in 2022 with 158 publications. This indicates that the number of papers is expected to continue to climb. This suggests that both researchers and practitioners are becoming increasingly interested in and attentive to this topic.

The breakdown of the papers into their respective types of publishing is presented in Fig. 3. We can observe that the majority of the papers were presented at conferences (82%) rather than being published in journals (18%). This shows that blockchain-based smart contracts are still a developing and active study subject that needs additional investigations that are more rigorous and mature before they can be published in journals.

Figure 4 illustrates the breakdown of the publications into several categories of research. It is clear that the majority of the publications (71%) were empirical, while only 29% of the studies were theoretical. This indicates that research on smart

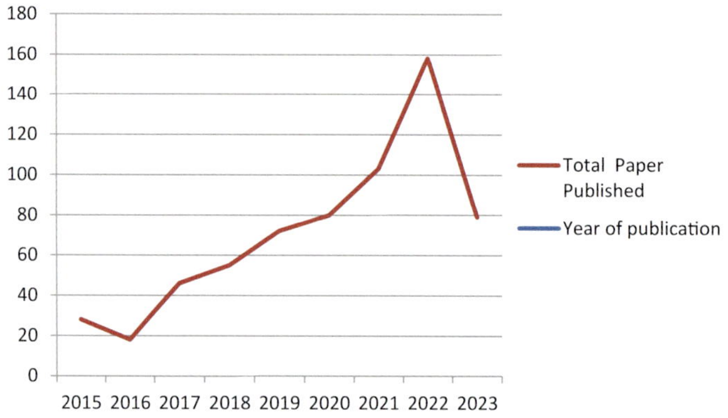

Fig. 2 Distribution of papers by year of publication

Fig. 3 Distribution of papers by type of publication

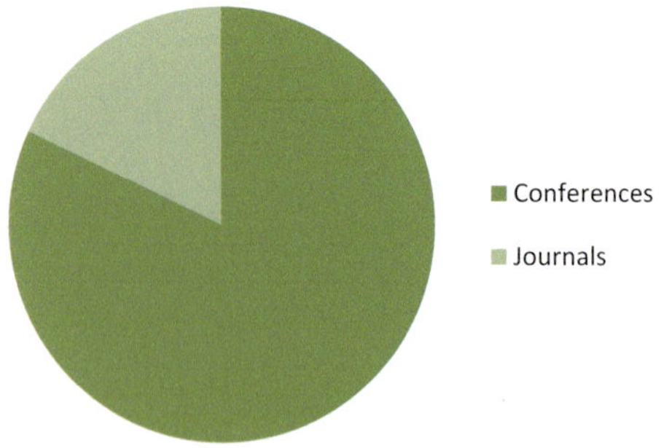

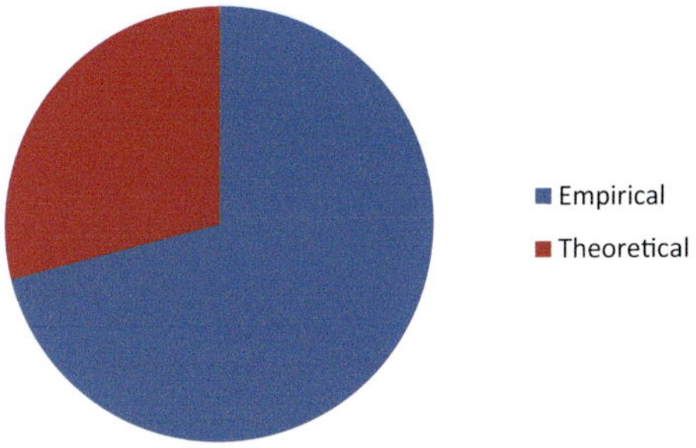

Fig. 4 Distribution of papers by type of research

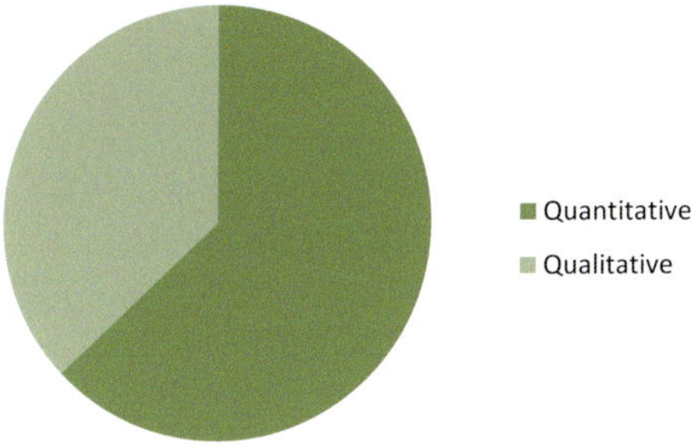

Fig. 5 Distribution of papers by type of approach

contracts that are based on blockchains is mostly conducted from a more applied viewpoint, employing experiments, case studies, surveys, or simulations to evaluate the performance, usability, or application of blockchain-based smart contracts.

The breakdown of the publications into their respective methods is presented in Fig. 5. It is clear that the majority of the publications (63%) focused on quantitative research, while just 37% of the papers explored qualitative topics. This suggests that the quality, efficacy, or efficiency of blockchain-based smart contracts is evaluated primarily through the use of numerical data, such as metrics, measurements, or statistics.

The distribution and frequency of the papers can be seen in Table 2, which is broken down by dimension and subcategory. We can observe that the majority of the papers (67%) utilized Ethereum as their blockchain platform of choice, followed by Hyperledger Fabric (16%), NXT (7%), and others (10%). The majority of the research articles (58%) made use of the smart contract programming language Solidity. This was followed by Java (15%), JavaScript (8%), and other languages (19%). The most common type of security issue that was discussed in the papers was code vulnerability, which accounted for 42% of the total, followed by transaction vulnerability (24%), network vulnerability (18%), and other types of vulnerability (16%). Sharding was the scalability approach that was proposed in the studies the most frequently (35%) followed by off-chain computation (25%), sidechains (20%), and other scalability solutions (20%). The technique of enhancing privacy with zero-knowledge proof was utilized by the majority of the papers (40%), followed by encryption (30%), ring signature (15%), and other privacy enhancement methods (15%). The majority

Table 2 Distribution and frequency of papers by dimension and subcategory

Dimension	Subcategory	Frequency	Percentage (%)
Blockchain platform	Ethereum	82	67
Hyperledger fabric	20	16%	
NXT	9	7%	
Others	12	10%	
Smart contract language	Solidity	71	58
Java	18	15%	
JavaScript	10	8%	
Others	23	19%	
Security issue	Code vulnerability	52	42
Transaction vulnerability	30	24%	
Network vulnerability	22	18%	
Others	19	16%	
Scalability solution	Sharding	43	35
Off-chain computation	31	25%	
Sidechains	25	20%	
Others	24	20%	
Privacy enhancement	Zero-knowledge proof	49	40
Encryption	37	30%	
Ring signature	18	15%	
Others	18	15%	
Programmability aspect	Formal verification	55	45
Testing	31	25%	
Debugging	18	15%	
Others	19	15%	

of the publications (45%) focused on formal verification as the primary method for evaluating programmability, followed by testing (25%), debugging (15%), and other methods (15%).

4.2 Themes and Patterns

Based on our thematic analysis of the papers, we identified four main themes and patterns that emerged from our literature review:

- **Theme A**: Smart contracts built on blockchain technology make it possible to create new applications and business models in a variety of industries, including cryptocurrency systems, supply chain management, agribusiness, real estate, and

energy trading, among others. These applications make use of the capabilities and advantages offered by smart contracts, such as automation, efficiency, safety, and transparency. On the other hand, they are also subject to a number of difficulties and constraints, including regulatory uncertainty, user uptake, interoperability, and so on.
- **Theme B**: Blockchain-based smart contracts require high-quality code to ensure their correct functionality and security. However, writing smart contract code is not easy due to its complexity, novelty, and irreversibility. Therefore, many papers propose methods and tools to improve the quality of smart contract code, such as formal verification, testing, debugging, etc.
- **Theme C**: Blockchain-based smart contracts suffer from scalability issues due to their high resource consumption and low throughput. Therefore, many papers propose solutions to enhance the scalability of smart contracts, such as sharding, off-chain computation, sidechains, etc. These solutions aim to reduce the load on the main blockchain or increase its capacity without compromising its security or decentralization.
- **Theme D**: Blockchain-based smart contracts raise privacy concerns due to their public visibility and traceability. Therefore, many papers propose techniques to enhance the privacy of smart contracts, such as zero-knowledge proof, encryption, ring signature, etc. These techniques aim to protect the identity or data of smart contract users or transactions without affecting their functionality or verifiability.

5 Future Research Directions

Following the completion of our literature study, we will now present several possible future research areas for smart contracts that are based on blockchain technology. We examine the possible benefits as well as the obstacles that may arise from following these areas, and we offer some suggestions or guidelines for future work.

5.1 Cross-Chain Smart Contracts

Cross-chain smart contracts are smart contracts that can interact with different blockchains or other external systems. This is one of the future research directions that will be pursued in order to enable cross-chain smart contracts. The interoperability and usefulness of blockchain-based systems, such as decentralized exchanges, cross-border payments, supply chain management, and other similar applications, can be improved with the help of cross-chain smart contracts. However, cross-chain smart contracts also present some technical and security challenges, such as how to ensure the consistency and atomicity of transactions across different blockchains, how to verify the validity and authenticity of data from external sources, how to prevent malicious attacks or frauds, and so on. These are just some of the challenges. As a result, future efforts ought to concentrate on building innovative protocols, structures, and standards with the purpose of making it easier to design, deploy, and carry out the execution of cross-chain smart contracts.

5.2 Machine Learning-Based Smart Contracts

Integration of machine learning methods with smart contracts, which are smart contracts that can learn from data and adapt to changing environments, is another potential study direction. Smart contracts are smart contracts that can learn from data. Applications built on blockchain technology, such as prediction markets, insurance policies, recommendation systems, and so on, can have their intelligence and efficiency improved by the use of smart contracts that are powered by machine learning. Nevertheless, machine learning-based smart contracts face a number of challenges and limitations. These include the following: how to ensure the privacy and security of data used for training and inference; how to deal with uncertainty and noise in data and models; how to balance the trade-off between complexity and performance of models; how to handle ethical and legal issues related to the outcomes of machine learning; and so on. As a result, work that will be done in the future should concentrate on establishing unique methodologies, tools, and frameworks that will enable the construction, deployment, and assessment of smart contracts that are based on machine learning.

5.3 Human-Centric Smart Contracts

The adoption of a human-centric approach to smart contracts, which are smart contracts that take into consideration the requirements, preferences, and actions of human users, is the third future research direction that will be pursued. The usability and general acceptability of blockchain-based applications, such as e-commerce platforms, social networking platforms, gaming platforms, and so on, can be improved with the help of human-centric smart contracts. However, human-centric smart contracts also raise a few challenges and questions, such as how to design user-friendly interfaces and interactions for smart contracts, how to incorporate human factors and emotions into the logic of smart contracts, how to align the outcomes of smart contracts with human values and norms, and how to resolve conflicts and disputes that arise from the execution of smart contracts, among other things. In light of this, work that will be done in the future should concentrate on establishing unique concepts, models, and approaches to support the design, implementation, and evaluation of human-centric smart contracts.

6 Conclusion

In this chapter, we have provided a comprehensive evaluation of the literature on blockchain-based smart contracts, both from a technical and an application standpoint. We have provided a taxonomy of the currently available blockchain-based smart contract solutions, categorized the included research papers, and discussed the various blockchain-based smart contract-based studies. In addition, we have identified and analyzed the major themes and patterns that emerged from our review of the pertinent literature. On the basis of our findings, we have suggested some potentially fruitful future research avenues for blockchain-based smart contracts and analyzed their potential benefits as well as potential drawbacks.

According to the findings of our study, blockchain-based smart contracts are an emerging and rapidly developing research topic that may be used in a variety of contexts and present a range of opportunities as well as difficulties. Nevertheless, blockchain-based smart contracts confront a number of challenges, both technological and usage-related, that need to be addressed by researchers and practitioners in the field. As a result, we have high hopes that our study will be able to operate as a helpful reference and guide for subsequent research on smart contracts based on blockchain technology.

Conflict of Interest There are no competing interests involved in this research.

Funding No funding is available for this research.

References

1. Petersen K, Feldt R, Mujtaba S, Mattsson M (2008) Systematic mapping studies in software engineering. EASE 8:68–77
2. Taherdoost H (2021) Smart contracts in blockchain technology: a critical review. Information 14(2):117
3. Khan SN, Loukil F, Ghedira-Guegan C, Benkhelifa E, Bani-Hani A (2021) Blockchain smart contracts: applications, challenges, and future trends. Peer-to-Peer Netw Appl 14(5):2901–2925
4. Nakamoto S (2009) Bitcoin: a peer-to-peer electronic cash system. http://www.bitcoin.org/bitcoin.pdf
5. Buterin V et al (2014) A next-generation smart contract and decentralized application platform. White Paper 3:37
6. Wang S et al (2018) An overview of smart contract: architecture, applications, and future trends. In: 2018 IEEE intelligent vehicles symposium (IV). IEEE, pp 108–113
7. Wright A, Filippi PD (2015) Decentralized blockchain technology and the rise of Lex Cryptographia. Available at SSRN 2580664
8. Alharby M, Moorsel AV (2017) Blockchain-based smart contracts: a systematic mapping study. arXiv:1710.06372
9. Udokwu C et al (2018) An exploration of blockchain enabled smart-contracts application in the enterprise. Technical report 2.2:36464.97287
10. Seijas PL, Thompson SJ, McAdams D (2016) Scripting smart contracts for distributed ledger technology. IACR cryptology ePrint, 1156

11. Aggarwal S et al (2019) Blockchain for smart communities: applications, challenges and opportunities. J Netw Comput Appl 144:13–48
12. Buterin V (2018) A next-generation smart contract and de-centralized application platform. https://github.com/ethereum/wiki/wiki/White-Paper/
13. Polyzos GC, Fotiou N (2022) Blockchain-assisted information distribution for the internet of things. In: 2017 IEEE International conference on information reuse and integration (IRI), pp 75–78
14. Wust K, Gervais A (2018) Do you need a blockchain? In: 2018 crypto valley conference on blockchain technology (CVCBT). IEEE, pp 45–54
15. Crawford M (2017) The insurance implications of blockchain. Risk Manag 64:24
16. Guo Y (2018) WISChain: an online insurance system based on blockchain and DengLu1 for web identity security. In: 2018 1st IEEE international conference on hot information-centric networking (HotICN), pp 242–243
17. Bogner A, Chanson M, Meeuw A (2016) A decentralised sharing app running a smart contract on the Ethereum blockchain. In: Proceedings of the 6th international conference on the internet of things. ACM, pp 177–178
18. Hans R, Zuber H, Rizk A, Steinmetz R (2017) Blockchain and smart contracts: disruptive technologies for the insurance market. In: 2017 Americas conference on information systems, pp 01–10
19. Bartoletti M, Pompianu L (2017) An empirical analysis of smart contracts: platforms, applications, and design patterns. In: International conference on financial cryptography and data security. Springer, pp 494–509
20. Peters GW, Panayi E (2016) Understanding modern banking ledgers through blockchain technologies: future of transaction processing and smart contracts on the internet of money. Banking beyond banks and money. Springer, pp 239–278
21. Sengupta J, Ruj S, Bit SD (2020) A comprehensive survey on attacks, security issues and blockchain solutions for IoT and IIoT. J Netw Comput Appl 149:102481
22. Sankar LS, Sindhu M, Sethumadhavan M (2017) Survey of consensus protocols on blockchain applications. In: 2017 4th international conference on advanced computing and communication systems (ICACCS). IEEE, pp 1–5
23. Singh A et al (2020) Sidechain technologies in blockchain networks: an examination and state-of-the-art review. J Netw Comput Appl 149:102471
24. Clack CD, Bakshi VA, Braine L (2016) Smart contract templates: essential requirements and design options. arXiv:1612.04496
25. Chen L et al (2017) Decentralized execution of smart contracts: agent model perspective and its implications. In: International conference on financial cryptography and data security. Springer, pp 468–477
26. Sousa J, Bessani A, Vukolic M (2018) A Byzantine fault-tolerant ordering service for the hyperledger fabric blockchain platform. In: 2018 48th annual IEEE/IFIP international conference on dependable systems and networks (DSN). IEEE, pp 51–58
27. Xu X et al (2017) A taxonomy of blockchain-based systems for architecture design. In: 2017 IEEE international conference on software architecture (ICSA). IEEE, pp 243–252
28. Marino B, Juels A (2016) Setting standards for altering and undoing smart contracts. In: International symposium on rules and rule markup languages for the semantic web. Springer, pp 151–166
29. Norta A (2016) Designing a smart-contract application layer for transacting decentralized autonomous organizations. In: International conference on advances in computing and data sciences. Springer, pp 595–604
30. Moyano JP, Ross O (2017) KYC optimization using distributed ledger technology. Bus Inf Syst Eng 59:411–423
31. Hu Y et al (2018) A delay-tolerant payment scheme based on the Ethereum blockchain. CoRR abs/1801.10295
32. Macrinici D, Cartofeanu C, Gao S (2018) Smart contract applications within blockchain technology: a systematic mapping study. Telemat Inform 35(8):2337–2354

33. Luu L (2017) Practical decentralized pooled mining. In: 26th fUSENIXg security symposium (fUSENIXg security 17), pp 1409–1426
34. Feng Q et al (2019) A survey on privacy protection in blockchain system. J Netw Comput Appl 126:45–58
35. Atzei N et al (2018) SoK: unraveling bitcoin smart contracts. In: Bauer L, Küsters R (eds) Principles of security and trust. POST 2018. Lecture notes in computer science, vol 10804. Springer, Cham
36. Manzoor Y (2018) A delay-tolerant payment scheme on the Ethereum blockchain. In: 2018 IEEE 19th international symposium on a world of wireless, mobile and multimedia networks (WoWMoM). IEEE, pp 14–16
37. Guo Y, Liang C (2016) Blockchain application and outlook in the banking industry. Financ Innov 2:24
38. Rosner MT, Kang A (2015) Understanding and regulating twenty-first century payment systems: the ripple case study. Mich L Rev 114:649
39. Hopwood D, Bowe S, Hornby T, Wilcox N (2016) Zcash protocol specification, Tech. rep. 2016–1.10. Zerocoin Electric Coin Company, Tech. rep
40. Rathore S, Kwon BW, Park JH (2022) Blockseciotnet: blockchain-based decentralized security architecture for IoT network. J Netw Comput Appl 143:167–177
41. Alphand O (2023) IoTChain: a blockchain security architecture for the internet of things. In: 2018 IEEE wireless communications and networking conference (WCNC). IEEE, pp 1–6
42. Nagothu D et al (2023) A microservice-enabled architecture for smart surveillance using blockchain technology. In: 2018 IEEE international smart cities conference (ISC2). IEEE, pp 1–4
43. Zheng Z et al (2018) Blockchain challenges and opportunities: a survey. Int J Web Grid Serv 14(2018):352–375
44. He P et al (2017) Survey on blockchain technology and its application prospect. Comput Sci 44(2017):1–7

An Impact of Cyber Security and Blockchain in Healthcare Industry: An Implementation Through AI

Monika Dandotiya and Indrajit Ghosal

Abstract The healthcare sector is intrigued by several aspects of AI and blockchain technology, including the immutability of data stored in a blockchain. Based on blockchain and AI Blockchain and AI will speed up clinical and biological research and develop biomedical and healthcare data ledgers. These assessments are based on fundamental blockchain and artificial intelligence (AI) capabilities, including decentralized management, data authenticity, immutable audit trails, resilience, enhanced security, and—most prominently—the restoration of charterers' rights. In this chapter, various blockchain-based solutions are examined, including frameworks and implementations to gauge the effectiveness of such frameworks, such as Composer, Docker Container, Hyperledger Fabric, and the Wire shark capture engine. Current limitations in healthcare systems are also discussed. In addition, this study suggests a Privacy Integration Algorithm among healthcare workers, aiding in a simulation of settings to implement hyperledger-based electronic healthcare record (EHR) distribution system that makes use of chain code idea. Round trip time (RTT), throughput, with other performance parameters in blockchain networks have also been adjusted for better outcomes. The suggested solution employs blockchain to increase efficiency and security in comparison to conventional EHR systems.

Keywords Healthcare · Cyber security · Blockchain · AI · EHR · Hyperledger fabric · Wireshark

M. Dandotiya (✉)
Poornima University, Jaipur, Rajasthan, India
e-mail: dandotiyamonika@gmail.com

I. Ghosal
Brainware University, Barasat, Kolkata, India

K. Kaushik and I. Sharma (eds.), *Next-Generation Cybersecurity*, Blockchain Technologies, https://doi.org/10.1007/978-981-97-1249-6_6

1 Introduction

Blockchain technology has gained popularity recently because of the clinical data's strong security method [1]. By automating Bitcoin transactions and providing decentralized, secure, and trusted access to a shared ledger of data, transactions, and records, blockchain technology works to secure data [2]. Through the use of intelligent contracts, it may also manage member interactions without the aid of a middleman or other reliable third party. Blockchain algorithms are used to assure security and secret storage of medical data because healthcare researchers are concerned about the security, storage, and privacy of clinical data [3] (Fig. 1).

Blockchain helps in the analysis and forecasting of health data. Technology helps in many different ways, according to research, including evaluating medical supply files, coming across computerized learning with pharmacologic warnings, successfully implementing health and treatment areas, and the potential for reassembling. Medical experts, healthcare professionals, and payers get fast updates, thanks to the collected medical data [5]. To take into account the aforementioned aspects, researchers have used a variety of strategies built on blockchain and AI technologies [6]. These methods enable the healthcare sector to analyze data at extraordinary rates while maintaining accuracy and data security, and they provide certain professionals temporary access to their data. The primary motivation for writing this chapter is to draw attention to the work that academics have done to inform us about the function and significance of employing AI Blockchain in EHR systems. The healthcare industry needs to reform badly, from infectious illnesses to cancer to radiography. There are various ways to use technology to provide more precise, trustworthy, and efficient therapies. When making a clinical decision, these treatments can be precise. A computer program with specific instructions called artificial intelligence is used to carry out tasks that would typically need human intellect. Algorithms are programmable rules that are programmed [7]. An algorithm may be continuously improved via machine learning. Huge amounts of data are used in the improvement process, which is carried out constantly to allow the algorithm to change to increase the artificial intelligence's accuracy. AI can comprehend and interpret language,

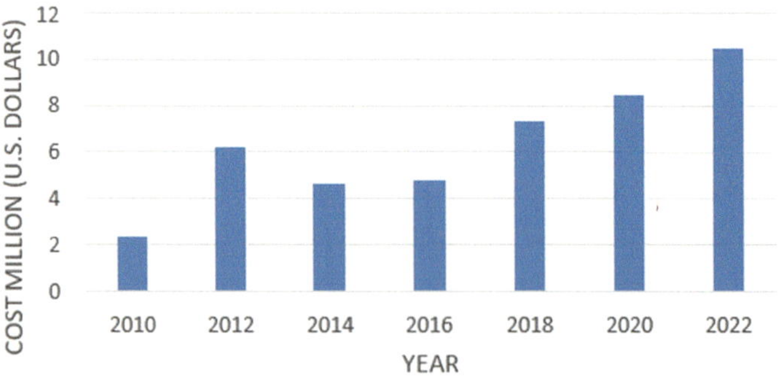

Fig. 1 Overall financial loss due to cyber-attacks on the healthcare industry (2010–2022) [4]

recognize things, hear noises, and learn patterns to carry out problem-solving tasks. The reliability and integrity of patient data are not guaranteed by existing medical data management system. This is a serious shortcoming. Under existing medical data management system, which is primarily targeted at medical institutions, there is no assurance of the validity and dependability of patient data. Medical data loss and hacking are unavoidable risks that must be recognized, and the data gathered is often vulnerable to data security breaches, violations of personal privacy, and other issues [8].

Blocks are immutable and auditable since they are cryptographically connected. Since the same copy of the ledger is replicated across all network nodes, the highest level of availability and transparency is achieved. Medical data may be verified via cryptographic linking, which can also provide a tamper-proof duplicate of it. Public, private, and consortium blockchains are the three kinds of blockchains that are now accessible. Anyone can join the network and take part in transactions when using a public blockchain. Private blockchain, in contrast, limits access without sufficient identification and verification. Public and private blockchain characteristics are combined in consortium blockchain. The movement of activities in the blockchain network is shown in Fig. 2.

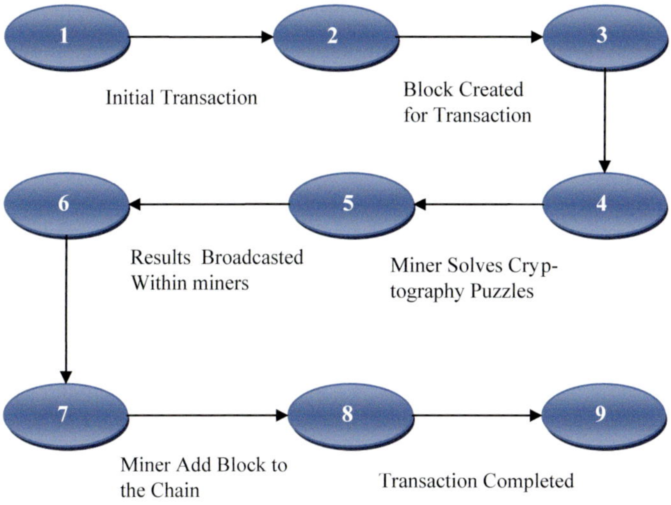

Fig. 2 Blockchain process

2 Literature Review

Clinical observatories, Doctors, patients, and labs, make up traditional healthcare systems. The effectiveness of the current healthcare system can be increased by integrating a variety of contemporary technologies, including AI, cyber security, blockchain, edge computing, and cloud computing, to provide the automation, intelligence, security, and low cost computational ecosystem that form the foundation of healthcare. The following are the main technologies that smart healthcare systems have included.

Costa et al. propose Sec-Health, a blockchain-based protocol that takes into account all of the key security and supplementary features outlined in existing rules. By examining Sec-Health under various attack scenarios and detailing how it resolves the issues with previous solutions, they demonstrate that it is an appropriate solution [9].

Ramachandran et al. presents a ground-breaking investigation into the security enabler function of blockchain in Industry 5.0. We were able to determine the most important drivers, and prospective applications, and give an institutional vision for Blockchain-based Industry 5.0 across a wide range of industries by using a subjective survey approach and study objectives. The poll aims to provide industry practitioners, academia, and researchers with recommendations for fresh even though solutions in Industry 5.0 verticals. One of the most promising technologies for increasing the security of Industrial 5.0 settings is blockchain technology, which has the inherent properties of absoluteness, chronological order, and integrity in industrial systems. Few books are recommended that provide an overview of Industry 5.0 applications that use blockchain technology [10].

Gunanidhi et al. suggested (E-PoW) consensus blockchain approach for IoT-based healthcare monitoring schemes. The primary issue with healthcare analytics systems is the patient data. Data privacy necessitates high-quality data security measures and calls for the finest design technique. Many people in the present situation have failed to ensure protection for data kept on servers that are accessible to other parties [11].

Rajora et al. discussed several mitigating techniques, such as Machine Learning, Blockchain, Deep Learning, HLE-SO hybrid, etc., and work is suggested for cyber attacks seen in the healthcare system, as well as the types of techniques used to prevent attacks in cyber systems. Even though there is no ideal solution, we should do everything in our ability to reduce cybercrime to guarantee a secure future for technology. Most security systems require weaknesses before inappropriate conduct can be investigated and remembered to prevent a repeat. It is futile to take such a reactive stance while protecting infrastructure, which includes our healthcare systems [12].

Zhang et al. provide a thorough examination of the security and privacy threats and requirements, along with technological solution methodologies and strategies, with a focus on the security and privacy requirements for medical data sharing using blockchain. They talk about the characteristics needed for electronic medical data sharing [13].

Omaretal. suggested an approach that combines decentralized storage and blockchain technology to increase transparency, improve take holder communication, and shorten the procurement cycle while preventing price differences and errors. They suggest leveraging smart contracts on the blockchain to automate the GPO contracting procedure. They provide a general architecture for the HCSC contracting process that includes intricate algorithms that show the different interconnections between HCSC stakeholders [14].

Charla et al. used the straightforward and widely accepted concept of cyber security as CIA to identify cyber risk in the context of blockchain. Based on the findings, a cyber-physical risk-based methodology was created, and blockchain implementation in particular cases was analyzed in order to find and evaluate the breaches following this framework. Distributed ledger technology, or blockchain, is still being used in industries like healthcare, energy, manufacturing, financial services, and retail. Because of its distinct technological features, blockchain proponents contend that this technology represents a new architectural foundation for establishing trust [15].

Egala et al. Uncompromising system security and privacy are maintained by the proposed blockchain-based new architecture, which offers a decentralized HER and smart-contract-based service automation. To discourse on the deficiencies of the blockchain-based cloud-centric IoMT healthcare system [16].

Im et al. It is suggested to use PHR based on the consortium blockchain, Hyperledger Fabric. In order to access private data, they also evaluate and compare performance in terms of ledger size and transaction delay. Personal health information should not be accessible to anyone because it is sensitive information. It should only be accessible to authorized users. Blockchain, a new middleware platform paradigm, can offer data sovereignty, secrecy, and transparency in addition to traceability [17].

3 Research Gaps

We look at how smart technologies are evolving and the security standards that must be met before they can be used in the healthcare sector. A description of blockchain technology is given, along with information on its advantages and potential uses in healthcare systems. Modular IT solutions have started to appear since the 1970s when healthcare services first began. This era is referred to as Healthcare 1.0. Due to a lack of resources, healthcare systems at this time were constrained and did not integrate with digital systems. In a similar vein, networked electronic devices did not yet include biomedical equipment that might be integrated. At healthcare companies at the time, paper-based prescriptions and reports were often employed, which raised expenses and took more time.

4 Objectives of the Study

The main objectives of the study aim to:

- To provide decentralized records of financial transactions independent of centralized institutions or authorities.
- To improve access to patient medical history, device tracking, hospital assets,
- and prescription databases with a whole life cycle of a device inside blockchain architecture.
- To change the interoperability of healthcare systems.
- To put forward a system architecture with a process for participants' access control policies to ensure patient data privacy and security in the HER system.

5 Research Methodology

Suggested electronic health record system is created using blockchain-based infrastructure hyperledger fabric with its sandbox hyperledger composer. The permission-based open-source distributed ledger technology known as hyperledger is (DLT). Since all users are aware of one another and the blockchain is permissioned and consortium-controlled, the network can be completely trusted and secured. Participants in the EHR sharing system fall into four categories: administrators, patients, physicians, and lab personnel. The exact administration of a blockchain network. The certifying authority is asked for admin's enrollment certificate. The administrator has complete control over the system, including the ability to write, read, update, and delete users. Whether doctors, patients, or lab workers are legitimate, the administrator may provide each participant with the appropriate ID to allow access to the blockchain network. If a participant's actions are deemed improper, the administrator may ban them by posting a message on the Hyperledger blockchain network.

Participants in the EHR sharing system fall into four categories: administrators, patients, physicians, and lab personnel. Of Algorithm, the administration of a blockchain network is executed precisely. The certify in authority is asked for admin's enrollment certificate. The administrator has complete control over the system, including the ability to write, read, update, and delete users. If doctors, patients, or lab workers are legitimate, the administrator may provide each participant with the appropriate ID to allow access to the blockchain network. If a participant's actions are deemed improper, the administrator may ban them by posting a message on the Hyperledger blockchain network as shown in Fig. 3. The healthcare service provider is associated with patients, doctors, caregivers, microbiological laboratory people, health service administrators, pharma people, insurance people, and other clinical authorities as participants of the network. In this scenario of a permissioned Blockchain e-health system, the patient records and all health-related information are transparent to all the participants of the network. Hence, a privacy-preserving Hybrid Encryption Technique is designed to protect the privacy of the data. Since

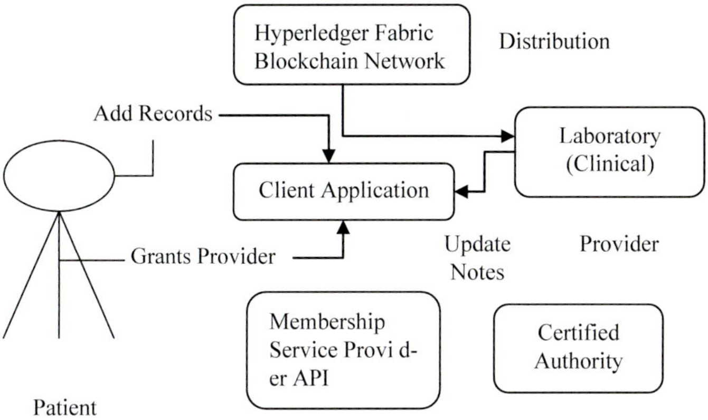

Fig. 3 Proposed architecture

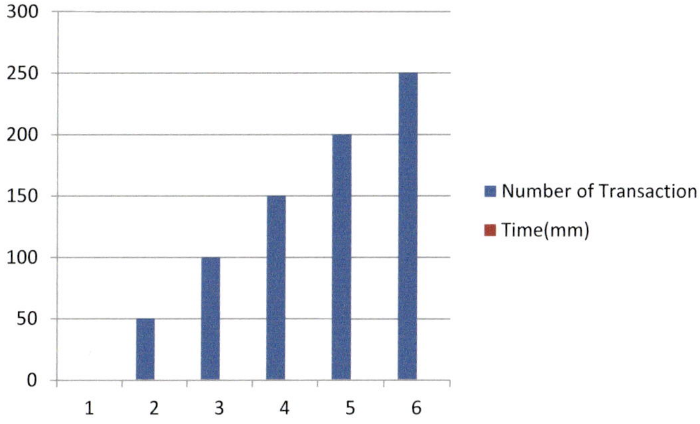

Fig. 4 Effect of block size on transactions

the proposed system is designed with a privacy-preserving Hybrid Encryption technique, the data in the e-Health system is not transparent to any participant of the network (Fig. 4).

5.1 Proposed Algorithm

Step 1: Procedure Admin(patientID,ClinicianID,LabID)

Step 2: While True(do)

  Step 3: If Clinician ID is valid then

Step 4: Add the Clinician to the blockchain Framework. Step 5: Add Clinician (GN, FID)

Step 6: else

Not exist(Clinician ID)

Step 7: end if

Step 8: if(Lab ID is valid)then

Step 9: Add Lab to blockchain Network

Step 10: Add Lab(Blockchain network,Lab ID) Step 11: end for

Step 12: end procedure

The suggested electronic health record system is created using blockchain-based infrastructure hyperledger fabric and its sandbox hyperledger composer. The permission-based open-source Distributed Ledger Technology (DLT) is known as the Hyperledger Project. The Linux Foundation created it to accommodate a variety of smart contracts and logic for creating several applications on the blockchain network. A sandbox called Hyperledger Composer is used to execute and test smart contracts by visualizing the network. Because all users are aware of one another and the blockchain is permissioned and consortium-controlled, the network can be completely trusted and secured. The endorsement policy in the fabric will select which peer nodes (endorsers) will receive the transaction proposal request.

The HER proposal is made up of a variety of parameters, including the patient identification data provided by the membership service provider, the transaction payload containing a list of operations to be carried out, the chain code identifier, a nonce value (counter or random value) that can only be used once by the user submitting the proposal, and the transaction proposal identifier as provided in the algorithm below. The method specifies a chain code in which various transaction actions carried out by the patient are shown as various functions. The proposal phase or the endorsement phase is what this stage is known as. Evaluation is essential for conducting targeted tests of system scalability and performance. Pre-processing is the first step. The Wire shark pcap file may be used to collect all network traffic. This scans all network traffic and only looks for TCP messages sent via the Hyperledger Fabric [18]. All communication inside the framework takes place through the gRPC protocol, which runs on top of TCP. To view the content, if any of the participants wish to view and access the content, it is important to get access permission. To get access permission, the proposed system designs a Patient based access control mechanism for smooth and automatic access with restricted users. To view the data, the proposed Patient based access control mechanism defines a set of permissions that contain access rights and transfer rights. Based on these permissions, the participants of the provider generate a request for viewing the content. In this way, the provider and the participants are receiving the permissions. The proposed system then automatically executes the request without intermediates. The content viewing request is permitted to only those users who are privileged through the access control mechanism [19].

The access permissions ensured in the proposed system are read, write, read-write, and delete. The permissions are provided based on the privileges given to the participants. In the same fashion, the resource owner grants access rights and transfer rights to the requested user.

The steps intricate in Granting Transfer Rights and Access Rights are as follows:

- Step 1: Resource holder of Health Service Provider 1 grants Access Rights to Health Service Provider 2.
- Step 2: Participants of Health Service Provide 2 generate a request for accessing data with the public key.
- Step 3: Resource holder of Health Service Provider 1 receives the public key of requester 1.
- Step 4: Resource Owner of Health Service Provider 1 grants Transfer Rights to request participant of Health Service Provider 2
- Step 5: Requester 1 of Health Service Provider 2 can access resource.

Local Database. In this subsection, the details of the local database of the proposed system are presented. In the proposed system, the original data is stored in the local database. But the hash values of the original data and the blocks generated during the hashing process are stored in the Blockchain. Utilization of the Blockchain repository, for all purposes like local storage and hashing process, automatically reduces the performance [20]. Since the Blockchain is positive, it does not store the original data; instead, it is reasonably the hash values of the original data to improve the performance of the proposed system. The proposed system utilizes a local database for rapid access to the original data and hence a local database is defined for keeping the data secure. This increases the performance of the proposed access control system.

Blockchain Node. The proposed system utilizes Blockchain as a distributed repository for storing the data. It reshapes data and historical access control system records in the permission Blockchain. On the Blockchain network, each provider is responsible for managing his Blockchain node. A new technology called blockchain enables dispersed and decentralized architectures to communicate among a wide network of unreliable parties. Transparency, immutability, cryptography, and operational resilience are built-in characteristics of blockchain technology that may enhance the storage capacity of access control systems, making blockchain technology more suited to decentralized design. Create, read, update, and delete activities are carried out by conventional database systems [21]. The Blockchain facilitates the addition of new transactions rather than updating or deleting existing ones; it is an append-only dataset range system. This append-only paradigm is applied in the planned Blockchain-based e-Health system. The suggested Blockchain technology is also a novel way to organize and keep track of the ownership of access rights. In this way, blockchain technology can be used to build a secure repository.

6 Patient-Based Access Control Mechanism in E-health System

In this section, the design of the Patient based Access Control Mechanism is emphasized in the e-health systems. In the e-Health system, considering a situation, there are four health service providers participating to share the data with permissions. The healthcare service provider is associated with patients, doctors, caregivers, microbiological laboratory people, health service administrators, pharma people, insurance people, and other clinical authorities as participants of the network. In this scenario of a permissioned blockchain health system, the patient record and health-related formation are transparent to all the participants of the network. Hence, a privacy-preserving Hybrid Encryption Technique is designed to protect the privacy of the data. Since the proposed system is designed with a privacy-preserving Hybrid Encryption Technique, the data in the e-Health system is not transparent to any participant of the network. It is further noted that the health records generated during the consultation process. The access rights could be defined as subscriptions for each health service and the resource owner of a Hospital defines how many times the clinical authority of another hospital can participate in the Blockchain network within a specified time limit. One of the important features of this proposed system is that any provider can transfer access rights to its internal departments such as microbiological groups, surgery groups, pharmaceutical groups, doctor groups, and patient groups. This process of transferring can continue from one provider to another and within the same provider departments.

7 Results

To be more specific, evaluation is essential for conducting targeted tests of systems capability and performance. Pre-processing is the first step. The Wire shark pcap file may be used to collect all network traffic. This examines all network traffic and only filters out TCP messages sent across the hyperledger Fabric. When the transaction rate rises during different rounds, the latency time also rises. More businesses and peers exhibit increasing latency. One of the most popular and extensively studied applications of blockchain technology in healthcare is electronic medical records (EMR). The majority of the studies that were chosen for the literature review also dealt with EMR. Multiple healthcare professionals can safely track patients using an EMR system that has been properly established. EMR also contributes to accurate and effective healthcare delivery. Data integrity and security are required by the nature of EMR systems, which call for the exchange of patient data among numerous healthcare providers.

The anaconda navigator-based spyder IDE is used for evaluation. It makes use of the statistical data sketching tool, matplotlib. Moreover, it imports pandas3, which provides data analysis and manipulation. Patient visits a doctor to get his/her ailment checked. After a thorough investigation, the doctor provides the prescription which then is added to the blockchain via a web application.

"Docker" is used for setup and initialization to interact with hyperledger fabric and composer. Developers and/or system administrators may utilize Docker, an operating system-level container. It helps develop, deploy, and operate hyperledger-based busi-container. It enables the developer to compile all requirements and features into a single container. The hyperledger fabric and composer network may operate within the container by utilizing Docker. Medical devices can charge for patient data, confirm that the intended patient is receiving the treatment, and exchange procedural data with patients and regulators in an anonymous manner. The use of blockchain technology in the healthcare industry is exciting. Recently notable advancements in medical science and high-quality medical treatments. It is a commonly used, transparent distributed digital ledger for transaction monitoring in numerous computers.

The above figure shows the number of clinical trials for different ORGPEERs. It is acknowledged for having a considerable impact on several sectors and industries. The existence of this technology addresses problems that cannot be solved by current methods. Confidence, protection, confidentiality, and data interchange among many systems are necessary. Blockchain presents the opportunity to approach it in healthcare in innovative ways.

It introduces the suggested method for sharing electronic health records based on the blockchain network. As a result, an architecture for an HER sharing system based on blockchain is suggested. The deployment to various techniques and configurations for block transactions in the network is depicted in Fig. 5. We employed a network model of three organizations, each with three peers, for our simulation phase. With thousand transactions per round at 200, 150, 100, 50, and 250 transactions per second, the experiment is run with basic writing transactions at varying rates. The experiment is carried out for the following organizations: 1 Org 1 Peer, 2 Org 2 Peer, and 3 Org 3 Peer. Each round of five calculations includes 1000 transactions at various transaction rates per second, with the total number of transactions in each round totaling 5000 (Fig. 6).

The number of transactions executed per second is measured in transaction per second (TPS) as shown in Fig. 4.

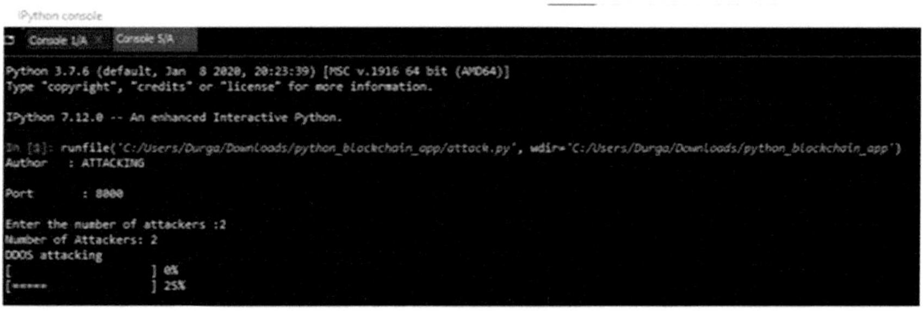

Fig. 5 DDoS injection

Fig. 6 The components that are for med along with spinning up the network

7.1 *CPU Utilization*

Another important parameter to look at how much CPU power the system uses. It is the most important operating system number to keep an eye on during the tuning process. Almost all operating systems show how much time the CPU is being used by people and by the system itself. These extra statistics make it easier to figure out what is being done on the CPU.

We employed a network model of three organizations with three peers in the simulation phase mentioned above. The experiment measured throughput and latency as the number of organizations, block size, and block time increased.

It has been observed that the latency increases with the scaling up of the network with the addition of new organizations and peers. On the other hand, latency for read or query operations is lower compared to write operations.

Block time increase reduces latency, which boosts the efficiency of the network model and increases throughput. The experiment demonstrates that when block time for write operations is increased from 250 ms to 2 s, throughput increases by up to 4 × in 250tps.

It is also observed that block size 20 has 50% lower latency than block size 40, increasing network performance. Since the network model's CPU consumption has remained stable over time, therefore, it is feasible to infer that a decrease in block size and an increase in block time will lead to a significant decrease in network latency, enhancing network performance, which may be used as a tool for performance enhancement.

It may result in all services being consumed on the targeted network. Using a themed-conserver overloads it, cluttering the network and causing the server to fail. The malicious code then injected a greater volume of DDoS packets onto the server sport as depicted.

7.2 Network Latency

Network latency is the amount of time it takes for data to be sent from one node to another. The lower the latency, the faster the connections between nodes in the network. Network latency is measured in milliseconds, and a connection is faster when the value is closer to zero. According to Eq. 1, latency may be determined.

$$\text{Latency} = \frac{\text{Data Size}}{\text{Band width}} \tag{1}$$

The attack codes are injected into the propagated chaos-based medicine to measure latency. Medical data is shared on an SHA and DES encrypt-based blockchain in order to ensure the validity of the data. The chaotic medichain network had a lower latency than a DES or SHA encrypted blockchain for 25, 50, 75, and 100% of the assaults. SHA and DES have a 14 and 18% greater latency than the suggested chaotic medichain at all attack levels. Connectivity and efficiency are improved when latency is reduced. Chaos medichain's decreased latency confirms its effectiveness.

7.3 Computational Time

The amount of time required by the specified architecture to process the data supplied by the user is referred to as computation all time. The propagated chaos-based medichain is injected with assaults in order to examine the computational time. SHA and DES encrypted blockchains are subjected to the same set of attacks to ensure the validity of the medical data. It depicts the research findings as a graph. Even after the assault, the suggested architecture still took 0.7 s to analyze the medical picture, but SHA and DES both took 0.8 s. It takes between 0.5 and 0.7 s to process a high-definition image. Consequently, the suggested design, which took 0.7 s to implement, demonstrates the network's ability to handle data more quickly. Various malicious attacks do not affect the behavior of the decentralized network-based chaos in the proposed architecture medichain, as seen by the extensive network latency and computing time experiments. It prevents an assault on the medichain and protects sensitive data from being accessed. In order to protect medical data, it is essential to do architectural quality tests. The server was injected with malicious code from a variety of attacks to test the proposed architecture. The results of the experiments show that the assaults sent to the chaotic medichain network had a lower latency of 25%, 50%, 75%, and 100% than the DES, SHA encrypted blockchain. The proposal demonstrates a 14% lower latency than SHA and an 18% lower latency than DES at all attack levels. The proposed approach was shown to be more secure than SHA and DES encryption-based blockchains in terms of network latency and computing time. As a result, there is a confidence that the chaotic medichain network presented in the experiment can handle medically sensitive data. There are two types of EHRs related to the patient i.e., on mandatory and other credentials which don't hold much sensitivity to the privacy or security of the patient. In the blockchain network, it is in charge of data storage and communication. The chained block of information, or set of records, is what creates the EHR Smart Contract. Every transaction that occurs throughout the whole blockchain network must be verified by the blockchain management module. The processing module makes sure that the transaction under examination has adhered to all necessary procedures and has received authentication from the Summary Contract. The authenticated data or records are transferred to the databases once all the prerequisites have been met. A framework has been created to create a Blockchain based distributed platform formed medical records that is safe and secure. With this strategy, it is easier to keep an eye on the lack of candour among medical practitioners, and only authorized users are given access to the EHR system. We concentrated on enhancing the security of health records data, which in turn opens up new vistas for future searches to satisfy other security needs. For the deployment of essential applications for any medical purpose, other security models that are more generalized and address the need for health data security as well as detecting the untrusted hosts will be useful in the future. In order to construct a predictive model using conventional AI approaches, data storage, and training are centralized, which increases computational complexity and compromises individual privacy. A federated blockchain-enabled system is proposed in this study as a solution to the issue, to improve the perception of numerous heterogeneous CT scans. It enables secure and private data exchange between hospitals without compromising patient privacy. Blockchain-based collaborative healthcare training also takes advantage of federated learning. This deep learning method has been thoroughly tested and compared to others in the field. A comparison was made between the proposed algorithm and that of previous federated deep learning modules powered by blockchain.

Participants could sign up via a client application or SDK, then obtain an enrollment certificate from a Membership Service Provider to send to the certificate authority. An MSP allows peers to sign off on endorsements and validate incoming transactions. For development purposes, auth.uidai.gov.in was used to connect to the host, and uidai provided java helper class was used to develop a client application to connect Aadhar API. Once verification is successful, Names, dates of birth, gender, and other identifying information can be retrieved from the Aadhar database. Patients can utilize the client application to update information such as their blood type, allergies, prescriptions, insurance information, and soon. In addition, the patient could have the capability to fetch data from the local hospital's EHR.

The anaconda navigator-based spyder IDE is used for evaluation. It makes use of the statistical data sketching tool, matlablib. Additionally, it imports pandas3, which provides data analysis and transformation. The evaluation phase involves the creation of graphs using the Python3 programming language. For better utilization, all network IP addresses have been substituted with node names from the Hyperledger caliper and peer organizational of the evaluation-related transactions, including transaction send rate, throughput, latency, organizations, peers, maximum CPU consumption, and memory usage, are taken from the caliper report file of HTML and processed for transformation. The data is then displayed using different angles in matlablib.

Table 1 Varied block time measurement

Parameter	Configuration
Round	10
Network size	1000 per round
Varied factor	Block time
Rate	150–250 tps

The anaconda navigator-based spyder IDE is used for evaluation. It makes use of the statistical data sketching tool, matlablib. Additionally, it imports pandas3, which provides data analysis and transformation. The evaluation phase involves the creation of graphs using the Python3 programming language. The Wireshark tool, which reads all destination port, sending times, source port, and TCP packets from the pcap file, also considers network data. For better visualization, all network IP addresses have been substituted with node names from the Hyperledger caliper and peer organization. All of the evaluation-related transactions, including transaction send rate, throughput, latency, organizations, peers, maximum CPU consumption, and memory used, are taken from the caliper report file of HTML and processed for transformation. The data is then displayed using different angles in matlablib. A blockchain network benchmarking tool is called Hyperledger Caliper. It is compatible with several Hyperledger frameworks, including Fabric, Composer, Sawtooth, Iroha, etc. In this study, the caliper device is used to check and test the system's performance as well as its many parameters, such as latency, throughput, CPU and memory utilization, disc write/read, network I/O, and others, as well as metrics for the system evaluation (Table 1).

The minimal transaction rates per second used in this case are 25, 50, 75,100, and 125. With this setup, block size 5 has produced better results with a roughly 1.75 × reduction in latency. Similar to this, an early1.75 × increase in transaction throughput is seen. Through investigation, it has been found that smaller block sizes and fewer Transactions Per Second (TPS) lead to better system performance. The blockchain system performs better when the Transaction Rate (TPS) is higher on a larger block size. Using Wireshark tcpdump, the network traffic and associated statistics are recorded as the caliper is run on the EHR system. In this, Wireshark is used to capture the packets during execution and save them to a pcap file. Information on one patient can be kept in a block linked in a chain. Its benefits include reduced costs, enhanced transparency, accurate tracking, a permanent ledger, a lack of intrinsic value, and the fact that it has no physical form and is not governed by a centralized authority. Because of advantages like real-time updates of shared data, distributed access with security, distributed encryption for data integrity, patient information protection, decreased transaction costs, and system performance, this technology is used in medical applications. Blockchain is a decentralized database with chronological connections between its data blocks. In the healthcare sector, a wide range of parties, including medical professionals, hospitals, insurer agencies, etc., need to handle personal EHRs blockchain cooperatively. Electronic Record Systems are centralized by design and are proprietary. This indicates that the programming base, dataset, and system outputs are all controlled by a single supplier, who also provides the monitoring tools. It is challenging for centralized systems to win the confidence of patients, physicians, and hospital administration. This problem is resolved by open-source, independently verifiable systems. EHR is laying the foundation for blockchain-based healthcare. It includes the specialty of employing blockchain technology. As it includes an updated patient medical report, blockchain technology makes it simple to track public health and identify dangers and patterns in the spread of any illnesses. This encourages patients around the world to receive adequate care. A single entity does not own it because it is decentralized and highly secure. After all, the data is kept using cryptography. The records are kept and preserved under the organization in the current system. To prevent the patient from accessing these documents for future comparisons. All records will be lost if the specific server (database) crashes. The proposed system has been created to address these issues. Electronic Health Records (EHRs) offer a valuable record-keeping service that encourages the electronic accessibility of paper-based patient medical records on the Internet. Patients currently disperse their electronic health records (EHRs) across many locations as life events occur, which causes the EHRs to shift from one service provider database to another. Therefore, while the service provider often retains primary management, the patient may loose custody of the current healthcare data. Patients often find it difficult to freely share these data with researchers because patient access permissions to EHRs are severely restricted. Compatibility issues between hospitals and diverse providers. The patient should be able to access his electronic health records (EHRs) for autonomous management and exchange. The application of transferring the patient's electronic health record without a centralized authority is the main topic of this chapter. Cryptography hash functions can be used to ensure security in this situation. The authorized party stores the patient data in blocks that can be connected in a chain using the private key. Bytes from the message can be joined with it to generate a digital signature that many nodes accept. By authenticating the public key image, anyone with a requirement for patient information can access the data. The current multi-signature algorithm can access up to 256 bytes and solely uses data as keys. Through the use of images that may be transformed into arrays of bytes, it was suggested in this work to increase the key size up to 512 bytes in order to address these problems. In terms of an array of bytes, time complexity, accuracy, throughput, and output hash bytes with multiple node accessibility, the proposed technique can then be compared to cryptography hash functions such as SHA224, SHA3-224, SHA256, SHA3-256, SHA384, SHA3-384, SHA512, SHA3-512, and MD-5.

References

1. Islam SMR, Kwak D, Kabir MH, Hossain M, Kwak K-S (2015) The Internet of Things for health care: a comprehensive survey. IEEE Access 3:678–708

2. Wang L et al (2010) A wireless biomedical signal interface system-on-chip for body sensor net- works. IEEE Trans Biomed Circuits Syst 4(2):112–117
3. Marrington A, Kerr D, Gammack J (2016) Management of security issues in wearble technology. IGI Global, Hershey, PA, USA
4. Khanuja SS, Garg S, Singh IP (2009) Method and apparatus for remotely monitoring the condition of a patient. U.S. Patent 12 259 905
5. Chan M, Estève D, Fourniols JY, Escriba C, Campo E (2012) Smart wearable systems: current status and future challenges. ArtifIntell Med 56(3):137–156
6. Clarke R Introduction to data veil lance and information privacy, and definitions of terms. http://www.rogerclarke.com/DV/Privacy.html. Accessed 10 Oct 2017
7. Vora J et al (2018) Blind signatures based secured e-healthcare system. 2018 International conference on computer, information, and tele-communication systems (CITS). IEEE
8. Bodkhe U et al (2020) Block chain for industry 4.0: a comprehensive review. IEEE Access 8:79764–79800
9. Costa LD, Pinheiro B, Cordeiro W, Araújo R, Abelém A (2023) SecHealth: a blockchain based protocol for securing health records. IEEE Access 11:16605–16620
10. Ramachandran KK, Nagarjuna B, Akram SV, Bharani J, Raju AM, Ponusamy R (2023) Innovative Cyber Security Solutions Built on Blockchain Technology for Industrial 5.0 Applications. 2023 International Conference on Artificial Intelligence and Smart Communication (AISC), Greater Noida, India, 643–650
11. Guanidhi GS, Krishnaveni R (2022) Improved Security Blockchain for IoTbased Healthcare monitoring system. Second international conference on artificial intelligence and smart energy (ICAIS). Coimbatore, India, pp 1244–1247
12. Rajora R, Kumar A, Malhotra S, Sharma A (2022) Data security breaches and mitigating methods in the healthcare system: a review. In: International conference on computational modelling, simulation and optimization (ICCMSO). Pathum Thani, Thailand, pp 325–330 (2022)
13. Zhang R, Xue R, Liu L (2022) Security and privacy for healthcare Block-chains. IEEE Trans Serv Comput 15(6):3668–3686
14. Omar A, Jayaraman R, Debe MS, Salah K, Yaqoob I, Omar M (2021) Automating pro cerement contracts in the health care supply chain using Blockchain smart contracts. IEEE Access 9:37397–37409
15. Charla GB, Karen J, Miller H, Chun M (2021) The human-side of emerging technologies and cyber risk: a case analysis of block chain across different verticals. 2021 IEEE technology and engineering management conference-Europe (TEMSCON-EUR). Dubrovnik, Croatia, pp 1–6
16. Egala BS et al (2021) Fortified-chain: a Blockchain based framework for security and privacy assured internet of medical things with effective access control. IEEE Internet Things J 8:11717–11731
17. Im H, Kim KH, Kim JH (2020) Privacy and ledger size analysis for healthcare blockchain. In: International conference on information networking (ICOIN). Barcelona, Spain, pp 825–829
18. Tekeste T, Saleh H, Mohammad B, Ismail M (2019) IoT for healthcare: ultra low power ECG processing system for IoT devices. International Publishing, Cham, Springer, pp 7–12
19. Zhang Y, Qiu M, Tsai C, Hassan MM, Alamri A (2017) Health-CPS: Healthcare cyber-physical system assisted by cloud and big data. IEEE Syst J 11(1):88–95
20. Xu X, Zhang X, Gao H, Xue Y, Qi L, Dou W (2020) BeCome: Blockchain enabled computation offloading for IoT in mobile edge computing. IEEE Trans Industr Inf 16(6):4187–4195
21. Biswas S, Sharif K, Li F, Mohanty SP (2020) Block chain for E-healthcare systems: Easier said than done. IEEE Comput 53(7):57–67

Deep Learning and Blockchain Applications in Healthcare Sector Using Imaging Data

Monika Sethi, Jatin Arora, Vidhu Baggan, Jyoti Verma, and Manish Snehi

Abstract The healthcare industry has witnessed the emergence of deep learning (DL) along with blockchain technologies as potent instruments with considerable potential for transforming the sector, in the domain of imaging analysis of information. The healthcare industry is highly significant to society as it is responsible for preserving and enhancing human health. It includes preventive medicine, diagnostics, rehabilitation, therapy, and palliative care offerings. In recent years, there have been notable advancements and improvements in the medical field regarding using images by implementing DL and blockchain applications. Integrating DL and blockchain technology in healthcare imaging has presented novel prospects for cutting-edge research and advancement. Scholars are investigating using DL models to examine voluminous imaging datasets, thereby revealing valuable insights and patterns that can facilitate progress in healthcare knowledge and treatment methodologies. The adoption of blockchain technology in clinical trials contributes to promoting transparency and consistency. This is owing to the inherent advantages of blockchain, which enable the creation of a transparent and visible system. As a result, the integrity of data collected for research is being protected and trust in the outcomes of clinical trials is being fostered. Integrating DL and the blockchain system presents an intriguing chance to transform the field of telemedicine by facilitating the safe and confidential transfer and retention of medical images, thereby enabling remote diagnosis and advice. This chapter aims to showcase the applications of DL along with the blockchain in healthcare using an imaging dataset. Using a colossal collection of data, combined with deep learning and blockchain techniques, it can be trained to exhibit the desired behaviour. Applications of DL and blockchain technology with imaging

M. Sethi (✉) · J. Arora · V. Baggan
Chitkara University Institute of Engineering & Technology, Chitkara University, Punjab, India
e-mail: sethi.monika4@gmail.com

J. Arora
e-mail: jatin.arora@chitkara.edu.in

V. Baggan
e-mail: vidhu.baggan@chitkara.edu.in

J. Verma · M. Snehi
Punjabi University, Patiala, India

K. Kaushik and I. Sharma (eds.), *Next-Generation Cybersecurity*, Blockchain Technologies, https://doi.org/10.1007/978-981-97-1249-6_7

dataset for various diseases such as cancer, diabetic retinopathy, Alzheimer's etc. can aid medical professionals in the early investigation and classification of diseases so that stricken are given effective therapies.

Keywords Deep learning · Alzheimer's disease · Diabetic retinopathy · Cancer

1 Introduction

The healthcare industry is pursuing novel technologies to augment patient care, enhance diagnostic precision, and advance healthcare. Over the past few years, DL and blockchain-based technologies have surfaced as potent instruments with substantial potential in diverse sectors, such as healthcare. DL algorithms, a constituent part of artificial intelligence, have shown exceptional abilities in examining intricate data, such as medical imaging. However, blockchain technology, renowned for its decentralized and impregnable characteristics, has attracted considerable interest due to its capacity to tackle data security and confidentiality predicaments [1]. With the availability of different kind of datasets such as biomedical imaging, genomics data, electronics health records, a novel paradigm in healthcare system is rapidly developing. In particular, highly precise therapy therefore in regards aims to "ensure that the effective medication is offered to the sufferer in the proper time" by considering a variety of individual information, such as variation in genetic features, surroundings, electronic health records (EHRs), and behavior [2]. The deployment of bio-medical data collection devices, the enormous improvement in image acquisition devices, and the rise in throughput have produced a continuous quantity of data. The aforementioned data is extremely varied, with large dimensions (MRI, PET, DTI, CT Scans etc.), and acquired from multiple (often incompatible) sources of information. This data being opted as consequence, particularly images, becomes intriguing for medical analysis. Medical experts put in their to keep up with the sector of medical scans. Automating the diagnostic process with machine learning (ML) approaches is a viable option, but conventional ML techniques are insufficient to handle such challenging situations [3] the development of effective healthcare strategies is greatly focused on data framework. ML is particularly constrained by the underlying issue of evaluating associations among all the many pieces of knowledge in such large sets of data [4]. As in recent years, DL a part of ML has experienced a remarkable boom; the main driving forces behind this progress are advancements in computational power and the ability to carry large and innovative databases with ease. The domain has already witnessed significant improvement in the potential of technology to interpret and operate with data, such as speech, languages, and imaging data [5]. DL techniques enable a machine to acquire the representations needed for recognition or classification from vast quantities of unstructured data. DL techniques involve various models of information using subsequent modifications that strengthen and eliminate insignificant divergences while amplifying elements of the input data which are essential to classification. It can happen under or without supervision i.e. supervised or unsupervised [6].

2 Applications of Deep Learning and Blockchain in Healthcare

A. Deep Learning and Blockchain in Alzheimer's Disease Classification

The utilization of blockchain technology has the potential to significantly contribute to the classification of Alzheimer's disease through the provision of safe and transparent administration of patient data [7]. The utilization of blockchain technology for storing medical records and diagnostic details can provide a secure means for healthcare professionals to access and analyze data, thereby enabling precise categorization and diagnosis of Alzheimer's disease (AD), one of the most prevalent kinds of dementia, is a fatal brain illness that primarily harms old people [8]. Dementia syndrome is mainly characterized by impairment in cognition and other intellectual abilities when compared to their past skills [9]. According to the survey reported in 2022, after COVID-19, AD ranks sixth among top 10 fatal factors in the United States in 2019 and the seventh-leading reason of fatalities in years 2020 and 2021. Amongst Americans aged over the age of 65, AD continues to be the fifth most common cause of mortality rates. Mortality from heart disease, HIV, and stroke all declined during 2000 and 2019, while recorded fatalities from AD grew by a rate of 145%. Approximately sixteen billion hours of attention were given to persons with AD or other kinds of dementia in 2021 by over ten million close relatives along with other caregivers. The medical assessment of individuals and connection to family members are all necessary for the identification of AD. The post-mortem investigation, which is not clinically vital, used to establish "a firmly grounded fact" diagnosis of AD for any scenario. Without access to real-world data, patients require another basis for confirming AD. Such criteria may help us better understand AD and make it simpler for live patients to determine it. AD has no known cure, but certain specific medications can slow down its progression [10]. Hence, identification of patients at early stage who are more likely to progress to AD is as vital as ever. This was emphasized in a noteworthy conversation delivered by the German neuropathology's and the psychiatrist Dr. Alois Alzheimer, who additionally explained the particular kind of dementia. Eventually, the disease was acknowledged and identified with the name of the person who first discovered it, Alzheimer's disease [11]. He revealed a Frankfurter lady, 51, that had passed away due to acute mental health issues. When Dr. Alzheimer studied her brain, he found amyloidal plaques and a mass of fibres which were likely caused by abnormal proteins aggregation in and around the neurons cells. Once healthy neurons in the cerebrums abruptly stop functioning properly, cells begin to lose their connections to other brain cells and eventually die [12, 13]. Computerized tomography (CT) scans, sMRI, functional MRI, and diffusion tensor imaging (DTI) are pinch of the diverse data collected from medical imaging modalities, which are currently exploited for diagnosing AD in different contexts incorporating distinct ML approaches [14]. The primary obstacle in interpreting these images is the vast amount of data they contain relative to the limited number of samples available, which results in high dimensionality. ML techniques could be used to deal with the aforementioned problems (limited number of samples as well as complex dimensionality) [15]. DL techniques contribute to a more accurate and independent strategy since they demand minimal prior understanding of additional complicated methods, which involves feature extraction and selection [16]. A wide range of methods of classification were applied across the past decade in order to classify AD based on one modality or an ensemble of modality. Support vector machines (SVM) and its enhanced architectures, which constitute traditional ML prediction models, have been extensively used for this area. The authors in [17] in order to predict the progression from the intermediate stage of Mild Cognitive Impairment (MCI) to Alzheimer's disease, a DL strategy was used that included a multi-modal neural network with a recurrent component, specifically a Recurrent Neural Network (RNN). Researchers devised an integrated paradigm that incorporates longitudinal cerebrospinal fluid (CSF) and cognitively functioning indicators gathered from the Alzheimer's disease Neuroimaging Initiative cohort (ADNI) in addition to cross-sectional brain scans indicators at baseline. The proposed design included data from several domains collected all through period. The results of study demonstrated that (i) when using only one modality of data separately, prediction model for MCI conversion to AD yielded up to 75% accuracy and (ii) and while integrating periodic multi-domain information; prediction system performed the best, achieving an accuracy rate of 81%. Thus, a multi-modal DL strategy could be implemented to assess clinical studies or to determine those at risk of experiencing AD who would get the best through these. An in-depth view of AD progression assessment could be provided with the combined analysis of several data modalities. To group participants into AD, MCI, and controls (CN), researchers thus employed DL to thoroughly evaluate brain imaging along with genetic and diagnostic test results. To derive features from medical and genetics facts, the authors build demonising auto-encoders; for images, employed three-dimensional convolutional neural networks (CNNs). Furthermore, they designed a cutting-edge data interpretation approach to locate the best features that the deep models learnt through clustered and perturbation evaluation. Researchers proved that deep models outperformed shallower designs, such as SVM, random forests, decision trees and k-nearest neighbours (KNN)', employing ADNI dataset [18]. Various ways and techniques utilized to classify AD into different categories are shown in Table 1.

B. **Deep Learning and Blockchain in Cancer**

DL algorithms can analyse copious amounts of cancer data, encompassing genome sequencing and medical imaging, to enhance the prompt identification, diagnosis, and individualized therapy of the disease. The utilization of blockchain technology guarantees a secure and transparent exchange of cancer-related information, which in

Table 1 Deep learning methods and approaches used for AD classification

Ref. year	Dataset	Single or multiple	Methods and approaches	Summary
2014 [19]	ADNI	Multiple	Stacked auto-encoders	With the goal to get over the barrier associated with the diagnosis of AD and its early stage MCI, researchers devised a DL framework in this research which comprises layered auto-encoders with a softmax at the output level
2016 [20]	ADNI	Single (patches of grey matter neuro images)	Ensembled deep belief networks (EDBN)	The study evaluates the development of DL-based classification methods specific to brain areas determined by AAL. The AAL atlas' areas were utilised to separate grey matter images from each brain region into 3D patches, and such patches were utilized for training multiple DBNs. The overall conclusion for classification is then generated by a method of voting. Two DL-based topologies and four distinct voting computations were built and compared with the objective to generate an efficient classification framework where discriminatory attributes are extracted unsupervised
2016 [21]	ADNI	Single	CNN	Identifying among the most effective distinctive characteristics remains the hardest part of categorizing healthcare information for neurological conditions like AD. Authors efficiently discriminated functional MRI data from individuals with AD sufferers from HC (Individuals who are normal) adopting the CNN with the prominent framework LeNet-5, with an accuracy of 97% on test data. The results of this study demonstrated which the most effective technique for separating clinical information with healthy data in fMRI is the shift and scaling variance features generated by CNN following DL classifier

(continued)

Table 1 (continued)

Ref. year	Dataset	Single or multiple	Methods and approaches	Summary
2017 [22]	ADNI	Single	Adapted auto encoder	In this research on the authors employed DL strategy on clinically relevant data along with neuro images to provide an early AD diagnoses. Gender, age, and the individual's ApoE genotype were all incorporated in the clinically relevant information. By applying data from resting-state functional-MRI, a neural network is formed by estimating the operational links among various regions of the brain. To distinguish between typical ageing and moderate cognitive impairment, an early stage of AD, an adapted auto encoder network was developed. The proposed approach precisely disclosed discriminatory neural network aspects and presented an effective AD classifier
2018 [23]	ADNI	Single	AlexNet	Researchers, however, found it challenging to identify various stages due to elderly individuals' and distinct stages' brain features are identical. The CNN model AlexNet was used in this research to classify diverse AD stages applying fMRI scans. With a DL technique, they were able to classify AD into five distinct stages. The method discriminated between AD, Early- MCI, Late-MCI, significant memory concern (SMC), and healthy controls
2019 [24]	ADNI	Multiple	FSBi-LSTM and 3D-CNN	In this study, a conceptual paradigm was created by the authors. The design they proposed specifically made use of advantages of FSBi-LSTM and 3D-CNN. In order to derive depth representation of features utilizing MRI and PET, they first developed 3D-CNN architecture. Then, in order to further enhance its efficacy, researchers applied FSBi-LSTM on the disguised spatial information obtained from deep mapping of features

(continued)

Table 1 (continued)

Ref. year	Dataset	Single or multiple	Methods and approaches	Summary
2020 [25]	ADNI	Clinical + single modality	CNN	The researchers developed an interpretable DL algorithm which identifies multiple signs associated with AD using heterogeneous sources of data such as sex, age group, and MMSE score as well as MRI neuro images. The framework connected a fully convolutional network, generating precise, comprehensible representations of each individual's accurate diagnosis, to a multiple-layer perceptron, that developed precise maps of disease probability from localised brain structures
2021 [18]	ADNI	Multiple	Stacked de-noising 3D CNNs	To derive characteristics from clinic and genetics samples, researchers used stacked de-noising auto-encoders; for imaging data, they employed 3D CNNs. Furthermore, they proposed an innovative data interpretation strategy for finding the most promising features associated with the deep models acquired using grouping and perturbations evaluation
2022 [26]	ADNI	Single	CNN	This work effectively established and analysed a CAD mechanism for distinguishing AD sufferers from healthy controlled individuals employing features from PET neuro scans. CNN was incorporated in the building process of the recommended CAD tool. The decomposition of the PET neuro scans into multiple 2D slices enabled an analysis of the features. The individual slices were then placed at short intervals avoiding overlapping
2022 [27]	ADNI	Single	Pre-trained models	Using the analysis of brain MRIs, the initial levels of AD were categorised this investigation as healthy cognitive, mild cognitive impairment, and AD. Multiple neuroimages from the ADNI database were recognised using several models based upon the CNN paradigm. The efficacy of 29 distinct pre-trained models on brain images has been examined and the research was laid out in an adequately complete comparative framework

(continued)

Table 1 (continued)

Ref. year	Dataset	Single or multiple	Methods and approaches	Summary
2022 [28]	Mayo clinic brain bank	Single	YOLOv3	To identify five different kinds of tau lesion, including neural inclusions, motor plaques, astrocytic deposits, and helical structures, researchers developed the YOLOv3 recognition system. During training, the team utilised 2522 digital sliding scans of CP13-immunostained motor cerebral slices comprising 10 cases. The dataset used for training was augmented to render it larger in size. The numerical loads for every tau stroke in the cerebellum, motor cortex, caudal core, and the superior temporal gyrus, derived by the object's recognition method, were subsequently leveraged to generate random forest models
2023 [29]	OASIS	Single	Shallow CNN	This study proposed a DL based network for reliable AD classification and diagnosis. The proposed analytic pipeline involved a shallow CNN framework on MRI neuro imaging dataset

turn promotes cooperation among researchers and facilitates the provision of patient-centred care. Integrating DL alongside blockchain technology holds promise for the progression of cancer research, improvement of treatment efficacy, and condition of patient empowerment in cancer management [30]. Globally, cancer is responsible for more deaths than any other cause. The obstacles of fighting cancer are being faced by both academics and clinicians. As per the American Cancer Society's 2019 Novel Cancer Released Research Report, there were approx. 18 thousands casualties from brain tumours in the year 2019, 142,670 being credited to lung cancer, 620 with skin cancer, 42,260 from breast cancer, and 31,620 stemming from prostate cancer [31]. One out of every ten adults in the US has been diagnosed of cancer disease. Kerala has India's largest cancer prevalence rate [32]. One in six casualties worldwide is caused by cancer disease [33]. Even while new methods could enhance cancer treatment and raise the survival rate, the purpose of cancer prognosis is to predict the evolution of the disease, provide life estimates, and assist with patient care [34]. Improved survival prediction based on the medical criteria and genetic profiles of individuals is a significant objective in cancer prognosis. Mathematical methods, such as log-ranks test [35], Cox proportionality hazards model [36], and Kaplan Meier estimator [37], have become cutting-edge analytical techniques in the field of cancer prognosis for longevity analysis. Healthcare data, such as detection of cancer, cancer categories, tumour levels, genetic background, etc., constitute to the significant input of data to support these approaches in cancer forecasting for survival forecasting. Different forms of data are currently available for comparing the current level of sickness to earlier decades. This data is multi-omics data from high throughput and multi-dimensional samples of patients [38]. The large amount of multi-omic input makes it hard for making predictions which believe in utilising statistical approaches. To deal with such challenges, several techniques including ML have been used or implemented. Principal component analysis (PCA), clustering, and auto encoder were all developed effectively for classifying various types of cancer [39]. Furthermore, cancer prognosis prediction has effectively employed techniques such as semi-supervised learning, SVM, Bayesian networks, and decision trees. The area of DL among other areas has benefited from significant growth in computational capacity and advancements in technology in the past decade. Table 2 displays several DL methodologies employed to classify cancer into a number of types.

C. **Deep Learning and Blockchain in Diabetic Retino**pathy

Diabetic retinopathy (DR), a rampant eye infliction, is a consequence of diabetes mellitus. It is characterized by abnormalities on the retina (back of the eye) which may hamper vision of an individual. If unacknowledged, it might result in blindness [51]. The Non-Proliferative DR (NPDR) and the Proliferative DR (PDR) are the two main categories of DR [52]. Early-stage DRs are generally referred to as NPDRs and can successfully be subdivided into three different categories: Mild, moderated, and extreme/severe stages [53]. Single aneurysm, which is a tiny red dot at the ends of blood arteries, exists in the mild phase. In the moderated stage, the aneurysm burst into beneath the retina to cause a retinal haemorrhage and the formation of an intricate structure that resemble a flame. In the most advanced stage, each of the four quadrants of the retina can exhibit over twenty intra-retinal haemorrhages, while there is apparent vascular bleeding besides apparent intra-retinal micro-vascular irregularities. The most extreme forms of DR, referred to as PDR, causes neovascularization, or the spontaneous development of new vessels of blood in the shape of operational micro-vascular systems across the innermost layer of the retinal area. Although various treatments are available to manage the condition, there is no known cure for DR at present. Detection and therapy of DR at initial stage may considerably minimize the chances of losing one's vision. Around 2025, there will be estimated 592 million DR afflicted worldwide, compared with 382 million presently. According to a survey completed by researchers in the Pakistani province of Khyber Pakhtunkhwa (KPK), 5.6% of diabetic individuals who developed DR were blinded. Whenever the mild NPDR is not controlled in the initial phases, it eventually matures into PDR [54]. According to reports, 26% of the identified individuals were having a diagnosis of PDR, reaching up 24% of the entire population of DR individuals identified in Sindh, Pakistan. Individuals with the DR remain undiagnosed in the initial stages, whereas later stages induce floaters, vision problems, and gradual degradation of the ability to see. Thus, it is challenging but essential to recognise DR in its initial phases with the objective to prevent the serious implications of its later stages. Coloured fundus images are used after diagnosing DR. The manual assessment is costly as well as time-consuming since it needs to be conducted by respective highly skilled experts. Consequently, it has become vital to adopt computer vision methods that

Table 2 Deep Learning Methods and Approaches used for Cancer Diagnosis

Ref. year	Dataset	Modality	Methods and approaches	Summary
2016 [40]	Histo.path	BreakHis	CNN	In this study, researchers obtained BreaKHis, a readily available database accessed at http://web.inf.ufpr.br/vri/breast-cancer-database, to perform initial experiments employing the DL approach for classifying breast cancer histological images. Authors developed an algorithm that involved extracting image patches for CNN training and integrating those patches towards final classification. By applying the higher definition histopathology scans via BreaKHis as feed to the present CNN, this strategy attempts to prevent model changes that could result in a more complicated and highly computational framework
2017 [41]	Histo.path	BreakHis	Structured deep neural network	In this study, the researcher's group employed a novel DL framework for classification a breast cancer images in to multiple categories. It is evident from the structured DNN framework's outstanding performance with an average of 93% accuracy on an extensive data that the approach is effective at offering a tool for breast cancer multi-classification in healthcare scenarios

(continued)

Table 2 (continued)

Ref. year	Dataset	Modality	Methods and approaches	Summary
2018 [42]	Histo.path	ICIAR 2018	Deep CNNs with gradients boosted tress	In the present research, researchers adopted a mathematical strategy for classifying breast cancer histology imaging dataset utilizing deep CNNS. The strategy made use of gradients boosted trees classifiers and multiple deep CNNs
2019 [43]	Histo.path	BreakHis	CNN with small SE-ResNet	In this research, researchers developed an innovative sort of CNN which is made up of a Conv Layers, a tiny SE-ResNet element, and lastly fully connected layer. They provided a lightweight SE-ResNet module that outperforms an amalgam of squeeze-and-excite units and residue sections alongside having fewer parameters in order to attain a comparable level of effectiveness. Also, they suggested an entirely novel scheduling for learning rates (LRs) that achieved exceptional results without requiring extensively tuning LRs. They implemented an algorithm to automatically classify breast cancer histological pictures into eight subgroups based on whether the disease was benign or malignant

(continued)

Table 2 (continued)

Ref. year	Dataset	Modality	Methods and approaches	Summary
2019 [44]	Colsanitas Colombia University's pathology department	Whole slides images	Ensemble DL (VGG 16 and VGG 19)	This study employed the combined DL approach to differentiate the histological pictures of breast cancer into two distinct classes: carcinoma and non-carcinoma. Using the transfer learning pre-trained frameworks named VGG 16 and 19, the authors trained four different models. Across all of the various designs, including the fully-trained and fine-tuned VGG16 and VGG 19, they first performed fivefold cross-validation methods. Subsequently adopting an ensemble technique that involved averaged forecasted probabilities, researchers discovered that the combined performance of the fine-tuned VGG16 and 19 algorithms was comparable in grouping, primarily with regard to the carcinoma category. The average effectiveness of the VGG16 and VGG19 classifiers was 95%, with a sensitivity of 98% for the malignancy cohort

(continued)

Table 2 (continued)

Ref. year	Dataset	Modality	Methods and approaches	Summary
2020 [45]	–	–	Hybrid convolution and recurrent neural network	For classification of breast cancer utilizing histopathology images, authors employed a novel hybrid convolutional and recurrent deep NN. The approach integrated the strengths of CNN and recurrent NN on the deeper hierarchical representation of features from histopathology patches of an image. Both the short and long-term spatially correlations across regions were retained. The experiment findings demonstrated with an overall performance of 91% for the multi class (4-class) classification
2021 [46]	–	–	Auxiliary classifier generative adversarial network (ACGAN)	The researcher's group adopted an ACGAN, which generated realistic images alongside their class labels, to expand the quantity of a confined dataset. By classifying breast cancer histopathology image data using deep CNN classifiers trained on our expanded dataset, they assessed the performance of ACGAN augmentation technique. They deployed a method known as transfer learning for building classifier, extracting the convolutional features from a pertained network and input these into multiple extreme gradient booster (XGBoost) classifiers

(continued)

Table 2 (continued)

Ref. year	Dataset	Modality	Methods and approaches	Summary
2022 [47]	Histo.path	BreakHis	AHoNet	This research designed a distinctive AHoNet by concurrent integrating attention strategy and high-order probabilistic estimation in remnant convolutional network in order to identify more discriminating hidden features from breast cancer pathologic imaging dataset. AHoNet at first applied a robust channel attention component without dimensionality reductions and locally cross-channel contact to identify localized significant deep features of pathological images of breast cancer. Later that, using matrix power standardization, the second-order covariance statistics were progressively computed, offered breast cancer pathological imaging an increasingly reliable overall feature depiction
2022 [48]	Histo.path	ICIAR 2018 Challenge datasets	Ensemble VGG-16 and DenseNet-201	Towards reliable breast histopathology imaging grouping, the proposed DEEP_Pachi integrated both local and global features. The ensemble VGG-16 and DenseNet-201 frameworks constituted the basis of the mentioned model. Whilst DEEP_Pachi gathered spatial or local features from patches, the ensembled model extracted global generic features from the whole image

(continued)

Table 2 (continued)

Ref. year	Dataset	Modality	Methods and approaches	Summary
2022 [49]	Histo.path	BreakHis	Efficient-Net framework with SVM classifier	For the diagnosis of breast cancer on histo-pathological imaging dataset, the study proposed a DL along with transfer learning model. The researchers applied transfer learning technique on only a few training histopathology samples to classify into classes without compromising efficiency. With the goal to derive features, patches/regions were initially extracted from Entire Slide Images and input to the CNN. The discriminative patches were chosen on the basis of these features before being input to an Efficient-Net framework which was already trained on the ImageNet data set. The Efficient-Net architecture's features were also exploited for training SVM classifier as well
2023 [50]	Histo.path	BreakHis	Arithmetic optimized DL technique for histological breast cancer (AOADL-HBCC)	In the present study, AOADL-HBCC is proposed by the authors. The AOADL-HBCC technique included a mechanism that improves contrast and eliminating background noise based on a median filter (MF). Furthermore, the AOADL-HBCC approach integrated an AOA alongside a Squeeze Net framework to extricate features. In order to classify breast cancer, a deep belief network (DBN) classifier that includes a hyper parameter optimizer (Adamax) was adopted

automatically interpret the fundus scans to aid radiologists and specialists. Hands-on feature engineering [55] and end-to-end learning that is DL [56] represents two categories of computer vision-based approaches. The conventional approaches adopted by the hands-on engineering procedures to extract features are time consuming and error prone. The potential applications of DL along with blockchain technology in the setting of diabetic retinopathy, a common cause of blindness among those who have diabetes, have demonstrated promising outcomes. The application of DL algorithms enables the evaluation of retinal images to identify symptoms of diabetic retinopathy, thereby facilitating timely treatment and diagnosis. The implementation of blockchain technology has the potential to safeguard the confidentiality and integrity of medical records, facilitating the efficient and protected exchange of retinal images between medical professionals. This can lead to improved precision in diagnosing and shrivelling diabetic retinopathy. The amalgamation of DL with blockchain computing holds promise for enhancing the management of diabetic retinopathy and averting vision loss among individuals with diabetes. Table 3 lists various DL techniques that were used to classify DR fundus images into different categories.

3 Conclusion

Integrating DL with blockchain technology has exhibited noteworthy potential in healthcare imaging. The utilization of DL algorithms in the analysis of medical images has demonstrated exceptional proficiency in facilitating precise diagnosis, computer-assisted detection, and individualized treatment planning. In contrast, blockchain technology ensures the security, privacy, compatibility, and transparency of data, thereby offering a solution to the significant challenges that arise in managing healthcare data. The impact of these technologies on healthcare imaging is significant, resulting in better patient outcomes, heightened diagnostic precision, efficient data sharing, and amplified healthcare advancement. Moreover, DL algorithms are effective in enhancing customary ML, enabling computers to acquire data in order to recognize strategies for advancing outcomes. Approaches ranging from natural language processing to computer vision, have had previously exploited these methods. Furthermore, considering the emphasis on representation learning, deep networks offer the ability to fuse assorted sets of data throughout distinctive types. Hierarchical learning design enables higher generality. Such in-depth representations can subsequently be applied for assisting medical procedures within areas, like health evaluation, individualised medicines suggestions, guidance on treatment, clinical study employment, and statistical data gathering and assessment. This chapter, focused on various DL approaches, can be applied in the healthcare industry. In light of this, we believe that DL has the power to usher in a new era of forecasting systems for healthcare, which may prove to be a game-changer for the field (i) scale to incorporate millions to a billion of medical details and (ii) employ only one, distributed individual's illustration to successfully assist medical professionals in everyday tasks—instead of multiple mechanisms through distinct patients data.

Table 3 Deep learning methods and approaches used for diabetic retinopathy image classification

Ref. year	Dataset	Modality	Methods and approaches	Summary
2016 [57]	Kaggle dataset (https://www.kaggle.com)	Fundus retinal images	CNN with augmentation	In this study, researchers suggested a CNN as a reliable methodology for identifying DR from digitised retinal ocular illustrations (images) and grading its level of severity. Researchers designed CNN network architecture along with augmented data that can recognise the complex elements needed for the classification task, such as micro-aneurysms, fluid, and haemorrhaging on retinas, and thereafter delivered an automatic diagnosis without requiring any human intervention
2017 [58]	FINDeRS dataset	Fundus retinal images	Deep-DR-Net	In this study, a cost-effective embedded framework driven by DL was developed that allows professionals estimate the extent of DR using retinal fundus images. Then, a simple embedded board-compatible DL model called Deep-DR-Net was recommended for employing in such a scenario. A cascaded codec-classifier framework was established up utilising residual approach at the core of Deep-DR-Net that ensured an optimal modelled size
2018 [59]	Ophthalmology Department, Health Management Center, and Endocrinology & Metabolism Department	Fundus retinal images	Deep CNN	The authors created an archive of DR fundus images featuring the relevant management approach labelled in order to automate the diagnosis of DR and offered appropriate recommendations to DR individuals. Researchers built deep CNN models on this dataset to classify the level of severity of DR fundus images

(continued)

Table 3 (continued)

Ref. year	Dataset	Modality	Methods and approaches	Summary
2019 [60]	Beijing Tongren Eye Center	Fundus retinal images	Integrated DL models	This study employs numerous appropriately trained DL models to illustrate an autonomous image-level DR recognition approach. Additionally, with the aim to lessen the biases of every individual framework, a number of DL models were combined via the Adaboost method. The weighted CAMs that could reveal the probable location of damages have been included in this study in order to clarify the DR findings. To expand the number of fundus images, eight image manipulation techniques were also incorporated during pre-processing
2019 [61]	Kaggle dataset	Fundus images	Ensemble of five deep CNN framework	In the present research, the contributors established a group of five deep CNN architectures namely Inceptionv3, DenseNet (121 and 169), Resnet50, and Xception to capture the meaningful features while improving classification accuracy across various phases of DR employing the freely accessible Kaggle database of retinal images
2020 [62]	Messidor-1 with the APTOS 2019 vision diagnosis	Fundus images	Hybrid CNN transfer learning framework	The paper discussed the challenge for automatically identifying DR and presented a novel DL hybrid model to fix the problem. In order to develop the hybrid framework, they used an additional blocks of CNN layering over top of the previously trained Inception-Res network utilizing the technique of transfer learning. Considering the Messidor-1 database for DR with the APTOS 2019 vision diagnosis (Kaggle dataset), researchers analysed the accuracy of the suggested model. Compared with other results reported, their approach performed better. Using the APTOS and Messidor-1 and datasets, the developers obtained test accuracy of 82% and 72% respectively

(continued)

Table 3 (continued)

Ref. year	Dataset	Modality	Methods and approaches	Summary
2020 [63]	Kaggle dataset (EyePACS dataset)	Fundus images	VGG-16 and VGG-19	This study introduced a computer-aided classification approach that used CNN, and two transfer learning VGG-16, and VGG-19 DL models to evaluate coloured fundus images involving various illumination and angles views and yielded a level of seriousness level for DR
2021 [64]	Kaggle dataset	Fundus images	–	In order accelerate the training process and convergence, researchers focused on identifying the DR's various stages based on the minimal learnable parameters needed. The VGG-Network in Network (NiN) model, a significantly nonlinear measure-invariant DL model, was built by stacking the VGG16, the spatially pyramidal layer for pooling (SPP), and the NiN. In addition to the advantages of the SPP layer, the recommended VGG-NiN model processed a DR image at any dimension. Furthermore, the stacking of NiN offered the model significant nonlinearity and enhanced classification results
2022 [65]	DRIVE and messidor datasets	Fundus images	CNN	The goal of this research was to establish a computer-based classification method for a set of retinal images used to recognize DR. CNN DL technique has been used to generate a multiclass classification framework which could automatically diagnose and classify sickness levels

(continued)

Table 3 (continued)

Ref. year	Dataset	Modality	Methods and approaches	Summary
2022 [66]	–	Fundus images	Hybrid CNN-SVD	In this study, the authors proposed an innovative a two-phase method for automatic DR categorization. Pre-processing and data augmentation techniques were utilised to improve the quality of the images and quantity because the asymmetric Optical Disc (OD) and blood vessels recognition technique has a small proportion of positive instances. In the initial phase, the segmentation of the OD and blood vessel was carried out employing two different U-Net algorithms. The next step involved building of the symmetrical hybrid CNN-SVD model, which observes DR by identifying retinal indicators like micro aneurysms, haemorrhages), and exudates, following initial processing to obtain and find the strongest discriminant features through Inception-V3
2022 [67]	Kaggle dataset (EyePACS t	Fundus images	ResNet and Alex Net	In this investigation, Multi-Resolution Analysis (MRA) and CNN architecture were utilised together for the input image features enhancement with no adding additional convolution filters. This study proposed a new HW step activated function having distinct properties associated with the wavelet channel-bands
2023 [68]	DIARETDB1 and APTOS 2019	Fundus images	Deep ensembled DenseNet101 and ResNeXt	For the DR recognition and DR, a fully automated ensemble DL model was presented in this research. For the purpose of detecting diabetic retinopathy, two DL algorithms, namely improved DenseNet101 and ResNeXt, are integrated

In this chapter, we showed how DL techniques can be applied in the healthcare industry using an imaging dataset. DL approaches can train it using comprehensive data. Applications of DL technology with imaging datasets for a range of diseases, including cancer, diabetic retinopathy, Alzheimer's disease, etc., can facilitate early investigation and classification of diseases so that patients can receive effective therapies.

References

1. Singh B, Tatiya M, Shrivastava A, Verma D, Pratap Srivastava A, Rana A (2022) Detection of Alzheimer's disease using deep learning, blockchain, and IoT cognitive data. In: 2022 2nd international conference on technological advancements in computational sciences (ICTACS), Oct 2022, pp 863–869.https://doi.org/10.1109/ICTACS56270.2022.9988058
2. Lyman GH, Moses HL (2016) Biomarker tests for molecularly targeted therapies—the key to unlocking precision medicine. N Engl J Med 375(1):4–6. https://doi.org/10.1056/NEJMp1604033
3. Razzak MI, Naz S, Zaib A (2018) Deep Learning for medical image processing: overview, challenges and the future. In: Lecture notes in computational vision and biomechanics, vol 26, pp 323–350
4. Miotto R, Wang F, Wang S, Jiang X, Dudley JT (2017) Deep learning for healthcare: review, opportunities and challenges. Brief Bioinform 19(6):1236–1246. https://doi.org/10.1093/bib/bbx044
5. Esteva A et al (2019) A guide to deep learning in healthcare. Nat Med 25(1):24–29. https://doi.org/10.1038/s41591-018-0316-z
6. Panch T, Szolovits P, Atun R (2018) Artificial intelligence, machine learning and health systems. J Glob Health 8(2):1–8. https://doi.org/10.7189/jogh.08.020303
7. Pilozzi A, Huang X (2020) Overcoming Alzheimer's disease stigma by leveraging artificial intelligence and blockchain technologies. Brain Sci 10(3):183. https://doi.org/10.3390/brainsci10030183
8. Bergstra J, Bengio Y (2012) Random search for hyper-parameter optimization. J Mach Learn Res 13:281–305
9. Chang CC, Lin CJ (2011) LIBSVM: A Library for support vector machines. ACM Trans Intell Syst Technol 2(3). https://doi.org/10.1145/1961189.1961199
10. Sethi M, Ahuja S, Rani S, Bawa P, Zaguia A (2021) Classification of Alzheimer's disease using gaussian-based Bayesian parameter optimization for deep convolutional LSTM network. Comput Math Methods Med 2021. https://doi.org/10.1155/2021/4186666
11. Yiğit A, Işik Z (2020) Applying deep learning models to structural MRI for stage prediction of Alzheimer's disease. Turkish J Electr Eng Comput Sci 28(1):196–210. https://doi.org/10.3906/elk-1904-172
12. Brückner G, Hausen D, Härtig W, Drlicek M, Arendt T, Brauer K (1999) Cortical areas abundant in extracellular matrix chondroitin sulphate proteoglycans are less affected by cytoskeletal changes in Alzheimer's disease. Neuroscience 92(3):791–805. https://doi.org/10.1016/S0306-4522(99)00071-8
13. Wang SH, Phillips P, Sui Y, Liu B, Yang M, Cheng H (2018) Classification of Alzheimer's disease based on eight-layer convolutional neural network with leaky rectified linear unit and max pooling. J Med Syst 42(5):85. https://doi.org/10.1007/s10916-018-0932-7
14. Azari NP et al (1993) Early detection of Alzheimer's disease: a statistical approach using positron emission tomographic data. J Cereb Blood Flow Metab 13(3):438–447. https://doi.org/10.1038/jcbfm.1993.58

15. Razavi F, Tarokh MJ, Alborzi M (2019) An intelligent Alzheimer's disease diagnosis method using unsupervised feature learning. J Big Data 6(1). https://doi.org/10.1186/s40537-019-0190-7
16. Lecun Y, Bengio Y, Hinton G (2015) Deep learning. Nature 521(7553):436–444. https://doi.org/10.1038/nature14539
17. Lee G et al (2019) Predicting Alzheimer's disease progression using multi-modal deep learning approach. Sci Rep 9(1):1–12. https://doi.org/10.1038/s41598-018-37769-z
18. Venugopalan J, Tong L, Hassanzadeh HR, Wang MD (2021) Multimodal deep learning models for early detection of Alzheimer's disease stage. Sci Rep 11(1):1–13. https://doi.org/10.1038/s41598-020-74399-w
19. Liu S, Liu S, Cai W, Pujol S, Kikinis R, Feng D (2014) Early diagnosis of Alzheimer's disease with deep learning. In: 2014 IEEE 11th int. symp. biomed. imaging, ISBI 2014, pp 1015–1018. https://doi.org/10.1109/isbi.2014.6868045
20. Ortiz A, Munilla J, Górriz JM, Ramírez J (2016) Ensembles of deep learning architectures for the early diagnosis of the Alzheimer's disease. Int J Neural Syst 26(7):1–23. https://doi.org/10.1142/S0129065716500258
21. Sarraf S, Tofighi G (2017) Deep learning-based pipeline to recognize Alzheimer's disease using fMRI data. In: FTC 2016 - proc. futur. technol. conf., no December, pp 816–820. https://doi.org/10.1109/FTC.2016.7821697.
22. Pan D, Huang Y, Zeng A, Jia L, Song X (2019) Early Diagnosis of Alzheimer's disease based on deep learning and GWAS. Commun Comput Inf Sci 1072(1):52–68. https://doi.org/10.1007/978-981-15-1398-5_4
23. Kazemi Y, Houghten S (2018) A deep learning pipeline to classify different stages of Alzheimer's disease from fMRI data. In: 2018 IEEE conf. comput. intell. bioinforma. comput. biol. CIBCB 2018, no Mci, pp 1–8. https://doi.org/10.1109/CIBCB.2018.8404980
24. Feng C et al (2019) Deep learning framework for Alzheimer's disease diagnosis via 3D-CNN and FSBi-LSTM. IEEE Access 7:63605–63618. https://doi.org/10.1109/ACCESS.2019.2913847
25. Qiu S et al (2020) Development and validation of an interpretable deep learning framework for Alzheimer's disease classification. Brain 143(6):1920–1933. https://doi.org/10.1093/brain/awaa137
26. Ahila A, Poongodi M, Hamdi M, Bourouis S, Rastislav K, Mohmed F (2022) Evaluation of neuro images for the diagnosis of Alzheimer's disease using deep learning neural network. Front Public Heal 10, no February, pp 834032. https://doi.org/10.3389/fpubh.2022.834032
27. Savaş S (2022) Detecting the Stages of Alzheimer's disease with pre-trained deep learning architectures. Arab J Sci Eng 47(2):2201–2218. https://doi.org/10.1007/s13369-021-06131-3
28. Koga S, Ikeda A, Dickson DW (2022) Deep learning-based model for diagnosing Alzheimer's disease and tauopathies. Neuropathol Appl Neurobiol 48(1):1–12. https://doi.org/10.1111/nan.12759
29. EL-Geneedy M, Moustafa HED, Khalifa F, Khater H, AbdElhalim E (2023) An MRI-based deep learning approach for accurate detection of Alzheimer's disease. Alexandria Eng J 63:211–221. https://doi.org/10.1016/j.aej.2022.07.062
30. Kumar R et al (2021) An Integration of blockchain and AI for secure data sharing and detection of CT images for the hospitals. Comput Med Imaging Graph 87:101812. https://doi.org/10.1016/j.compmedimag.2020.101812
31. Munir K, Elahi H, Ayub A, Frezza F, Rizzi A (2019) Cancer diagnosis using deep learning: a bibliographic review. Cancers (Basel) 11(9):1–36. https://doi.org/10.3390/cancers11091235
32. Kaushal C, Singla A (2020) Automated segmentation technique with self-driven post-processing for histopathological breast cancer images. CAAI Trans Intell Technol 5(4):294–300. https://doi.org/10.1049/trit.2019.0077
33. Zhu W, Xie L, Han J, Guo X (2020) The application of deep learning in cancer prognosis prediction. Cancers (Basel) 12(3):1–19. https://doi.org/10.3390/cancers12030603
34. Kaushal C, Kaushal K, Singla A (2021) Firefly optimization-based segmentation technique to analyse medical images of breast cancer. Int J Comput Math 98(7):1293–1308. https://doi.org/10.1080/00207160.2020.1817411

35. Society RS (2013) Asymptotically efficient rank invariant test procedures author (s): Richard Peto and Julian Peto Reviewed work (s): Source : Journal of the Royal Statistical Society . Series A (General), 135(2) (1972), pp Published by : Wiley for th, vol 135, no 2, pp 185–207
36. Ahmed FE, Vos PW, Holbert D (2007) Modeling survival in colon cancer: a methodological review. Mol Cancer 6:1–12. https://doi.org/10.1186/1476-4598-6-15
37. Kaplan EL, Meier P (1958) Nonparametric estimation from incomplete observations. J Am Stat Assoc, vol 53, no March 2013, pp 457–81
38. Goossens N, Nakagawa S, Sun X, Hoshida Y (2015) Cancer biomarker discovery and validation. Transl Cancer Res 4(3):256–269. https://doi.org/10.3978/j.issn.2218-676X.2015.06.04
39. Tan M et al (2014) Lysine glutarylation is a protein posttranslational modification regulated by SIRT5. Cell Metab 19(4):605–617. https://doi.org/10.1016/j.cmet.2014.03.014
40. Jiang Y, Chen L, Zhang H, Xiao X (2019) Breast cancer histopathological image classification using convolutional neural networks with small SE-ResNet module. PLoS One 14(3). https://doi.org/10.1371/journal.pone.0214587
41. Han Z, Wei B, Zheng Y, Yin Y, Li K, Li S (2017) Breast cancer multi-classification from histopathological images with structured deep learning model. Sci Rep 7(1):1–10. https://doi.org/10.1038/s41598-017-04075-z
42. Kumar ES, Bindu CS, Madhu S (2020) Deep convolutional neural network-based analysis for breast cancer histology images, vol 1. Springer International Publishing
43. Jiang Y, Chen L, Zhang H, Xiao X (2019) Breast cancer histopathological image classification using convolutional neural networks with small SE-ResNet module. PLoS ONE 14(3):1–21. https://doi.org/10.1371/journal.pone.0214587
44. Hameed Z, Zahia S, Garcia-Zapirain B, Aguirre JJ, Vanegas AM (2020) Breast cancer histopathology image classification using an ensemble of deep learning models. Sensors (Switzerland) 20(16):1–17. https://doi.org/10.3390/s20164373
45. Yan R et al (2020) Breast cancer histopathological image classification using a hybrid deep neural network. Methods 173(2019):52–60. https://doi.org/10.1016/j.ymeth.2019.06.014
46. Gheshlaghi SH, Nok Enoch Kan C, Ye DH (2021) Breast cancer histopathological image classification with adversarial image synthesis. In: 2021 43rd annual international conference of the IEEE engineering in medicine & biology society (EMBC), pp 3387–3390. https://doi.org/10.1109/EMBC46164.2021.9630678
47. Zou Y, Zhang J, Huang S, Liu B (2022) Breast cancer histopathological image classification using attention high-order deep network. Int J Imaging Syst Technol 32(1):266–279. https://doi.org/10.1002/ima.22628
48. Ukwuoma CC, Hossain MA, Jackson JK, Nneji GU, Monday HN, Qin Z (2022) Multi-Classification of breast cancer lesions in histopathological images using DEEP_Pachi: multiple self-attention head. Diagnostics 12(5). https://doi.org/10.3390/diagnostics12051152
49. Ahmad N, Asghar S, Gillani SA (2022) Transfer learning-assisted multi-resolution breast cancer histopathological images classification. Vis Comput 38(8):2751–2770. https://doi.org/10.1007/s00371-021-02153-y
50. Obayya M, et al (2023) Hyperparameter optimizer with deep learning-based decision-support systems for histopathological breast cancer diagnosis. Cancers (Basel) 15(3). https://doi.org/10.3390/cancers15030885
51. Alyoubi WL, Shalash WM, Abulkhair MF (2020) Diabetic retinopathy detection through deep learning techniques: a review. Informatics Med Unlocked 20:100377. https://doi.org/10.1016/j.imu.2020.100377
52. Raja Memon DW, Lal DB, Aziz Sahto DA (2017) Diabetic retinopathy; frequency at level of hba1c greater than 6.5%. Prof Med J 24(2):234–238. https://doi.org/10.17957/tpmj/17.3616
53. Wu L (2013) Classification of diabetic retinopathy and diabetic macular edema. World J Diabetes 4(6):290. https://doi.org/10.4239/wjd.v4.i6.290
54. Jan S, Ahmad I, Karim S, Hussain Z, Rehman M, Shah MA (2018) Status of diabetic retinopathy and its presentation patterns in diabetics at ophthalomogy clinics. J Postgrad Med Inst 32(1):24–27

55. Seoud L, Hurtut T, Chelbi J, Cheriet F, Langlois JMP (2016) Red lesion detection using dynamic shape features for diabetic retinopathy screening. IEEE Trans Med Imaging 35(4):1116–1126. https://doi.org/10.1109/TMI.2015.2509785
56. Quellec G, Charrière K, Boudi Y, Cochener B, Lamard M (2017) Deep image mining for diabetic retinopathy screening. Med Image Anal 39:178–193. https://doi.org/10.1016/j.media.2017.04.012
57. Pratt H, Coenen F, Broadbent DM, Harding SP, Zheng Y (2016) Convolutional neural networks for diabetic retinopathy. Procedia Comput Sci 90(July):200–205. https://doi.org/10.1016/j.procs.2016.07.014
58. Ardiyanto I, Nugroho HA, Buana RLB (2017) Deep learning-based Diabetic Retinopathy assessment on embedded system. In: Proc. annu. int. conf. IEEE eng. med. biol. soc. EMBS, pp 1760–1763. https://doi.org/10.1109/EMBC.2017.8037184
59. Gao Z, Li J, Guo J, Chen Y, Yi Z, Zhong J (2019) Diagnosis of diabetic retinopathy using deep neural networks. IEEE Access 7(c):3360–3370. https://doi.org/10.1109/ACCESS.2018.2888639
60. Jiang H, Yang K, Gao M, Zhang D, Ma H, Qian W (2019) An interpretable ensemble deep learning model for diabetic retinopathy disease classification. In: Proc. annu. int. conf. IEEE eng. med. biol. soc. EMBS, pp 2045–2048.https://doi.org/10.1109/EMBC.2019.8857160
61. Qummar S et al (2019) A deep learning ensemble approach for diabetic retinopathy detection. IEEE Access 7:150530–150539. https://doi.org/10.1109/ACCESS.2019.2947484
62. Gangwar AK, Ravi V (2021) Diabetic retinopathy detection using transfer learning and deep learning, vol 1176. Springer, Singapore
63. Nguyen QH, et al (2020) Diabetic retinopathy detection using deep learning. In: ACM int. conf. proceeding ser., pp 103–107. https://doi.org/10.1145/3380688.3380709
64. Khan Z et al (2021) Diabetic retinopathy detection using VGG-NIN a deep learning architecture. IEEE Access 9:61408–61416. https://doi.org/10.1109/ACCESS.2021.3074422
65. Sebti R, Zroug S, Kahloul L, Benharzallah S (2022) A Deep Learning Approach for the Diabetic Retinopathy Detection. In: 2022 2nd international conference on intelligent technologies, CONIT 2022, no April, 2022, pp 459–469
66. Bilal A, Zhu L, Deng A, Lu H, Wu N (2022) AI-based automatic detection and classification of diabetic retinopathy using u-net and deep learning. Symmetry (Basel) 14(7). https://doi.org/10.3390/sym14071427
67. Chandrasekaran R, Loganathan B (2022) Retinopathy grading with deep learning and wavelet hyper-analytic activations. Vis Comput. https://doi.org/10.1007/s00371-022-02489-z
68. Mondal SS, Mandal N, Singh KK, Singh A, Izonin I (2023) EDLDR: an ensemble deep learning technique for detection and classification of diabetic retinopathy. Diagnostics 13(1):1–14. https://doi.org/10.3390/diagnostics13010124

Healthcare Data Security Using AI and Blockchain: Safeguarding Sensitive Information for a Safer Society

Jyoti Upadhyay, Sunil Kumar Singh, Naresh Kumar Kar, Manoj Kumar Pandey, Priyanka Gupta, and Preeti Tiwari

Abstract The synergy of AI and blockchain presents transformative opportunities, such as enabling secure sharing of medical data across institutions, facilitating research collaborations, and streamlining insurance claims and billing processes. Blockchain technology is known for its secure and transparent data management addresses critical challenges in healthcare data interoperability, security, and privacy. By creating a tamper-proof and decentralized ledger of patient records, medical practitioners can access a patient's comprehensive medical history, leading to more informed decision-making. However, several hurdles including regulatory compliance, technical integration, and ethical concerns must be overcome to fully harness the potential of this combined approach. This chapter explores the current state, usage of blockchain and AI in healthcare, trends of utilizing AI, benefits, challenges, and future prospects of employing AI and blockchain in healthcare, highlighting their potential to reshape the industry's landscape and enhance patient care quality.

J. Upadhyay
GD Rungta College of Science and Technology, Bhilai, India
e-mail: upadhyaydrjyoti@gmail.com

S. K. Singh
CMREC, Hyderabad, India
e-mail: kumarsunil@cmrec.ac.in

N. K. Kar
GITAM Deemed to be University, Hyderabad, India
e-mail: nkar@gitam.edu

M. K. Pandey (✉)
Pranveer Singh Institute of Technology, Kanpur, India
e-mail: manoj.pandey@psit.ac.in

P. Gupta
Research Scholar, Guru Ghasidas Central University, Bilaspur, India
e-mail: priyanka13666@gmail.com

P. Tiwari
IBITF, India Institute of Technology, Bhilai, India
e-mail: preetit@iitbhilai.ac.in

K. Kaushik and I. Sharma (eds.), *Next-Generation Cybersecurity*, Blockchain Technologies, https://doi.org/10.1007/978-981-97-1249-6_8

Keywords Blockchain · AI · Healthcare · Security · Privacy · Electronic health record

1 Introduction

With increasing amount of data on the web, it is very essential to come up with some technologies that can ensure the security and privacy on digital data. It is also very important to note that the E-data related to the healthcare is utmost important to keep it protected from unauthorized users. The form and sensitivity of digital data has been transformed day by day and along with that the technology to misinterpret that data is also improved so there is a need to improve the security of digital data using the latest technology. The digital data is also very much aligned with the security of the society and the nation therefore it is essential to provide maximum security to the digital data. If we look into the past decades, our necessity is also increased to work with the digital data due to the evolvement of artificial intelligence (AI) and with the help of AI now we are more powerful and sophisticated to achieve some good results and decision as compare to our traditional methods so both the terminology i.e. AI and security of digital contents are equally important in order to be more effective and precisely in any field [1]. The machine learning is term associated with the AI that enables machine to think like human brains and act accordingly but to be more effective and precise it needs enough digital data to train a model through which a decision or result can be made [2]. When machine learning is not enough to reach a decision and shows that it is less effective then deep learning is applied to draw a decision by applying more complex calculations [3]. It is very clear as of now that AI and security of digital data both are equally important for any society and nation. Looking at the two or three decades back one can observe that the privacy and integrity of digital data has become more vulnerable to digital attacks as the decades passes because of the improvement in the technology so it is very clear the with increasingly usages of the technology, the security of digital content is also at high risk. This chapter includes the various aspects of how integration of AI and blockchain can provide good infrastructure to build up good health infrastructures for any nation.

There are various types of electronic data that is associated with a patient like X-Ray data, Blood test data, MRI scan data, personal data, and many more, and all these data need to be secured and processed effectively for better decision making. Nowadays machine vision technique also plays an important role in the healthcare industry by recognizing various objects and acting upon them like human does. Machine vision is a technique which enables a machine to mimic computer as human vison and to make them understand what object is placed in front of the machine [4]. Machine vision can promote the automation of system in case of emergency that can save a human life. A patient data can be partition into text, image, audio and videos, and for all these types of data different machine learning and deep learning based application need to be developed or enhanced in order to automate the diagnosis whenever needed. The digital data produced by patient in form of text and image like prescription, written document, summary, lab report, and clinical reports are very much important and these data can be effectively used for providing good health advice and decision making if one can process it effectively. Natural language processing (NLP) is a sub field of AI, which enable a machine to comprehend written documents like a human does. With use of machine vision and NLP machine can interpret, deduce, summarize, translate, and synthesize exact text for better human diagnosis [5]. Image segmentation is also a type of processing in which medical image is portioned into different regions in order to find region of interest [6]. Segmentation plays an important role in medical diagnosis, and semantic segmentation is a type of segmentation in which pixel belonging to one object is given one colour and pixel belonging to another object is given other colour. Figure 1 shows the various requirements of healthcare industry which shows that security and remote monitoring are among the most requirements of the industry.

Audio data related to patient like heart-beat, crying, breathing, coughing etc. plays an excellent role in diagnosis respiratory diseases, pulmonary diseases, and cardiac diseases. AI based medical system can assist to analyse these sounds and classifying audio according to their characteristics in order to diagnosis patient. There is various state-of-art methods are present for audio signal processing that can aid in medical industry. Though all these types of automation need the usages of AI and machine learning concepts where the training data plays a significant role and also the integration of these data is also an important concern. Collecting, analysing, integrating and protecting all these forms of medical digital data are also an important

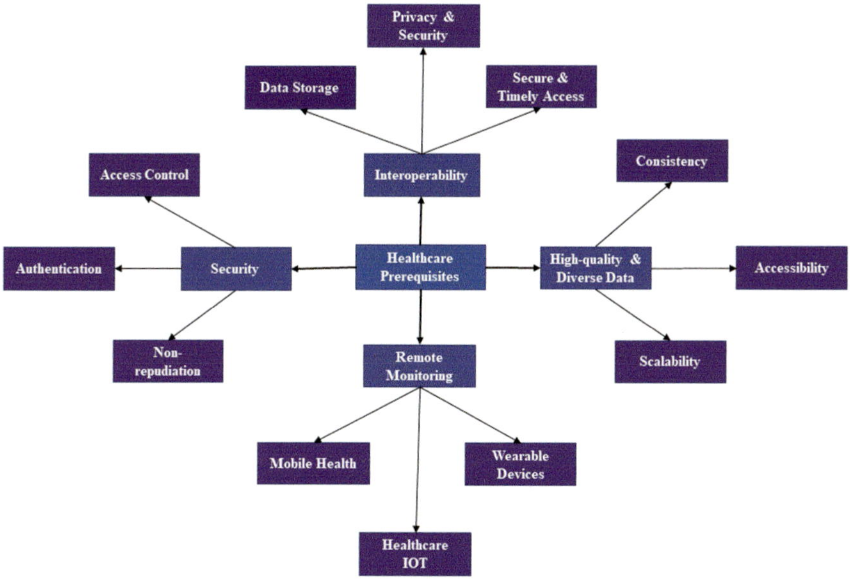

Fig. 1 Healthcare industry requirements

concern for the industry. There are few threats to enabling AI in the medical industry are like, firstly, a patient data is very precious for any medical centre and securing data and maintain the integrity of data needs specialized work force and resources. Secondly integrating data from various resources in order to maximize the quantity of training data is also a challenging task. It is also very obvious that these data may be stolen or vulnerable to different attacks for different reasons therefore protecting it along with the integration is a serious issue. Third, let us assume that training is done for models and now AI is ready to support the clinical decision, but AI based result are black box in nature and one cannot take risk of human life because of AI recommendation therefore the simultaneous intervention of specialized doctors is also needed. Fourth, there should be secure resource sharing for overcoming the threat of the rouge devices. Fifth the sharing of knowledgeable data among the researchers and clinical staff has also a security issues and threats. Hence we need to work to overcome these threats collectively to make AI to assist medical diagnosis. Basic terminologies of the chapter are as follows, and Sect. 1 covers the introduction part, Sect. 2 covers the block chain in healthcare, Sect. 3 covers the related study, Sect. 4 covers the analysis of various related work, and Sect. 5 contains the conclusion and future work.

1.1 Artificial Intelligence (AI)

AI is a technique which enable a machine to think and mimic like human brains, and it has now been a part of everyone life. The term artificial intelligence (AI) describes how computers, particularly computer systems, may simulate human intelligence processes. Learning, reasoning, problem-solving, perception, and language comprehension are some of these processes. Artificial Intelligence (AI) seeks to create machines that are capable of tasks that normally require human intelligence, like pattern recognition, decision-making, and situational adaptation. In the context of medical diagnosis, AI involves using advanced algorithms to analyse medical data and assist healthcare professionals in diagnosing diseases, predicting outcomes, and recommending treatments. The idea of building intelligent machines that could express human intelligence was first investigated by researchers in the 1950s, which is when artificial intelligence (AI) first emerged. The Dartmouth Workshop in 1956 is credited with coining the phrase "artificial intelligence" [7], which officially established AI as a discipline of research. Early AI research focused on symbolic AI, using rules and logic to model human cognitive processes. However, progress was slower than initially anticipated during the 1970 and 1980s, characterized by reduced funding and enthusiasm due to unmet expectations. The field regained momentum in the 1990s with advancements in machine learning and neural networks. After the year 2000, with the rise of electronic health records and digital medical imaging, AI began to make strides in medical diagnosis. Computer-aided detection systems were developed to assist radiologists in identifying abnormalities in medical images. In the year 2010, deep learning, a subset of AI, gained prominence due to improved computational resources. Deep learning algorithms demonstrated remarkable capabilities in image and speech recognition, which paved the way for breakthroughs in medical image analysis. At present, AI in medical diagnosis has expanded to various domains, including radiology (detecting tumours and fractures), pathology (identifying cellular abnormalities), and genomics (analysing genetic data for disease risks). AI-driven tools assist in early detection of conditions like diabetic retinopathy and lung diseases, improving disease management and patient outcomes.

1.2 Blockchain Technology

Blockchain is a digital ledger that has no central administrator and records transactions in an open, secure, and permanent manner. Since its initial introduction in 2008 as the core technology behind Bitcoin, it has been embraced by numerous other industries for a variety of purposes [8]. Blockchain is a combination of peer-to-peer communication and cryptographic technology. It is made up of a number of blocks connected together by cryptographic hash functions. The blockchain is a simple but brilliant way to distribute and receive data securely and automatically. Block production by one of the parties is required to start a transaction. Thousands of devices dispersed over the internet are verifying this block. After that, the verified block is added to a chain and maintained online, producing a record that is distinct, unique, and has a clear past. Therefore, on a blockchain, a transaction is regarded valid when a block's consensus through a contract is reached. Every member of a blockchain network has a copy of the ledger because blockchain technology is decentralized, which also means that trust is decentralized. Each block in a blockchain network is made up of a collection of transactions that have been approved and added to the network by a consensus mechanism. A chain of data that cannot be changed is created by connecting the blocks in chronological order. Blockchain is extremely secure because it is decentralised; no single entity controls the network. The unchangeable nature of the ledger allows for speedy and effective transaction verification, processing, and tracking, as well as convenient auditing of transactions.

Blockchain has established the ground work [9–12] for cryptocurrencies such as Ripple, Bitcoin, and Ethereum, among others. Notably, Bitcoin is acclaimed as the inaugural cryptocurrency. Given the contemporary reliance on digital authentication for business transactions, blockchain introduces prospects for decentralized platforms and services that are accessible to all. This innovative technology furnishes avenues for conducting a diverse array of imaginative financial tools like micropayments and peer-to-peer lending, streamlining transactions while reducing associated costs. Each transaction functions as a digital block, necessitating consensus from numerous participants within the network to be verified and incorporated. In response to the susceptibilities witnessed by conventional databases, blockchain has emerged as a safeguard, ensuring data security through cryptographic validation. Blockchain technology [13–15] has undergone distinct evolutionary versions and each version marking significant advancements. In its first iteration, Blockchain 1.0, the revolutionary concept of decentralized ledgers birthed cryptocurrencies like Bitcoin, leveraging Proof of Work for consensus. Blockchain 2.0 followed, introducing smart contracts and platforms like Ethereum, enabling self-executing agreements. The third version, Blockchain 3.0, emphasized interoperability and scalability through platforms like Polkadot. Privacy and security took the forefront in Blockchain 4.0, exemplified by privacy-focused coins. The quest for sustainability led to Blockchain 5.0, championing energy-efficient consensus mechanisms. Presently, Blockchain 6.0 is emerging, integrating Artificial Intelligence and the IoT into the blockchain based ecosystem, ushering in a new era of enhanced capabilities and applications across industries.

While artificial intelligence may improve our ability to recognize and react promptly to disease diagnoses, blockchain technology, which was initially developed to support the cryptocurrency ecosystem, is now being used in many other industries to achieve extraordinary levels of security [16], improve the security of medical records, and protect the privacy of record owners [17]. Blockchain technology is applicable to a number of industries, including e-commerce, cross-border payments, digital identities, healthcare, IoT, and online voting. When utilised in data management systems, blockchain technology can help solve important problems with data transparency, traceability, immutability, auditing, safe data provenance, and healthcare system confidence. Additionally, this can be effectively used to enhance the management of medical records [18]. Figure 2 shows the various versions of blockchain with its technology.

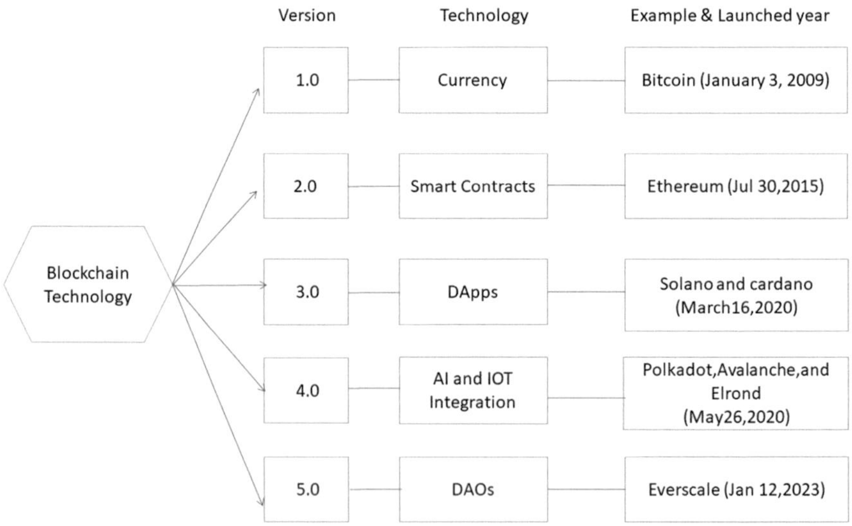

Fig. 2 Various versions of blockchain

2 Blockchain in Healthcare

Healthcare industry is one of the most prominent and significant area for any nation and the failure of traditional methods has significantly leads to the adaption of blockchain technology in the healthcare. The healthcare sector is going through a revolutionary change thanks to blockchain technology, which has ushered in a new era of data security, interoperability, transparency, and patient-centric treatment. Traditionally plagued by fragmented systems, data silos, and security vulnerabilities, healthcare has found a potent ally in blockchain's decentralized and tamper-proof architecture [19]. One of the most significant impacts of blockchain lies in data security. Patient records, medical histories, and sensitive health information are vulnerable to breaches and unauthorized access in centralized databases. Blockchain's cryptographic techniques offer a robust solution by ensuring that patient data remains encrypted and accessible only to authorized individuals. Interoperability, a long-standing challenge in healthcare, is another realm revolutionized by blockchain. Health systems are often hindered by incompatible data formats and disparate information sources, resulting in inefficient care coordination. Blockchain's distributed ledger eliminates this obstacle by providing a standardized, transparent platform where healthcare providers, insurance companies, pharmacies, and even patients can securely share and access information [20]. While blockchain's potential in healthcare is vast, challenges remain. Integration with legacy systems, scalability concerns, and regulatory alignment require careful navigation. Collaborative efforts among stakeholders, including healthcare providers, technology companies, and policymakers, are essential to fully harness blockchain's potential. Figure 3 shows the usages of blockchain in healthcare industry. Here one can notice that the stack holders can utilize the cloud system for storing and retrieving of medical records using the interface called blockchain hand-shaker, which ensure the secure way of communication by applying blockchain mechanism and cryptography if required. The interface blockchain hand-shaker is also connected with the distributed ledger and smart contact in bidirectional way for fetching any required details.

It is very clear that the involvement of AI and blockchain in healthcare industry revolutionized the medical industry and the continuous development in these sectors can boost the medical for the wellbeing of human kind. The convergence of Artificial Intelligence (AI) and blockchain technology holds the promise of a revolutionary transformation in the medical healthcare system, addressing critical challenges and unlocking new avenues for improved patient care, data management, research, and operational efficiency. This synergistic approach harnesses the strengths of both technologies to create a more secure, transparent, and patient-centric healthcare ecosystem. Figure 4 shows the AI usages in healthcare system. Here one can observe that the electronic patient record which is composed of audio, image and text can be processed very efficiently using various AI technology. This figure also shows the common applications and common challenges associated with healthcare system using AI.

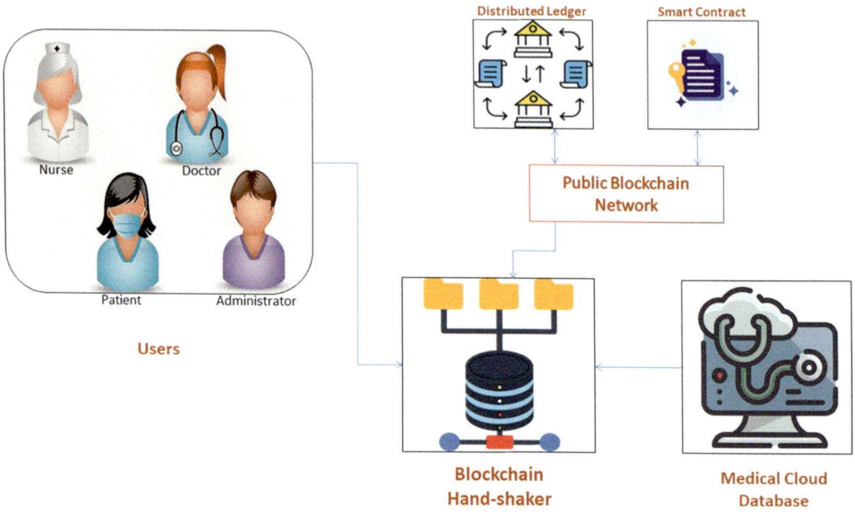

Fig. 3 Usages of Blockchain in healthcare

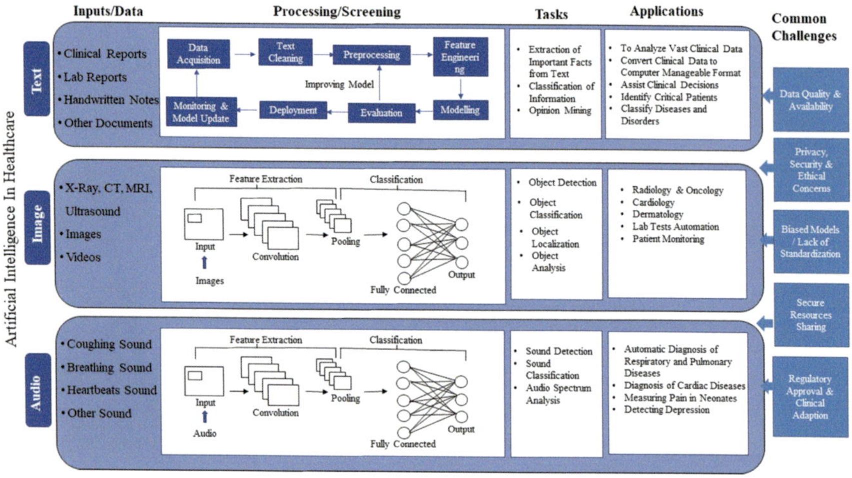

Fig. 4 Various usages of AI in healthcare

Blockchain technology combined with artificial intelligence (AI) has the potential to transform the healthcare industry, overcoming obstacles and changing the game. Through enhanced data security, interoperability, diagnostic precision, personalized treatment, and streamlined research processes, this convergence empowers patients, improves outcomes, and ushers in an era of efficient, transparent, and patient-centred healthcare.

3 Related Study

In this segment, we will address several recent research investigations concerning the security of healthcare data using the assistance of AI methodologies and blockchain technologies. Challenges prevalent within the healthcare sector encompass issues like interoperability, unavailable medical records, and the absence of thorough and protected population health information. This chapter aims to identify cutting-edge approaches utilizing AI techniques and blockchain technology to ensure healthcare data security.

Shinde et al. [4] had delved into the notion that utilizing blockchain can enhance the dependability and credibility of AI-driven healthcare. The proposed work indicates that blockchain having the ability to tackle privacy and security concerns within the healthcare community. Additionally, implementing blockchain could enable verifying medical data and user origin. Furthermore, they identified specific defensive strategies designed to counter certain adversarial attacks on healthcare data and also mentioned how AI and blockchain technology together can be used for the healthcare industry using various AI techniques.

Andrew et al. [21] have detailed several attributes and practical applications of blockchain across various contexts, including its relevance to achieving interoperability in the healthcare sector. The discussion encompassed an overview of blockchain architecture, platforms, and categorizations to aid in selecting an appropriate forum for healthcare purposes. They provided a thorough analysis to demonstrate the blockchain technology's importance to the healthcare sector from both an application and a technological standpoint. Furthermore, the presentation contained a comprehensive examination of security breaches targeting blockchain protocols, including the classification of threat models. They also provided a comparative assessment of detection and safeguarding methods. Various perspective of using blockchain and AI along with is mentioned and safeguard measure of using blockchain is also mentioned. Finally, they proposed several strategies to bolster the security and confidentiality of blockchain networks.

Ali et al. [22] provided an architectural proposal comprising three distinct environments: doctor's, patient's, and the meta-verse environment. Within this meta-verse setting, doctors and patients engage with the support of blockchain technology, ensuring the security, safety, and privacy of their information. The meta-verse environment stands as the central component of the proposed architecture. By registering on the blockchain, doctors, patients, and nurses enter this environment and assume avatar forms. Every interaction and consultation between doctors and patients are meticulously documented, encompassing diverse data such as images, speech, text, videos, and clinical records. These data are then collected, transmitted, and securely stored using blockchain technology. Explainable artificial intelligence (XAI) models use the collected data to predict and diagnose diseases. Blockchain strengthens the security of patient data while enabling information to be transparent, traceable, and unchangeable. These intrinsic attributes of blockchain cultivate patient trust in their data's integrity, ultimately ensuring transparency, trustworthiness in disease diagnosis, and robust patient data security within the scope of this proposed architectural framework.

Attaran [23] suggested the utilization of Blockchain solutions to enhance data security and manage identities. This technology offers the potential to mitigate threats and safeguard private information from unauthorized access. As data is added to the blockchain, it becomes encrypted, rendering it immutable and indecipherable. Transactions are validated through a private key, which is known exclusively to the all the individual person. Unlike contemporary healthcare data systems, this approach would restrict a healthcare provider's access to a patient's medical data unless granted explicit permission through the blockchain record. Enhanced collaboration of data among providers enhances the accuracy of diagnoses, increases treatment effectiveness, and enables healthcare facilities to provide cost-efficient care. Through Blockchain, patient information can remain secure while still allowing sharing with selected service providers. The technology establishes proof of medical records ownership and ensures authenticity to counteract counterfeit techniques.

Farouk et al. [24] suggested novel attributes inherent to blockchain and delved into the potential advantages and future trajectories of its integration into the healthcare domain. Healthcare entails combining blockchain with Internet of Things (IoT) technologies, offering healthcare establishments a robust and precise system for record management, a pivotal requirement. The suggested framework is segmented into three contract categories: Registrar Contract (RC), Patient-Provider Relationship Contract (PPR), and Summary Contract (SC). RC serves the purpose of associating participants' identification with their corresponding Ethereum address identity. PPR is established between system nodes to facilitate the storage and administration of patients' medical records. SC retrieves patients' complete medical history, showcasing all past and ongoing interactions between participants and other nodes within the system.

De Moraes Rossetto [25] put forward a framework for managing Healthcare Data Privacy using Blockchain. They introduced a structure aimed at safeguarding the confidentiality of medical information. This data is stored and exchanged within a decentralized blockchain network. This protection is achieved by employing encryption methods such as RSA, ECC, and AES. Tests were conducted to assess how cryptography influenced the suggested framework regarding computational workload, memory consumption, and processing duration. The findings reveal that the primary impacts were on execution time and heightened computational burden when transmitting data to the blockchain. These effects are reasonable, given the enhanced privacy and security offered by the architectural design and encryption techniques.

Haddad et al. [26] presented a classification of suggested AI-Blockchain remedies for EHR challenges. They extensively examined the advantages of implementing Blockchain technology for health record administration. They also included a recent examination of privacy and security concerns linked to EHR management. Furthermore, they proposed prospective avenues for research, aiming to establish a proficient and dependable AI-Blockchain EHR ecosystem that surpasses existing centralized models. This approach envisions a patient-focused data-sharing platform, granting individuals full authority over their data.

Singh et al. [27] suggested strategies to enhance blockchain security by consolidating pivotal aspects that could be leveraged to create diverse blockchain systems and security utilities, thereby mitigating security susceptibilities. They elucidated a range of blockchain based attacks prevalent in IoT based networks and presented remedies for such vulnerabilities. Their paper systematically laid out a methodical exploration of security constraints, weak points, hurdles, and complexities related to blockchain technology. The same thorough treatment was given to security challenges within blockchain enterprises. Existing security resolutions for blockchain in varied contexts were also elucidated. The discourse further delved into security tools capable of addressing these vulnerabilities. A comprehensive analysis was undertaken to underscore both the concerns related to security and advantages of blockchain components, notably decentralization, that poses significant obstacles concerning data privacy and transparency, leading to intricacies within the network.

Tanwar et al. [28] introduced a system framework and algorithm designed to adopt a patient-centred strategy, facilitating access control policies through symmetric key cryptography to diverse healthcare providers. They additionally suggested the utilization of blockchain for a permission-oriented EHR sharing system, incorporating the chain code concept. The proposed system was assessed to evaluate its alignment with the requisites of patients, healthcare providers, and other stakeholders. Lastly, an optimal strategy for enhancing the performance of the blockchain system was identified, addressing metrics such as latency and throughput, network scalability, and security considerations.

Hylock and Zeng [29] have presented a conceptual framework detailing the application of blockchain technology to ascertain patients' health status. This framework emphasizes the integration of concurrent execution and AI within healthcare networks. The approach focuses on evaluating the comprehensive well-being of patients, including diagnosis and recovery procedures. Moreover, the study delves into the analysis of pertinent surgical interventions, employing parallel operations and computational assessments of clinical decision-making. By examining patient care quality and diagnostic feasibility, both within real-world and simulated healthcare systems, the proposed methodology underwent practical evaluation.

Feng et al. [30] provided a detailed analysis of the distributed blockchain database using state machine replication, with atomic database changes known as transactions organised into blocks and the transaction log's integrity and tamper-resistance guaranteed by hash links between blocks. The idea of a blockchain was first suggested in relation to Bitcoin and decentralised electronic money. The core blockchain building blocks have been the topic of countless reports, and with the success of Bitcoin, blockchain technology may now be utilised to conduct secure and reliable transactions over unreliable networks without the need for a third party.

By integrating blockchain into several sources, Andoni et al. [31] analyses the need for sharing of electronic healthcare record (EHR). It is designed primarily for personal health data and electronic medical records, but it also investigates many other types of data. They combined the application with on-chain and off-chain authentication procedures to guarantee that it matched good standards for validity and privacy. The adaption of blockchain for the interchange of EHR, obscurity, and shield between clinical staff and organisations can be greatly increased.

Khurshid [32] has suggested adopting a similar mechanism, proposing the implementation of an inventive and organized infrastructure using Blockchain technology. This innovative approach aims to enhance private patient records' security and address core data security concerns. Additionally, it seeks to establish a foundational blockchain software for hospitals. The potential of Blockchain technology extends to the medical and therapeutic domains as well. By employing blockchain solutions, there is a possibility to store all clinical authorizations, schedules, and protocols on a blockchain before the commencement of a clinical trial or examination. This would result in more up-to-date, secure, and timestamped clinical trial data, which can be readily accessed.

Wahl et al. [33] utilized a method to address or reply to an inquiry; these computational systems employ a diverse array of suggested algorithms, decision-making skills, and extensive information. The technologies capable of assisting us in our day-to-day existence encompass our surroundings. The efficiency of autonomous automobiles, progress in medical investigations, and wearable gadgets tailored for individuals have showcased the immense capability of technology and the remarkable strides it has taken in the recent past. Jaiswal et al. [34] developed a GLSTM model. The strength of this research lies in its creative combination of GANs and LSTM, providing a robust framework for handling diverse datasets. The use of GANs to generate synthetic PID diabetes data is particularly commendable, as it addresses the challenges associated with limited real-world data.

4 Analysis of AI and Blockchain Technique Used for Healthcare

The present progression of AI's application in healthcare is expected to form a cautious and well-balanced framework. Han et al. [35] have examined several aspects of healthcare provision, indicating that AI, when combined with collaboratively trained medical professionals, is likely to follow the current trajectory for a considerable period. This trend demonstrates AI's capacity for accurate patient diagnosis. As depicted in Fig. 5, the year-by-year trends in AI's healthcare utilization reveal a continuous enhancement in AI technology, undeniably bolstering the healthcare sector's performance. However, this technological advancement comes with an elevated risk to the security of medical record data, necessitating robust quality measures.

The integration of AI and blockchain has undoubtedly strengthened every aspect of the medical healthcare system as the evolvement of AI has brought various technologies together to process various electronic medical records like X-Ray, medical

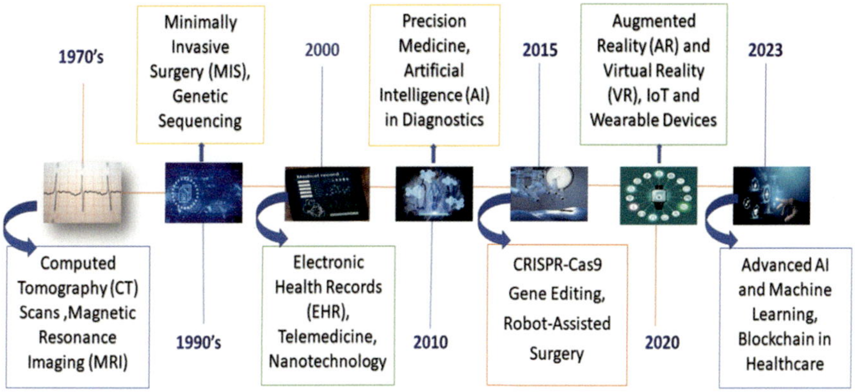

Fig. 5 Trends of AI in healthcare year wise

prescription, ECG data, reports etc., in order to extract some useful pattern and information related to the patient for automating the suggestion in case of emergency. It also helps during various decision-making process related to the healthcare. As we all know that the lots of data is being generated and every data related to the medical healthcare is of utmost important and significant so blockchain is an emerging technology which ensures the exchange of data in efficient manner. There are various related research papers published every year by Indian authors and authors from other countries than then the India and it is seen that every year the number of publication in this field is being increased rapidly. Figure 6 shows the worldwide publication in the domain of healthcare using AI and blockchain.

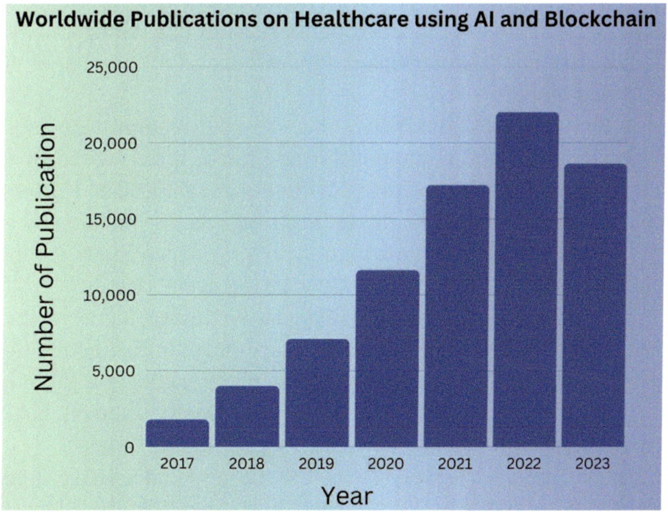

Fig. 6 Worldwide publication details year-wise

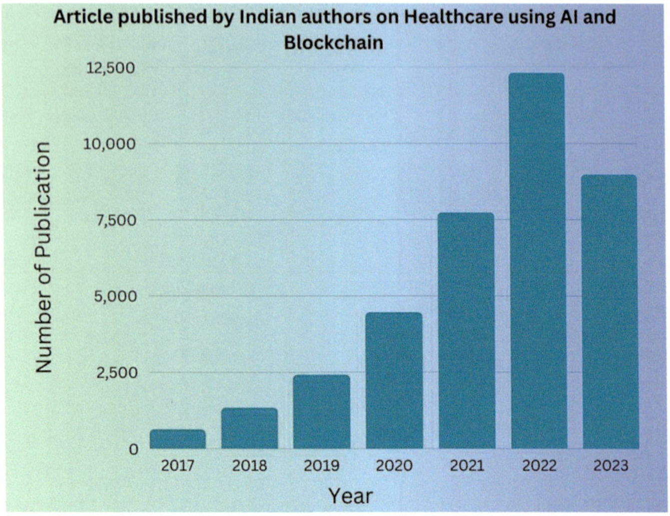

Fig. 7 Publication by India authors year-wise

After observing Fig. 6 it is very clear that the number of publication in this domain is continuously increasing worldwide and one can notice that in year 2017 the total publication in this domain was 1780, but in 2018 it is increased by 124% than the year 2017. Likewise for the year 2019 it is increased by 77% than the 2018 to reach 7100 publication worldwide. For year 2020 the number of publication worldwide is 11600 with 63% of growth than the previous year. For year 2021 the total publication is 17200, which shows 48% growth than the previous year. Like for year 2022, the total number of publication is 22000, which is 27% more than the previous year. For year 2023, till 25 August the total number of publication in this domain is 18600 and it is expected to be continuously increasing. Figure 7 shows the number of publication done by the Indian authors.

Looking at Fig. 6 one can easily say that the contribution done by the Indian authors are good as well and continuously the growth is very good by the Indian authors and for each year it is showing growth of more than 60%. Figure 8 shows a comparison between the publication by India authors and publication by worldwide authors and it shows a significant growth in every year which shows that publication and research in this domain is continuously growing every year.

Table 1 contains the recent state-of-art related to healthcare industry which utilizes the blockchain and the AI for providing better healthcare infrastructure. After going through various literature and recent literature as mentioned in the Table 1, it is very clear that there is room for the improvement in the technology and AI, but along with this securing digital health related data is also a challenging issue.

Referencing to Table 1 it is very clear that the researchers have used different blockchain technology like Consensus mechanisms specifically, Blockchain technology with smart contracts—Modified SHA256 for data security, with AI and IOT.

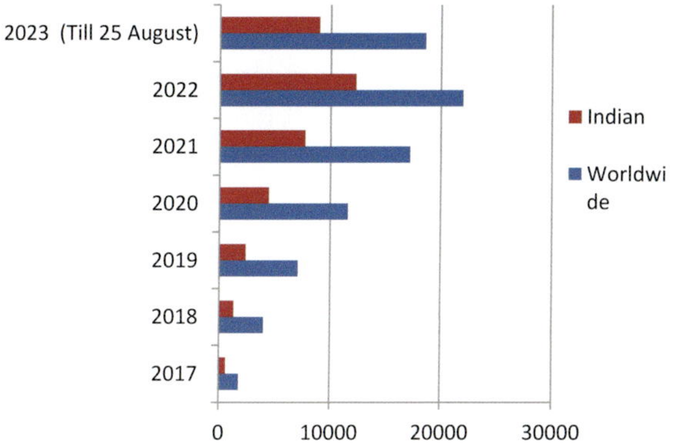

Fig. 8 Comparison between numbers of publication

As mentioned by Farahat et al. [43] the version 2.0 of blockchain is very fast to create data block. We can say that Blockchain technology is an effective method for enhancing the security of healthcare data and the use of blockchain in healthcare ensures that sensitive data is protected from hackers and unauthorized access. The integration of blockchain technology with IoT devices in the healthcare sector has significantly impacted security, privacy, and efficiency.

5 Conclusion and Future Scope

The integration of AI and blockchain technologies has emerged as a promising solution to address the critical issue of healthcare data security, fostering a safer society in the process. The exponential growth of sensitive healthcare data and the increasing sophistication of cyber threats necessitate innovative approaches to safeguarding this information. AI algorithms offer advanced data monitoring, anomaly detection, and predictive analytics, enhancing the real-time identification of potential breaches. Concurrently, blockchain's decentralized and tamper-resistant nature ensures the integrity, traceability, and confidentiality of healthcare data, thereby building trust among stakeholders. The synergy of AI and blockchain not only bolsters the protection of patient records, diagnoses, and treatments but also streamlines data interoperability across healthcare institutions. This interoperability enhances the continuity and quality of patient care, enabling timely and well-informed medical decisions. Additionally, the collaboration between these technologies holds the potential to mitigate the administrative burden, reduce costs, and improve the efficiency of healthcare systems globally. Future scope includes utilizing latest technology and security mechanism to ensure good healthcare infrastructure.

Table 1 Comparison of recent state-of-the-art on healthcare using blockchain

Literature	Objective	Method used	Result	Limitation
Mahammad et al. [36]	Blockchain technology for secure storage and maintenance of healthcare data but does not mention the use of AI for healthcare data security	Blockchain security with smart contracts Integration of EHR data	The paper proposes a Blockchain-based architectural framework for storing EHR (Electronic health data) data in Bigdata storage systems	Security and storage challenges with EHR data—excessive administrative and operative costs for data management
Alruwaill et al. [37]	Blockchain based healthcare system using IoMT devices for continuous patient monitoring and addresses security concerns with the use of symmetric and asymmetric keys	Symmetric and asymmetric keys for data concealment Hash function SHA256 for data alteration detection	Blockchain-based smart healthcare system using IoMT devices - Location-based authentication for authenticity and data source	Not available
Dayana et al. [38]	The paper discusses healthcare data security using blockchain technology, but does not mention the use of AI	Use of decentralized blockchain technology Implementation of permissioned blockchain with authorized access	Solutions for healthcare data security using blockchain technology Benefits include secure data storage and reduced fear of data theft	Not available
Devi et al. [39]	The paper discusses enhancing healthcare data security using blockchain technology, but does not mention the use of AI	Blockchain method for data security Consensus mechanisms: PoW (Proof Of Work), PoS (Proof of Stake), DPoS, (Delegated Proof of Stake)	Enhanced proof of stake increases security in blockchain Protects medical records from hackers	Not available
Sharma et al. [40]	Blockchain technology enhances security of healthcare data by providing features like confidentiality, decentralization, and privacy	Blockchain-based Proposed Application (PA), Smart contracts for confidentiality, authentication, and access control	Blockchain-based IoT systems enhance security and privacy Proposed Application (PA) provides more effective solution	Enhances security, privacy, transparency, and efficiency in healthcare Offers a remedy that is more efficient than current plans

(continued)

Table 1 (continued)

Literature	Objective	Method used	Result	Limitation
Abdulatif et al. [41]	This work proposes an AI and blockchain-based secure architecture to analyze malware and network attacks on wearable devices in the smart healthcare system	Conducting a comprehensive review of security challenges Proposing an AI and blockchain-based secure architecture	Identification of security challenges in smart healthcare systems Proposal of AI and blockchain-based secure architecture	Lack of integration of modern technologies in healthcare security Research gap in exploring countermeasures for security challenges
Ali et al. [42]	The proposed architecture combines AI and blockchain in the metaverse to ensure data security and privacy in healthcare	Amalgamation of AI and blockchain in the metaverse Use of explainable AI models for disease prediction and diagnosis	Proposed architecture integrates AI, blockchain, and metaverse Ensures trust, security, and transparency in healthcare	Not available
Farahat et al. [43]	The paper discusses the use of blockchain technology to secure patient records and ensure privacy in healthcare applications	Amalgamation of AI and blockchain in the metaverse Use of explainable AI models for disease prediction and diagnosis	Proposed architecture for integrating AI and blockchain in the metaverse for healthcare Ensures transparency, trust, and data security in healthcare	Standard privacy techniques are not secure enough Blockchain version 2.0 performs better than subsequent versions
Sarker et al. [44]	It shows the utilization of blockchain technology in healthcare applications, but does not mention the use of AI in conjunction with blockchain	Review of blockchain technology in healthcare Evaluation of healthcare technologies based on blockchain	Review of blockchain technology in healthcare Identification of limitations and future research direction	Limitations of previous approaches in healthcare blockchain technology
AlGhamdi et al. [45]	The healthcare industry is adopting AI and blockchain technologies to enhance patient outcomes, reduce costs, and improve operational efficiencies	AI, IoMT, and blockchain technologies in healthcare Role, applications, obstacles, and future research areas are mentioned	Adoption of AI, IoMT, and blockchain in healthcare Advantages and challenges in implementing these technologies	Implementation requires collaboration between multiple stakeholders Ensuring patient privacy and data security

Conflict of Interest There is no conflict of interest and all authors agree for the publication of the chapter.

References

1. Nilsson N (2009) The quest for artificial intelligence: a history of ideas and achievements. Cambridge University Press, New York. ISBN 978-0-521-12293-1
2. Ethem A (2020) Introduction to machine learning, 4th edn. MIT, pp xix, 1–3, 13–18. ISBN 978–0262043793
3. Ciresan D, Meier U, Schmidhuber J (2012) Multi-column deep neural networks for image classification. In: 2012 IEEE conference on computer vision and pattern recognition, pp 3642–3649. arXiv:1202.2745, https://doi.org/10.1109/cvpr.2012.6248110
4. Shinde R, Patil S, Kotecha K, Potdar V, Selvachandran G, Abraham A (2022) Securing AI-based healthcare systems using blockchain technology: a state-of-the-art systematic literature review and future research directions. https://arxiv.org/pdf/2206.04793
5. Bengio Y, Ducharme R, Vincent P, Janvin C (2003) A neural probabilistic language model. J Mach Learn Res 3:1137–1155—via ACM Digital Library
6. Jaiswal S, Pandey MK (2021) A review on image segmentation. In: Rathore VS, Dey N, Piuri V, Babo R, Polkowski Z, Tavares JMRS (eds) Rising threats in expert applications and solutions. advances in intelligent systems and computing, vol 1187. Springer, Singapore. https://doi.org/10.1007/978-981-15-6014-9_27
7. McCarthy J, Minsky M, Rochester N, Shannon C (1955) A proposal for the dartmouth summer research project on artificial intelligence. AI Mag 27:12–14. http://www-formal.stanford.edu/jmc/history/dartmouth/dartmouth.html
8. Narayanan A, Bonneau J, Felten E, Andrew M, Steven G (2016) Bitcoin and cryptocurrency technologies: a comprehensive introduction. Princeton University Press, Princeton, New Jersey. ISBN 978-0-691-17169-2
9. Armknecht F, Karame GO, Mandal A, Youssef F, Zenner E (2015) Ripple: overview and outlook. Lect Notes Comput Sci 9229:163–180. https://doi.org/10.1007/978-3-319-22846-4_10
10. Radziwill N (2018) Blockchain revolution: how the technology behind bitcoin is changing money. Bus World 25(1). https://doi.org/10.1080/10686967.2018.1404373
11. Dannen C (2017) Introducing ethereum and solidity: foundations of cryptocurrency and blockchain programming for beginners, vol 1. Springer. https://doi.org/10.1007/978-1-4842-2535-6
12. Nakamoto S (2008) Bitcoin: a peer-to-peer electronic cash system. Decentralized Bus Rev Article 21260
13. Lindman J, Rossi M, Tuunainen VK (2017) Opportunities and risks of blockchain technologies in payments—a research agenda. In: Proceedings of the 50th Hawaii international conference on system science, pp 1533–1542. https://doi.org/10.24251/HICSS.2017.185
14. Lundqvist T, Blanche A, De Andersson HRH (2017) Thing-to-thing electricity micro payments using blockchain technology. In: Global Internet of Things Summit (GIoTS), Geneva, Switzerland, pp 1–6. https://doi.org/10.1109/GIOTS.2017.8016254
15. Tosh DK, Shetty S, Liang X, Kamhoua CA, Kwiat KA, Njilla L (2017) Security implications of blockchain cloud with analysis of block withholding attack. In: Proceedings 17th IEEE/ACM international symposium on cluster, cloud and grid computing, CCGRID, pp 458–467. https://doi.org/10.1109/CCGRID.2017.111
16. Yaeger K, Martini M, Rasouli J, Costa A (2019) Emerging blockchain technology solutions for modern healthcare infrastructure. J Sci Innov Med 2(1):1. https://doi.org/10.29024/jsim.7
17. Kumar R, Arjunaditya SD, Srinivasan K, Hu Y-C (2023) AI-powered blockchain technology for public health: a contemporary review, open challenges, and future research directions. Healthcare 11:81. https://doi.org/10.3390/healthcare11010081

18. Singh S, Sharma SK, Mehrotra P, Bhatt P, Kaurav M (2022) Blockchain technology for efficient data management in healthcare system: opportunity, challenges and future perspectives. https://doi.org/10.1016/j.matpr.2022.04.998
19. McFarlane TD, Dixon BE, Grannis SJ (2016) Client registries: identifying and linking patients. In: Health Information Exchange (HIE): navigating and managing a network of health information systems, pp 163–182. https://doi.org/10.1016/B978-0-12-803135-3.00011-6
20. Neelakandan S, Beulah JR, Prathiba L, Murthy GLN, Irudaya Raj EF, Arulkumar N (2022) Blockchain with deep learning-enabled secure healthcare data transmission and diagnostic model. Int J Model Simul Sci Comput 13(04):2241006. https://doi.org/10.1142/S1793962322410069
21. Andrew J, Isravel DP, Sagayam KM, Bhushan B, Sei Y, Eunice J (2023) Blockchain for healthcare systems: architecture, security challenges, trends and future directions. J Netw Comput Appl. https://doi.org/10.1016/j.jnca.2023.103633
22. Ali S, Abdullah, Armand TPT, Athar A, Hussain A, Ali M, Yaseen M, Joo MI, Kim HC (2023) Metaverse in healthcare integrated with explainable AI and blockchain: enabling immersiveness, ensuring trust, and providing patient data security. Sensors (Basel, Switzerland) 23(2):565. https://doi.org/10.3390/s23020565
23. Attaran M (2022) Blockchain technology in healthcare: challenges and opportunities. Int J Healthc Manag 15(1):70–83. https://doi.org/10.1080/20479700.2020.1843887
24. Farouk A, Alahmadi A, Ghose S, Mashatan A (2020) Blockchain platform for industrial healthcare: vision and future opportunities. Comput Commun 154:223–235. ISSN 0140-3664, https://doi.org/10.1016/j.comcom.2020.02.058
25. De Moraes Rossetto AG, Sega C, Leithardt VRQ (2022) An architecture for managing data privacy in healthcare with blockchain. Sensors 22(21):8292. https://doi.org/10.3390/s22218292
26. Haddad A, Habaebi MH, Islam MH, Hasbullah NF, Zabidi SA (2022) Systematic review on AI-blockchain based e-healthcare records management systems. IEEE Access 10:94583–94615. https://doi.org/10.1109/ACCESS.2022.3201878
27. Singh S, Hosen ASMS, Yoon B (2021) Blockchain security attacks, challenges, and solutions for the future distributed IoT network. IEEE Access 9:13938–13959. https://doi.org/10.1109/ACCESS.2021.3051602
28. Tanwar S, Parekh K, Evans R (2020) Blockchain-based electronic healthcare record system for healthcare 4.0 applications. J Inform Secur Appl 50:102407, ISSN 2214-2126. https://doi.org/10.1016/j.jisa.2019.102407
29. Hylock RH, Zeng X (2019) A blockchain framework for patient-centered health records and exchange (healthChain): evaluation and proof-of-concept study. J Med Internet Res 21
30. Feng Q, He D, Zeadally S, Khan MK, Kumar N (2019) A survey on privacy protection in blockchain system. J Netw Comput Appl 126:45–58. https://doi.org/10.1016/j.jnca.2018.10.020
31. Andoni M et al (2019) Blockchain technology in the energy sector: a systematic review of challenges and opportunities. Renew Sust Energ 21(100):143–174. https://doi.org/10.1016/j.rser.2018.10.014
32. Khurshid A (2020) Applying blockchain technology to address the crisis of trust during the COVID-19 pandemic. JMIR Med Inform 8(9):e20477. https://doi.org/10.2196/20477
33. Wahl B, Cossy-Gantner A, Germann S, Schwalbe NR (2018) Artificial intelligence (AI) and global health: how can AI contribute to health in resource-poor settings? BMJ Glob Health 3(4):e000798. https://doi.org/10.1136/bmjgh-2018-000798
34. Jaiswal S, Gupta P (2023) GLSTM: a novel approach for prediction of real & synthetic PID diabetes data using GANs and LSTM classification model. Int J Exp Res Rev 30:32–45. https://doi.org/10.52756/ijerr.2023.v30.004
35. Han C, Rundo L, Murao K et al (2020) Bridging the gap between AI and healthcare sides: towards developing clinically relevant AI-powered diagnosis systems. In: IFIP advances in information and communication technology, vol 584. Springer, Cham, pp 320–333. https://doi.org/10.1007/978-3-030-49186-4_27

36. Mahammad AB, Kumar R (2023) Scalable and security framework to secure and maintain healthcare data using blockchain technology. In: International conference on computational intelligence and sustainable engineering solutions (CISES), Greater Noida, India, pp 417–423. https://doi.org/10.1109/CISES58720.2023.10183494
37. Alruwaill AM, Mohanty SP, Kougianos E (2023) HChain: blockchain based healthcare data sharing with enhanced security and privacy location-based-authentication. In: Proceedings of the Great Lakes symposium on VLSI 2023 (GLSVLSI '23). Association for Computing Machinery, New York, NY, USA, pp 97–102. https://doi.org/10.1145/3583781.3590255
38. Dayana R, Vadivukkarasi K (2023) Healthcare data security using blockchain technology. https://doi.org/10.1109/ICISCoIS56541.2023.10100572
39. Devi T, Kamatchi SB, Deepa N (2023) Enhancing the security for healthcare data using blockchain technology. In: International conference on computer communication and informatics (ICCCI), Coimbatore, India, pp 1–7. https://doi.org/10.1109/ICCCI56745.2023.10128545
40. Sharma P, Namasudra S, Crespo RG, Parra-Fuente J, Trivedi MC (2023) EHDHE: enhancing security of healthcare documents in IoT-enabled digital healthcare ecosystems using blockchain. Inf Sci. https://doi.org/10.1016/j.ins.2023.01.148
41. Alabdulatif A, Khalil I, Saidur Rahman M (2022) Security of blockchain and AI-empowered smart healthcare: application-based analysis. Appl Sci. https://doi.org/10.3390/app122111039
42. Ali S, Abdullah, Armand TPT, Athar A, Hussain A, Ali M, Yaseen M, Joo M-I, Kim HC (2023) Metaverse in healthcare integrated with explainable AI and blockchain: enabling immersiveness, ensuring trust, and providing patient data security. https://doi.org/10.3390/s23020565
43. Farahat IS, Aladrousy W, Elhoseny M, Elmougy S, Tolba AE (2022) Improving healthcare applications security using blockchain. Electronics. https://doi.org/10.3390/electronics11223786
44. Sarker B, Sharif N, Rahman MA, Parvez AHM (2023) AI, IoMT and blockchain in healthcare. J Trends Comput Sci Smart Technol 5:30–50. https://doi.org/10.36548/jtcsst.2023.1.003
45. AlGhamdi R, Alassafi MO, Alshdadi AA, Dessouky MM, Ramdan RA, Aboshosha BW (2022) Developing trusted IoT healthcare information-based AI and blockchain. Processes. https://doi.org/10.3390/pr11010034

Future of Electronic Healthcare Management: Blockchain and Artificial Intelligence Integration

Parag Verma, Chakka Mohana Rao, Prudhvi Kumar Chapalamadugu, Rajeev Tiwari, and Shuchi Upadhyay

Abstract Blockchain and artificial intelligence (AI) technologies have emerged as innovative solutions in the healthcare sector. Using a thorough risk management strategy and a generalizable analytical technology, this research study explores how the integration of these technologies may have a transformational effect on healthcare. Many reliable sources, like the Web of Science and Google polls carried out by regulatory authorities, were examined to compile statistics on healthcare indices. The evaluation concentrates on several facets of blockchain and AI and emphasizes the advantages of merging these technologies in the healthcare industry. The study addresses how to create trustworthy artificial intelligence models for e-Health by utilizing blockchain, an open network for secure information sharing and authorization. The integration of these cutting-edge technologies can result in improved service efficiency, lower costs, and democratized healthcare by giving healthcare professionals access to the blockchain to display patient medical records and using AI's proposed algorithms and decision-making capabilities, along with vast amounts of data. Additionally, blockchain makes storing the encrypted data needed for AI applications easier. By utilizing the breakthroughs in blockchain and AI, this research offers a vision for the future of healthcare delivery, paving the road for a more effective and inclusive healthcare system.

Keywords Healthcare · Risk management · Service efficiency · Cost reduction · Democratized healthcare

---

P. Verma (✉) · C. M. Rao · P. K. Chapalamadugu
Datafoundry Private Limited, Bengaluru, Karnataka 560080, India
e-mail: parag_verma@yahoo.com

R. Tiwari
IILM University, Greater Noida 201306, India

S. Upadhyay
Department of Allied and Health Sciences, School of Health Science and Technology, UPES, Dehradun 248007, India

K. Kaushik and I. Sharma (eds.), *Next-Generation Cybersecurity*, Blockchain Technologies, https://doi.org/10.1007/978-981-97-1249-6_9

1 Introduction

1.1 A Subsection Sample

According to Alotaibi and Federico [1], healthcare systems are complicated and comprise a variety of procedures, workflows, and patient-care activities. For healthcare providers to be successful, internal controls must be improved, performance, compliance, and consistency must be improved, and risk, workload, and overhead must be decreased [2]. This study suggests a healthcare smart contract structure to handle these issues based on cutting-edge healthcare blockchain analysis and a solid method of healthcare administration [3].

Integration of technology like blockchain and artificial intelligence (AI) is becoming increasingly important as governments and corporate sectors digitize healthcare systems [4, 5]. This integration intends to promote medical research, achieve patient centricity, and push novel data processing methodologies for better results [6]. AI may be very useful in medication monitoring, prioritizing essential patients, and data-driven decision-making [7]. The repurposing of pharmaceuticals and accurate dosage measurement in clinical trials are also made possible by AI and numerical drug design methodologies [8]. Governments must make use of their resources and push for change in the fast-changing healthcare industry while also assuring uniformity, compliance, and data protection [9]. Medical professionals, healthcare providers, and payers may receive fast updates thanks to blockchain technology's safe and automated mechanism for storing health-related data.

Additionally, the integration of blockchain and AI algorithms allows AI to discover and study patterns and trends in the field of health. AI can efficiently analyze unstructured data and collect data from a variety of sources, including radiologists and patients. Despite the potential advantages of AI, there are worries among healthcare professionals about its effects on patient well-being and the need for prudence in its use [10]. While AI has proven its worth in a number of fields, such as autonomous cars and fraud detection, its maturity level in the healthcare sector is still being investigated [11]. This study explores how blockchain technology and artificial intelligence may be used to enhance patient outcomes, data security, and healthcare administration.

2 The Blockchain: Advancements and Applications in Healthcare

The blockchain has drawn a lot of interest since it was first introduced as the foundational technology behind Bitcoin, decentralized electronic money. It offers integrity, tamper resistance, and trust in a distributed context, presenting a novel data management paradigm. The blockchain functions as a distributed database that makes use of replication using state machines. Blocks are collections of transactions, which represent atomic modifications to the database. A chain of transactions is then created by connecting these blocks together using cryptographic hash connections. The main goal of the blockchain is to create a transparent, decentralized system for recording and validating transactions. The blockchain does not rely on a single authority or middleman for transaction confirmation and security, in contrast to conventional centralized databases. Instead, it uses a network-wide consensus technique to reach a consensus on the state of the database. The blockchain's capacity to guarantee the consistency and tamper-resistance of the transaction log is a key feature. A strong cryptographic connection is made possible by the fact that each block in the chain holds a hash of the one before it. Any modification to a prior block would cause a mismatched hash, instantly alerting the network of possible meddling. Any modification to a prior block would cause a mismatched hash, instantly alerting the network of possible meddling. The immutability and openness of the blockchain contribute to its high level of security and trust. The blockchain has the potential to transform several industries, including banking, supply chain management, and healthcare, in addition to its use in decentralized electronic currency. It is the perfect option for ensuring data integrity, enabling safe transactions, and boosting confidence in digital ecosystems because of its transparency and immutability.

The essential components of blockchain technology have been the subject of substantial research and reporting [12]. A sequential collection of blocks that include complete and accurate transaction data make up the core of the blockchain. A connection between these blocks and the preceding block, which is commonly represented by a hash value, connects them and creates a chain-like structure [13]. The block of Genesis is the first block in the chain, and the block header is used to identify each succeeding block [14, 15]. Due to blockchain's adaptability, there is increasing interest in a wide range of applications, including data storage, financial markets, computer security, the Internet of Things (IoT), nutritional science, healthcare, and brain research. Blockchain technology has gained a lot of attention, especially in the healthcare industry, and has made major strides in ensuring the safe and trustworthy monitoring of medical records. By providing individualized, dependable, and secure access to real-time clinical records and enabling a thorough and current perspective of a patient's well-being inside a secure healthcare ecosystem, it has the potential to change healthcare in the future [16].

By integrating artificial intelligence healthcare networks and running many transactions concurrently, the use of a blockchain platform in healthcare intends to evaluate the health state of patients' ailments. The suggested strategy entails assessing the general health of patients, providing correct diagnoses, assessing the recovery system, and researching pertinent surgical treatments through concurrent procedures and computer studies in clinical decision-making. With the aid of this thorough technique, it is possible to evaluate the effectiveness of the patient care given and the viability of making correct diagnoses. The efficiency and practical applicability of the suggested strategy have been tested in both actual and simulated healthcare systems [17].

A customized blockchain-oriented healthcare knowledge-sharing network called Blo CHIE was created by researchers [18]. By combining blockchains from many sources, this platform focuses on assessing the requirements for transferring healthcare data, such as personal health data, electronic medical reports, and several other forms of data. The program combines both on-chain and off-chain authentication procedures to guarantee the accuracy and privacy of the data transferred. Blockchain technology makes it possible to improve data safety, anonymity, and medical record interchange between clinical specialists and healthcare organizations. Further evidence for the possibility of substantial advancements in the application of blockchain in the healthcare sector can be found in the study undertaken by Andoni et al. [19].

The adoption of a related mechanism that emphasizes a systematic and cutting-edge infrastructure utilizing blockchain technologies to improve the security of private patient records, address fundamental data protection concerns, and establish a thorough blockchain software framework inside a hospital setting was proposed by Cryan, M.A. [20]. Blockchain technology has shown a lot of promise in the domains of pharmaceuticals and biological science. It is now feasible to use blockchain technology to keep all clinical clearances, schedules, and protocols on a blockchain even before clinical research or trial begins. This strategy makes sure that crucial information about clinical trials is current, safe, time-stamped, and transparent, which improves the efficiency and openness of the research process.

3 Blockchain in Healthcare: e-Health Use Cases

Blockchain in healthcare offers promising e-Health use cases, revolutionizing the industry. Its decentralized and secure nature ensures patient data integrity, streamlined medical record management, and efficient supply chain tracking. Smart contracts enable automated insurance claim processing, while telemedicine platforms benefit from secure and transparent patient-doctor interactions. Figure 1 showcases high-potential blockchain use cases in the healthcare industry.

3.1 Clinical Trials and Research

By securely storing and distributing trial data, blockchain has the potential to simplify the process of conducting clinical trials. It enables effective and transparent data gathering, participant consent management, and data exchange across researchers, assuring data integrity and lowering fraud. For instance, Patientory is a blockchain-based platform that provides a complete solution for storing and managing patients' medical records and data. Patientory uses blockchain technology to store patient data in a decentralized, tamper-proof way, ensuring the confidentiality and privacy of the data. The platform gives patients more control over their own health data and offers them individualized medical care based on their unique requirements. Additionally, Patientory can help with the integration of health insurance programs, enabling users

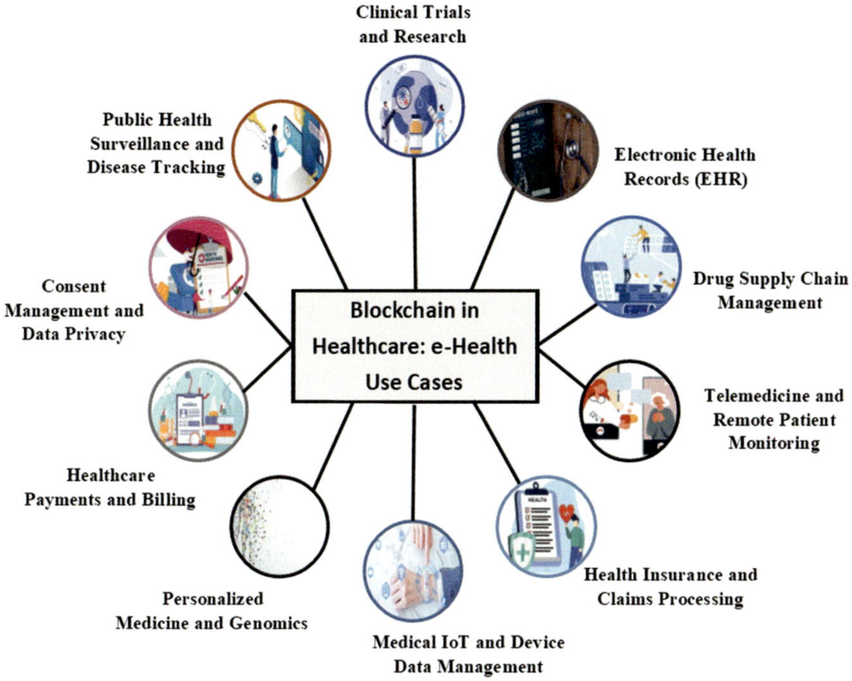

Fig. 1 High-potential blockchain use cases in the healthcare industry

to view and manage their insurance data without any difficulty. Another platform powered by blockchain that focuses on storing patient medical records and other data on the blockchain is MediBloc. MediBloc guarantees the safety and integrity of patient data by utilizing the built-in security features of blockchain technology. In addition to serving as a safe place to store medical documents, the platform also provides users with assistance and other services like medical consultations and insurance alternatives. By giving patients access to a complete platform that satisfies their medical requirements, data security concerns, and access to multiple healthcare providers, this integrated approach improves the overall healthcare experience. The transformational potential of blockchain technology in healthcare is highlighted by both Patientory and MediBloc. These systems improve data security and privacy, give patients authority over their health information, and provide individualized medical services and insurance alternatives by utilizing the blockchain's decentralized and transparent nature. These developments support better patient outcomes and experiences by developing a healthcare ecosystem that is more patient-centric and effective. The review articles by Shah and Garg [21] and Pegoraro et al. [22] give a general summary of how digital technologies have affected clinical trials. The researchers talk about how to better data collecting, patient monitoring, and remote data analysis by wearable technology, mobile health applications, and electronic data recording. They draw attention to the difficulties and possibilities that digital technology brings for improving the efficacy and efficiency of clinical trials. A systematic study that looks at the ethical issues in clinical research was done by Alemayehu et al. [23] and de Jongh et al. [24]. Key difficulties such as informed permission, data privacy, participant recruitment, and conflict of interest are identified by the researchers through analysis of diverse studies and ethical frameworks. They shed light on the moral dilemmas that confront scientists and make suggestions for upholding moral principles in clinical trials. A study on novel methods for patient recruitment in clinical trials was undertaken by Peipert et al. [25]. To reach a larger and more varied patient group, the researchers talk about using social media, internet platforms, and tailored advertising. They discuss the benefits and drawbacks of each approach to patient recruitment and offer doable suggestions for enhancing patient recruitment and retention in clinical trials. An investigation of the regulatory frameworks governing clinical trials in various nations was done by Welch et al. [26]. The researchers examine the legal procedures, moral standards, and times for approval in various countries. They talk about the regulatory standards' parallels and variations and provide information on the difficulties and chances associated with performing international clinical trials.

3.2 Electronic Health Records (EHR) Management

Blockchain improves EHR systems by collecting pertinent data, identifying mistakes, and automating repetitive processes. The decision-making process enhances data accuracy, enables data integration, and helps with clinical documentation and coding. For instance, Ciox Health employs blockchain to increase the effectiveness of medical record retrieval. Their AI-powered technology examines EHRs to find pertinent patient records, saving time and effort compared to manual record retrieval. Studies on the difficulties and factors to be taken into account when integrating blockchain technology into EHR systems have been done by researchers [27, 28]. Scalability, interoperability, regulatory compliance, and data governance are just a few of the important topics they explore. The authors address the challenges of implementation by offering insightful research into potential fixes and suggestions for the successful integration of blockchain in EHR administration. Researchers [10, 29] launch Medledger, a blockchain-based platform created with the express purpose of connecting EHR and pharmaceutical supply chain management systems. The benefits of using blockchain technology to guarantee the authenticity and traceability of pharmaceutical items are discussed by the writers. Additionally, they go through how Mediledger may improve patient safety and drug tracking while addressing important facets of pharmaceutical supply chain integration. The systematic review of the use of blockchain technology for EHR interoperability is carried out by Schmeelk et al. [28] and Maitra et al. [30]. The authors analyze the potential of blockchain in promoting smooth data interchange and sharing among healthcare institutions via the examination of numerous research and projects. In tackling interoperability issues within EHR systems, the assessment highlights the importance of blockchain, which presents a viable alternative for enhanced data interoperability and cooperation in healthcare settings.

3.3 Drug Supply Chain Management

Blockchain enhances supply chain transparency and traceability for pharmaceuticals. It makes it possible to follow a drug's path from its maker to the patient, lowering the chance that it will be counterfeited and verifying the genuineness and safety of medicines. An illustration of blockchain-based medicine supply chain management is IBM's medicine Supply Chain Integrity Solution. The application of blockchain technology in managing drug recalls was the subject of a systematic review by Han et al. [31]. They looked at numerous research papers and initiatives that investigated how blockchain technology may be used to enhance the speed and accuracy of drug recall procedures. The assessment focused on how blockchain technology may improve coordination, communication, and traceability during medication recall incidents.

Blockchain technology was the subject of a study [32] that concentrated on safe medicine authentication. They suggested a blockchain-based system that guarantees the integrity and validity of pharmaceutical items all the way through the supply chain. The benefits of adopting blockchain to thwart medicine fraud and raise patient safety were outlined by the researchers. Additionally, they implemented the suggested system as a proof-of-concept and assessed it. Clauson et al. [33] gave a summary of the application of blockchain technology in the management of the medication supply chain. The advantages of using blockchain to improve traceability, transparency, and security in the pharmaceutical supply chain were outlined by the researchers. They looked at a few blockchain applications, including supply chain visibility, inventory management, and product verification. The report also discussed the difficulties and potential outcomes of using blockchain technology in the management of the medication supply chain. The difficulties and factors to be considered while integrating blockchain technology into the pharmaceutical supply chain were examined by Humayun et al. [34]. The researchers looked at things including scalability, regulatory compliance, stakeholder confidence, and data interoperability. They offered perceptions of prospective remedies and suggestions for the effective application of blockchain in the management of the medication supply chain.

The relevance of blockchain in allowing track and trace systems in pharmaceutical supply chains was investigated by researchers [35, 36]. The benefits of using blockchain to provide real-time transparency and accountability in drug trafficking were outlined by academics. To integrate blockchain into current supply chain management systems, they examined the implementation problems and potential solutions. Collectively, these research projects advance our knowledge of the potential uses of blockchain technology in the management of the medication supply chain.

They address a few issues, including managing medicine recalls, supply chain problems, secure authentication, transparency, and traceability. These research results and insights offer important information and pointers for developing blockchain technology's use in the pharmaceutical sector.

3.4 Telemedicine and Remote Patient Monitoring

For telemedicine consultations and remote patient monitoring, blockchain makes it possible to securely and decentrally share patient health data. Between patients and healthcare providers, it assures data protection, interoperability, and consent management. Blockchain telemedicine platform examples include Gem's GemOS. Using a literature study, Liu et al. [37] created a summary of telemedicine and remote patient monitoring in the medical field. They investigate the advantages of using these technologies as well as the difficulties that may arise. The researchers look at a few studies and programs that have looked into how telemedicine and remote patient monitoring might be used to increase patient outcomes, cost-effectiveness, and access to healthcare. The paper emphasizes how telemedicine and remote patient monitoring have the potential to revolutionize healthcare delivery by increasing accessibility, better patient outcomes, and maximizing resource use. A study report on the ethical issues of telemedicine and remote patient monitoring was released by Field and Grigsby [38]. They examine the ethical issues raised by telemedicine activities in relation to patient autonomy, data security, privacy, and informed permission. To guarantee responsible and patient-centered use of telemedicine and remote patient monitoring, the researchers talk on the significance of ethical frameworks, rules, and policies. The study offers insights into the most effective ways to uphold moral norms in telemedicine, safeguard patient rights, and preserve patient-provider confidence.

A study was undertaken by researchers [39] to look into the use of telemedicine in rural and underserved regions. The difficulties that patients and healthcare professionals experience while trying to receive medical care in rural areas are highlighted by the study. They talk about how telemedicine might improve access to treatment, lower healthcare inequities, and close the healthcare gap in rural regions. The paper offers suggestions for the effective adoption and implementation of telemedicine in rural healthcare settings, taking into account issues with infrastructure, education, and funding. A literature study on the financial effects of telemedicine and remote patient monitoring was undertaken by Lloyd and Lee [40] and Niu et al. [41]. The economic effects, cost-effectiveness, and reimbursement practices of utilizing these technologies in healthcare are all examined by the researchers. From an economic standpoint, they talk about the difficulties and advantages that telemedicine and remote patient monitoring have for healthcare practitioners and systems. The analysis sheds light on the advantages and factors that should be taken into account in terms of cost reductions, resource allocation, and long-term financial viability.

3.5 Health Insurance and Claims Processing

Blockchain simplifies procedures related to health insurance, such as fraud detection and claims handling. Insurers, healthcare providers, and patients may share patient data in a safe and transparent manner, which lowers administrative costs and boosts productivity. Health insurance options based on blockchain have been investigated by Aetna and Humana. A literature review that [42] did gives an overview of health insurance and claims processing in the healthcare sector. The difficulties and inefficiencies in the current health insurance systems and claim-processing practices were investigated by the researchers. They talked about how new technology, shifting government regulations, and policy changes might affect how quickly, accurately, and transparently health insurance and claims are processed. To promote affordable and patient-centered healthcare, the assessment also looked at the possible advantages of value-based payment models and alternative payment strategies. A study on the application of blockchain technology to the processing of health insurance claims was released in 2021 by Abekah-Nkrumah et al. [43]. They talked about the usage of smart contracts, claims adjudication, and safe, transparent transactions on the blockchain in the insurance sector. The potential of blockchain to lower fraud, simplify billing procedures, and increase the precision of claims processing was investigated by the researchers. The study highlighted the potential for blockchain to alter existing procedures by shedding light on the opportunities and problems related to its implementation in health insurance and claims processing.

Electronic Health Records (EHRs) and systems for processing health insurance claims were the subject of a study by researchers [44, 45]. The researchers looked at the possible advantages of integrating EHR data with claims processing practices, including strengthening payment accuracy, lowering billing mistakes, and increasing coding accuracy. Data privacy, interoperability, and regulatory compliance were some of the topics they covered while talking about the difficulties and factors involved in integrating EHRs and claims processing systems. The study shed light on the possibility of EHR integration in expediting the handling of health insurance and claims, resulting in more accurate and efficient billing operations.

A study on the use of data analytics in the processing of medical claims was published by a researcher [46]. The researchers looked at how data analytics methods may be used to examine huge amounts of insurance data, find fraud, and enhance billing accuracy. They emphasized how data analytics may improve revenue cycle management, spot billing mistakes, and boost financial performance in healthcare enterprises. The study revealed the prospects and difficulties of using data analytics in health insurance and claims processing, highlighting its potential to raise the general efficacy and precision of these procedures. A study of the literature was undertaken by Ho et al. [47] with a focus on the changing environment of health insurance and claims processing. The researchers looked at how technology improvements, payment reform initiatives, and legislative changes affected health insurance systems and claims processing practices. They talked on the transition to value-based payment models, the development of alternative payment options, and the significance of pricing transparency in the healthcare industry. The review offered vital light on the reasons underlying the evolution of these processes while revealing patterns and future directions in health insurance and claims processing.

3.6 Medical IoT and Device Data Management

The integrity and security of data produced by medical Internet of Things (IoT) devices are improved by blockchain. It allows for the safe storage and transfer of device data, preserving data security and integrity, and promoting device interoperability. One example of a blockchain-based system for managing medical IoT data is Chronicled's MediLedger. A literature study that offers an overview of Medical IoT (Internet of Things) and device data management was undertaken by Baumfeld Andre et al. [48] and Pradhan et al. [49]. The adoption of IoT devices in healthcare and its effects on data management procedures were investigated by the researchers. They talked about how IoT can improve remote patient monitoring, enable real-time data collecting, and monitor patient health. The evaluation also covered the issues with interoperability, data integration, data security, and privacy in medical IoT contexts. A study article concentrating on the security and privacy issues in Medical IoT and device data management was released by Frikha et al. [50] and Azbeg et al. [51]. Their analysis focused on the security and privacy of patient data as well as other risks and vulnerabilities linked to the usage of IoT devices in healthcare. To protect patient data in Medical IoT contexts, the researchers talked about the significance of secure communication protocols, data encryption, authentication systems, and access control. The study included recommendations for the safest ways to protect the privacy and security of medical IoT data. The interoperability issues in medical IoT and device data management were investigated in a study by researchers [52]. The necessity for standardized communication protocols and data formats was investigated to facilitate smooth integration and data sharing across various IoT devices and systems. To facilitate data interoperability in Medical IoT contexts, they explored the role of healthcare standards and interoperability frameworks. To facilitate efficient data exchange and usage throughout the healthcare ecosystem, the research offered solutions for tackling interoperability difficulties in device data management. The application of data analytics in medical IoT and device data management was covered in a paper by Abounassar et al. [53]. The researchers investigated the use of data analytics tools to examine the massive amounts of data produced by IoT devices in the healthcare industry. They talked about how data analytics may be used to identify abnormalities, forecast health outcomes, and extract insightful information. The study demonstrated how data analytics may enhance decision-making, customized medicine, and patient outcomes in Medical IoT contexts, allowing healthcare practitioners to get insightful knowledge and enhance patient care. A literature study on the ethical issues of medical IoT and device data management was undertaken by Pradhan et al. [49]. The researchers looked at the ethical issues surrounding data privacy, permission, ownership, and openness in IoT contexts for medical purposes.

To ensure the acceptable and ethical use of Medical IoT data, they emphasized the significance of informed permission, data governance systems, and ethical norms. The evaluation shed light on the ethical implications and issues that should be considered while managing device data in the healthcare industry, placing emphasis on the necessity of giving patient privacy and data protection a priority.

3.7 Personalized Medicine and Genomics

Blockchain enables the safe management and exchange of genetic data for individualized medication. While protecting privacy and data ownership, it enables patients to manage their genetic data, share it with researchers, and take part in genomic research projects. An illustration of a blockchain-based genomics platform is Nebula Genomics. Rajeswari and Ponnusamy [54] and Hong and Oh [55] did a study of the literature to give an overview of genomics and personalized medicine. The use of genetic information to customize medical interventions and therapies for specific individuals was investigated by the researchers. The development of genomic technology, such as next-generation sequencing, and its implications for personalized treatment were highlighted. The study looked at several research and programs that addressed the use of genetics in illness diagnosis, prognosis, and therapy choice, underlining the potential of genomics to change healthcare by enabling more targeted and precise treatments. Beccia et al. [56] and Offit [57] released a study on the ethics of genetics and customized medicine. They looked at the moral issues raised by genetic discrimination, data privacy, and genetic testing. In order to ensure the responsible and fair use of genetic data in personalized medicine, the researchers stressed the significance of ethical frameworks, rules, and regulations. The study highlighted the necessity for patient autonomy, well-informed decision-making, and privacy protection while offering insights into best practices for upholding ethical norms and defending patient rights in the context of genomics. A study on the use of genomes and clinical data for customized medicine was carried out by Santaló and Berdasco [58]. They looked at the difficulties and possibilities of fusing genetic data with clinical data and electronic health records. The researchers talked about how integrated data analysis may help with illness risk assessment, therapy response assessment, and medication development. To enable the application of genomics in personalized medicine and enable the smooth integration of genetic information into clinical practice, the study underlined the necessity for interoperability and data integration standards. A study of the literature was undertaken by researchers [59] with a focus on the financial effects of genomics and customized medicine. They looked at the economic effects, cost-effectiveness, and reimbursement practices of integrating genomics into clinical practice. From an economic standpoint, the researchers explored the difficulties and advantages of implementing customized medical techniques, taking into account things like healthcare costs, resource allocation, and reimbursement models. In order to maximize the value of genomic medicine, the study offered insights into the possible advantages and considerations for adopting personalized medicine and genomics in healthcare systems. It also highlighted the necessity for evidence-based decision-making and effective resource allocation.

3.8 Healthcare Payments and Billing

Blockchain reduces administrative expenses and eliminates middlemen in the billing and payment of medical services. It makes it possible for payers, providers, and patients to transact in a secure and open manner, increasing productivity and lowering billing mistakes. A blockchain-based platform for healthcare provider payments and credentialing is called ProCredEx. A literature study undertaken by Veenstra et al. [60] offers a summary of the procedures for paying for and billing for medical services. The difficulties and inefficiencies in the present healthcare payment systems were investigated by the researchers. They talked about how electronic payment systems and blockchain, two new technologies, can increase the speed, accuracy, and security of healthcare transactions. The review also looked at the potential advantages of value-based payment models and alternative payment strategies in promoting affordable and patient-centered healthcare, highlighting the need for creative methods of billing and reimbursement to better align incentives and advance the entire healthcare system. A study on the use of blockchain technology in healthcare billing and payments was released by Meyer [61]. They talked about smart contracts, the processing of medical claims, and the usage of blockchain for safe and transparent transactions. The possibility of using blockchain to lessen fraud, reorganize billing procedures, and boost claims adjudication accuracy was investigated by the researchers. The study highlighted the significance of data integrity, privacy, and security in the digital healthcare ecosystem and offered insights into the prospects and problems of integrating blockchain in healthcare payments. The integration of electronic health records (EHRs) with healthcare payment and billing systems was the subject of research by researchers [62]. They looked at the possible advantages of integrating EHR data with billing procedures, including increased coding precision, decreased billing mistakes, and improved reimbursement accuracy. The researchers talked about interoperability, data privacy, and regulatory compliance as well as the difficulties and factors to be considered when integrating EHRs and billing systems.

To increase billing efficiency and accuracy, the study emphasized the necessity for seamless data connectivity between clinical and financial systems. An essay on the application of data analytics in healthcare billing and payments was written by Britton [63]. They looked at ways to evaluate billing data, spot fraud, and increase billing accuracy using data analytics approaches. The advantages of data analytics in enhancing revenue cycle management, detecting billing problems, and enhancing financial performance in healthcare businesses were underlined by the researchers. The study shed light on the difficulties and possibilities of using data analytics for healthcare billing and payments, with particular emphasis on the necessity of strong data governance and analytical skills in managing financial operations. A literature study on the changing nature of healthcare billing and payments was undertaken by Yaqoob et al. [64]. They looked at how technology developments, legislative changes, and payment reform initiatives affected healthcare payment systems. The researchers talked about how healthcare is moving toward value-based payment models, the emergence of alternative payment options, and the significance of pricing transparency. The analysis highlighted the need for flexibility and creativity to suit the shifting demands of the healthcare business and offered insightful information on current trends and potential directions for healthcare payments and billing.

3.9 Consent Management and Data Privacy

Blockchain allows for granular consent management and gives individuals authority over their health data. Patients have more control over who has access to their sensitive information and can give or cancel access to their data, preserving privacy. An illustration of a blockchain-based consent management system is the Health Nexus platform. In a literature study, Kommunuri [65] presented an overview of permission management and data privacy in healthcare. The significance of informed consent in preserving patient confidentiality and privacy was investigated by the researchers. They talked about the difficulties and factors to be considered when handling consent, as well as consent models, consent documents, and patient involvement. The analysis highlighted the necessity for enterprises to adhere to legal obligations and safeguard patient rights by examining the effects of data privacy laws like GDPR and HIPAA on consent management procedures in healthcare. A study on the application of blockchain technology to consent management and data privacy was released in 2019 by Asghar et al. [66]. They talked about how to leverage blockchain for data exchange, auditing, and safe, open consent management. The researchers looked at how blockchain may improve data privacy procedures and increase patient ownership over their health data.

In addition to demonstrating the value of decentralized and verifiable permission systems in preserving patient confidence, the study offered insights into the difficulties and prospects of integrating blockchain in consent management and data protection. In research published in 2023, Rantos et al. [67] looked at patient involvement in consent management and data privacy. They looked at how crucial it is for consent procedures to involve patients in decision-making and empower them. The necessity for clear and easily available information was emphasized as the researchers explored the difficulties and factors to be considered when properly conveying privacy rules and consent alternatives to patients. The study offered insights on tactics for increasing openness and patient-centered treatment while also promoting patient engagement and improving data privacy standards. A paper on the effects of data privacy laws on consent management in healthcare was written by Kommunuri [65]. They looked at the demands and effects of laws like the GDPR, HIPAA, and CCPA on consent procedures. The researchers highlighted the necessity for businesses to set up strong consent frameworks and privacy policies as they explored the difficulties and factors to be considered when coordinating consent management procedures with legal obligations. By highlighting the significance of privacy by design and data governance, the study offered insights into best practices for ensuring compliance and sustaining data privacy in healthcare settings. A literature study on consent management and data privacy ethics was undertaken by Maher et al. [68]. Informed permission, data sharing, secondary use of data, and data ownership were all topics of discussion by the researchers. They talked about the value of ethical guidelines, procedures for getting informed consent, and openness in preserving patient autonomy and safeguarding data privacy. The assessment shed light on the ethical issues and concerns surrounding permission management and data protection, highlighting the necessity for companies to give patient rights and moral behavior priority in their data management procedures.

3.10 Public Health Surveillance and Disease Tracking

Blockchain enhances disease monitoring and public health surveillance by safely gathering and distributing epidemiological data. It enables safe data exchange between healthcare institutions and public health organizations, real-time monitoring, and early disease epidemic identification. Blockchain technology has been applied to public health surveillance through Solve.Care platform. A literature study by Martin et al. [69] gave an overview of disease monitoring and public health surveillance. The significance of surveillance systems in observing and limiting the spread of infectious illnesses was investigated by the researchers. They reviewed several public health monitoring techniques and tools, including electronic health records, syndromic surveillance, and data fusion.

The evaluation emphasized the necessity of effective surveillance systems to safeguard people's health by highlighting the importance of monitoring in early detection, response planning, and public health interventions. A study concentrating on the application of big data analytics in public health surveillance was released by Zeng et al. [70]. They presented how to analyze sizable datasets for disease monitoring using data mining, machine learning, and predictive modeling approaches. The advantages of big data analytics in detecting epidemics, identifying risk factors, and enhancing public health decision-making were investigated by the researchers. The study highlighted the potential for data-driven techniques to improve public health monitoring skills while underlining the limitations and opportunities of using big data analytics in disease tracking.

An investigation of the use of mobile health technology for illness tracking and surveillance was done by Khoury et al. [71]. They looked at how mobile apps, wearable technology, and sensors may be used to gather real-time health data for monitoring. The researchers talked about how mobile health technology may improve data collecting, track illness patterns, and encourage people to report their own health conditions. In addition to highlighting the necessity for privacy and security measures to secure sensitive health data, the study offered insights into the difficulties and factors to be considered when using mobile health technology for illness tracking. In a 2018 publication, MacKinnon and Brittain [72] discussed the significance of data security and privacy in public health monitoring.

They looked at the issues and factors involved in safeguarding private health information gathered for monitoring. The ethical and legal concerns around data privacy, permission, and sharing in public health surveillance were highlighted by the researchers. The report included recommendations for the best ways to maintain data security and privacy in surveillance systems, highlighting the necessity of extensive privacy regulations, safe data storage, and stringent access restrictions. Iwaya et al. [73], Aiello et al. [74] and Chiou et al. [75] performed a literature analysis with an emphasis on the moral issues involved in disease monitoring and public health surveillance. The researchers looked at the moral issues surrounding data sharing, privacy, informed permission, and public health measures. To ensure responsible and ethical surveillance techniques, they talked about the significance of ethical frameworks, openness, and community involvement. The assessment highlighted the need to strike a balance between public health goals, individual private rights, and ethical standards by offering insightful information on the ethical implications and concerns of public health surveillance.

4 Artificial Intelligence in Healthcare: e-Health Use Cases

AI in healthcare, as shown in e-Health use cases, offers significant potential for transforming the industry. Figure 2 showcases the high potential of AI use cases in the healthcare industry.

4.1 Public Health Surveillance and Disease Tracking

e-Health uses artificial intelligence (AI) to help in medical condition diagnosis. Large patient data sets, including medical history, symptoms, and test results, may be analyzed by machine learning algorithms to produce precise and rapid diagnoses. AI-based diagnostic tools can aid medical personnel in improving patient outcomes and decision-making. Researchers [76] ran two online tests to determine if people should trust AI to make medical diagnoses. For high-risk disorders, participants trusted human doctors more than AI doctors. Surprisingly, participants' expectations of AI and human doctors were similar, and their level of confidence did not rise when AI performed better than humans. However, when participants were given the option of choosing between human and AI doctors, their trust in AI diagnoses considerably improved, highlighting the significance of patient control in fostering trust. The use of molecular imaging in identifying and treating inflammatory illnesses was highlighted by Juravle et al. [77]. Diabetes, cardiovascular disease, and immune-mediated inflammatory disorders are just a few of the ailments that include inflammation as a major contributing role. However, molecular imaging of inflammation

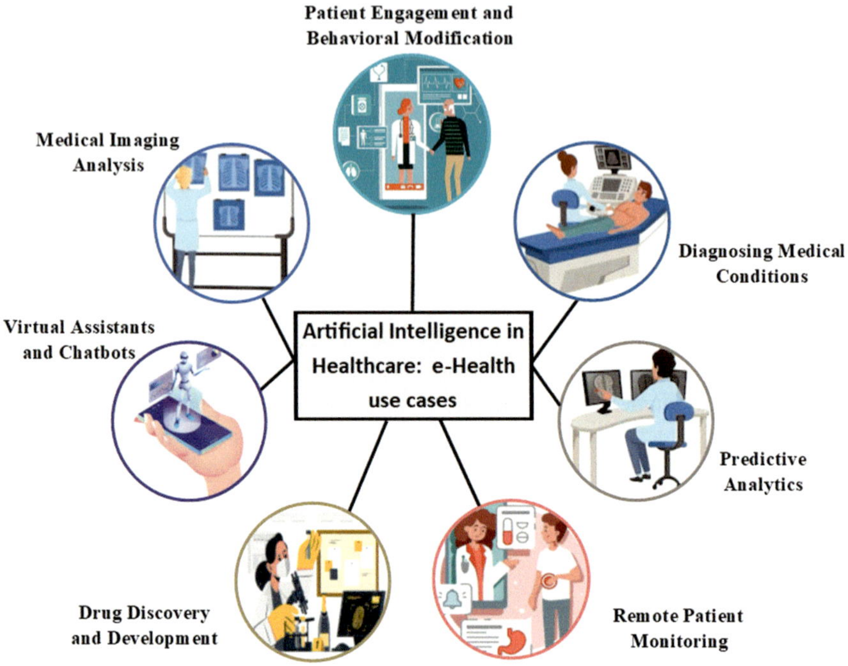

Fig. 2 AI high potential of use cases in the healthcare industry

is not frequently performed using existing clinical imaging methods. To improve illness diagnosis, direct therapeutics, and identify new research possibilities, the review investigates the potential of known and developing imaging methods to image inflammatory biomarkers, pathways, and cells. MacRitchie et al. [78] examined how data science and digital technology have revolutionized diagnostic categorization, with an emphasis on autism.

The study emphasizes data-driven approaches, including computer vision, feature selection algorithms, and early indicators of autism. Concerns about privacy and the use of people in the diagnosis process are also covered. The evaluation also covers digital therapy treatments for data collection and result monitoring, paving the way for more accurate and scalable psychiatric diagnostic methods. Washington et al. [79] updated summary of categorization, etiopathogenesis, diagnosis, and therapy for insomnia problems. Hyperarousal driven by a number of variables characterizes insomnia, which is categorized as an "insomnia disorder." It is divided into chronic, transient, and other subtypes. Modified diagnostic criteria are needed, as well as an assessment of related variables and co-existing disorders. Pharmacotherapy and cognitive-behavioral therapy for insomnia (CBTI) are both available as treatments. The use of phytochemicals and complementary alternative medicines (CAMs) such as digital cognitive behavioral therapy and mindfulness-based techniques has shown potential. Paul et al. [80] examined prenatal diagnostic methods' ethical considerations, focusing on preimplantation genetic testing (PGT). PGT gives those who are more at risk the chance to examine embryos for genetic abnormalities. PGT raises questions about embryo safety and increased expenses even while it improves embryo selection and reproductive results. The paper examines the obstacles, ethical ramifications, and demand for patient-centered decision-support technologies. It highlights the value of making informed decisions, considering the concerns of patients, and addressing ethical issues during counseling sessions.

4.2 Predictive Analytics

Predictive analytics, which uses previous data to estimate future health outcomes, is made possible by AI in e-health. AI algorithms may identify people who are at risk of contracting diseases by examining patient data such as demographics, lifestyle variables, and genetic markers. Early intervention and individualized preventative care are both made possible by this proactive strategy. To address the restricted use of radio frequency identification (RFID) technology in the healthcare supply chain, researchers [81] performed a study. The study sought to forecast RFID adoption by combining the unified theory of acceptance and use of technology (UTAUT) with individual variations, such as personality traits and demographic variables. Neural network analysis was used to gather and examine data from 252 doctors and nurses.

The study suggested 11 criteria, with an emphasis on individual characteristics, to forecast RFID adoption. The results showed that, in comparison to factors generated from UTAUT, individual differences were more useful in predicting RFID adoption. This study stresses the significance of considering individual aspects in technology adoption and advances our understanding of RFID acceptability in the healthcare sector. The hard issue of customer churn prediction (CCP) in the telecom business was tackled by Chong et al. [82]. Six steps make up their suggested methodology: pre-processing the data, feature analysis, gravitational search feature selection, and separating the data into train and test sets. On the train set, several prediction models were applied, including boosting and ensemble methods, logistic regression, naive Bayes, support vector machines, random forests, and decision trees. For hyperparameter tuning, K-fold cross-validation was utilized, and the AUC curve and confusion matrix were used to assess the outcomes. Adaboost and XGBoost classifiers outperformed other models, achieving the greatest accuracy of 81.71 and 80.8%, respectively, with an AUC score of 84%. This work advances machine learning methodologies for forecasting customer turnover in the telecom sector. A predictive model for the discontinuation of antihyperglycemic medication in patients with type 2 diabetes after laparoscopic metabolic surgery was created and validated by Coussement et al. [83].

In two big US healthcare databases, they employed machine learning techniques and a shared data model. The model's ability to help with patient selection and enhance outcomes was demonstrated by its high accuracy in predicting drug discontinuation. However, such models would need to be implemented in real-world decision support, which would require a sufficient technology foundation. To create predictive models for clinical decision-making, this study shows the potential of machine learning with real-world healthcare data. Birth Match is described by Johnston et al. [84] as an innovative policy approach to reduce newborn maltreatment by utilizing data systems to forecast future risk. With this method, a child protection response is started using information from birth certificates and child welfare records. The research explores the moral implications of Birth Match and emphasizes the significance of openness and responsibility in its execution. While technology has the potential to stop newborn maltreatment that is deadly, ethical issues and trade-offs need to be carefully considered. The study highlights the requirement for moral frameworks to direct the use of policy innovations and guarantee the avoidance of baby abuse. Lanier et al. [85] investigates the use of artificial intelligence and machine learning to detect high-risk suicidal patients using predictive analytics. They stress the necessity for extensive, sensitive patient data to underpin difficult medical choices. To illustrate its difficulties, the research compares suicide prediction to non-medical and medical forecasts. A risk–benefit paradigm is used to examine clinical and ethical issues, with an emphasis on the possible drawbacks of misclassifying suicide risk. The authors urge a thorough evaluation of the hazards and advantages in healthcare populations and offer useful guidelines to safeguard patient rights and improve the therapeutic value of suicide prediction analytics technologies.

4.3 Patient Monitoring

Remote patient monitoring systems use AI technology to collect and analyze patients' vital signs and health data in real time. Wearable technology using AI algorithms and sensors can identify unusual trends, notify healthcare professionals, and enable quick treatments. AI-powered remote patient monitoring increases accessibility to healthcare, promotes disease management, and lowers hospital readmissions. For instance, Biofourmis, a firm that develops digital therapies, employs AI to monitor patients with heart failure. Their wearable gadget gathers physiological data, which AI systems then analyze to forecast and identify heart failure exacerbations, allowing for prompt therapies. Artificial intelligence (AI) is discussed in papers by researchers [86, 87] that include telemedicine and remote patient monitoring. The researchers look at how AI algorithms may be used to interpret data from remote patient monitoring, speed up diagnostic choices, and enhance individualized patient care. They look at how AI may improve telemedicine practices' effectiveness, accuracy, and patient outcomes. The study sheds light on the difficulties and possible possibilities of using artificial intelligence (AI) in telemedicine and remote patient monitoring, emphasizing how AI could change healthcare delivery and enhance patient care. A thorough analysis of remote patient monitoring (RPM) systems in healthcare is provided by Jeddi and Bohr [87] with an emphasis on the use of artificial intelligence (AI). RPM is a useful tool for keeping tabs on patients in a variety of contexts, and AI has the ability to improve it. The paper examines the effects of AI on RPM, taking into account cutting-edge technological applications, difficulties, and new trends.

The assessment has a focus on patient-centric RPM systems that make use of wearable gear, sensors, and cutting-edge technologies like blockchain, fog, and edge. In several facets of RPM, including activity categorization, chronic illness monitoring, and vital sign tracking, AI plays a crucial role. The results underscore the revolutionary possibilities of AI-enabled RPM, including tailored monitoring, early diagnosis of health decline, and learning human behavior patterns. The paper also covers the difficulties and implementation problems that come with incorporating AI into RPM systems and offers predictions about the future uses of AI in RPM applications based on new developments and issues. Shaik et al. [88] emphasizes the Internet of Things' (IoT) considerable influence on healthcare, notably in the tracking of important health indicators. IoT devices have completely changed how health monitoring is done by measuring factors like blood pressure, body temperature, pulse rate, and blood oxygen saturation. Additionally, these gadgets pick up on physical movements to spot dangers like falls and injuries. For such gadgets, mobility, minimal weight, and user-friendliness are essential design elements. An inbuilt CPU is used in the system's architecture to pre-process sensor signals and gather data. A shared cloud platform is used to execute feature extraction, recognition algorithms, and the data presentation of vital indicators, along with emergency call-outs as needed. Elango et al. [89] examines how machine learning (ML) and artificial intelligence (AI) are transforming social media in the healthcare industry. Telehealth, remote patient monitoring, and general well-being may all benefit from the appropriate management of the massive amounts of data created on social media platforms thanks to AI and ML algorithms. The paper identifies key trends in the adoption of AI-ML, such as the use of sentiment analysis for improved social media marketing, the use of social media as a tool for data collection with privacy protections, and the use of chatbots and personalized content to build long-term relationships with stakeholders.

In the context of telehealth and remote patient monitoring, the study identifies research gaps, offers a conceptual framework to maximize AI-ML application, handle ethical issues, and counteract false information on social media platforms. The research focused on deploying wearable Internet of Things devices to monitor COVID-19 patients was carried out by Leung [90]. The technology monitors vital signs and uses real-time GPS data to notify medical authorities of potential confinement breaches. A layer of wearable IoT sensors, a layer of mobile Android applications for alerts, and a cloud layer for data processing make up the proposed system's three tiers. The paper also presents a CNN-UUGRU deep neural network model for identifying human activities that outperforms other models on the Kaggle dataset with accuracy, precision, and F-measures of 97.75, 96.8, and 97.8%, respectively.

4.4 Drug Discovery and Development

AI is essential for expediting the processes of drug discovery and development. Large datasets of genomic and proteomic data may be analyzed by machine learning algorithms to pinpoint new therapeutic targets and forecast the potency of medication candidates. AI-powered platforms can cut expenses, accelerate the creation of novel drugs, and streamline the drug discovery process. Researchers reviewed the most recent approaches to medication development for uncommon disorders [91]. There is a need to close the gap between fundamental research and therapeutic therapies since there are millions of people in the United States who suffer from one of the 7,000 rare illnesses that are known to exist. The review focuses on whole genome sequencing and pharmacogenetics for determining causes and creating therapies for genetic uncommon disorders. High throughput screening, medication repurposing, and the utilization of biologics like gene therapy and recombinant proteins are all covered in this article. Explored are many disease models, such as induced pluripotent stem cells and animal models. The importance of biomarkers in the development and discovery of new drugs is also underlined. Sarkar et al. [7] emphasizes how Artificial Intelligence (AI) has the potential to revolutionize drug research. In computer-facilitated drug development, the application of machine learning, in particular deep learning (DL), in conjunction with massive data and improved processing capacity, has shown promising outcomes. Artificial neural networks, for example, enable the automated extraction of characteristics from input data and the establishment of nonlinear connections.

The early pessimism regarding the use of AI in pharmaceutical discovery is dissipating since it is anticipated that AI would speed up the search for new and better medications when combined with contemporary experimental approaches. The growth of AI-driven drug discovery will be greatly aided by technological breakthroughs and open data sharing. In this paper, the potential uses of AI to speed up the drug discovery process are examined. Various methods used to increase pharmaceutical research and development success rates are discussed in Sun et al. [92]. The review entails a discussion of genomes and proteomics, target-based and phenotypic screening, drug repurposing, collaborative research, underdeveloped therapeutic domains, outsourcing, and artificial intelligence-assisted pharmaceutical modeling.

These methods have improved target identification and validation, rational drug design, cost, and time savings, higher returns on investment, cross-fertilization of ideas, resource sharing, niche drug discovery, and effective computer-aided drug design, all of which have resulted in the discovery of successful drugs. The application of these tactics, either singly or in combination, has the potential to stimulate pharmaceutical research and development. Kiriiri et al. [93] draws attention to the substantial expansion of biological data and the expanding usage of machine learning (ML) and artificial intelligence (AI) techniques in data mining for drug development. Deep learning (DL) is one of the AI techniques that has shown promise in a variety of tasks, including the production of chemical structures, scoring of binding affinities, position prediction, and molecular dynamics.

The paper evaluates the present state-of-the-art in AI-assisted drug development, examines obstacles to be overcome, and tackles problems. It also looks at possible directions for this quickly developing profession in the future. The difficulties in drug design and development, such as low effectiveness, off-target delivery, time requirements, and high costs, are highlighted by Cerchia and Lavecchia [94]. The discipline has undergone a revolution thanks to the introduction of artificial intelligence (AI) and machine learning (ML) technologies, which have made strides in peptide synthesis, virtual screening, toxicity prediction, medication monitoring, and other fields. With the use of data mining and management strategies, AI and deep learning algorithms have shown to be successful in a few drug development procedures. These developments have the potential to have a significant influence on human health and present exciting options for rational medication design. The intricate interaction of genomic, proteomic, and metabolic networks in regulating human biological processes is discussed in Patel and Shah [95]. The discovery of biological processes that are being reprogrammed and the creation of network-based models to comprehend these intricate systems are both made possible by advancements in the omic sciences. A comprehensive interpretation of the system is made possible by the integration of molecular networks from many angles, which offers promise for precision medicine. Pharmaco-omics studies, which combine pharmacology with omics data, can produce treatment plans that are specifically suited to a patient's needs. The review emphasizes the fundamental concepts of network-based systems biology, the incorporation of multi-omics data, and the value of network-based drug discovery techniques for the treatment of breast cancer.

AlzGPS, a complete systems biology tool created to assist in the process of drug discovery and development for Alzheimer's disease (AD), is introduced by Turanli et al. [96]. AlzGPS combines heterogeneous biological networks, multi-omics data, clinical databases, and more to make it easier to identify targets and create AD preventive and treatment plans that work. The platform features clinical trial information, FDA-approved/investigational medication information, literature references for therapeutic candidates, curated AD multi-omics datasets, endophenotype disease modules, and a user-friendly online interface for browsing and displaying data. AlzGPS offers useful methods for advancing AD treatment research by utilizing systems pharmacology and network-based analyses. An overview of machine learning techniques for incorporating multi-omics data in cancer molecular biology investigations may be found in Zhou et al. [97]. The review divides the tools into approaches that are general-purpose and task-specific, including supervised and unsupervised learning methods. Utilizing information from the Cancer Cell Line Encyclopedia, benchmarking is done to assess runtime effectiveness, mean absolute error in drug response prediction, and accuracy of cancer type categorization. The results emphasize the need for continued study and development of the fresh methodology to improve drug discovery, clinical trial design, and individualized therapies. They also provide direction to researchers in choosing effective machine-learning methods for their applications.

4.5 Virtual Assistants and Chatbots

AI-based chatbots and virtual assistants are used in e-health to schedule appointments, respond to questions, and deliver individualized health information. These conversational interfaces utilize machine learning and natural language processing to comprehend user questions and provide relevant answers. Virtual assistants improve accessibility to healthcare services and patient involvement For instance, Buoy Health has created an AI-powered chatbot that assists users in evaluating their symptoms and offers tailored health recommendations. It leads users through a series of questions to identify probable reasons and suggests the best course of action, such as engaging in self-care or seeing a doctor. To pinpoint key trends and knowledge gaps in the area of chatbots and stakeholder interactions, researchers [98] undertook a thorough literature study.

The review analyzed 62 peer-reviewed English articles using a code book and inductive analysis. Findings revealed that existing studies primarily focused on the technical aspects of chatbots and their language skills, with limited consideration for organizational and societal perspectives. The study emphasized the need for corporate communication scholars to contribute more to the discussion of chatbot-stakeholder interactions. The results provide valuable insights into the organizational capabilities and affordances of chatbots, highlighting their importance in stakeholder engagement strategies. Syvänen and Valentini [99] analyzes the advantages, limitations, ethical considerations, and future prospects of ChatGPT and artificial intelligence (AI) in healthcare. ChatGPT, a powerful language model, generates human-like responses using deep learning techniques. 62 Peer-reviewed English publications were examined using inductive analysis and a code book. Findings showed that previous research mostly concentrated on the technical features of chatbots and their linguistic capabilities, with little attention paid to organizational and societal viewpoints. The study highlighted the need for corporate communication experts to add more to the conversation on chatbot-stakeholder interactions. The findings underscore the significance of chatbots in stakeholder engagement strategies by offering insightful information on the organizational capabilities and affordances of these tools. Syvänen and Valentini [99] examines the benefits, drawbacks, moral issues, and potential applications of ChatGPT and AI in the medical field. Deep learning methods are used by ChatGPT, a potent language model, to produce replies that are human-like.

It has several uses in medicine, including supporting patient care, clinical diagnosis, medical education, and research. Copyright infringement, medico-legal problems, and transparency in AI-generated information are examples of ethical challenges, nevertheless. The study analyses these issues, highlighting ChatGPT's promise and difficulties in the healthcare industry. The Leora model is one example of an AI-powered platform that [100] discusses as having the potential to help mental health. These systems, which include conversational agents like Leora, can offer tailored and easily available mental health help to people who are only mildly to moderately affected by anxiety and sadness. However, for the appropriate development and use of AI in mental health treatment, ethical concerns relating to trust, transparency, bias, and potential negative repercussions must be addressed. To guarantee the efficacy of such models, intensive user testing is required for validation. A study was undertaken to determine how eager people are to interact with AI-driven health chatbots [101]. Participants' responses to an online survey and semi-structured interviews were gathered for data. Three themes—"Understanding of chatbots," "AI hesitancy," and "Motivations for health chatbots"—came out of the interviews, stressing worries about accuracy, cyber-security, and the perceived lack of empathy in AI-led services.

The poll found a modest level of acceptance for health chatbots, with characteristics impacting adoption including IT proficiency, attitude, and perceived benefit. Intervention designers should address patients' concerns and improve user experience by utilizing a user-centered and theory-based approach to guarantee effective uptake and use. Nadarzynski et al. [102] addresses the demand for proactive asthma treatment as well as the difficulties in maintaining ongoing asthma monitoring and control in conventional clinical settings. They unveil kBot, a customized chatbot system made to help children with asthma. Through an Android app, kBot continually collects pertinent health and environmental data while tracking medication adherence.

It makes use of contextualization through obtaining patient feedback and domain expertise, as well as customization through surveys and regular interactions. KBot's excellent acceptability and utility during preliminary testing with physicians and researchers suggested that it has the potential to be an important tool in the management of asthma. The goal of [103] is to create a virtual caregiver system for individualized healthcare for the elderly. The system uses a mobile chatbot to communicate with the user and gather details about their physical and emotional well-being. The technology incorporates a rule-based virtual caregiver system dubbed "Mind Monitoring" with physical, mental, and social questions within the chat application, in contrast to conventional health monitoring techniques. The elderly individual answers a question from the chatbot every day by pressing buttons or speaking via a microphone. The technology quantifies the responses, creates illustrative graphs, and offers summaries and recommendations that are personalized for each user. Positive outcomes from an experimental examination with eight senior participants and 19 younger participants over a 14-month period included a response rate above 80% and useful feedback messages. To better examine and enhance health outcomes, interviews were also performed.

4.6 Medical Imaging Analysis

For the examination of medical imaging, artificial intelligence is widely employed in the healthcare industry. Medical pictures like X-rays, CT scans, and MRIs may be analyzed by AI algorithms to help in the identification and diagnosis of a variety of illnesses. AI-powered image analysis tools can accurately identify abnormalities, cancers, and other anomalies, assisting radiologists with their interpretation and enhancing diagnosis efficiency. For instance, Google's DeepMind created an AI system that can examine retinal scans to find signs of eye conditions like diabetic retinopathy. The device helps ophthalmologists diagnose and treat patients in real-world clinics. Deep learning approaches are being looked at by researchers [104] for the analysis of medical pictures and the localization of areas of interest (ROI). Due to these photos' complicated look, conventional techniques are limited. The study focuses on using a powerful deep learning architecture, especially the FCN-UNET approach, to segment MRI images of skin lesions, retinal lesions, and brain lesions. The suggested approach exhibits effective ROI recognition, making it possible to diagnose conditions including skin cancer, eye abnormalities, and brain tumors. Evaluation using databases that are open to the public reveals good accuracy and a dice coefficient that exceeds 90%. Adegun et al. [105] suggests a deep learning-based multi-mode technique for fusing medical images. This technique enables batch processing of diverse multi-modal medical pictures while addressing the drawbacks of MRI, CT, and SPECT image fusion. It enhances time effectiveness, image detail clarity, and fusion effect. Because of quantitative assessment criteria, the suggested method outperforms existing techniques when applied to multimodal medical pictures. This study advances medical picture fusion using deep learning and makes contributions to pattern identification, image processing, and practical implementation in medical diagnostics. A deep learning-based radiomics model named DRTOP is presented by Li et al. [106] for forecasting time-to-event clinical outcomes in patients with lung cancer. DRTOP use raw pictures as opposed to manually created radiomics and determines image-based chances of mortality or recurrence.

The model's predictive ability for outcomes such as overall survival (OS), distant control (DC), and local control (LC) outcomes using computed tomography (CT)-based characteristics was assessed on a dataset of 132 patients. Features from positron emission tomography (PET) were used to predict both overall survival (OS) and recurrence-free survival (RFS). In terms of predicting clinical outcomes, the deep learning-based radiomics model performed better than hand-crafted radiomics, potentially providing advantages in the management of individual cases of lung cancer. The significance of explainable machine learning (XML) techniques in medical image analysis is covered in Afshar et al. [107]. Even though machine learning models can predict clinical parameters with accuracy, their lack of interpretability raises questions. ML algorithms may be explained and important predictors in a prediction can be identified using XML techniques like Shapley values. The integration of XML techniques into medical software solutions might increase confidence in ML predictions and help doctors make wise judgments. The work presents a unique XML imaging pipeline for explaining result predictions from sophisticated medical imaging models using radiomics data and Shapley values. The outcomes are shown on a dashboard designed specifically for clinicians and used to demonstrate how to forecast genetic mutations in glioma patients using MRI data. An overview of the development and uses of deep learning in medical image processing is given in Severn et al. [108]. The paper focuses on the neurological system, cardiovascular system, digestive system, and skeletal system as particular examples of therapeutic applications for convolutional neural network-based methodologies. Although deep learning models have demonstrated remarkable performance, the research notes that the practical usefulness of these models is constrained by the limited size of the medical datasets. In order to advance the area, the future direction advises investigating federated learning, gathering benchmark datasets, and adding domain-specific topic knowledge. Overall, cutting-edge deep learning methods have shown significant promise in the interpretation of medical images, but further development is required to handle data issues and improve scalability.

4.7 Patient Engagement and Behavioral Modification

Artificial intelligence (AI) technologies are used to increase patient engagement and encourage behavioral change for better health outcomes. Applications and platforms driven by AI may give individualized health recommendations, perform focused interventions, and provide feedback and encouragement to people so they follow their treatment programs, adopt better lives, and effectively manage chronic illnesses. The Patient Health Engagement Scale (PHE scale) was produced through thorough conceptualization and psychometric techniques, according to research [109]. The PHE model was used to develop the scale, which was then tested on a large national sample of chronic patients and showed to have strong psychometric characteristics and reliability. A deeper understanding of the role that patient participation plays in healthcare quality, outcomes, and cost containment is made possible by this valid and reliable metric. The PHE scale shows potential in terms of helping to adapt interventions and evaluate improvements following patient engagement efforts. An expert agreement on digital behavior change interventions is presented by Graffigna et al. [110], with an emphasis on conceptualizing and evaluating engagement. The report places more emphasis on "effective engagement" than it does on merely increasing involvement to achieve desired results. It emphasizes the necessity for reliable and effective metrics to create multifaceted models of interaction. In addition, the article advocates for an iterative, user-centered approach to intervention design, combining mixed techniques and qualitative research to fine-tune treatments in accordance with user needs and context.

Yardley et al. [111] emphasizes the significance of taking patient involvement into account while designing rehabilitation trials to enhance the outcomes during the implementation phase. The study suggests that by including patient participation as a key factor in trials, we move from a therapist-focused to a patient-focused strategy. Exercise prescription is used as an example to demonstrate how the Behavior Change Wheel may be used to examine obstacles and facilitators relating to patients' skills, opportunities, and motivations. In order to guarantee realistic, significant, and transferable findings that enable actual effects on rehabilitation treatments at the primary care level, the study suggests a framework based on the COM-B model. The adoption and use of artificial intelligence (AI) technology in healthcare settings are highlighted as being critically important by Zheng et al. [112]. Twelve Canadian patients participated in semi-structured interviews as part of the study to learn more about their opinions on what skills healthcare professionals should possess for the future of AI-enabled healthcare and how to better involve patients in the usage of AI technology. In addition to establishing data governance and validating AI technology, the panelists emphasized the necessity to create trust and patient involvement. To address health disparities and improve the quality of care, the study focused on the role of healthcare providers in including patients.

To capture various patient viewpoints, more study is required. The necessity of comprehending and enhancing adolescent involvement with behavior modification programs is emphasized by Jeyakumar et al. [113]. To evaluate and improve treatments aimed at teenagers, the article advises combining artificial intelligence (AI) and process-level data. The approach that has been developed focuses on four main objectives: assessing engagement, modeling engagement, improving existing interventions, and developing new interventions. The framework is built on the illustration of the INSPIRE narrative-centered intervention for hazardous alcohol consumption. However, when using AI in treatments for children, ethical issues, such as privacy concerns, must be considered. The report ends by outlining the many possibilities for additional research in this quickly changing sector.

5 Blockchain-Based Various Tools Used in Healthcare

Blockchain technology was being explored and implemented in healthcare for various purposes. Here is a list of some existing blockchain-based tools used in healthcare.

5.1 MedicalChain

It is a blockchain-based platform created to overcome the difficulties associated with the safekeeping and exchange of patient data and medical records in the healthcare sector. It makes use of blockchain technology to offer a decentralized and unchangeable mechanism for handling private medical data. Patients may safely exchange their health information with approved healthcare professionals, researchers, and organizations using MedicalChain, giving them ownership over their personal health information. The technology uses the immutability of blockchain to record every transaction and access to medical records, ensuring data integrity, privacy, and transparency. Advanced encryption and smart contracts are used to preserve patient privacy and permission for data access. It makes it easier for healthcare practitioners to interact seamlessly, cutting down on redundant testing and enhancing care coordination. MedicalChain also makes it possible for patients to consult with medical experts securely, remotely, and in real-time through telemedicine services. This function is especially helpful in enhancing accessibility to medical knowledge and lessening the strain of in-person appointments. MedicalChain benefits patients and healthcare providers alike by strengthening data security and expediting medical data sharing, enabling a more effective and patient-centric healthcare ecosystem.

5.2 *SimplyVital Health*

A blockchain-based platform called SimplyVital Health aims to revolutionize healthcare by improving care coordination between healthcare professionals and securely exchanging patient data. The platform uses blockchain technology to build a tamper- and immutable-proof ledger of patient records, assuring the security and integrity of the data. Healthcare practitioners may safely communicate patient data, such as medical records, test results, and treatment plans, with other authorized parties with SimplyVital Health. The smooth coordination of treatment is promoted by this safe and open data interchange, which also improves patient satisfaction. Incorporating smart contracts into the platform also enables healthcare providers to join into value-based care agreements and simplify payment procedures. Smart contracts ensure fast and correct payment for services by automatically executing when certain criteria are satisfied. SimplyVital Health lessens the administrative complexity associated with data exchange and improves data interoperability among healthcare systems by utilizing blockchain's decentralized nature. This results in better patient outcomes, lower healthcare expenses, and greater ecosystem efficiency. SimplyVital Health's overall goal is to improve the healthcare sector by promoting teamwork, data security, and patient-centered treatment.

5.3 *Patientory*

A blockchain-based platform called Patientory focuses on giving patients ownership over their health data while assuring a secure and effective transmission of that information. Patients may save their test results, medical records, and other health-related data on the platform in a decentralized, unhackable way. Patients may use it to provide healthcare practitioners, researchers, and other authorized parties access to their health data, facilitating safe and secure data exchange within the healthcare ecosystem. By utilizing blockchain technology, the platform provides data security, privacy, and integrity, lowering the possibility of data breaches and illegal access. Smart contracts, which allow for automatic and open communication between patients and healthcare professionals, are also incorporated into Patientory. These smart contracts simplify payment procedures and support value-based care agreements, improving care coordination and lowering administrative burdens. Patientory enables people to make educated decisions about their health and take an active role in their care by placing patients at the heart of their health data management. The platform's blockchain-based methodology promotes data interoperability, cooperation, and trust among all stakeholders, improving healthcare outcomes and making the healthcare system more patient-centric.

5.4 Nebula Genomics

Nebula Genomics is a revolutionary platform that leverages blockchain technology to empower individuals with control over their genomic data. Users may choose how they wish to share their genetic data with academics, healthcare professionals, or other interested parties via the platform, which securely stores their data. By offering a decentralized and unchangeable storage solution using blockchain, Nebula Genomics protects the confidentiality and security of its consumers' genetic data. This promotes openness and trust by allowing users to retain ownership of their data while providing authorization for access to parties. The opportunity for consumers to profit from their genetic data is one of Nebula Genomics' distinctive features. Users can receive prizes in the form of cryptocurrency for taking part in research projects or clinical trials, encouraging data sharing while retaining their privacy. The platform also makes it simple for people and research organizations to share data, which speeds up genetic analysis and advances in precision medicine. The blockchain-powered ecosystem of Nebula Genomics intends to transform the genomics sector by fostering data privacy, ownership, and cooperation for a more individualized and efficient approach to healthcare.

5.5 Healthereum

Healthereum is a pioneering blockchain-based platform designed to enhance patient engagement and incentivize positive healthcare actions. The platform uses blockchain technology to provide a safe and open environment where stakeholders such as patients, healthcare professionals, and others may communicate easily. Patients are encouraged to actively participate in their healthcare journey through Healthereum. For fulfilling specific tasks, such as making appointments, following treatment programs, or taking part in health-related surveys, they are rewarded with Halthereum tokens. On the platform, these tokens may be used for a variety of things, like getting access to upscale medical treatment or getting deals on hospital bills. The blockchain-based technology used by Healthereum guarantees the security and immutability of patient data and interactions, fostering a culture of trust. The technology also makes it easier for patients and healthcare professionals to communicate, which improves patient-provider connections and allows for real-time feedback. Healthereum aims to encourage individuals to adopt a more pro-active attitude to their health by gamifying healthcare encounters and offering concrete benefits. This cutting-edge platform's ultimate goals are to increase patient involvement, enhance healthcare results, and promote constructive behavior changes within the healthcare sector.

5.6 *Solve.Care*

Solve.Care is a comprehensive blockchain platform that revolutionizes the way healthcare is administered and coordinated.

It makes use of blockchain technology to raise the level of efficiency, security, and transparency throughout the healthcare industry. By establishing connections between patients, healthcare providers, insurers, and other stakeholders through a decentralized network, Solve.Care's primary goal is to enhance care coordination. All stakeholders engaged in a patient's treatment path may communicate and collaborate easily because to the platform's secure patient data sharing capabilities. By automating activities including appointment scheduling, claims processing, and insurance verifications, Solve.Care simplifies administrative procedures. Through automation, administrative tasks are simplified, mistakes are reduced, and processing times are shortened. SOLVE, the platform's native utility token, is employed to streamline internal rewards and payments. Incentives for following treatment plans may be given to patients, and the token can be used by providers and insurers to speed up payment procedures. Solve.Care establishes data integrity, privacy, and auditability using blockchain technology, fostering confidence in the healthcare system. The platform's cutting-edge method of managing and coordinating care has the ability to greatly raise the standard of healthcare delivery while lowering costs and inefficiencies.

5.7 *MedRec*

MedRec is a blockchain-based system developed by researchers at the Massachusetts Institute of Technology (MIT) to address the challenges in managing electronic health records (EHRs). It makes use of blockchain technology to ensure that patient data is stored securely and impervious to tampering, giving patients and healthcare professionals transparent and auditable access to medical information. The platform improves privacy and security by giving patients more control over their health data and the decision of who may access it. By streamlining data exchange and enhancing interoperability between healthcare organizations, MedRec hopes to produce healthcare services that are more effective and patient-focused (Table 1).

Table 1 Comparative table of blockchain tools support the healthcare system

Platform	Purpose and features	Medical records storage	Data sharing and coordination	Genomic data sharing	Patient engagement	Care coordination
MedicalChain	Blockchain platform for secure storage and sharing of medical records and patient data	✓	✓	✗	✗	✗
SimplyVital Health	Blockchain-based platform for healthcare providers to securely share patient data and coordinate care	✓	✓	✗	✗	✓
Patientory	Blockchain-based platform for patient data management and health information exchange	✓	✓	✗	✗	✗
Nebula Genomics	Blockchain-powered genomic data sharing platform that allows individuals to control and monetize their genetic data	✗	✗	✓	✗	✗
Healthereum	Blockchain-based platform for patient engagement and incentivizing healthcare actions	✗	✗	✗	✓	✗
Solve.Care	Blockchain platform for care coordination, administration, and payments in healthcare	✓	✓	✗	✗	✓
MedRec	Blockchain system developed by MIT for managing electronic health records and providing secure data access	✓	✓	✗	✗	✗

6 Integration of AI and Blockchain in Healthcare

To overcome the issues with both technologies, it has been suggested that blockchain and AI integration in healthcare be used [114]. Data is a major component of how AI systems learn, understand, and make decisions. Machine learning algorithms work best when the data they use comes from a reliable, trustworthy, and secure platform. The blockchain acts as a distributed ledger, guaranteeing that data is kept and shared by all participating nodes in an unhackable and cryptographically secure way. The outcomes may be trusted and validated with high integrity and resilience by using smart contracts for decision-making and analytics using machine learning algorithms. With improved security and dependability, this connection has the potential to transform healthcare data management and analytics. For handling sensitive data gathered, saved, and used by AI-driven techniques, the integration of blockchain with AI offers a safe, unchangeable, and decentralized solution [115]. The medical, personal, banking, financial, trade, and legal sectors, among others, all benefit greatly from these developments in data and information security [116]. According to Mamoshina et al. [117], this integration enables intelligent decentralized autonomous agents, or DAOs, to evaluate data, value, and asset transfers autonomously and quickly among several authorities. As a result, the combination of AI with blockchain technology has the promise of revolutionizing data management and transactional procedures while providing unmatched efficiency and security. To analyze incoming data, spot trends, and enable more efficient operations and insights into patients' habits and health, AI is essential. Analytics may be strategically applied by AI to improve decision-making. AI systems can identify life-threatening illnesses and prescribe treatments and length of hospital stays for patients by studying algorithms and making difficult judgments. With the help of AI and Google's enormous computing capacity, massive datasets can be processed with astounding precision. Hanover, a Microsoft AI product, is also used to mine healthcare data and unlock the promise of AI for improvements in the healthcare industry. The use of AI in healthcare has a lot of potential to enhance patient care and healthcare outcomes in general. With the use of cutting-edge technology, machines are now able to read, understand, and retain medical research articles, providing prospective treatments specific to each patient [118]. Pharmaceutical firms use blockchain technology to protect the integrity of their supply chains, collect data in real time, and share it effectively to improve patient outcomes [119]. Blockchain aids in the fight against medicine fraud, which claims millions of lives annually, by issuing distinctive serial numbers to pharmaceuticals. By prohibiting unlawful swaps and assuring prompt delivery of high-quality drugs, our serialization technology guarantees precise drug tracking. For pharmaceutical firms, the full traceability of pharmaceuticals utilizing blockchain technology has proven advantageous since it improves drug management and safety [120].

7 Advancing Healthcare with AI and Blockchain Integration: Overcoming Challenges

Blockchain and AI work well together to improve safety, integrity, and treatment alternatives that are specifically suited to each patient. Healthcare will improve and accuracy will increase because of AI's superior capacity to detect abnormalities in medical imaging. With the goal of using patient data, improving diagnoses, and providing suggestions based on evidence, major tech giants like Google, Microsoft, Apple, and Amazon, as well as startups, are aggressively investigating AI's uses in healthcare. Robotic surgery and digital counseling via smartphone applications are innovations that go hand in hand with one another to bring about considerable cost reductions in the healthcare industry. Security and interoperability issues must be resolved for healthcare AI to completely realize its potential. The fusion of complex algorithms, enormous databases, and effective machines offers a cutting-edge and revolutionary method for integrating healthcare [121].

8 Enhancing Healthcare Data Storage: Integrating Blockchain and AI

The use of blockchain technology and artificial intelligence in the storage of healthcare data presents intriguing answers to the problems posed by the administration of massive amounts of medical records and data. The blockchain can be used to store medical data uploaded via Electronic Health Record (EHR) systems, however, processing such data directly within the blockchain network has computational cost and storage constraints owing to its tiny block sizes. Additionally, privacy issues come up and put data breaches at danger. A strong architecture is used to solve these challenges, relying on trustworthy third parties to manage enormous volumes of sensitive data while blockchain is used for secure on-chain storage. In managing healthcare information, this strategy achieves a balance between effectiveness, security, and privacy [122, 123].

9 Decentralizing Biomedical Research: Blockchain and AI Integration

Blockchain and AI technology integration is changing how biological research and healthcare are conducted. Ahuja [124] showed the effectiveness of deep learning, a crucial AI method, in a variety of fields, including medication categorization and medical diagnosis. Its potential as a fruitful tool for drug development and research was proved by their study. In this situation, the MolAICal software is essential for producing 3D structural ligands inside protein targets using both conventional and deep learning methods. In the first of its two modules, MolAICal uses fragments of FDA-approved pharmaceuticals to create a deep learning model based on WGANs. The protein pocket's 3D ligands are then made using these produced fragments. In the second module of the software, WGAN-based deep learning models are trained using drug-like compounds from the zinc database, and then molecular docking is used to calculate the affinities between produced compounds and proteins. Testing on the membrane target GCGR and the non-membrane target SARSCoV-2 Mpro showed the effectiveness of MolAICal's drug design functionalities. Notably, the method was effective in producing ligands that differed in their degree of 3D structural similarity to the crystal ligands of GCGR and SARS-CoV-2 Mpro, providing helpful resources for researchers looking for novel drug candidates.

Additionally, the integration of blockchain and AI in customized cardiovascular therapy was investigated by Bai et al. [125]. This integration offers the potential for better results by expanding the amount of data available for AI training, sharing confidential AI techniques, and decentralizing databases. However, other concerns about implementation, such as technical trust, data transfer among competing organizations, compensation, and ethical implications, need more research. Disseminating information for lung cancer patients across a decentralized network is another use for blockchain and AI. Organizations may jointly create global models to find novel patterns and symptoms by employing locally taught models for big data analytics. With this novel strategy, patient care might be improved while maintaining privacy concerns.

10 Conclusion

The use of blockchain and AI technology in the healthcare sector has enormous potential to transform patient care and medical procedures. Healthcare workers now have quick access to detailed patient information thanks to the usage of blockchain to securely store and display medical records, enabling more informed and effective decision-making. On the other side, AI's sophisticated algorithms and data analysis skills are essential for processing huge amounts of patient data and gleaning insightful information. The medical system may gain from improved service efficiency, simpler processes, and cost savings by integrating the benefits of various technologies.

Blockchain technology's decentralized structure protects data confidentiality and integrity, while AI's capacity to decode and comprehend complicated medical data supports individualized healthcare, diagnosis, and treatment planning. Additionally, the combination of blockchain technology and artificial intelligence democratizes the healthcare industry by giving individuals more ownership over their health data and enabling frictionless data exchange between healthcare providers. As a result, patient outcomes and overall healthcare quality are enhanced via collaboration and interoperability. But in order to successfully use blockchain and AI in healthcare, a number of issues must be resolved, including interoperability, data protection, and ethical considerations. To overcome these obstacles and realize the full potential of this integration, stakeholders such as healthcare practitioners, academics, politicians, and technology experts must work together. A more patient-centric, effective, and accessible healthcare ecosystem will be made possible by embracing the transformational power of blockchain and AI as the healthcare sector continues to develop. To continuously improve the capabilities of these technologies and take use of their potential to improve healthcare in the future, further research and development in this area are essential. Overall, the combination of blockchain and AI is a significant step toward a more sophisticated, secure, and egalitarian healthcare environment.

Acknowledgements Authors would like to extend their appreciation to every member of the Datafoundry Pvt. Ltd. team who contributed to this endeavor. The dedication and commitment of Datafoundry Pvt. Ltd. have played a pivotal role in enhancing the depth and breadth of our research. The seamless access to data, technical guidance, and collaborative discussions have enriched our understanding and enabled us to present a comprehensive and well-informed study.

References

1. Alotaibi YK, Federico F (2017) The impact of health information technology on patient safety. Saudi Med J 38(12):1173
2. Chapuis C et al (2010) Automated drug dispensing system reduces medication errors in an intensive care setting. Crit Care Med 38(12):2275–2281
3. Campanella P et al (2016) The impact of electronic health records on healthcare quality: a systematic review and meta-analysis. Eur J Public Health 26(1):60–64
4. Wong ZSY, Zhou J, Zhang Q (2019) Artificial intelligence for infectious disease big data analytics. Infect Dis Health 24(1):44–48
5. Bragazzi NL, Dai H, Damiani G, Behzadifar M, Martini M, Wu J (2020) How big data and artificial intelligence can help better manage the COVID-19 pandemic. Int J Environ Res Public Health 17(9):3176
6. Sahoo MS, Baruah PK (2018) Hbasechaindb—a scalable blockchain framework on Hadoop ecosystem. In Supercomputing frontiers: 4th Asian conference, SCFA 2018, Singapore, March 26–29, 2018, proceedings 4. Springer, pp 18–29
7. Sarkar C et al (2023) Artificial intelligence and machine learning technology driven modern drug discovery and development. Int J Mol Sci 24(3):2026
8. Cha Y et al (2018) Drug repurposing from the perspective of pharmaceutical companies. Br J Pharmacol 175(2):168–180
9. Siyal AA, Junejo AZ, Zawish M, Ahmed K, Khalil A, Soursou G (2019) Applications of blockchain technology in medicine and healthcare: challenges and future perspectives. Cryptography 3(1):3
10. Hang L, Choi E, Kim D-H (2019) A novel EMR integrity management based on a medical blockchain platform in hospital. Electronics (Basel) 8(4):467
11. Akkiraju R et al (2020) Characterizing machine learning processes: a maturity framework. In: Business process management: 18th international conference, BPM 2020, Seville, Spain, September 13–18, 2020, proceedings 18. Springer, pp 17–31
12. Feng Q, He D, Zeadally S, Khan MK, Kumar N (2019) A survey on privacy protection in blockchain system. J Netw Comput Appl 126:45–58
13. Lin C, He D, Huang X, Khan MK, Choo K-KR (2020) DCAP: a secure and efficient decentralized conditional anonymous payment system based on blockchain. IEEE Trans Inf Forensics Secur 15:2440–2452

14. Ahmad SS, Khan S, Kamal MA (2019) What is blockchain technology and its significance in the current healthcare system? A brief insight. Curr Pharm Des 25(12):1402–1408
15. Zhang P, White J, Schmidt DC, Lenz G, Rosenbloom ST (2018) FHIRChain: applying blockchain to securely and scalably share clinical data. Comput Struct Biotechnol J 16:267–278
16. Linn LA, Koo MB (2016) Blockchain for health data and its potential use in health it and health care related research. In: ONC/NIST use of blockchain for healthcare and research workshop. ONC/NIST, Gaithersburg, Maryland, United States, pp 1–10
17. Bryatov SR, Borodinov AA (2019) Blockchain technology in the pharmaceutical supply chain: researching a business model based on Hyperledger fabric. In: Proceedings of the international conference on information technology and nanotechnology (ITNT), Samara, Russia, pp 1613–1673
18. Singh M, Kim S (2018) Branch based blockchain technology in intelligent vehicle. Comput Netw 145:219–231
19. Andoni M et al (2019) Blockchain technology in the energy sector: a systematic review of challenges and opportunities. Renew Sustain Energy Rev 100:143–174
20. Khurshid A (2020) Applying blockchain technology to address the crisis of trust during the COVID-19 pandemic. JMIR Med Inform 8(9):e20477
21. Shah VN, Garg SK (2015) Managing diabetes in the digital age. Clin Diabetes Endocrinol 1:1–7
22. Pegoraro V et al (2023) Cardiology in a digital age: opportunities and challenges for e-Health: a literature review. J Clin Med 12(13):4278
23. Alemayehu C, Mitchell G, Nikles J (2018) Barriers for conducting clinical trials in developing countries—a systematic review. Int J Equity Health 17:1–11
24. de Jongh D et al (2022) Early-phase clinical trials of bio-artificial organ technology: a systematic review of ethical issues. Transpl Int 35:10751
25. Peipert BJ, Spinosa D, Howell EP, Weber JM, Truong T, Harris BS (2021) Innovations in infertility: a comprehensive analysis of the ClinicalTrials.gov database. Fertil Steril 116(5):1381–1390
26. Welch MJ et al (2015) The ethics and regulatory landscape of including vulnerable populations in pragmatic clinical trials. Clin Trials 12(5):503–510
27. Kassab M, DeFranco J, Malas T, Laplante P, Destefanis G, Neto VVG (2019) Exploring research in blockchain for healthcare and a roadmap for the future. IEEE Trans Emerg Top Comput 9(4):1835–1852
28. Schmeelk S, Kanabar M, Peterson K, Pathak J (2022) Electronic health records and blockchain interoperability requirements: a scoping review. JAMIA Open 5(3):ooac068
29. Maitra S, Yanambaka VP, Puthal D, Abdelgawad A, Yelamarthi K (2021) Integration of Internet of Things and blockchain toward portability and low-energy consumption. Trans Emerg Telecommun Technol 32(6):e4103
30. Han Y, Zhang Y, Vermund SH (2022) Blockchain technology for electronic health records. Int J Environ Res Public Health 19(23):15577
31. Agrawal D, Minocha S, Namasudra S, Gandomi AH (2022) A robust drug recall supply chain management system using hyperledger blockchain ecosystem. Comput Biol Med 140:105100
32. Clauson KA, Breeden EA, Davidson C, Mackey TK (2018) Leveraging blockchain technology to enhance supply chain management in healthcare: an exploration of challenges and opportunities in the health supply chain. Blockchain Healthc Today
33. Humayun M, Jhanjhi NZ, Niazi M, Amsaad F, Masood I (2022) Securing drug distribution systems from tampering using blockchain. Electronics (Basel) 11(8):1195
34. Ghadge A, Bourlakis M, Kamble S, Seuring S (2022) Blockchain implementation in pharmaceutical supply chains: a review and conceptual framework. Int J Prod Res 1–19
35. Dasaklis TK, Voutsinas TG, Tsoulfas GT, Casino F (2022) A systematic literature review of blockchain-enabled supply chain traceability implementations. Sustainability 14(4):2439
36. Liu X, Barenji AV, Li Z, Montreuil B, Huang GQ (2021) Blockchain-based smart tracking and tracing platform for drug supply chain. Comput Ind Eng 161:107669

37. Field MJ, Grigsby J (2002) Telemedicine and remote patient monitoring. JAMA 288(4):423–425
38. Pirtle CJ, Payne K, Drolet BC (2019) Telehealth: legal and ethical considerations for success. Telehealth Med Today
39. Lloyd J, Lee CJ (2022) Use of telemedicine in care of hematologic malignancy patients: challenges and opportunities. Curr Hematol Malig Rep 17(1):25–30
40. Niu B, Mukhtarova N, Alagoz O, Hoppe K (2022) Cost-effectiveness of telehealth with remote patient monitoring for postpartum hypertension. J Matern Fetal Neonatal Med 35(25):7555–7561
41. De Guzman KR, Snoswell CL, Taylor ML, Gray LC, Caffery LJ (2022) Economic evaluations of remote patient monitoring for chronic disease: a systematic review. Value Health 25(6):897–913
42. Abekah-Nkrumah G, Antwi M, Attachey AY, Janssens W, Rinke de Wit TF (2022) Readiness of Ghanaian health facilities to deploy a health insurance claims management software (CLAIM-it). PLoS One 17(10):e0275493
43. Thenmozhi M, Dhanalakshmi R, Geetha S, Valli R (2021) WITHDRAWN: implementing blockchain technologies for health insurance claim processing in hospitals. Elsevier
44. Deluca JM, Enmark R (2000) E-health: the changing model of healthcare. Front Health Serv Manage 17(1):3–15
45. Desai RJ et al (2021) Broadening the reach of the FDA sentinel system: a roadmap for integrating electronic health record data in a causal analysis framework. NPJ Digit Med 4(1):170
46. Ho CWL, Ali J, Caals K (2020) Ensuring trustworthy use of artificial intelligence and big data analytics in health insurance. Bull World Health Organ 98(4):263
47. Baumfeld Andre E, Reynolds R, Caubel P, Azoulay L, Dreyer NA (2020) Trial designs using real-world data: the changing landscape of the regulatory approval process. Pharmacoepidemiol Drug Saf 29(10):1201–1212
48. Pradhan B, Bhattacharyya S, Pal K (2021) IoT-based applications in healthcare devices. J Healthc Eng 2021:1–18
49. Frikha T, Chaari A, Chaabane F, Cheikhrouhou O, Zaguia A (2021) Healthcare and fitness data management using the IoT-based blockchain platform. J Healthc Eng 2021
50. Azbeg K, Ouchetto O, Andaloussi SJ (2022) BlockMedCare: a healthcare system based on IoT, blockchain and IPFS for data management security. Egypt Inform J 23(2):329–343
51. Esposito C, De Santis A, Tortora G, Chang H, Choo K-KR (2018) Blockchain: a panacea for healthcare cloud-based data security and privacy? IEEE Cloud Comput 5(1):31–37
52. Abounassar EM, El-Kafrawy P, Abd El-Latif AA (2022) Security and interoperability issues with internet of things (IoT) in healthcare industry: a survey. In: Security and privacy preserving for IoT and 5G networks: techniques, challenges, and new directions, pp 159–189
53. Rajeswari S, Ponnusamy V (2022) AI-based IoT analytics on the cloud for diabetic data management system. In: Integrating AI in IoT analytics on the cloud for healthcare applications. IGI Global, pp 143–161
54. Hong K-W, Oh B-S (2010) Overview of personalized medicine in the disease genomic era. BMB Rep 43(10):643–648
55. Beccia F et al (2022) An overview of personalized medicine landscape and policies in the European Union. Eur J Public Health 32(6):844–851
56. Offit K (2011) Personalized medicine: new genomics, old lessons. Hum Genet 130:3–14
57. Santaló J, Berdasco M (2022) Ethical implications of epigenetics in the era of personalized medicine. Clin Epigenet 14(1):1–14
58. McGowan ML, Settersten RA Jr, Juengst ET, Fishman JR (2014) Integrating genomics into clinical oncology: ethical and social challenges from proponents of personalized medicine. In: Urologic oncology: seminars and original investigations. Elsevier, pp 187–192
59. Veenstra DL, Mandelblatt J, Neumann P, Basu A, Peterson JF, Ramsey SD (2020) Health economics tools and precision medicine: opportunities and challenges. In: Forum for health economics and policy. De Gruyter, p 20190013

60. Meyer MA (2023) A patient's journey to pay a healthcare bill: it's way too complicated. J Patient Exp 10:23743735231174760
61. Al Barazanchi I et al (2022) Blockchain: the next direction of digital payment in drug purchase. In: 2022 International congress on human-computer interaction, optimization and robotic applications (HORA). IEEE, pp 1–7
62. Britton JR (2015) Healthcare reimbursement and quality improvement: integration using the electronic medical record: comment on "fee-for-service payment—an evil practice that must be stamped out?" Int J Health Policy Manag 4(8):549
63. Yaqoob I, Salah K, Jayaraman R, Al-Hammadi Y (2021) Blockchain for healthcare data management: opportunities, challenges, and future recommendations. Neural Comput Appl 1–16
64. Kommunuri J (2022) Artificial intelligence and the changing landscape of accounting: a viewpoint. Pac Account Rev 34(4):585–594
65. Asghar MR, Lee T, Baig MM, Ullah E, Russello G, Dobbie G (2017) A review of privacy and consent management in healthcare: a focus on emerging data sources. In: 2017 IEEE 13th international conference on e-Science (e-Science). IEEE, pp 518–522
66. Rantos K, Drosatos G, Kritsas A, Ilioudis C, Papanikolaou A, Filippidis AP (2019) A blockchain-based platform for consent management of personal data processing in the IoT ecosystem. Secur Commun Netw 2019:1–15
67. Maher M, Khan I, Prikshat V (2023) Monetisation of digital health data through a GDPR-compliant and blockchain enabled digital health data marketplace: a proposal to enhance patient's engagement with health data repositories. Int J Inf Manag Data Insights 3(1):100159
68. Martin C et al (2022) The ethical considerations including inclusion and biases, data protection, and proper implementation among AI in radiology and potential implications. Intell Based Med 100073
69. Zeng D, Cao Z, Neill DB (2021) Artificial intelligence–enabled public health surveillance—from local detection to global epidemic monitoring and control. In: Artificial intelligence in medicine. Elsevier, pp 437–453
70. Khoury MJ, Armstrong GL, Bunnell RE, Cyril J, Iademarco MF (2020) The intersection of genomics and big data with public health: opportunities for precision public health. PLoS Med 17(10):e1003373
71. MacKinnon GE, Brittain EL (2020) Mobile health technologies in cardiopulmonary disease. Chest 157(3):654–664
72. Iwaya LH, Fischer-Hübner S, Åhlfeldt R-M, Martucci LA (2018) mhealth: a privacy threat analysis for public health surveillance systems. In: 2018 IEEE 31st international symposium on computer-based medical systems (CBMS). IEEE, pp 42–47
73. Aiello AE, Renson A, Zivich P (2020) Social media- and internet-based disease surveillance for public health. Annu Rev Public Health 41:101
74. Chiou H, Voegeli C, Wilhelm E, Kolis J, Brookmeyer K, Prybylski D (2022) The future of infodemic surveillance as public health surveillance. Emerg Infect Dis 28(Suppl 1):S121
75. Mello MM, Wang CJ (1979) Ethics and governance for digital disease surveillance. Science 368(6494):951–954
76. Juravle G, Boudouraki A, Terziyska M, Rezlescu C (2020) Trust in artificial intelligence for medical diagnoses. Prog Brain Res 253:263–282
77. MacRitchie N, Frleta-Gilchrist M, Sugiyama A, Lawton T, McInnes IB, Maffia P (2020) Molecular imaging of inflammation—current and emerging technologies for diagnosis and treatment. Pharmacol Ther 211:107550
78. Washington P et al (2020) Data-driven diagnostics and the potential of mobile artificial intelligence for digital therapeutic phenotyping in computational psychiatry. Biol Psychiatry Cogn Neurosci Neuroimaging 5(8):759–769
79. Paul S, Vidusha K, Thilagar S, Lakshmanan DK, Ravichandran G, Arunachalam A (2022) Advancement in the contemporary clinical diagnosis and treatment strategies of insomnia disorder. Sleep Med 91:124–140

80. Kaye DK (2023) Addressing ethical issues related to prenatal diagnostic procedures. Matern Health Neonatol Perinatol 9(1):1–9
81. Chong AY-L, Liu MJ, Luo J, Keng-Boon O (2015) Predicting RFID adoption in healthcare supply chain from the perspectives of users. Int J Prod Econ 159:66–75
82. Coussement K, Lessmann S, Verstraeten G (2017) A comparative analysis of data preparation algorithms for customer churn prediction: a case study in the telecommunication industry. Decis Support Syst 95:27–36
83. Johnston SS, Morton JM, Kalsekar I, Ammann EM, Hsiao C-W, Reps J (2019) Using machine learning applied to real-world healthcare data for predictive analytics: an applied example in bariatric surgery. Value Health 22(5):580–586
84. Lanier P, Rodriguez M, Verbiest S, Bryant K, Guan T, Zolotor A (2020) Preventing infant maltreatment with predictive analytics: applying ethical principles to evidence-based child welfare policy. J Fam Violence 35:1–13
85. Luk JW, Pruitt LD, Smolenski DJ, Tucker J, Workman DE, Belsher BE (2022) From everyday life predictions to suicide prevention: clinical and ethical considerations in suicide predictive analytic tools. J Clin Psychol 78(2):137–148
86. Jeddi Z, Bohr A (2020) Remote patient monitoring using artificial intelligence. In: Artificial intelligence in healthcare. Elsevier, pp 203–234
87. Shaik T et al (2023) Remote patient monitoring using artificial intelligence: current state, applications, and challenges. Wiley Interdiscip Rev Data Min Knowl Discov 13(2):e1485
88. Elango S, Manjunath L, Prasad D, Sheela T, Ramachandran G, Selvaraju S (2023) Super artificial intelligence medical healthcare services and smart wearable system based on IoT for remote health monitoring. In: 2023 5th International conference on smart systems and inventive technology (ICSSIT). IEEE, pp 1180–1186
89. Leung R (2023) Using AI–ML to augment the capabilities of social media for tele-health and remote patient monitoring. In: Healthcare, MDPI, p 1704
90. Palanisamy P, Padmanabhan A, Ramasamy A, Subramaniam S (2023) Remote patient activity monitoring system by integrating IoT sensors and artificial intelligence techniques. Sensors 23(13):5869
91. Sun W, Zheng W, Simeonov A (2017) Drug discovery and development for rare genetic disorders. Am J Med Genet A 173(9):2307–2322
92. Kiriiri GK, Njogu PM, Mwangi AN (2020) Exploring different approaches to improve the success of drug discovery and development projects: a review. Futur J Pharm Sci 6(1):1–12
93. Cerchia C, Lavecchia A (2023) New avenues in artificial-intelligence-assisted drug discovery. Drug Discov Today 103516
94. Patel V, Shah M (2022) Artificial intelligence and machine learning in drug discovery and development. Intell Med 2(3):134–140
95. Turanli B, Karagoz K, Gulfidan G, Sinha R, Mardinoglu A, Arga KY (2018) A network-based cancer drug discovery: from integrated multi-omics approaches to preci-sion medicine. Curr Pharm Des 24(32):3778–3790
96. Zhou Y et al (2021) AlzGPS: a genome-wide positioning systems platform to catalyze multi-omics for Alzheimer's drug discovery. Alzheimers Res Ther 13(1):1–13
97. Cai Z, Poulos RC, Liu J, Zhong Q (2022) Machine learning for multi-omics data integration in cancer. iScience
98. Syvänen S, Valentini C (2020) Conversational agents in online organization–stakeholder interactions: a state-of-the-art analysis and implications for further re-search. J Commun Manag 24(4):339–362
99. Dave T, Athaluri SA, Singh S (2023) ChatGPT in medicine: an overview of its applications, advantages, limitations, future prospects, and ethical considerations. Front Artif Intell 6:1169595
100. van der Schyff EL, Ridout B, Amon KL, Forsyth R, Campbell AJ (2023) Providing self-led mental health support through an artificial intelligence-powered chat bot (Leora) to meet the demand of mental health care. J Med Internet Res 25:e46448

101. Nadarzynski T, Miles O, Cowie A, Ridge D (2019) Acceptability of artificial intelligence (AI)-led chatbot services in healthcare: a mixed-methods study. Digit Health 5:2055207619871808
102. Kadariya D, Venkataramanan R, Yip HY, Kalra M, Thirunarayanan K, Sheth A (2019) kBot: knowledge-enabled personalized chatbot for asthma self-management. In: 2019 IEEE international conference on smart computing (SMARTCOMP). IEEE, pp 138–143
103. Miura C, Chen S, Saiki S, Nakamura M, Yasuda K (2022) Assisting personalized healthcare of elderly people: developing a rule-based virtual caregiver system using mobile chatbot. Sensors 22(10):3829
104. Adegun AA, Viriri S, Ogundokun RO (2021) Deep learning approach for medical image analysis. Comput Intell Neurosci 2021:1–9
105. Li Y, Zhao J, Lv Z, Li J (2021) Medical image fusion method by deep learning. Int J Cognit Comput Eng 2:21–29
106. Afshar P et al (2020) DRTOP: deep learning-based radiomics for the time-to-event outcome prediction in lung cancer. Sci Rep 10(1):12366
107. Severn C, Suresh K, Görg C, Choi YS, Jain R, Ghosh D (2022) A pipeline for the implementation and visualization of explainable machine learning for medical imaging using radiomics features. Sensors 22(14):5205
108. Liu X et al (2021) Advances in deep learning-based medical image analysis. Health Data Sci 2021 (2021)
109. Graffigna G, Barello S, Bonanomi A, Lozza E (2015) Measuring patient engagement: development and psychometric properties of the Patient Health Engagement (PHE) scale. Front Psychol 6:274
110. Yardley L et al (2016) Understanding and promoting effective engagement with digital behavior change interventions. Am J Prev Med 51(5):833–842
111. Zheng Z et al (2022) Patient engagement as a core element of translating clinical evidence into practice-application of the COM-B model behaviour change model. Disabil Rehabil 1–10
112. Jeyakumar T et al (2023) Preparing for an artificial intelligence-enabled future: patient perspectives on engagement and health care professional training for adopting artificial intelligence technologies in health care settings. JMIR AI 2(1):e40973
113. Giovanelli A et al (2023) Supporting adolescent engagement with artificial intelligence—driven digital health behavior change interventions. J Med Internet Res 25:e40306
114. Campbell D (2018) Combining ai and blockchain to push frontiers in healthcare. http://www.macadamian.com/2018/03/16/combining-ai-andblockchain-in-healthcare/, vol online
115. Marwala T, Xing B (2018) Blockchain and artificial intelligence. arXiv preprint arXiv:1802.04451
116. Mamoshina P et al (2018) Converging blockchain and next-generation artificial intelligence technologies to decentralize and accelerate biomedical research and healthcare. Oncotarget 9(5):5665
117. Magazzeni D, McBurney P, Nash W (2017) Validation and verification of smart contracts: a research agenda. Computer (Long Beach Calif) 50(9):50–57
118. Bohr A, Memarzadeh K (2020) The rise of artificial intelligence in healthcare applications. In: Artificial intelligence in healthcare. Elsevier, pp 25–60
119. Khezr S, Moniruzzaman M, Yassine A, Benlamri R (2019) Blockchain technology in healthcare: a comprehensive review and directions for future research. Appl Sci 9(9):1736
120. Behner P, Hecht M-L, Wahl F (2017) Fighting counterfeit pharmaceuticals: new defenses for an underestimated and growing menace, vol 12, p 2017
121. Tran V-T, Riveros C, Ravaud P (2019) Patients' views of wearable devices and AI in healthcare: findings from the ComPaRe e-cohort. NPJ Digit Med 2(1):53
122. Jennath HS, Anoop VS, Asharaf S (2020) Blockchain for healthcare: secur-ing patient data and enabling trusted artificial intelligence
123. Ahuja AS (2019) The impact of artificial intelligence in medicine on the future role of the physician. PeerJ 7:e7702

124. Bai Q, Tan S, Xu T, Liu H, Huang J, Yao X (2021) MolAICal: a soft tool for 3D drug design of protein targets by artificial intelligence and classical algorithm. Brief Bioinform 22(3):bbaa161
125. Krittanawong C et al (2020) Integrating blockchain technology with artificial intelli-gence for cardiovascular medicine. Nat Rev Cardiol 17(1):1–3

Impact of Neural Network on Malware Detection

Aaeen Alchi, Kiranbhai Dodiya, and Varayogula Sai Niveditha

Abstract Attacks by malware have significantly increased during the last several years, endangering the security of computer systems and networks. The continually shifting landscape of malware assaults makes it challenging for traditional techniques like a rule- and signature-based detection to stay up. Consequently, researchers are now looking at more cutting-edge methods, including neural networks, to increase the precision of malware detection. In assessing new malware by comparable traits, neural networks may gather data from previously discovered malware. This method makes it possible to quickly and reliably identify malware, which is essential for stopping infestation spread. Finding patterns and connections among massive data sets is one of the main benefits of utilising neural networks to detect malware. However, there are some issues with employing neural networks to find viruses. One of the main difficulties is that the network needs a lot of labelled data training. Effective evasion attacks are also feasible, in which fraudsters try to control neural networks by feeding them harmful material designed to elude detection. Despite these difficulties, researchers are still exploring neural networks' potential for virus detection. This book chapter provides a detailed analysis of how neural networks affect malware detection. Along with an outline of the current status of the research area, updates on neural network topologies and training methods are provided. We'll discuss possible advancements and future possibilities in the industry, including reinforcement learning and generative adversarial networks for malware detection. The first section of the chapter summarises the essential characteristics of neural networks, including their capacity to identify patterns and abnormalities in massive amounts of data. The benefits and drawbacks of using neural networks to detect malware are highlighted. Finally, the extensive usage of neural networks in this book chapter's malware detection analysis is highlighted. Researchers may greatly enhance computer and network security and better protect against the rising danger of assaults by malware by using neural networks' complicated capabilities.

A. Alchi (✉) · K. Dodiya
Department of Biochemistry and Forensic Science, Gujarat University, Ahmedabad, Gujarat, India
e-mail: alchiaaeen7@gmail.com

V. S. Niveditha
Digital Forensics, Rashtriya Raksha University, Lavad, Gandhinagar, Gujarat, India

K. Kaushik and I. Sharma (eds.), *Next-Generation Cybersecurity*, Blockchain Technologies, https://doi.org/10.1007/978-981-97-1249-6_10

Keywords Malware · Signature-based · Detection · Examination · Large data sets · Neural Network

1 A Summary of Malware Detection's Current State

The complexity and volume of malware assaults are swellings, which pose a grim threat to the safety of computer structures and networks and are a contemporary hallmark of malware detection. A broad spectrum of harmful software, including viruses, worms, Trojan horses, and ransomware, are called Malware. Malware attacks may cause data loss, system outages, financial losses, and brand damage.

The ever-evolving nature of malware attacks has rendered conventional malware detection methods like rule-based and signature-based detection ineffective. Experts may identify malicious software and stop using signature-based detection, which relies on known malware patterns but can easily be evaded by attackers using new or updated malware versions. Rule-based detection finds and disables Malware using predetermined criteria but may produce false positives and negatives [13].

As a result, researchers and security professionals have turned to more cutting-edge technology to improve the precision and efficacy of malware detection. However, these approaches rely on massive extents of data to teach computers to spot patterns and anomalies in Malware, and they may also react to new, unforeseen threats.

However, there are still challenges in malware detection, including the necessity for extensive and diverse training datasets, the possibility that attackers would evade detection by using cunning tactics, and the need to strike a balance concerning detection accuracy and processing speed. In addition, researchers and security specialists must keep coming up with fresh ideas and methods to spot and stop malware attacks as they increase in frequency.

1.1 More Complex Techniques Are Required

Experts must develop malware detection systems with ever-greater sophistication due to the complexity and culture of malware attacks. Malware developers continuously formulate innovative strategies to avoid detection and entering computer systems and networks. The upshot is that conventional signature-based and rule-based detection techniques often must be in sync with the changing threat environment.

More complicated technologies like machine learning and neural networks have emerged to satisfy this need. These technologies can improve malware detection accuracy and efficacy by utilising enormous datasets and training algorithms to uncover patterns and irregularities associated with Malware.

Furthermore, the number of possible entry points for malware attacks grows as connected devices and the Internet of Things (IoT) increase. As a result, there is a particular need for more complicated detection algorithms capable of identifying and stopping threats across several platforms and devices.

In addition to these technological challenges, regulatory and compliance demands necessitate enhanced detection approaches. In Europe, for example, the General Data Protection Regulation (GDPR) requires commerce to put in place appropriate technological and organisational protections to guarantee the security of personal data. As a result, powerful malware detection techniques must be used to avoid data breaches and other security calamities.

Overall, the rising complexity and sophistication of malware assaults and modern information security's legal and regulatory requirements drive the demand for more sophisticated malware detection technologies.

1.2 Artificial Neural Networks

Neural networks are machine learning algorithms that simulate the human brain's structure and function. They comprise layers of linked nodes called neurons that process input data and anticipate output. Each neuron in the preceding layer gets input from other neurons, computes, and then sends the result to the next layer.

Since they can understand patterns and abnormalities associated with Malware even in vast and complicated datasets, neural networks have become popular in detection. They can also constantly adapt to new and undiscovered dangers by learning from new data.

Feedforward neural networks, recurrent neural networks, and convolutional neural networks are among the neural networks utilised in malware detection. Liable on the nature of the data and the detection job, each kind has different strengths and disadvantages (Jakub Gotthans et al., n.d.).

One of the benefits of neural networks is their capacity to do feature extraction, which entails finding essential data properties relevant to the detection job. This is especially beneficial in detecting Malware since characteristics like code obfuscation and packing can make it challenging to detect Malware using standard approaches.

Yet, there are several drawbacks to using neural networks for malware detection. One of the difficulties is the requirement for extensive and diverse training datasets, which might be challenging when dealing with uncommon or novel malware strains. Moreover, training and evaluating neural networks may be computationally costly, which might be a limiting issue for real-time detection systems.

Overall, neural networks have emerged as a viable solution to malware detection, promising more remarkable accuracy and the capability to adapt to new and developing threats. Their success, however, depends on the quality of the training data and the neural network's unique design and configuration.

1.3 Neural Network Overview

Neural networks are machine learning algorithms that mimic the human brain's structure and function. They comprise layers of linked nodes called neurons that process input data and anticipate output.

Each neuron in the preceding layer takes input from other neurons in that layer, computes using an instigation function, and then transmits the outcome to the next layer. The neural network's prediction is the output of the last layer.

Classification, regression, and pattern recognition are just a few tasks that neural networks can do. Therefore, professions that involve managing intricate and non-linear relationships between input and output variables can be significantly enhanced by individuals who possess exceptional skills in this area.

There are several forms of neural networks, each with its structure and function. Among the most regularly utilised kinds are (Fig. 1):

Feedforward neural networks are the most basic neural networks, with information flowing in just one technique, from the input to the output layer. They are often employed in classification and regression problems [10].

Backpropagation neural networks: Recurrent networks have feedback connections that allow them to store and comprehend consecutive input across time. They are often used for time-series data applications, for instance, speech recognition and natural dialectal processing.

Convolutional neural networks analyse image data using a two-dimensional pixel input matrix. They may be used for object identification and picture classification and employ filters to extract photo information [12].

Backpropagation is a technique for training neural networks that involves changing the weights and Preferences of the neurons to close the gap between predicted and observed output. To do this, one must first calculate the gradient of the loss function concerning the weights and biases before changing the parameters [12].

Some areas where neural networks have shown to be very successful are image identification, natural language processing, and voice commands. Nevertheless,

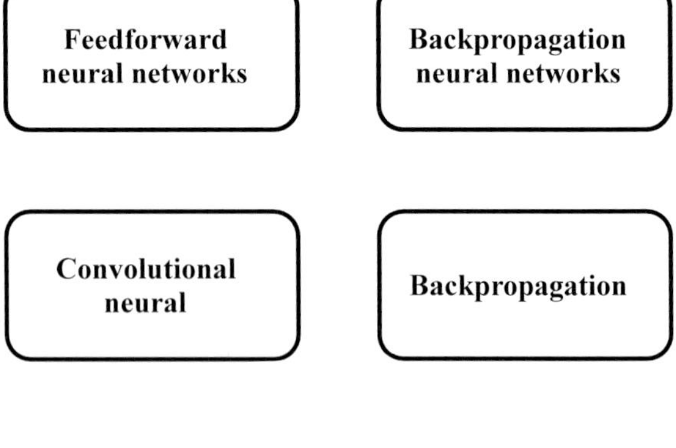

Fig. 1 Form of neural network

they need a substantial amount of training data to achieve high accuracy and are computationally costly to train.

1.4 Types of Neural Networks

Neural networks come in many different forms, each with its specific structure and function. Listed below are a few of the most typical types (Fig. 2):

1. **Feedforward neural networks**: A primary type of neural network that allows information to flow in a single direction, from the input to the output layer. They are often used to solve classification and regression problems (Jakub Gotthans et al., n.d.).
2. **Recurrent neural networks**: Thanks to their feedback connections, these networks can store and analyse sequential input across time. They are often used in time-series data applications, including natural language processing and speech recognition (Jakub Gotthans et al., n.d.).
3. **Convolutional neural networks**: These networks interpret picture data and use a two-dimensional matrix of pixel values as input.
4. **Autoencoders**: The purpose of these networks is to learn a condensed version of the input data, which can then be applied for tasks such as data compression, denoising, and detecting anomalies. These networks acquire a compact understanding of the input data, which can be effective until serious charges such as data compression, denoising, and anomaly detection. They do this by collecting information from images using filters, which can be used for functional classification and object detection (Jakub Gotthans et al., n.d.).
5. **GANs (generative adversarial networks)**: A GAN consists of a generator and a discriminator. The discriminator separates the created samples from the actual data, while the generator produces new examples comparable to the original data. Among other things, GANs may be used to create images and transfer fashion.

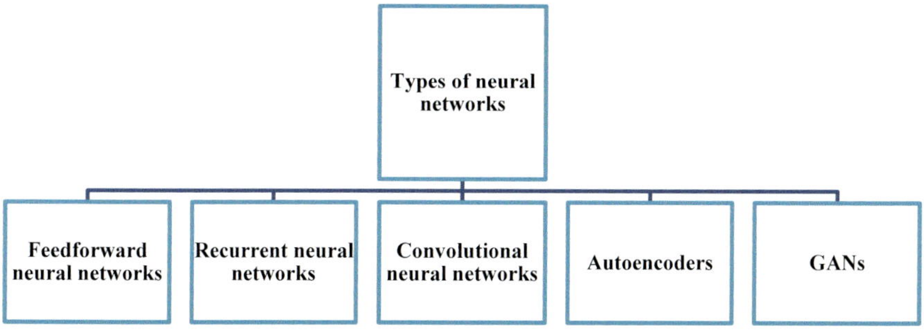

Fig. 2 Types of neural networks

6. LSTM networks: These networks are a form of the recurrent neural network intended to handle consecutive input over extended periods. They're typically employed for voice recognition, language translation, and sentiment analysis [4].

Deep belief networks are feedforward neural networks constructing a hierarchical input data representation. They are commonly applied in feature extraction, unsupervised learning, and recommendation systems.

Each neural network style has its benefits and weaknesses and is best suited for specific tasks and data. Therefore, the neural network utilised is dictated by the problem's nature and the data's characteristics.

1.5 Advantages of Employing Neural Networks to Identify Malware

The use of neural networks for malware detection has various advantages (Fig. 3):

1. **Learning patterns**: As neural networks can learn ways and identify abnormalities in big datasets, they are highly suited for malware detection. Even when standard signature-based approaches fail, they may uncover minute changes in code or behaviour that may suggest Malware.
2. **Adaptability**: Neural networks can adjust to new kinds of Malware and evolving threats. They may learn from new instances and increase their detecting skills.
3. **Speed**: Since neural networks can handle enormous volumes of data fast and effectively, they are well suited for real-time detection and reaction.
4. **Scalability**: Since neural networks can be scaled up or down based on the amount and complexity of the data collection, they can be used in large-scale systems.
5. **Accuracy**: When trained on vast and diverse datasets, neural networks may reach significant levels of accuracy in malware detection.
6. **Automated feature extraction**: Neural networks can extract features from data automatically, removing manual feature engineering requirements. This can save time and lower the likelihood of human mistakes.

Fig. 3 Advantages of Employing Neural Networks to Identify Malware

Advantages	Learning patterns
	Adaptability
	Speed
	Scalability
	Accuracy
	Automated feature extraction
	Low false positive rate

7. **Low false positive rate**: since neural networks have a low false positive rate, they are less likely to mistake innocuous software for Malware.

Overall, using neural networks for malware detection is a viable way to tackle the rising sophistication of malware assaults and enhance computer system and network security.

1.6 Neural Network Limitations for Malware Detection

While neural networks provide various advantages for malware detection, there are several limits to be aware of (Fig. 4):

1. **Black box model**: Neural networks are sometimes called "black box" models since it might be challenging to comprehend how the network makes choices. This makes interpreting the data and identifying the exact traits utilised to detect Malware challenging.
2. **Data bias**: A neural network model's accuracy depends on the training data's quality and representativeness. If the training data are skewed or inadequate, the model may fail to recognise new varieties of Malware.
3. **Overfitting**: When training data are excessively fed to neural networks, they become overly specific and incapable of adapting to new information. This can ultimately result in poor detection of unique forms of Malware that were not present during their initial training.
4. **Adversarial attacks**: Malware makers can build their software to avoid detection by neural networks, potentially resulting in an arms race between malware creators and malware detection systems.

Fig. 4 Neural network limitations for malware detection

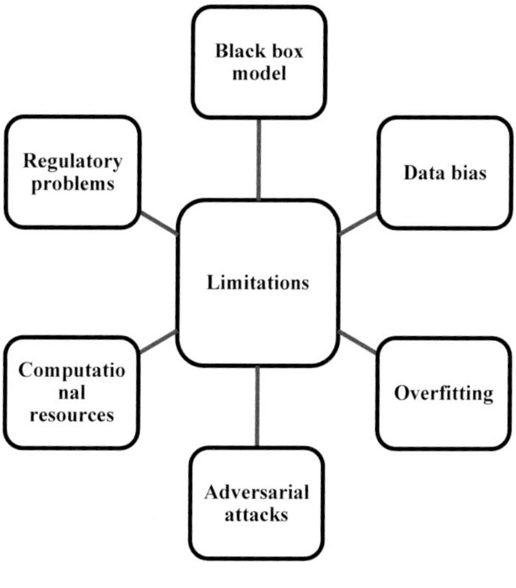

5. **Computational resources**: Since neural networks are computationally expensive, significant resources are required for training and deployment, which can be a hurdle for smaller firms with low resources.
6. **Regulatory problems**: Using neural networks to identify Malware may raise worries about data privacy and security, as well as ethical questions regarding the use of AI in security-related applications.

Thus, while neural networks hold great potential for enhancing malware detection, they also have limits that must be examined and addressed to be successful and ethical.

2 The Use of Neural Networks in Malware Detection

2.1 Data Pre-Processing for Neural Network Input

Pre-processing is required to ensure the quality and relevance of the input data before feeding it to a neural network for malware detection. Pre-processing is a sequence of procedures that clean, standardise, and convert input so the neural network can use it. Some popular neural network input pre-processing processes include (Fig. 5):

Data cleaning includes deleting unneeded or superfluous data and dealing with missing or incorrect values.

Data normalisation is scaling the data to a specified range or distribution to guarantee that each attribute is given equal weight.

Feature selection entails picking a subset of essential characteristics from the data to decrease the model's complexity and enhance performance.

Data transformation entails changing the data into a more acceptable format for neural network input. Text data, for example, could need to be tokenised and transformed to a numerical representation, while image data would need to be scaled or converted to grayscale.

Data augmentation is new synthetic data generated by applying changes to the original data, such as rotation or cropping. Data augmentation can assist in increasing the neural network's accuracy and generalisation [3].

The pre-processing stages for neural network input will vary depending on the type of data utilised for malware detection and the neural network model's needs. Therefore, it is critical to carefully evaluate the pre-processing stages to ensure the

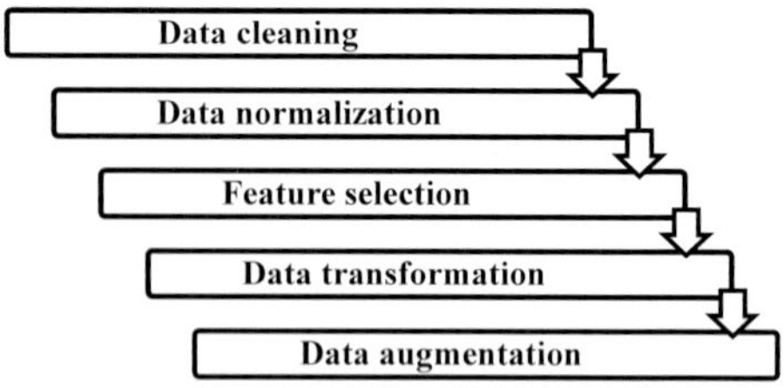

Fig. 5 Data pre-processing for neural network input

input data's quality and relevance and enhance the neural network's performance for malware detection.

2.2 Neural Network Training for Malware Detection

Many steps are involved in training a neural network for malware detection, including:

Designing the network architecture entails selecting the proper type of neural network, calculating the number of layers and nodes, and deciding on the activation functions for each layer.

- Preparing the training data entails pre-processing the data to ensure that it is in a format appropriate for training the neural network.
- Dividing the data into training and validation sets entails partitioning it into two groups: training the neural network and assessing its performance during training.
- Setting hyperparameters entails adjusting the learning rate, batch size, and other training-related factors.
- Training the network entails entering training data into the neural network and changing the network's weights and biases to minimise the error between expected and actual output.
- Assessing the network entails running the trained neural network on the validation set to evaluate its performance and, if necessary, make improvements.
- Testing the network entails running the trained neural network on a different test set to assess its performance on previously unknown data.
- Tuning the network entails adjusting the hyperparameters and network design to optimise performance on test data.

To ensure the accurate detection of Malware and prevent false positives, it is essential to train a neural network, even though it may require significant time and computational resources. Therefore, the quality and relevance of the training data, as well as the choice of appropriate hyperparameters and network design, are all crucial elements that can substantially influence the neural network's performance for malware detection.

2.3 Performance Evaluation of Neural Networks

Assessing neural network performance is essential to ensuring the network detects Malware effectively while avoiding false positives. Many measures are used to determine a neural network's effectiveness in malware detection, including:

Accuracy: This parameter measures the neural network's proportion of correct classifications.

Precision is the fraction of successfully detected malware samples among all malware samples categorised by the neural network.

Remember that this statistic counts the proportion of successfully recognised malware samples among all malware samples in the dataset.

The F1 score calculates the harmonic mean of accuracy and recall, providing a fair evaluation of both factors. The ROC curve is a graph that shows the effectiveness of a neural network in different classification scenarios. The AUC, or area under the ROC curve, is a widely used overall performance measure.

Confusion matrix: The breakdown of categorisation findings is presented in this matrix, providing a detailed account of the true positives, true negatives, false positives, and false negatives.

The false positive rate (FPR) and false negative rate (FNR) are metrics that assess the proportion of false positives and false negatives among all malware samples categorised by the neural network [1].

The malware detection application's unique requirements determine the selection of relevant assessment metrics. In specific applications, for example, avoiding false positives may be more essential than maximising recall, but in others, the converse may be true. It is critical to carefully consider the assessment criteria and employ suitable approaches to analyse the neural network's performance for malware detection effectively [11].

2.4 Evaluation of Neural Network Performance in Comparison to Alternative Malware Detection Approaches

Neural networks have exhibited promising results in malware detection, outperforming standard malware detection approaches in several circumstances. Following are some performance comparisons of neural networks to different malware detection techniques:

1. Signature-based detection compares a file's code or behaviour to a database of known malware signatures. While signature-based detection can be successful against known Malware, it is vulnerable to zero-day assaults and new malware variants. By understanding the fundamental patterns of malware activity, neural networks can detect new and undiscovered viruses, making them more successful than signature-based detection [14].
2. Heuristic-based detection entails studying a file's activity and determining its maliciousness based on specified criteria or heuristics. While heuristic detection can be helpful, it can also generate many false positives. By learning from a massive collection of benign and malicious samples, neural networks can minimise the number of false positives [14].
3. Machine learning-based detection: This method entails developing a model to identify files based on their attributes. While machine learning can be helpful,

it frequently necessitates extensive feature engineering, which can be time-consuming and may fail to collect all essential information. Neural networks can learn characteristics from data automatically, making them more effective than standard machine learning approaches [5].
4. Behaviour-based detection to assess a file's maliciousness; behaviour-based detection analyses its activity in a sandboxed environment. While behaviour-based detection can be helpful, it is computationally demanding and may miss Malware that can elude detection in a sandbox. Without a sandboxed environment, neural networks can evaluate file activity and highly accurately detect malware [14].

Overall, neural networks outperformed traditional malware detection algorithms in several circumstances. On the other hand, the success of neural networks is determined by the quality and relevancy of the input data and the design of the neural network architecture. Therefore, it is critical to thoroughly examine the efficacy of various malware detection algorithms and select the best solution for the individual application.

2.5 The Present Status of Neural Network-Based Malware Detection Research

Neural network-based malware detection is an active study area, with notable breakthroughs in recent years. Among the current research areas are:

Transfer learning is employing a previously trained neural network for a different job and fine-tuning it for malware detection. This method can minimise the training data necessary while improving neural network performance.

Adversarial attacks entail altering the input to a neural network to avoid detection or generate a false positive. As a result, researchers are looking for ways to increase the robustness of neural networks in malware detection against adversarial assaults.

Explainability and interpretability: Neural networks may be difficult to interpret, which can be an issue in malware detection applications where understanding the reasoning for a particular categorisation result is critical. Researchers are working on techniques to increase the explainability and interpretability of neural network models used to identify Malware.

Hybrid models enhance malware detection performance by combining various detection methods, including signature-based and neural network-based detection. Edge computing is processing at the network's edge, such as on a mobile device or IoT device, rather than in a centralised server. Researchers are researching using neural networks for edge-based malware detection to increase the detection process's efficiency and scalability.

The present level of research in neural network-based malware detection is focused on enhancing neural network models' accuracy, speed, and robustness. As neural networks progress, they more successfully detect and prevent malware assaults.

2.6 Current Developments in Neural Network Designs and Training Methods

Recent advances in neural network topologies and training methodologies have helped to produce more powerful and effective malware detection models. Among the noteworthy advancements in this field are:

CNNs (convolutional neural networks): CNNs are a neural network that has proven highly successful in image recognition applications. They have been used to identify Malware by portraying binary data as pictures and extracting characteristics from these images using CNNs. This method has been demonstrated to be highly efficient in identifying Malware, with the bonus of viewing the features used by the model for categorisation [8].

RNNs are neural networks that can process sequential data, such as time series or text data. They have been used to identify Malware by expressing a program's activities as a series of API or system calls. RNNs may then be trained on this data to detect aberrant activity that may indicate Malware.

GANs (generative adversarial networks) are neural networks that can produce synthetic data comparable to actual data. They've been used to identify Malware by creating simulated malware samples that may be used to train and evaluate neural network models. This method could create more diverse and representative data for neural network training, improving their performance (Maria Rigaki & Sebastian Garcia, 2018).

Transfer learning is employing a previously trained neural network for a different job and fine-tuning it for malware detection. This method can minimise the training data necessary while improving neural network performance.

Adversarial training adds adversarial samples to the training data to increase the neural network's robustness against hostile attacks. This method has been found to improve the resilience of neural network models for malware detection.

Overall, advances in neural network topologies and training methodologies have aided in creating increasingly powerful and effective malware detection models. However, more complex and successful models are expected to be completed as research in this field develops.

2.7 Current Research's Challenges and Limits

Despite the encouraging outcomes of neural network-based malware detection, several obstacles and limits in current research must be solved. Among these difficulties are:

Malware data are frequently uneven, with a small number of malware samples outnumbering many benign ones. This might result in skewed models and lower detection rates for malware samples. Data augmentation and class weighting must be used to b data and enhance model performance to address this issue.

Malware makers can employ adversarial tactics to change malware code to avoid detection by neural networks. This can lead to false negatives and lower detection rates. There is continuing research to construct robust neural network models capable of detecting adversarial assaults and improving detection rates.

Interpretability: Because neural networks are generally regarded as "black box" models, it might be challenging to grasp how they make their predictions. This lack of interpretability can make identifying and correcting model faults easier. Techniques for improving the interpretability of neural network models for malware detection are being developed.

Overfitting: Neural networks can overfit specific datasets, resulting in poor performance on new, previously unknown data. Creating neural network models that generalise effectively to new data is an ongoing issue in malware detection research.

Scalability: Since neural networks take significant computer resources and time to train, they might be challenging to scale to massive datasets or real-time detection applications. Researchers are working to build more efficient neural network topologies and training methodologies to solve this difficulty.

Overall, resolving these issues and constraints is crucial for progressing in neural network-based malware detection and increasing computer system and network security.

2.8 Case Studies of Effective Neural Network-Based Malware Detection Systems

Numerous effective neural network-based malware detection systems have been created in recent years. Here are a couple of such examples:

DeepMal: DeepMal is a deep learning-based malware detection solution that analyses portable executable (PE) files using a convolutional neural network (CNN). The method beat numerous classic machine learning-based malware detection systems in detecting known and unknown malware samples [2].

N-BaIoT is a neural network-based solution that detects Malware in IoT devices. To monitor network traffic and detect Malware, the system employs a hybrid neural network design that combines a CNN and a long short-term memory (LSTM) network. As a result, the system detected Malware with great accuracy and beat numerous state-of-the-art malware detection methods.

MalwareNet is a deep neural network-based solution that detects Malware in Android apps. Our system uses a deep neural network to detect malware by analysing permissions, API calls, and file actions. It's more effective than traditional methods, identifying known and unknown samples.

DeepHunter: DeepHunter is a malware detection solution that uses deep learning to detect new and undiscovered malware samples. The system analyses the binary properties of the malware sample and sees similarities with known malware families using a stacked autoencoder neural network. The approach surpassed various state-of-the-art malware detection techniques to detect unknown malware samples [9].

Overall, these neural network-based malware detection systems indicate the promise of deep learning in enhancing malware detection accuracy and efficacy, and they serve as a foundation for future study in this field.

3 Future Directions and Possible Developments

GANs (generative adversarial networks) have lately been investigated as a viable solution for virus detection. GANs comprise two neural networks: a generator and a discriminator.

The generator produces new data samples that mimic the distribution of actual malware samples, while the discriminator differentiates between authentic and fabricated models. Both networks are trained together, with the generator striving to create convincing malware samples that can deceive the discriminator and the discriminator trying to detect which examples are legitimate and artificially generated. The use of GANs for malware detection offers various possible benefits. One advantage is that the generator may generate synthetic malware samples comparable to real-world Malware even if the training dataset is restricted. Another advantage of GANs is that they can detect new forms of Malware that were not included in the training dataset by producing samples that were never seen before (Maria Rigaki & Sebastian Garcia, 2018).

Unfortunately, there are significant obstacles and limits to using GANs for malware detection. One disadvantage is that GANs can be challenging to train since the two networks must be properly balanced to avoid one network overwhelming the other. Another issue is that GANs are vulnerable to adversarial assaults, in which an attacker creates samples tailored to deceive the discriminator and avoid detection.

Notwithstanding these obstacles, research into the use of GANs for malware detection is underway, and this technique can potentially increase the accuracy and efficacy of malware detection systems.

3.1 Detecting Malware via Reinforcement Learning

The process of educating an agent to make choices in situations where rewards and penalties are given is known as reinforcement learning or RL. The viability of using RL to enhance malware detection has been examined. Using RL-based malware detection, an agent learns to identify malicious software or processes based on their behaviour and the environment in which they operate.

The agent receives bonuses for correctly identifying Malware and suffers consequences for false positives or negatives. The objective is to teach the agent to accurately determine if a specific program is Malware, even if the infection is intricate and capable of eluding conventional detection techniques.

RL-based malware detection has the benefit of being adaptable to new and changing disease strains. The agent may adapt its decision-making process when the infected environment changes due to its experiences.

RL for malware detection has a variety of drawbacks and restrictions, which is unfortunate. For example, as the agent must investigate various speculative possibilities and learn from its experiences, RL-based systems may be challenging to train. Another issue is that RL-based systems could be susceptible to adversarial assaults, in which a perpetrator tries to influence the surroundings to deceive the agent into doing the wrong action.

Notwithstanding these difficulties, research into the use of RL for malware detection is ongoing, and this te potentially improves the precision and effectiveness of malware detection systems.

3.2 Prospects for Neural Network-Based Malware Detection

Future developments in neural network-based malware detection might increase the accuracy and effectiveness of these systems.

Advanced neural network structures, such as deep neural networks and convolutional neural networks, are a crucial area of study in this field. They are specifically crafted to handle the complex tasks of malware detection with utmost efficiency. In addition, the efficiency of neural network-based malware detection may be enhanced by similar architectures, which have shown promising results in other areas, including picture and voice recognition.

Another area of research is using more challenging training approaches, such as transfer learning and meta-learning, which may allow neural networks to incorporate knowledge gained from earlier tasks or experiences to improve their ability for malware detection. To increase its ability to detect Malware, a neural network trained on a large dataset of benign software, for instance, would be able to transmit part of that knowledge. With neural network-based malware detection systems, combining several data types, such as network traffic data, system call data, and file information, is possible.

By mixing several data types, neural networks could spot patterns and anomalies that aren't apparent from a single data form.

Lastly, using reinforcement learning and machine learning to create adaptive malware detection systems that can react to threats shows promise.

To constantly changing virus types. However, with ongoing learning and adaptation, these systems can stay ahead of the always-evolving danger scenario.

Overall, neural network-based malware detection has several potential advancements, and ongoing research in this area is expected to result in future improvements in the precision and efficacy of these systems.

3.3 Neural Network Forensic Application for Malware Detection

The forensic use of neural networks for malware detection entails employing these networks to examine digital data and identify malware risks. In addition, forensic analysis is critical for determining the breadth and effects of a malware attack and implementing appropriate mitigation solutions.

Neural networks may be trained on enormous datasets of known Malware and benign files to understand the traits and properties of each type of file. Once introduced, these networks may categorise new files as harmful or harmless. This method can assist forensic investigators in swiftly identifying potentially dangerous data and prioritising their investigation.

Neural networks may also be used for behavioural analysis, which entails monitoring system activity to detect malicious conduct. For example, neural networks can recognise abnormal movements and highlight possible malware risks by evaluating patterns in system activity. This method can assist investigators in identifying Malware that has eluded standard detection techniques and give insights into attacker behaviour and strategies.

The network must be trained on a diverse and comprehensive dataset to ensure successful malware detection in forensic applications using neural networks. Outdated or insufficient datasets may hinder the network's ability to accurately identify new or advanced threats due to the constant evolution of malware. Nevertheless, neural networks are simply one tool in the forensic investigator's toolbox, and its output must be understood in the context of other evidence and information.

Overall, the forensic use of neural networks for malware detection shows promise as a solid and efficient technique for identifying and reducing malware risks.

As technology advances, neural networks are anticipated to play an increasingly crucial role in digital forensics.

4 Conclusion

4.1 *A Description of the Effect of Neural Networks on Malware Detection*

Finally, neural networks have had a significant impact on malware detection. Traditional malware detection methods can no longer keep up with the increasing sophistication of malware attacks. Because of their ability to learn patterns and detect irregularities in massive datasets, neural networks have emerged as a potential method for finding and detecting Malware. When identifying malware, one sound choice is to deploy neural networks, given its numerous benefits above traditional methods, including efficient management of large datasets through adaptation when encountering new threats. Although there are a few drawbacks to deploying this algorithm l, like exacting superior quality training data while still being prone to generating false positive or negative outcomes, it is still more efficient compared with common approaches for detecting malicious activities online. Furthermore, the recent advancements made through improved tactics and enhanced topologies depict positive signs concerning the future effectiveness in identifying different types of malware.

This trend is anticipated to continue as researchers develop new strategies and methods to improve the precision and effectiveness of neural network-based malware detection systems.

4.2 *Future Prospects and Potential Effects on the Computer System and Network Security.*

1. **Improved Detection Accuracy**: Neural networks are exceptionally good at identifying intricate patterns and separating essential elements from big datasets. This feature enables neural networks to analyse massive volumes of data, including file structures, behavioural patterns, network traffic, and code fragments, for malware identification. Furthermore, neural networks may learn complex correlations between these features by training on various malware samples, leading to the development of precise models for recognising malicious software. As a result, detection accuracy increases, making it possible to identify known and new malware strains. In malware detection, neural networks have greatly improved detection accuracy. Signature-based approaches traditionally used predetermined patterns and signatures to identify known malware variants. These techniques, however, could not locate polymorphic and obfuscated malware, which may change its signature or display variations in its code structure. Neural networks have overcome this restriction thanks to their ability to learn intricate patterns and extract valuable characteristics from big datasets. Furthermore, neural networks

may identify subtle markers and associations within the data that suggest malicious behaviour by training on various malware samples. Thanks to this, they can accurately identify malware versions that are both known and undiscovered. The ability of neural networks to analyse multiple data sources, including file structures, behavioural patterns, network traffic, and code fragments, is one of their most vital points. Neural networks can find hidden patterns and correlations by analysing these factors that might not be visible through manual analysis or conventional methods. As a result, malware detection is more successful and has better detection accuracy. Additionally, neural networks are excellent at managing massive datasets, enabling them to handle enormous volumes of data quickly. Given the exponential growth in malware samples, this capacity is vital for malware identification. Since neural networks can analyse this data quickly, they can identify malware more quickly and precisely.

Furthermore, improved detection accuracy results from neural networks' capacity to adapt to and learn from new data. Neural networks may update their models and incorporate this further information into their detection algorithms as new malware strains appear. Because of their flexibility, neural networks can efficiently recognise even the most recent and sophisticated malware strains and stay current with the most recent threats.

2. **Robust Feature Extraction**: Conventional signature-based approaches mainly rely on preset patterns and signatures, which limits their efficacy against polymorphic and obfuscated malware. On the other hand, neural networks may automatically discover and extract pertinent features from unprocessed data, negating the need for intentional feature engineering. By picking up on tiny signs of harmful behaviour, this adaptability enables neural networks to adapt to new malware types and significantly increases detection rates. The capacity of neural networks to perform reliable feature extraction is one of their primary advantages in malware detection. In this case, malicious and benign software, significant traits or patterns must be extracted from raw data to distinguish between distinct groups or categories. Traditionally, selecting specific qualities considered indicative of malware required manual selection and subject knowledge. The complexity and diversity of malware, as well as the laborious process of feature selection, can put a limit on this technique. On the other hand, neural networks are excellent at automatically learning and extracting features from unprocessed data. As a result, there is no longer a requirement for explicit feature engineering, and the neural network can detect subtle and complicated patterns that human-designed feature sets would miss. Neural networks can analyse a wide range of data in the context of malware detection, including binary file formats, API calls, network traffic, system logs, and behaviour patterns. They can record data at various abstraction levels, from simple byte sequences to complex behaviour sequences. Neural networks analyse the input data and derive meaningful representations using layers of interconnected nodes (neurons). These networks modify the weights and biases of the neurons through a process known as backpropagation to reduce the discrepancy between the projected output and the actual output. This optimisation process lets the network learn and recognise

the most pertinent features for differentiating between malware and good software. The ability of neural networks to automatically learn intricate, hierarchical representations of data gives them an edge in feature extraction. As a result, the network can recognise clear indicators and abstract, higher-level features that reveal malware's fundamental makeup and behaviour. Neural networks, for instance, can spot recurrent code patterns, strange system calls, or unusual network traffic that point to malicious activities. Additionally, neural networks can modify their feature extraction skills to account for newly emerging and changing malware kinds. Neural networks may generalise their learned features and use them to identify malware samples that had not previously been observed by training on various datasets that cover a wide range of malware families and variants. Nevertheless, it is important to remember that the calibre and representativeness of the training data significantly impact how well feature extraction in neural networks performs. Sufficient and diverse training datasets are essential to guarantee that the network learns pertinent properties that may generalise successfully to real-world malware samples.

3. **Behavioural Analysis**: Neural networks are excellent at analysing malware's intricate behavioural patterns. Neural networks may learn to distinguish between legitimate and harmful behaviour by watching how malware interacts with a system, including file alterations, network communications, and process executions. This behavioural analysis method works exceptionally well in spotting complex malware that may use sophisticated evasion strategies or display zero-day traits. Anomalies that deviate from typical system behaviour can be detected using neural networks, which can also provide alerts. Neural networks have successfully tackled the vital component of malware detection known as behavioural analysis. Behavioural analysis entails watching and examining how software or code interacts and behaves to spot trends or deviations from expected behaviour. It focuses on comprehending the behaviour displayed by malware in the context of malware detection to distinguish it from legitimate software.

The capacity of neural networks to learn and recognise intricate patterns in data makes them particularly effective in behavioural analysis. They can be trained on big datasets that record both malware and good software behaviours, allowing them to comprehend the everyday interactions and behaviours of various software kinds.

Here is how neural networks use behavioural analysis to detect malware:

Various aspects of a system's behaviour, including file modifications, registry changes, network communications, process executions, and API requests, can be observed using neural networks. Neural networks can detect abnormalities or variations from typical behaviour that can suggest the presence of malware by recording and examining these behavioural occurrences.

Anomaly Detection: Neural networks are good at spotting patterns and can spot changes in how a system behaves typically. In the training phase, neural networks build a baseline of typical behaviour to identify unexpected or suspicious behaviours that deviate from the expected patterns. This helps them to identify malware that is zero-day-aware or malicious variants that display novel behaviours.

Evasion Methods: Advanced malware frequently uses sophisticated evasion methods to prevent detection. These evasive behaviours, such as code obfuscation, anti-analysis tactics, or attempts to conceal or modify their presence, can be identified by neural networks with practice. Furthermore, neural networks can improve the detection of such evasive malware strains by examining the behavioural patterns linked to evasion.

Dynamic Analysis: Based on continuing behavioural analysis, neural networks can process real-time data and make judgements. They can respond quickly to developing malware threats thanks to their ability to adapt to changing circumstances. Neural networks can instantly detect and respond to dangerous behaviours in real-time by continuously observing and analysing system behaviour, offering a proactive defence against malware.

Contextual Understanding: Neural networks can record and examine the context of system behaviour. This comprises elements like the chronology of events, timing, interdependencies between activities, and connections between various parts. As a result, neural networks can gain a more thorough grasp of the behaviour displayed by malware and improve the accuracy of their detections by considering these contextual elements.

4. **Adaptability and Generalisation**: Neural networks can apply their understanding to other virus families. They can accurately detect new and undiscovered malware types after being trained on various datasets. This adaptability is essential in today's quickly changing threat environment, as attackers continuously create new malware strains to circumvent established defences. Neural networks provide a proactive defence system that can adapt to new threats.

In malware detection, neural networks' primary advantages include adaptability and generalisation. These characteristics allow neural networks to efficiently identify and categorise known and unidentified malware variants and adapt to changing cybersecurity threats.

Neural networks can modify and update their knowledge in response to new input. As a result, neural networks can be retrained on updated datasets to include understanding these new dangers as new malware types are found. This adaptability guarantees that the network stays current with the most recent malware trends and may precisely identify emerging threats that may have eluded conventional signature-based detection techniques.

1. 2. **Generalising Across Malware Families**: Neural networks gain knowledge from various datasets covering various malware families and variants. By training on multiple samples, neural networks can identify underlying patterns and traits shared by several malware strains. By identifying similarities and shared features between malware families, they can use this generalisation to find new malware variants with similar structures or behaviours.
2. **Determining Unknown Malware**: Neural networks can identify zero-day malware, software that has yet to be discovered. This is accomplished through their ability to pick up on the traits and actions displayed by well-known malware

strains. In addition, neural networks may detect departures from the norm even without detailed knowledge about a specific malware strain by comprehending the underlying patterns and traits that separate malicious software from benign software.

3. **Transfer Learning**: To adapt their expertise from one malware detection task to another, neural networks can use transfer learning. Transfer learning uses skills developed during training on one dataset to enhance performance on another related dataset. This indicates that, in the context of malware detection, a neural network trained on a substantial and varied malware dataset can transfer its learned features, representations, and behaviours to enhance detection accuracy on different malware detection problems.
4. **Variability Resistance**: Neural networks are renowned for their capacity to deal with erratic and noisy input. This robustness is helpful for malware identification when dealing with polymorphic malware that can alter its structure or behaviour to avoid detection. In addition, neural networks can capture the constant underlying patterns and characteristics across many malware strain changes, allowing them to correctly categorise such malware even when it manifests in various ways.

Neural networks offer a proactive and adaptable malware detection method by fusing flexibility with generalisation. They can manage variances in malware strains, generalise their knowledge across many malware families, detect unknown malware, and use transfer learning. They can also adapt to new threats. Because of these features, neural networks are valuable weapons in the continuous conflict against sophisticated and constantly changing malware threats.

1. **Limitations and Challenges**: While neural networks have many benefits, detecting malware presents particular difficulties. The requirement for sizable and representative training datasets is a significant obstacle. Gathering and tagging various malware samples can take time and resources. Additionally, when harmless files are mistakenly labelled as malware or vice versa, false positives and negatives may occur when using neural networks. The neural network models must be continuously monitored and improved to lessen these difficulties. While neural networks have much to offer regarding malware detection, several restrictions and challenges must be considered. The following are some significant drawbacks and problems related to the application of neural networks in this field:
1. **Large and Representative Training Datasets**: For neural networks to perform at their best, large volumes of training data are necessary. Gathering and categorising a broad and representative array of malware samples might take time and effort. The quantity and calibre of training data can significantly impact how well neural networks detect malware.
2. **False Positives and False Negatives**: Although resistant, neural networks are not impervious to them. False positives happen when suitable files or actions are mistakenly categorised as malicious, resulting in pointless alarms and potential disruptions. False negatives, on the other hand, occur when malware slips through

the cracks and lets dangerous activity continue unchecked. Therefore, the neural network models must be adjusted and fine-tuned to reduce false alarms and missed detections.

3. **Interpretability and Explainability**: Because neural networks are frequently viewed as "black boxes," it might be challenging to understand how they make decisions. This lack of interpretability can make determining if a file or action is malicious or benign complex. The capacity to defend a choice is crucial in several circumstances, including legal inquiries and regulatory compliance.
4. **Avoiding Adversarial Attacks**: Neural networks can be subject to adversarial attacks, in which evildoers purposefully alter or create inputs to trick the web and avoid detection. For example, malware samples can be changed to evade detection or adversarial instances that look harmless but are mistakenly identified as malicious can be made. As a result, neural network-based malware detection faces continual challenges in creating effective defences against such assaults.
5. **Resource Needs**: Neural network training and deployment for malware detection can be computationally demanding, needing significant computational time and resources. Due to their intricacy, deep learning models and other neural network topologies might require a lot of processing power and memory. For deployment to be realistic, it is essential to implement effective and scalable ways to handle these resource requirements.
6. **Dynamic and Evolving Danger Environment**: As new malware variants and attack strategies often appear, the dangerous climate for malware is constantly changing. To respond to evolving threats effectively, neural networks must be regularly updated and trained on new datasets. It's a constant challenge to stay on top of malware developments and ensure that neural network models accurately represent current malicious behaviours.

Cyber security experts, data scientists, and machine learning specialists must continue researching, developing, and collaborating to address these constraints and obstacles. However, these challenges can be overcome, and malware detection capabilities can be strengthened by ongoing enhancement and refinement of neural network-based malware detection systems.

References

1. Narayanan BN, Djaneye-Boundjou O, Kebede TM (2016) Performance analysis of machine learning and pattern recognition algorithms for Malware classification
2. Yang C, Xu J, Liang S, Wu Y, Wen Y, Zhang B, Meng D (2021) DeepMal: maliciousness-preserving adversarial instruction learning against static malware detection
3. Catak FO, Ahmed J, Sahinbas K, Khand ZH (2021) Data augmentation-based malware detection using convolutional neural networks
4. Goodfellow IJ Pouget-Abadie J, Mirza M, Xu B, Warde-Farley D, Ozair S, Courville A, Bengio Y (2014) Generative adversarial networks
5. Singh J, Singh J (2021) A survey on machine learning-based malware detection in executable files. J Syst Arch

6. Gotthans J, Gotthans T, Novak D (n.d.) Improving TDOA radar performance in jammed areas through neural network-based signal processing
7. Rigaki M, Garcia S (2018) Bringing a GAN to a knife-fight: adapting malware communication to avoid detection
8. Recent advances in convolutional neural networks (2018)
9. Wei R, Cai L, Zhao L, Yu A, Meng D (2021) DeepHunter: a graph neural network based approach for robust cyber threat hunting
10. Rita (2022) Neural networks for pattern recognition
11. Anandhi V, Vinod P, Menon VG, Aditya KM (2022) Performance evaluation of deep neural network on malware detection: visual feature approach
12. Wise A, Kruglyak KM (2019) Variant classifier based on deep neural networks
13. Lee Y-S, Lee J-U, Soh W-Y (2018) The trend of malware detection using deep learning
14. Bazrafshan Z, Hashemi H, Fard SMH, Hamzeh A (2013) A survey on heuristic malware detection techniques

Evaluating Different Malware Detection Neural Network Architectures

Harinadh Varikuti and Valli Kumari Vatsavayi

Abstract In today's digital world, cutting-edge technologies in machine learning, cybersecurity, data science, and blockchain are emerging rapidly. Every day billions of devices are connected together through internet for data exchange. Due to the existence of some vulnerabilities present in the device, attackers perform malicious activities through such devices. Detecting the malwares using anti-malware engines and signature-based methods is becoming more and more complex day by day. Various techniques are used by the attackers in the generation of malwares such as encryption, obfuscation, etc., to escape the detection. Few years ago, in traditional machine learning-based malware detection algorithms, feature engineering was the most complex activity to perform. Inefficient extraction of features gives poor performance in the detection of malware samples. Neural network techniques in deep learning give better results in the identification of malware data. This chapter presents a survey on the performance of various deep learning-based neural networks for malware detection and classification. It is also noted that these models require a lot of computational time during training. Transfer learning is a technique used to reduce the training time by using pre-trained models such as VGG-16, ResNet-50, Inception V3, etc. This chapter explains how the evolution of neural networks changed the malware analysis giving exceptional results during detection of malware samples. It classifies, compares, and evaluates several deep learning-based malware detection approaches published in literature elaborating their advantages, disadvantages, applicability, assumptions, and scope.

Keywords Neural networks · Malware · Detection · Classification · Machine learning · Deep learning · Transfer learning

H. Varikuti (✉) · V. K. Vatsavayi
Andhra University, Visakhapatnam, India
e-mail: varikutiharinadh@gmail.com

V. K. Vatsavayi
e-mail: vallikumari@gmail.com

K. Kaushik and I. Sharma (eds.), *Next-Generation Cybersecurity*, Blockchain Technologies, https://doi.org/10.1007/978-981-97-1249-6_11

1 Introduction

The cybersecurity industry is growing rapidly with the growth in the number of users and connectivity. A large number of devices are interconnected for the exchange of data, operations, etc. Insecure connection of devices causes lot of vulnerabilities and security issues in the network. Meanwhile, attackers have been trying to gain access to targeted systems by sending malicious files. With the new technologies at hand, attackers attempt new obfuscation techniques to create a variety of malware files that escape detection using traditional malware detection systems. Attackers perform cybercrimes which include data theft, data destruction, money theft, denial of service, hacking the operations of the system, etc. According to cybercrime reports prediction [1], 8 trillion USD will be lost annually due to cybercrime damage in 2023. The total loss is more than most of the developed countries' economies. The growth in cyber threat damage demands focus on research towards overcoming security vulnerabilities and building resilient secured systems. A telecommunication company named T-Mobile suffered data breach [2] recently in January 2023. In this attack, over 37 million customers' data was stolen by the attacker through Application Programming Interface (API). Due to some vulnerability identified in the API, the attacker retrieved customer personal information through it. These types of data breaches affect the customer's trust in the organization. Ransomware attacks are the popular malware attacks that happen more frequently throughout the world. Attacker [3] takes advantage of vulnerabilities present in the security layer of organizations and encrypts all the files related to it. Organizations include private businesses, healthcare, and government institutions. Recent ransomware attackers used triple extortion techniques [4] in the attacks to get more profits.

According to the malware analysis market [5], the predicted global malware analysis market size is likely to reach USD 24.1 billion by 2026 with a Compound Annual Growth Rate (CAGR) of 28.5% between 2019 and 2026.

1.1 Malware Analysis

Detecting the malware is done by analyzing the malware samples. Malware analysis depends on the functionality and behavior of the code present in it. Daily a large number of new malwares are generated by the attackers using various techniques such as obfuscation, compression, etc. to escape detection. Malware analysis is performed in three ways (i) static analysis, (ii) dynamic analysis, and (iii) hybrid analysis.

(i) Static analysis is about studying the code present in malware file, for example, in the Windows executable file without its execution. It examines the code structure related to functions, libraries, and operations used by the executable file. This type of analysis initially converts executable files into low-level assembly language code files to get information about what exactly the code performs. Sometimes, extraction of strings from the executables gives hints about the file functionality. Strings include

DLL files, domain names, attacker file paths, IP addresses, etc. Tools used in the process of static analysis are Dependency Walker, PEView, IDA freeware, IDA pro, and Ghidra, etc. The dependency walker scans the entire executable file and represents a hierarchical tree with all dependent modules. PEView is used to view portable executable files with header information of the file. IDA freeware is an open tool to perform analysis on malware binaries with limited functionality, whereas IDA Pro is an enterprise tool, which is commonly used by reverse engineers and security analysts. Attackers use obfuscation techniques in the malware generation to evade detection. In the obfuscation method, encryption and compression have been used to generate the code which is difficult to get the information using traditional tools mentioned. Static analysis depends mostly on the domain knowledge of the analyst and also misinterpretation of data may lead to errors in detecting the malware.

While considering the limitations of static analysis, dynamic analysis gives reliable results in the detection of malicious files. Dynamic analysis is the process of performing the execution of malware executable files in a safe isolated environment. It tells the way in which the malicious code interacts with the system. This type of analysis has been performed by isolated physical systems or virtual machines. Besides those two environments, dynamic analysis has been performed in a sandbox environment to detect the malware. Tools used for analyzing malware dynamically are Process Monitor, Process Explorer, Regshot, NetCat, Wireshark, etc. The dynamic analysis gives better results in identifying the zero-day malware when compared with the static analysis approach. The major limitation of performing the dynamic analysis is providing a safe environment, because erroneous setting of the environment will infect the system and damage it badly.

Hybrid analysis includes the combination of both static and dynamic analysis to perform malware detection. Information obtained from both approaches has been efficiently used to perform the malware analysis.

1.2 Machine Learning in Malware Detection

A few years ago, anti-malware engines have been used to perform malware detection. These engines used signature and heuristic-based techniques to detect the malware. In the signature-based method, known malwares are identified by matching the signatures present in the anti-malware database. Unknown malware is not detected by the engines which limits its capability. The heuristic-based method is the process of investigating the source code for apprehensive action. Polymorphic malware and zero-day malware have been detected using the heuristic methods. But there is a problem of getting false positives and false negatives with this method. False positive, which identifies benign files as malware and doesn't affect costly, whereas false negatives identify malware files as benign which destructs badly to the system.

Nowadays with the increase of digital data, it is difficult for humans to continuously monitor and analyze to take decisions. To overcome this issue, machine learning has been used to train the model on viable data and apply the model to similar data items for taking decisions. Attackers also apply various technologies in generating malicious code to perform illegal operations throughout the world. Digitalization increased the opportunities for attackers with more volume availability. On the other side, the difficulty increases in detecting malicious structures manually for huge volumes of data. Here machine learning techniques are used to perform malware analysis and detect malicious files. Along with the detection of malware files, classifying the malware class is also important to know the functionality of the malware file. Machine learning approaches initially perform feature engineering steps to extract the features. Features are extracted based on the user domain knowledge. Here the features identified are used in building the model using trained dataset to perform the required operation. Based on the trained model, test data is used to find the performance of the model. Various malware detection models built using machine learning are discussed in the related work section.

1.3 Artificial Neural Networks

Traditional machine learning methods believe in user-extracted features and their analysis. But these hand-crafted feature models sometimes lead to inefficient models due to inappropriate feature selection. To overcome the problem that occurred by the manual feature selection, deep learning models perform automatic deep feature extraction using neural networks. Here artificial neurons are created and connected to form artificial neural networks. Every neuron is associated with weight when it has been given as input to the next layer. An artificial neural network includes an input layer that takes source data, one or more hidden layers which analyze and perform computation on the data to extract features, and an output layer that predicts the output of the source data provided. Applications, where neural networks play an important role, include chatbots, pattern recognition, machine vision, speech recognition, medical, telecommunication, transport, and time series prediction, etc. unlike hand-crafted features used in machine learning models, neural networks extract meaningful patterns from the data to give better performance when compared to traditional machine learning models. Neural networks have been performing exceptionally well in the field of malware detection and classification. Detection of malware samples relies on the behavior and structure of the malware. To detect the malware samples, patterns are identified as feature vectors by the neural network.

2 Related Work

In this section, various research works published in literature on malware detection and classification are discussed. Using machine learning and neural network models, efficient identification of malware is possible. Traditional machine learning approaches extract the hand-crafted features in the feature generation process. With limited domain expert knowledge, feature extraction is difficult to perform. By observing the immense growth in the machine learning field, malware analysis has advanced toward deep learning. Using deep learning techniques, an end-to-end model takes input of malware sample with little preprocessing and directly detects its related malware family as output. Malware detection also is efficiently performed by the neural network models by differentiating the behavior patterns of malware and benign files. To deal with malware bytes files, various approaches [10–16] use grayscale images for malware classification. Narayanan et al. [11] used principal component analysis (PCA) for getting reduced features from the images, then various machine learning algorithms such as K-nearest neighbor, support vector machines, and artificial neural network (ANN) were applied. K-nearest neighbor gives the best performance with the PCA features. In the paper [12], the authors proposed deep learning model which uses autoencoders [16] in the multi-layer architecture.

As per the studies, Convolutional Neural Network (CNN) models performed well on the image dataset. Gibert et al. [14] presented a CNN architecture for the classification of malware images, which performs better with minimal prediction time. Gibert et al. [13] represented the malware executable as an entropy stream [17], which gives randomness of the system. Haar wavelet transform was used on the entropy times series to reduce the noise and size, and CNN model was applied for malware classification. The author concluded that wavelet features performed strongly against obfuscation techniques. Singh et al. [15] performed malware analysis on Android malware images using feature fusion methods. Malware APK files have been used to generate grayscale images. Hand-crafted feature extraction methods such as Gray-Level Co-occurrence Matrix (GLCM), Global Image descriptors (GIST), and Local Binary Pattern (LBP) uproot features. Traditional machine learning approaches such as support vector machine, K-nearest neighbor, and random forest have been applied on the above hand-crafted features. Feature fusion applied with CNN and hand-crafted features to perform malware classification. Feature fusion method gives better results when compared with traditional methods. Xiao et al. [16] used structural entropy graphs for extracting features using CNN model and then support vector machine method was applied to perform malware classification. Yamashita et al. [10] provided an overview of CNN model, also gave how to overcome overfitting using various techniques such as data augmentation, batch normalization, dropout, etc. Khan et al. [18] classified the malware using deep neural networks using binary files, later a CNN model was built using grayscale images as input. CNN model gives better results emphasizing the fact that convolution models are more efficient in classifying the image data. M. Alam et al. [19] used standard neural network models such as VGG-16, ResNet-50, and Inception v3 and proposed convolutional neural network for classifying the malware data which gives a significant performance. Depuru et al. [20] performed a comparative analysis of malware classification using neural network models and machine learning models. Shinde et al. [21] constructed four neural networks with malware data image sizes of 224×224, 240×240, 260×260, and 300×300, respectively. Using a large dataset, the neural network which uses high image pixel size gives better results when compared with low image pixel size.

Yamashita et al. [10] used neural network models to perform malware detection on the EMBER dataset and observed better performance. Akthar et al. [22] performed malware detection using machine learning models and CNN model. Among the machine learning models used, decision tree gives good results along with neural network system. Pavithra et al. [23] applied malware analysis using various machine learning classifiers in detecting the adware malware files. In their work, random forest, SVM, and Naïve Bayes algorithms have been used, in which random forest gives better results with lesser time complexity. Aslan et al. [24] proposed a hybrid model which uses pre-trained models such as AlexNet and ResNet-50 for feature engineering and combined features have been applied to fully connected layers and final output layer. Xing et al. [25] used autoencoders in the detection of Android malwares. Kumari VV et al. [26] proposed an efficient model for malware detection and also gives impact of neural networks in malware detection.

From the above-discussed papers, it is observed that mostly deep learning models give good results when compared with the machine learning models in classifying and detecting the malware samples for both balanced and unbalanced data. This paper provides implementation of machine learning and neural network models for the task of malware detection and classification and also gives better results when compared to the state-of-art models discussed above.

3 System Overview

3.1 Datasets

In this work, three benchmark datasets are used. They are (1) the MALIMG dataset for malware classification, (2) the Microsoft malware dataset for malware classification, and (3) the BODMAS dataset for malware detection and classification.

Malimg **Dataset**: The Malimg [7] dataset contains 9339 malware samples of grayscale images which belong to 25 malware families. The malware family distribution of the Malimg dataset is shown in Table 1. Allaple.A has a greater number of samples with a count of 2949 and Skintrim.N has the least number of samples with a count of 80.

Microsoft Malware Dataset: Microsoft [8] provided the malware dataset with 10,868 samples belonging to 9 malware families. It provides both bytes file and assembly code file for every malware sample. The data has been launched for the challenge hosted on Kaggle. The malware family distribution of Microsoft malware data has been shown in Table 2. Kelihos_ver3 has a greater number of samples with a count of 2942 and Simda has the least number of samples with a count of 42. It is considered as the benchmark dataset to evaluate the performance of machine learning algorithms in malware classification.

BODMAS Dataset: BODMAS [6] stands for Blue Hexagon Open Dataset for Malware Analysis. The researchers performed joint work with Blue Hexagon to

Table 1 Class distribution of the malimg dataset

Malware family	Number of samples
Allaple.A	2949
Allaple.L	1591
Yuner.A	800
Instantaccess	431
VB.AT	408
Fakerean	381
Lolyda.AA1	213
C2LOP.gen!g	200
Alueron.gen!J	198
Lolyda.AA2	184
Dialplatform.B	177
Dontovo.A	162
Lolyda.AT	159
Rbot!gen	158
C2LOP.P	146
Obfuscator.AD	142
Malex.gen!J	136
Swizzor.gen!I	132
Swizzor.gen!E	128
Lolyda.AA3	123
Adialer.C	122
Agent.FYI	116
Autorun.K	106
Wintrim.BX	97
Skintrim.N	80

Table 2 Class distribution of the microsoft malware dataset

Malware family	Number of samples
Kelihos_ver3	2942
Lollipop	2478
Ramnit	1541
Obfuscator.ACY	1228
Gatak	1013
Vundo	475
Tracur	751
Kelihos_ver1	398
Simda	42

Table 3 Malware category distribution of BODMAS dataset

Malware category	Number of samples
Trojan worm	29,972
Worm	16,697
Backdoor	7331
Downloader	1031
Ransomware	821
Dropper	715
Information stealer	448
Virus	192
Pua	29
Cryptominer	20
P2p-worm	16
Exploit	12
Trojan-game thief	6
Rootkit	3

produce timestamped malware data and related malware family information. The dataset includes 57,293 malware samples and 77,142 benign samples with a total of 134,435 samples. The total malware samples constitute 14 malware categories such as virus, worm, Trojan, downloader, dropper, ransomware, backdoor, information stealer, rootkit, p2p-worm, cryptominer, trojan-game thief, pua, and exploit. The malware samples cover 581 malware families. Here the data samples include the SHA-256 hash of the malware file and pre-extracted features. The total malware samples category distribution is shown in Table 3.

3.2 *Malware Representation and Features*

Malware files can be represented in different ways such as hex view (bytes files) and assembly view (assembly language files). In the bytes file, all the machine code which represents opcodes, functions, jump statements, control statements, texts, images, etc., are represented in hexadecimal format shown in Fig. 1a. The left-most value (e.g., 00401000) represents the starting address of the machine code in the memory and consecutive 16-byte values represent opcode or data. Malware samples are also represented as grayscale images shown in Fig. .b from the Malimg dataset. In the assembly language files, all the malware sample data has been represented in the form of mnemonic codes to represent machine codes, function calls, register information, etc., as shown in Fig. 2.

```
00401000 E8 0B 00 00 00 E9 16 00 00 00 90 90 90 90 90 90
00401010 B9 25 2B 56 00 FF 25 80 23 41 00 90 90 90 90 90
00401020 68 30 10 40 00 E8 70 03 01 00 59 C3 90 90 90 90
00401030 B9 25 2B 56 00 FF 25 74 23 41 00 90 90 90 90 90
00401040 E8 0B 00 00 00 E9 16 00 00 00 90 90 90 90 90 90
00401050 B9 24 2B 56 00 FF 25 78 23 41 00 90 90 90 90 90
00401060 68 70 10 40 00 E8 30 03 01 00 59 C3 90 90 90 90
00401070 B9 24 2B 56 00 FF 25 7C 23 41 00 90 90 90 90 90
00401080 B0 9F C2 04 00 90 90 90 90 90 90 90 90 90 90 90
00401090 83 05 C4 24 56 00 20 8B 15 A4 24 56 00 85 D2 56
004010A0 B9 7B 00 00 00 75 10 A1 18 24 56 00 85 C0 75 07
004010B0 B8 FD 00 00 00 EB 37 8B 35 C4 24 56 00 B8 60 00
004010C0 00 00 2B C6 8B 35 60 24 56 00 83 EA 7B 83 FE 5C
004010D0 89 15 A4 24 56 00 B9 7F 00 00 00 7D 11 83 E8 39
004010E0 83 C6 3B B9 41 1F 00 00 89 35 60 24 56 00 8B 35
004010F0 74 24 56 00 83 FE 7B 7E 08 29 05 70 24 56 00 EB
00401100 0A 81 05 1C 24 56 00 FD 00 00 00 85 F6 5E 74 06
00401110 83 F0 30 8D 48 85 83 F9 34 7E 11 A1 24 24 56 00
00401120 83 F8 59 7F 07 2B C2 A3 24 24 56 00 B0 06 C3 90
00401130 51 53 33 C0 A0 F4 23 56 00 56 57 BE 91 00 00 00
```

(a)

(b)

Fig. 1 **a** Hexadecimal representation of sample malware file from the Microsoft malware dataset, **b** Grayscale image of sample malware file from the Malimg dataset

```
.text:0040102B CC CC CC CC CC                          align 10h
.text:00401030 56                              push    esi
.text:00401031 8B F1                              mov     esi, ecx
.text:00401033 C7 06 08 BB 42 00                      mov     dword ptr [esi], offset off_42BB08
.text:00401039 E8 13 1C 00 00                         call    sub_402C51
.text:0040103E F6 44 24 08 01                         test    byte ptr [esp+8], 1
.text:00401043 74 09                              jz      short loc_40104E
.text:00401045 56                              push    esi
.text:00401046 E8 6C 1E 00 00                         call    ??3@YAXPAX@Z    ; operator delete(void *)
.text:0040104B 83 C4 04                           add     esp, 4
.text:0040104E
.text:0040104E                         loc_40104E:                ; CODE XREF: .text:00401043↑j
.text:0040104E 8B C6                              mov     eax, esi
.text:00401050 5E                              pop     esi
.text:00401051 C2 04 00                           retn    4
.text:00401051                         ; ---------------------------------------------------------------
.text:00401054 CC CC CC CC CC CC CC CC CC CC CC CC             align 10h
.text:00401060 8B 44 24 08                        mov     eax, [esp+8]
.text:00401064 8A 08                              mov     cl, [eax]
.text:00401066 8B 54 24 04                        mov     edx, [esp+4]
.text:0040106A 88 0A                              mov     [edx], cl
.text:0040106C C3                              retn
```

Fig. 2 Assembly language code of sample malware file

To perform the malware classification or detection, features have been extracted from the malware binary files. Feature extraction plays a significant role in the performance of the model. Various techniques like N-Gram analysis and image-based analysis have been used to extract the features.

N-Gram Analysis

N-gram analysis is the standard technique to extract the features for various problems along with malware detection and classification. N-grams are the sequence of n data items considered as a feature and count the frequency of the sequence in the entire document. In malware bytes file, bytes N-grams are considered. Opcode, registers, keywords, prefixes n-grams are considered from the assembly language source files.

1-g, 2-g, 3-g, etc., are examples of n-grams. Features extracted using 1-g are discussed below.

Byte_histogram. It counts the frequency of each byte value in the file. Byte values range from 00 to ff which is 256 values. Same family malware files have nearly matched the frequency of byte values.

Prefixes. These features are extracted from assembly files. In assembly language files, various sections are present such as data section, bss section, text section, etc. Features extracted include HEADER, text, pav, idata, data, bss, rdata, edata, rsrc, tls, reloc, BSS, and CODE.

Opcodes. Assembly code contains various operation codes which give behavioral aspect of the file. Opcode features considered are rol, jnb, jz, rtn, lea, movzx, push, pop, dec, add, imul, xchg, or, xor, retn, nop, sub, inc, shr, cmp, call, shl, ror, jmp, mov, and retf.

Keywords. It includes std, dword, and dll.

Registers. It stores data values used for processing in assembly files. Registers used in assembly code are eax, edx, esi, eip, ecx, ebx, ebp, esp, and edi.

The BODMAS dataset used for malware detection contains 2351 pre-extracted 1-g features of malware and benign files shown in Table 4.

Image-Based Analysis

In image-based analysis, malware binaries are converted into image files such as grayscale images of size 128 × 128, 256 × 256, etc. Every hexadecimal view of byte values is represented as one pixel in the image. By using neural networks, features are extracted automatically from malware images. To extract these types of features, no additional knowledge is required in contrast to the handcrafted features which use n-gram analysis. Malware files belonging same family have similar grayscale image structures. Bytes data has been converted into 2D array, thereby array values are represented as pixels. Neural networks take a large computation time to extract

Table 4 Feature groups present in the BODMAS dataset

Feature groups	No of features
Header information	62
General file information	10
Byte histogram	256
Byte entropy histogram	256
Imported functions	1280
Exported functions	128
String information	104
Section information	255
Total	2351

the features when compared with handcrafted features. We considered two datasets named Malimg and Microsoft malware datasets to perform image-based analysis for feature extraction.

4 Architectures

4.1 Machine Learning Models

Machine learning models learn knowledge from either labeled data or unlabeled data and perform predictions or decisions. In this work, multiclass malware classification and malware detection have been performed. Both these problems come under the classification method of supervised learning.

In this chapter, machine learning models such as KNN, Random Forest, Support Vector Machine, Decision Tree, and Extra Tree classifiers are used in the detection and classification of malware data.

K-Nearest Neighbor

K-nearest neighbor (KNN) [9] is one of the most commonly used classification algorithms in detecting malware samples. KNN is a supervised machine learning algorithm used for both classification and regression problems. It is an instance-based learning technique which doesn't construct the generalized function. New instances are classified by mapping with already stored trained data points using distance measures.

Random Forest

Random forest is a supervised machine learning model which uses an ensemble technique that assumes predictions from multiple decision trees to perform classification. It is the most used algorithm due to its effective handling of high-dimensional and imbalanced datasets. Hyperparameter tuning has been performed in the model to get the optimal results. Hyperparameters include a number of decision trees, depth of the tree, number of samples, minimum sample split, number of features, etc. Grid search and random search have been used to identify the best hyperparameter values.

Support Vector Machines (SVMs)

Support vector machines are classification models which use the hyperplane to separate the data points of different classes in the high-dimensional space. The hyperplane represents the boundary line which differentiates the data points in feature space. Support vectors are used to identify the margin, which is the distance from the hyperplane to the closest data points. Kernel function has been used to convert the data points into feature vectors. Some of the kernel functions used are linear, radial bias function, sigmoid, and polynomial. Based on the characteristics of the data points, one of the kernel functions is used. SVM performs classification on linearly separable as well as non-linear separable data points.

Decision Trees

Decision trees are tree-like structures which are used to perform classification. Each internal node defines the selected feature which either splits into different branches or it represents the target class based on the outcome of the feature for all the data points. Each leaf node represents the target class label. Features are selected based on the entropy and information gain parameter values. Tree pruning technique has been used if an overfitting situation occurs in the decision tree. Ensemble methods such as AdaBoost and gradient boosting have been used to increase the performance of the decision tree systems.

Extra Trees Classifier

Extra trees classifier is an ensemble machine learning model which uses the decision trees. It considers entire data points to construct the trees which increases the variance. Extra tree classifier trains multiple number of decision trees and aggregates all the tree results to predict the output. It randomly selects the feature for splitting the node into child nodes. It is robust against overfitting and gives a more generalized model. It is similar to random forest algorithm but computationally faster than it.

4.2 Deep Learning Models

Deep learning models extract high-level abstract features and give improved performance when compared with traditional machine learning models. It automatically extracts the standard features from data samples. Unlike deep learning models, machine learning models need more domain knowledge and computational time for feature extraction. As per the study from related work, convolutional neural networks perform better with image data. Deep learning models use large datasets and neural networks for training. Classification has been performed by directly training the model using images, text, etc.

Grayscale Image-Based Convolutional Neural Network for Malware Classification

Convolutional neural network (CNN) is a widely used architecture to perform image classification and detection. In this work, CNN architecture is implemented as shown in Fig. 3. to extract features from grayscale images for the malware classification. Initially, grayscale images of all malware files have been resized to 256×256. These resized images are given to convolution layer. The first two blocks of the architecture are the convolution layer followed by the max-pooling layer. The third block contains the dropout layer immediately after the convolution layer which cuts some features to reduce the computational time, and then the max-pooling layer. Now the low dimensional space has been flattened and given as input to a fully connected layer, which is followed by the output layer.

At output layer, malware image datasets such as Malimg and Microsoft malware gave 25 and 9 units, respectively, as per the count of malware families present. Malware samples have been detected based on the extracted features. The above CNN model is a shallow model with three convolution layers, one fully connected layer, and one output layer.

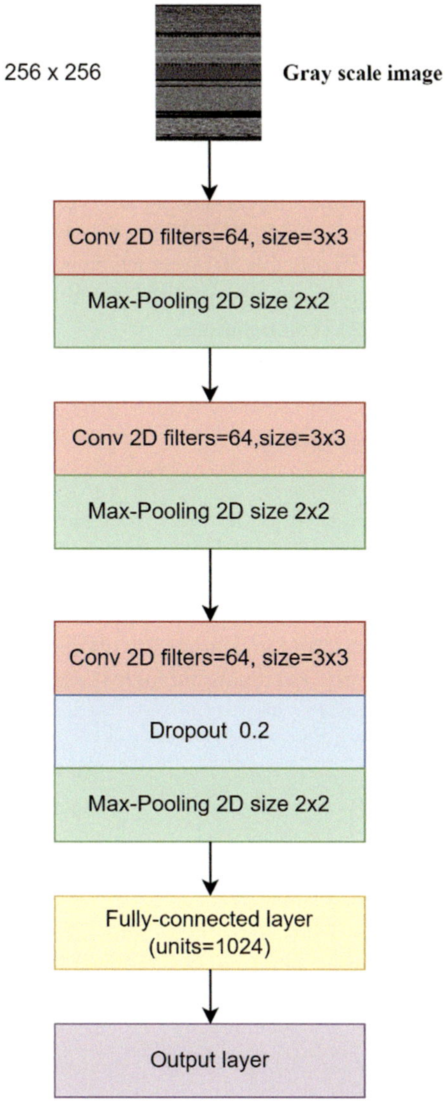

Fig. 3 Convolutional neural network architecture for malware classification using grayscale images

4.3 Transfer Learning

Transfer learning is the one type of machine learning method which uses the knowledge of pre-trained model and applies on new tasks. Construction of deep learning models from scratch takes more amount of time in training the data and also adjusting the weights to extract the features. Pre-trained models use large datasets for training the model so that smaller datasets also give good performance with it. Meanwhile, usage of pre-trained model knowledge reduces the computational time and resources.

A large number of pre-trained transfer learning models, which are trained on large-scale datasets, have been available. The pre-trained models such as VGG-16, Resnet-50, and InceptionV3 for initial layer training as shown in Fig. 4. Firstly, malware images are provided to pre-trained models. These pre-trained models extracted the features based on the kernel filters provided. Output of pre-trained models is connected to one fully connected layer (dense layer) with 1024 units and one output layer (softmax layer) which classifies the data. Inner layer training depends on the pre-trained model, VGG16 has 13 convolution layers and 3 fully connected layers. ResNet-50 contains 50 layers which include convolution layer, pooling layer, normalization layer, etc., InceptionV3 contains factorized convolutions, regularization, and parallelized computations.

5 Evaluation

This section provides the experimental evaluation of various machine learning models and deep learning models discussed in Sect. 4. We considered benchmark malware datasets such as Malimg, Microsoft malware data, and BODMAS as discussed in Sect. 3 to perform malware classification and detection. To build the models, all the datasets are divided into train set and test set with 75 and 25%, respectively. Accuracy and logloss are the two metrics used for analyzing the model performance.

$$\text{Accuracy} = \frac{\text{number of correct predictions}}{\text{total number of predictions}}$$

Accuracy is the performance metric used to get the correctly predicted samples among all the samples. Accuracy is not enough to assess the performance of the models, because of the class imbalance that exists in the datasets. Logloss gives the closeness of the predicted state and target state.

$$\text{Logloss} = -\frac{1}{N}\sum_{i=1}^{N}\sum_{j=1}^{M} y_{i,j}\log(p_{i,j})$$

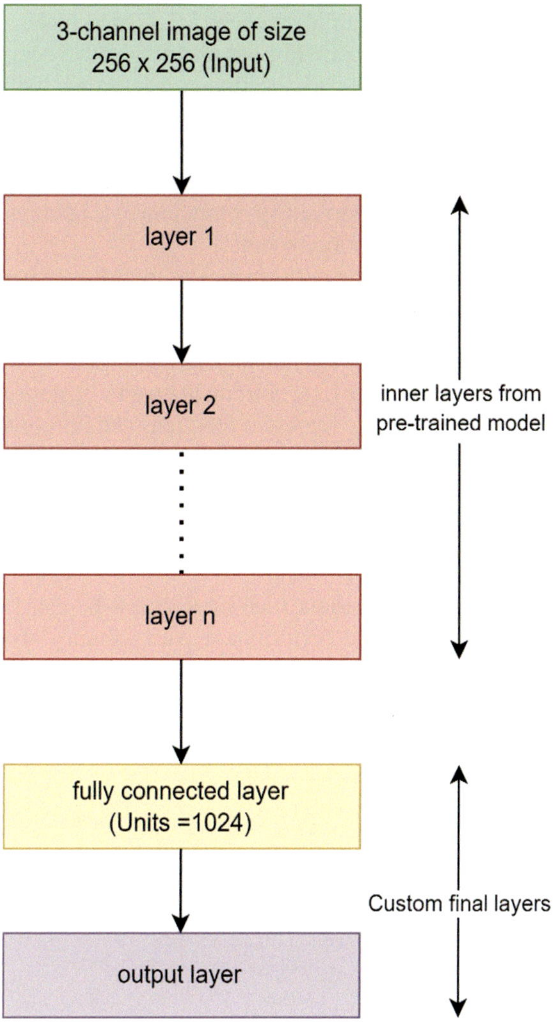

Fig. 4 Transfer learning model to perform malware classification

where N represents the number of samples, M represents the number of class labels, $y_{i,j}$ is 1 if the sample i belongs to class j and 0 otherwise, and $p_{i,j}$ is the prediction probability that observation i is in class j.

5.1 Feature Performance Analysis

In this work, 256 features from bytes files and 48 features from assembly code files of 10,868 malware samples of the Microsoft malware dataset are extracted. From the BODMAS dataset, 2351 pre-extracted features have been provided as shown in

Table 4. Machine learning models (ML models) such as KNN, Random Forest, SVM, Decision Trees, and Extra trees classifiers have been applied to the above datasets to perform malware classification and detection. Table 5 shows the performance of malware classification on the Microsoft malware dataset. Byte features and Assembly features (ASM) individually are applied on the machine learning models, among them ASM features attain better performance when compared with byte features. Feature fusion has been performed by combining byte features and ASM features and applied to the machine learning models.

Table 6 shows the performance of malware detection on the BODMAS dataset. Here each feature category labeled as G1, G2, … given below.

- G1 represents the byte histogram features.
- G2 represents the byte entropy histogram features.
- G3 represents the imported functions and library features.
- G4 represents the string information features.
- G5 represents the general file information features.
- G6 represents the header information features.
- G7 represents the combined features of all the groups specified in Table 4.

Tree-based classifiers such as Random Forest, Decision tree, and Extra Trees classifiers give good performance with accuracy above 98% in malware detection as well as classification, when compared with KNN and SVM (linear). SVM gives

Table 5 Malware classification performance based on feature category of the Microsoft dataset

Machine learning models	Accuracy		
	Byte features (256)	ASM features (48)	Combined Byte + ASM features (304)
KNN	92.5	99.08	97.6
Random Forest (RF)	98.11	99.63	99.07
SVM (Linear)	79.6	88.5	89.73
Decision Tree	95.6	98.75	98.71
Extra Trees Classifier	98.7	99.3	99.1

Table 6 Malware detection performance based on feature category of the BODMAS dataset

Machine learning models	Accuracy						
	G1	G2	G3	G4	G5	G6	G7
KNN	96.5	98.03	97.7	86.4	97.4	98.1	98.1
Random Forest (RF)	98.2	98.5	98	98.4	99.1	99	99.5
SVM (Linear)	77.2	83	97	86	75.7	87.9	70.3
Decision Tree	95.5	96.6	97	96.36	98.6	98.8	98.9
Extra Trees Classifier	98.3	98.6	97.9	98.4	99.09	98.9	99.5

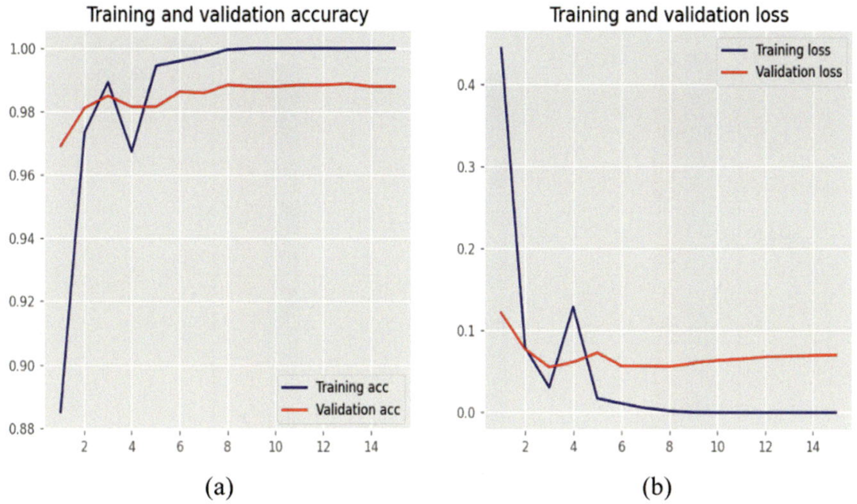

Fig. 5 **a** Training and validation accuracy of malimg data and **b** training and validation logloss of malimg dataset

better performance with imported library functions when compared with other feature groups.

5.2 *Image Based Model Analysis*

Two datasets Malimg and Microsoft malware (discussed in Sect. 3) are taken, with gray scale images to perform malware classification. In Sect. 4.2, a shallow CNN model has been proposed to classify the malware samples. The CNN model gives us the training accuracy of 1.00 and validation accuracy of 0.988 on malimg dataset as shown in Fig. 5a. The logloss values of the model on malimg data is shown in Fig. 5b.

5.3 *Transfer Learning Model Analysis*

In the malware classification, pre-trained models such as VGG16, ResNet-50 and InceptionV3 are used. Pre-trained models reduce the training time, as they are already trained on large Image dataset. Features obtained from pre-trained model are given to two customized dense layers to classify the malware data. In Sect. 4.3, transfer learning model has been proposed and applied on the malimg and Microsoft data, which gives accuracy as shown in Figs. 6 and 7. As per the observation, from the bar plotting, among the three pre-trained models, VGG16 performs betters and gives

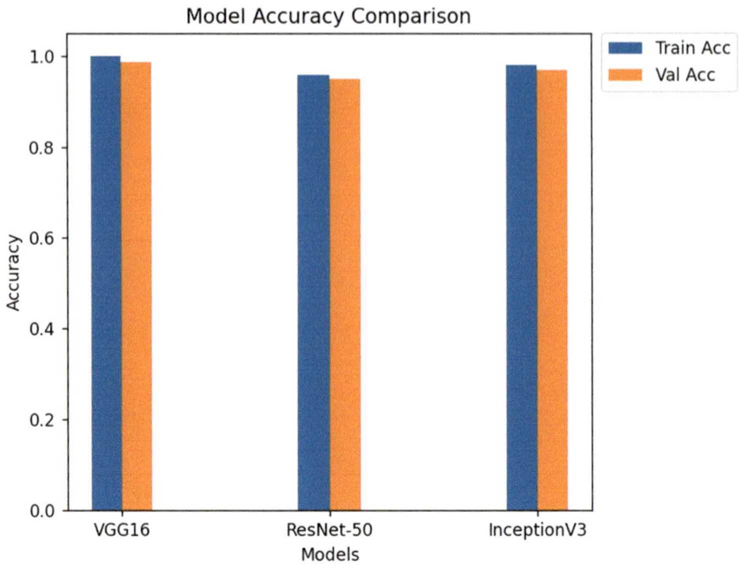

Fig. 6 Accuracy scores of models using pre-trained architectures such as VGG16, ResNet-50, and InceptionV3 in the Malimg data

good training accuracy and validation accuracy for both malimg and Microsoft malware datasets.

5.4 Comparison of Architectures

This section presents a performance analysis of all the models discussed in the previous sections of machine learning models and neural network models.

Feature Engineering

Machine learning models are highly dependent on the hand-crafted features. Feature engineering has been performed by the domain expert with much-needed knowledge to get the best feature set. On the other side, neural networks perform automatic extraction of features from raw data. For malware analysis, attackers always tend to use new techniques to perform malicious activities. So, if we apply similar hand-crafted features to the latest malwares, it may not give good performance.

Handling Complex and Non-linear Data

Smaller datasets with less complexity in feature extraction can be easily handled by the machine learning models, whereas with high dimensionality datasets, non-linearity in the data and unstructured data such as images can be handled by the neural

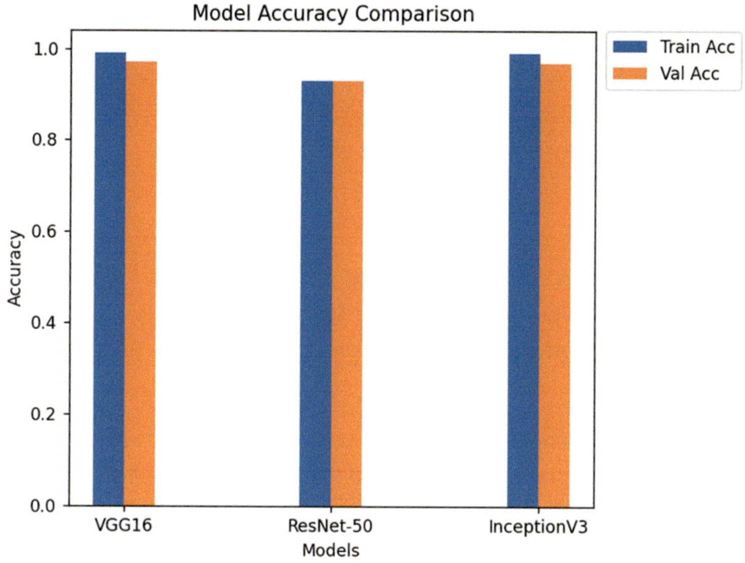

Fig. 7 Accuracy scores of models using pre-trained architectures such as VGG16, ResNet-50, and InceptionV3 in the Microsoft malware data

network models and give good performance. Complex patterns or relationships can be easily extracted by the neural network layers.

Time Complexity and Resource Consumption

Training machine learning models takes small amount of time when compared with deep learning models, because of a smaller number of parameters. Neural networks extract deep features in the training stage of large datasets. Sometimes neural networks need GPUs to run the large and complex datasets. In deep learning training time is reduced by using pre-trained neural network models. These pre-trained models transfer the knowledge.

Table 7 represents the comparison of theproposed neural network model and transfer learning models which use various pre-trained models such as VGG16, ResNet-50, and InceptionV3 with previous bytes file implementations. Except model pre-trained with ResNet-50, all the proposed models give better performance with 99% accuracy in classifying and detecting the malware data.

5.5 Concept Drift in Malware Analysis

Concept drift is one of the significant challenges to come across when we are performing malware detection or classification. Models trained on historic malware data may/may not detect current or future malware variants, due to differences in the

Table 7 Bytes file approach comparison of various neural networks in the Microsoft malware data

Method	Accuracy	Macro F1-score
PCA + K-NN [11]	0.9660	0.9102
CNN IMG [14]	0.975	0.940
CNN ENTROPY [13]	0.9828	0.9636
Deep neural network using autoencoders	0.9915	–
Proposed CNN model	**0.993**	**0.98**
Neural network model with pre-trained VGG16	**0.99**	–
Neural network model with pre-trained ResNet-50	**0.93**	–
Neural network model with pre-trained InceptionV3	**0.99**	–

characteristics and behavior of the malware samples. Malware creators continuously change the techniques used in new malware variant creation to escape detection. Sometimes, adding new malware variant data to an already existing dataset may change the overall data distribution. To overcome the concept drift challenge, continuous updation of models has been taking place. Ensemble methods have been used to combine the different time series models to get more generalized models which reduces the concept drift issue. Sustained monitoring of updated malware signatures helps in detecting the trends in the new malware variants.

6 Conclusion and Future Work

In this paper, malware detection and classification are performed on various datasets using machine learning and neural network models. As per the observation, feature fusion is effective in detecting and classifying the malware data. A novel neural network architecture has been used to perform malware classification and get better results than the state-of-the-art model's performance. Obfuscated malware files, which use encryption and code masking, are detected efficiently using neural networks. To reduce the training time of the neural networks, a transfer learning model is applied and gives good performance. Results show that the proposed shallow neural network model outperforms the machine learning models and performs as better as state-of-the-art models in the task of malware classification. Transfer learning models also give a good way of performing malware classification and detection by reducing model training time. A future direction of research work is to perform malware analysis on the assembly language files by getting graph features such as control flow graphs, which gives a technical perspective of different patterns used in malware logic and also provides the broad domain knowledge in depth. Concept drift is a significant issue in malware detection because models trained on historical data are less efficient when new malware variants emerge. Generally, as times change, new types of malwares also emerge rapidly with different structures and patterns.

References

1. https://cybersecurityventures.com/cybercrime-to-cost-the-world-8-trillion-annually-in-2023/
2. https://www.bleepingcomputer.com/news/security/t-mobile-hacked-to-steal-data-of-37-million-accounts-in-api-data-breach/
3. https://heimdalsecurity.com/blog/companies-affected-by-ransomware/
4. https://www.akamai.com/blog/security/defeating-triple-extortion-ransomware#:~:text=This%20combination%20of%20encryption%20and,as%20an%20additional%20extortion%20technique.
5. https://www.alliedmarketresearch.com/malware-analysis-market-A05963
6. Yang L, Ciptadi A, Laziuk I, Ahmadzadeh A, Wang G (2021) BODMAS: an open dataset for learning based temporal analysis of PE Malware. In: 2021 IEEE security and privacy workshops (SPW), San Francisco, CA, USA, pp 78–84. https://doi.org/10.1109/SPW53761.2021.00020
7. Nataraj L et al (2011) Malware images: visualization and automatic classification. In: Proceedings of the 8th international symposium on visualization for cyber security
8. Ronen R, Radu M, Feuerstein C, Yom-Tov E, Ahmadi M (2018) Microsoft Malware classification challenge
9. Choi S (2020) Combined kNN classification and hierarchical similarity hash for fast Malware detection. Appl Sci 10:5173. https://doi.org/10.3390/app10155173
10. Yamashita R, Nishio M, Do RKG et al (2018) Convolutional neural networks: an overview and application in radiology. Insights Imaging 9:611–629. https://doi.org/10.1007/s13244-018-0639-9
11. Narayanan BN, Djaneye-Boundjou O, Kebede TM (2016) Performance analysis of machine learning and pattern recognition algorithms for Malware classification. In: 2016 IEEE national aerospace and electronics conference (NAECON) and Ohio innovation summit (OIS), pp 338–342, July 2016
12. Kebede TM, Djaneye-Boundjou O, Narayanan BN, Ralescu A, Kapp D (2017) Classification of Malware programs using autoencoders based deep learning architecture and its application to the microsoft Malware Classification challenge (BIG 2015) dataset. In: 2017 IEEE national aerospace and electronics conference (NAECON), Dayton, OH, USA, 2017, pp 70–75. https://doi.org/10.1109/NAECON.2017.8268747
13. Gibert D, Mateu C, Planes J, Vicens R (2018) Classification of malware by using structural entropy on convolutional neural networks. In: Proceedings of the thirty-second AAAI conference on artificial intelligence, (AAAI-18), the 30th innovative applications of artificial intelligence (IAAI-18), and the 8th AAAI symposium on educational advances in artificial intelligence (EAAI-18), New Orleans, Louisiana, USA, February 2–7, 2018, pp 7759–7764
14. Gibert D, Mateu C, Planes J, Vicens R (2018) Using convolutional neural networks for classification of malware represented as images. J Comput Virol Hacking Tech, August 2018
15. Singh J, Thakur D, Gera T, Shah B, Abuhmed T, Ali F (2021) Classification and analysis of android malware images using feature fusion technique. IEEE Access 9:90102–90117. https://doi.org/10.1109/ACCESS.2021.3090998
16. Xiao G, Li J, Chen Y, Li K (2020) Malfcs: an effective malware classification framework with automated feature extraction based on deep convolutional neural networks. J Parallel Distrib Comput 141:49–58
17. Lyda R, Hamrock J (2007) Using entropy analysis to find encrypted and packed Malware. In: IEEE security & privacy, vol 5, no 2, pp 40–45, March–April 2007. https://doi.org/10.1109/MSP.2007.48
18. Khan M, Baig D, Khan US, Karim A (2020) Malware classification framework using convolutional neural network. In: 2020 international conference on cyber warfare and security (ICCWS), Islamabad, Pakistan, pp 1–7. https://doi.org/10.1109/ICCWS48432.2020.9292384
19. Alam M, Akram A, Saeed T, Arshad S (2021) DeepMalware: a deep learning based malware images classification. In: 2021 international conference on cyber warfare and security (ICCWS), Islamabad, Pakistan, pp 93–99. https://doi.org/10.1109/ICCWS53234.2021.9703021

20. Depuru S, Santhi K, Amala K, Sakthivel M, Sivanantham S, Akshaya V (2023) Deep learning-based malware classification methodology of comprehensive study. In: 2023 international conference on sustainable computing and data communication systems (ICSCDS), Erode, India, pp 322–328. https://doi.org/10.1109/ICSCDS56580.2023.10105027
21. Shinde S, Dhotarkar A, Pajankar D, Dhone K, Babar S (2023) Malware detection using efficientnet. In: 2023 international conference on emerging smart computing and informatics (ESCI), Pune, India, pp 1–6. https://doi.org/10.1109/ESCI56872.2023.10099693
22. Akhtar MS, Feng T (2022) Malware analysis and detection using machine learning algorithms. Symmetry 14(11):2304. https://doi.org/10.3390/sym14112304
23. Pavithra J, Selvakumara Samy S (2022) A comparative study on detection of Malware and benign on the internet using machine learning classifiers. Math Prob Eng 2022, Article ID 4893390, 8 p. https://doi.org/10.1155/2022/4893390
24. Aslan Ö, Yilmaz AA (2021) A new malware classification framework based on deep learning algorithms. IEEE Access 9:87936–87951. https://doi.org/10.1109/ACCESS.2021.3089586
25. Xing X, Jin X, Elahi H, Jiang H, Wang G (2022) A Malware detection approach using autoencoder in deep learning. IEEE Access 10:25696–25706. https://doi.org/10.1109/ACCESS.2022.3155695
26. Kumari VV, Jani S (2023) An effective model for Malware detection. In: Rao BNK, Balasubramanian R, Wang SJ, Nayak R (eds) Intelligent computing and applications. Smart innovation, systems and technologies, vol 315. Springer, Singapore. https://doi.org/10.1007/978-981-19-4162-7_35

Protecting Your Assets: Effective Use of Cybersecurity Measures in Banking Industries

Sulagna Das and Devraj Ganguly

Abstract The Indian banking system, originating from the first Bank of Hindustan, has evolved through nationalization, liberalization, and mergers, enhancing serviceability, automation, and operational cost reduction. Banks are focusing on digitization, launching digital payment systems like UPI and BHIM, and promoting a cashless society through NEFT, ECS, RTGS, and Prepaid Cards. The digital transformation aims to enhance operational efficiency, client experiences, and competitiveness. Key areas include digital channels, process automation, AI, machine learning, robotics, personalized customer experiences, sustainable banking, agile workforce, and remote employment. Technological advancements are increasing cyber-attacks on banks and financial organizations, posing privacy and phishing issues. Cybersecurity technology is crucial in protecting financial information, preventing fraud, and ensuring system reliability. Key factors include advanced threat detection, collaboration, quantum-safe encryption, AI, machine learning, a cybersecurity workforce, a zero-trust framework, and continuous monitoring for incident response and security.

Keywords Indian banking sector · Cybersecurity · Digitalization · Blockchain · Cybercrime

1 Evolution of the Indian Banking System

The traditional banking system in India has evolved since the 1770s when the first Indian bank (Bank of Hindustan) was established. Over the last five centuries, the banks have been nationalized, then liberalized, and recently have gone through bank mergers. Before this, the banks were performing manually, the account-related information was maintained in hard copy, through ledger books, transactions were limited, with limited transactional accuracy, and customers had to visit branches physically to conduct transactions. The Indian banking industry understood the importance

S. Das (✉) · D. Ganguly
Department of Management Studies, JIS University, Kolkata, India
e-mail: sulagna.das@jisuniversity.ac.in

K. Kaushik and I. Sharma (eds.), *Next-Generation Cybersecurity*, Blockchain Technologies, https://doi.org/10.1007/978-981-97-1249-6_12

Fig. 1 Evolution of banking. *Source* Author

of computerization in the 1980s to improve serviceability and the recording and accounting of financial transactions.

Banking automation began with the computerization of key departmental functions, such as Account-related banking operations [9]. Banks improved customer service by automating branches and providing "Single Window Service" facilities, which reduced transaction processing times and enabled interbank communication. This revolution was welcomed by the banking industry [2]. The multichannel banking strategy prioritized the creation of a centralized data infrastructure capable of consolidating databases from various banks into a single large database. ATMs fundamentally changed the customer–bank relationship, allowing customers to conduct basic financial transactions without entering a bank. The implementation of a new customer service interface "Core Banking Applications" benefited banks by significantly lowering operational costs. (Fig. 1, gives a clear idea of the Evolution of Banking, from branch banking to Block Chain).

The Institute for Development and Research in Banking Technology (IDRBT) was founded in 1996 to advance banking technology in India. The Indian Financial Network (INFINET), Structured Financial Messaging System (SFMS), and National Financial Switch (NFS) were established by the organization. It deployed high-end PKI-based services and solutions to protect transactions across INFINET in India's banking and financial sectors. PayPal, Google Pay, and Amazon Pay have significantly impacted the banking industry by automating processes and enabling computerized financial transfers through the widespread availability of the Internet.

2 Present Scenario of the Indian Banking System

Banks seek to give their consumers a banking experience that is rapid, accurate, and of high quality. For Indian banks, digitization is a primary priority, and the NPCI's (National Payments Corporation of India) launch of United Payments Interface (UPI) and Bharat Interface for Money (BHIM) are two major steps for innovation. Digital payment innovation and popularity are driving up digital banking transactions. Indian banks are increasingly adopting electronic payment systems like National Electronic Fund Transfer (NEFT), Electronic Clearing Service (ECS), Real Time Gross Settlement (RTGS), Cheque Truncation System, Mobile Banking, Debit Cards, Credit Cards, and Prepaid Cards. Online banking has transformed the banking industry, with low-cost methods like UPI and BHIM. NEFT is India's most common electronic payment mechanism, with half-hourly batches and 23 settlements. Real-Time Gross Settlement (RTGS) is used for high-value transactions with a minimum sum

Fig. 2 Trends in digital banking. *Source* Author

of Rs. 2 lakhs. Prepaid payment instruments like gift cards, travel cards, corporate cards, and mobile wallets have become increasingly popular, leading to a cashless society. Prepaid payment instruments like gift cards, travel cards, corporate cards, and mobile wallets have become increasingly popular, leading to a cashless society. Indian banks are offering innovative products and features to attract more clients, benefiting both banks and making banking easier for customers. a few examples are listed below (Fig. 2, illustrates the modern trends in digital banking):

- **Online and Mobile Banking**: With the rise of the internet and mobile technology, users may now access their accounts, conduct transactions, and manage their money from any location at any time. The use of physical branch visits is decreased by the convenience, real-time updates, and self-service capabilities offered by online and mobile banking systems.
- **Digital Payments and Wallets**: Technology has enabled the development of digital payment systems such as mobile wallets, contactless payments, and peer-to-peer payment platforms. The use of these solutions encourages financial inclusion and lessens dependency on the physical currency by providing quick, safe, and practical substitutes for conventional cash and card transactions.
- **Blockchain and Cryptocurrencies**: The financial industry has taken notice of blockchain technology, which enables safe and open transactions. It facilitates the growth of cryptocurrencies like Bitcoin, allows for quicker and more secure cross-border payments, and streamlines procedures like trade finance and supply chain management.

- **Open Banking and APIs**: Application Programming Interfaces (APIs) have facilitated cooperation and innovation in the banking industry. Open Banking efforts have done the same. Financial technology (fintech) firms can safely get client data from banks, enabling the creation of new services and fostering competition.
- **Enhanced Security and Fraud Prevention**: As the number of digital transactions has increased, banks' key priorities now include cybersecurity. Security measures have been strengthened by the use of cutting-edge technology like multi-factor authentication and biometric authentication (such as fingerprints and face recognition). Systems for detecting fraud that is driven by AI pore over enormous volumes of data to spot suspicious activity and stop fraudulent transactions.
- **Disruption and Innovation**: Technology has paved the road for novel financial solutions and industry disruptions. Blockchain technology provides smart contracts, safe and transparent transactions, and facilitates cross-border payments. Cryptocurrencies have developed as new forms of digital currency, posing a threat to existing financial institutions. Fintech firms have provided new financial goods and services, increased competition, and forced banks to adapt and innovate.

Banks are undergoing digital transformation to improve operational effectiveness, improve client experiences, and maintain competitiveness. Digital channels, process automation, data analytics, and cutting-edge technology such as AI, machine learning, and robotic process automation are essential for success.

- **Personalization and Customer Experience**: Banks are putting a lot of effort into providing consumers with personalized experiences. Insights into client preferences, behavior, and requirements may be gained by banks by utilizing data analytics and AI, enabling the creation of specialized product offerings, specialized suggestions, and proactive assistance.
- **Sustainable Banking**: Environmental, social, and governance (ESG) issues are becoming increasingly important in the banking industry. Banks are making more and more decisions about their investments and operations based on sustainable practices. This involves using responsible lending and risk assessment techniques, delivering green financial products, and financing renewable energy projects.
- **Agile Workforce and Remote Employment**: The epidemic has caused a movement in the banking sector toward remote employment. To increase efficiency, recruit talent, and swiftly adjust to shifting market conditions, banks are embracing flexible work arrangements and using agile approaches.
- **Application Programming Interfaces (APIs)** are used by banks to securely exchange client data with third-party suppliers as part of the open banking initiatives that are gaining traction globally. This pattern supports collaboration between banks and fintech firms, fosters innovation, and permits individualized financial services.

Artificial intelligence is helping the banking sector to offer a higher degree of value, reduce risks, and provide better chances. It aids in the development of new ideas and the transformation of customer service. The banking and financial industries' key aims are to improve performance, increase profitability, and reduce risk. Data

Analytics is essential for lowering costs, developing new products, and expanding client base, it has also enabled banks to lend at a reduced interest rate to key industries such as agriculture, housing, and education.

Cyber-attacks are increasing as technology advances, and criminals are actively seeking victims for catastrophic cyber-attacks on sensitive data stored by banks and financial organizations. Most banks have been pushed to go online, leading to privacy issues and phishing efforts. Cybercriminals use customer and employee information to steal bank data and money. It is important to understand the importance of cybersecurity in the banking sector before delving into cybersecurity concerns.

3 Status of Cybersecurity in the Indian Banking System

Cybersecurity is the act of preventing unauthorized access to, damage to, and theft of computer systems, networks, and data. Systems and procedures must be implemented to protect information from cyber threats. Cyber threats such as hacking, malware, phishing, and ransomware attack people, businesses, and even entire countries, posing a threat to sensitive information and financial and reputational harm. Banks need to protect their users' assets, such as debit and credit cards, from cyber-attacks. Banks must implement security protocols to protect data from cyber-attacks [17]. Banks must prioritize cybersecurity to protect their data from cyber-attacks, and they must implement cybersecurity measures to protect their customers' data from cyber-attacks, which can have a negative impact on their reputation and assets. Card fraud is usually recoverable, but data infringement can take time to recover funds. This results in the loss of customers. Data infringement is a critical issue for banks [6], as it results in the loss of user data and makes it difficult for customers to trust the bank. Banks must have cybersecurity requirements to evaluate current security measures and protect critical data. Cyber-attackers have found various ways to attack and steal data, as everything turns digital these days, and the country goes for "Digital India". Banks are vulnerable to organized criminals and hackers. In the recent past, A cyber-attacker attacked and vandalized Canara Bank's website by inserting a malicious page and blocking e-payments. Union Bank of India also suffered a significant loss due to a cybersecurity attack in India's banking sector (Fig. 3, illustrates the Key Cyber Considerations for Banks to Maintain Cybersecurity).

Hackers gained access by impersonating an RBI employee, and one of the bank's employees clicked on a malicious link, allowing the malware to manipulate the system. Regarding the report of Money Control [19] there were around 248 successful cybercrime cases of data breaches by hackers and criminals in the year 2022. According to the data shared by Statista, in India, there were 1,343 cybercrime cases related to online banking in the year 2016 and 2,095, 968, 2,093, 4,047, and 4,823, in 2017, 2018, 2019, 2020, and 2021, respectively (Fig. 4, Graph illustrates the above discussion on cybercrime cases in Indian Banking).

Worldwide around four out of ten internet users have experienced cybercrime in the year 2022, and India was the most likely country to experience cybercrime,

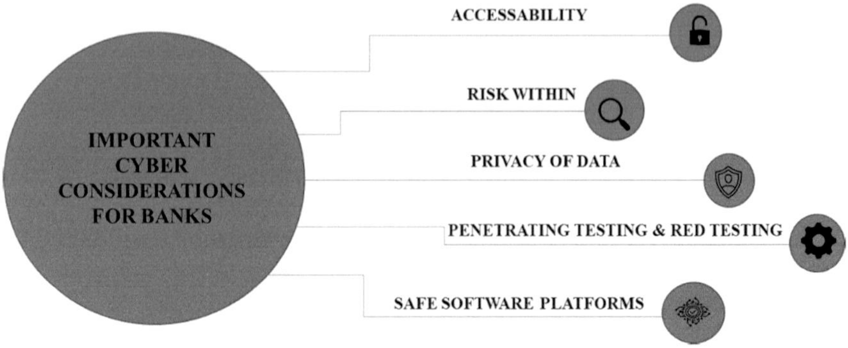

Fig. 3 Key cyber considerations for banks to maintain cybersecurity. *Source* Author

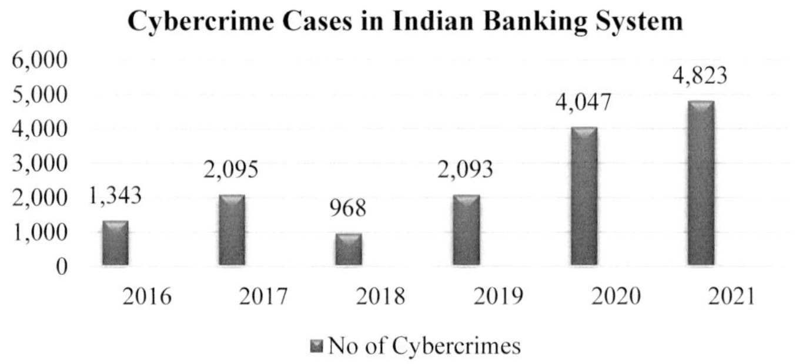

Fig. 4 Cybercrime cases related to online banking across India. *Source* Statista [28]

with nearly 70% of internet users reporting having experienced it. The United States came in second, with 49% reporting having been victims. Australia is third with 40%, followed by New Zealand with 38%, the United Kingdom, France with 33%, Germany with 30%, and Japan with only 21%. Cybercrime has become a major threat to the banking industry, as the technology and expertise used by hackers are becoming increasingly advanced, making it impossible to prevent attacks consistently. Banks face a variety of threats to their cybersecurity, which have been elaborated on below (Fig. 5, illustrates the country specific experience of cybercrime).

4 Cybersecurity Threats Faced by the Banking System

Cybercrime is the use of digital instruments to commit illegal acts, such as fraud, invasion of privacy, and identity theft, which can be carried out from afar, increasing the risk for both the bank and the consumer [23], with financial institutions and banks being the most common targets due to their sensitive client data and potential financial gain. Banks and financial organizations have prioritized cybersecurity by

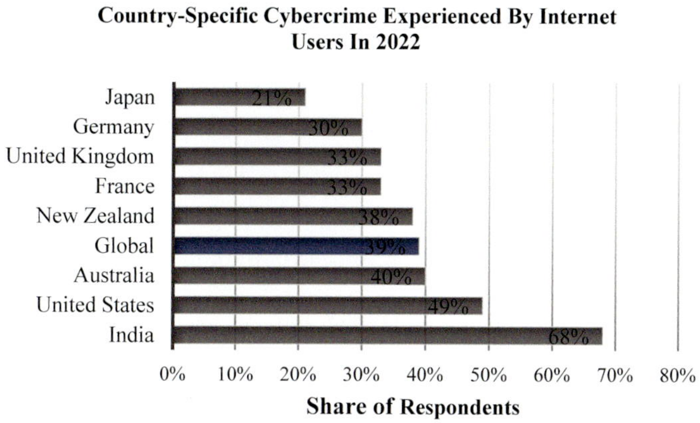

Fig. 5 Cybercrime experienced by internet user's country wise. *Source* Statista [31]

investing in security infrastructure and educating clients. The Reserve Bank of India has conducted financial awareness campaigns, but cybercriminals are constantly changing their strategies [8]. Some of the most common strategies they use are listed below:

- **Reverse Engineering of Mobile Apps**: A scammer may reverse engineer an app to examine its source code and parts to create malware or tamper with it. For example, attackers could create their malicious app to exploit vulnerabilities discovered while reverse engineering the banking app [21]. If a user has both applications installed on their device, the malicious app can redirect banking deposits to the account without the user being aware of the breach (Fig. 6, gives a clear image of the list of cybercrimes).
- **Screen Overlay Attack**: Overlay attacks are used to hijack data entry and trick users into installing additional malware or performing unsafe tasks on their mobile devices. They consist of an attacker-generated screen that appears on top of the legitimate application UI and is programmed to send any information entered into it directly to the attacker. Overlay strikes are also used to trick users into installing other malware or performing insecure tasks on their mobile devices, such as giving a malware app complete control of the user's phone [33].

Fig. 6 List of cybercrimes. *Source* Author

- **Fraudulent Screen Sharing**: Screen sharing fraud is a recent scam that involves a fraudster posing as an employee of an online gaming company or a bank and asking for remote access to the victim's phone under a false guise. The victim is then prompted to share the screen-sharing app's code and complete a transaction for a small fee. The catch is that once the victim gives the scammer the code, they can see what they are typing on their screen, their bank account number, and all of the information they are seeing in real time. To avoid this, one should always check the authenticity of the official website of the company from which they get a call.
- **Keylogging/Screen Reading**: Keyboard apps are available in the app store to replace the original keyboards on mobile devices. Users usually download these apps in an innocent attempt to personalize their gadgets. People may enjoy the new keyboard's hue and functionality, but some programs are rogue and contain code that can steal personal information or do other dangerous operations.
- **Message App Banking Fraud**: Cheaters contact bank customers on behalf of the bank with which the customer has an account and request that the customer download an app in exchange for increased security or rewards. When the app is installed, the customer is prompted to enter sensitive information, which is then relayed back to the fraudster on the other end of the line. To avoid this, one must be cautious when answering calls from unknown numbers and immediately notify his branch if he has installed such an app.
- **Malicious Applications Fraud**: Malicious application fraud is a scam in which the victim receives a call offering a freelance/Work-from-home job in exchange for installing an app suggested by them. To avoid becoming a victim of this scam, one should contact the company directly on whose behalf they were offered the job [35].
- **Sim Swap Fraud**: The idea of a "sim swap" involves exchanging a defective SIM for a new one, but the effects of this fraud can be extremely detrimental. The fraudster can make multiple money transfers from the victim's account without his knowledge when they switch his SIM. To prevent this, one should maintain a two-step verification process and set a withdrawal amount cap. Additionally, one should contact their bank and block all transactions if their SIM suddenly stops working.
- **QR Code Scams**: This fraud mostly affects people who are attempting to sell their goods on internet marketplaces, typically valuable items such as automobiles or cell phones. The con artist contacts the consumer and offers to buy the item for sale. Once accepted, the fraudster will request account information through which the payment can be paid. The buyer will claim that they are unable to send the payment and will request that the seller scan a QR code and enter their UPI PIN. The seller then scans the code and enters the pin, allowing the fraudster to withdraw any amount of money from the seller while remaining anonymous. Only scan QR codes from reliable sources to avoid this. Furthermore, the UPI PIN is never used to receive payments.

- **Phishing/Smishing/Vishing/Spoofing Attacks**: Phishing is a scam where private information is stolen through emails from an authentic source, such as bank officials. Hoaxers create fake links to direct customers to a fake page, which is used for fraudulent transactions. SMS phishing is called Smishing, while voice phishing is a social engineering attack where the attacker calls the victim to obtain information, pretending to be representative of financial institutions. They often ask the victim to install screen-sharing apps to gain access to their device.
- **Spurt in Spy Loan Apps**: Spy loan apps are instant loans with high interest rates and fees. They collect personal information before approval and use it to harass the borrower during loan recovery. This has caused harm to victims, such as sending scandalous messages and blackmailing them [25].
- **Crime-as-a-Service**: Cybercriminals are marketing their services in the form of Crime-as-a-Service (CaaS) and Malware-as-a-Service (MaaS), allowing novice hackers to launch targeted attacks.

The Indian banking system has faced many real-life challenges in the recent past, which have devastated the financial sector, and impacted a lot. These types of incidents have been discussed in the below section.

4.1 Cyber-Attack Cases in the Indian Banking Sector

- **Cosmos Bank Cyber-Attack in Pune**: The Cosmos Cooperative Bank in Pune was the target of a cyber-attack in 2018 that stole Rs. 94.42 crores. Hackers gained access to the bank's ATM server and stole personal information from debit cardholders and visa holders. Hacker gangs from 28 countries withdrew funds as soon as they were notified. To prevent this type of fraud, toughening surveillance measures and assisting approved individuals are needed [27].
- **ATM System Hacked**: Canara Bank's ATM servers were targeted for a cyber-attack in 2018, resulting in the transfer of 20 lakh rupees from several bank accounts. According to sources, cybercriminals gained access to ATM information for over 300 individuals, resulting in the deaths of 50 people. Hackers acquired information from debit cardholders using skimming devices, with transactions utilizing stolen information ranging in value from Rs. 10,000 to Rs. 40,000. To prevent data misuse, ATM security methods must be strengthened [32].
- **RBI Phishing Scam**: Criminals attacked the Reserve Bank of India in an audacious phishing attempt, promising Rs.10 lakhs in prize money within 48 h if the recipient clicked on a link to a website that looked like the RBI's official website. The user was asked to give personal information such as their password, I-pin, and savings account number. The RBI had issued a warning about its bogus email address and official website [35].
- **The Bank NSP Case**: A bank management trainee was engaged to be married and used company computers to exchange emails. After a while, the young lady made up false email addresses and sent them to the boy's international clientele,

leading to the boy's company losing many customers and going to court against the bank. The court determined that the bank was accountable for the emails sent through the bank's system [27].

- **UIDAI Aadhaar Software Hacked**: In 2018, the UIDAI released an official notification on a data breach that exposed 1.1 billion Indian Aadhaar card info. Anonymous merchants on WhatsApp were selling Aadhaar information for Rs. 500 and Aadhaar card copies for Rs. 300. The UIDAI received notification of the hacking of 210 Indian government websites [27].
- **SIM Swap Scam**: In August 2018, two Navi Mumbai hackers deceptively obtained the SIM card information of the people and illegally transferred Rs. 4 crores from their bank accounts, through online banking. In this respect, organizations must implement cybersecurity measures and adhere to the security guidelines outlined below, and financial sectors should be aware of the risks of cyber threats and take steps to protect themselves. One should be alert and restrict himself from sharing his personal information with any unknown domains, as this can help in reducing the risk of malicious content reaching people. The cyber-attacks in India should serve as a warning to all vulnerable individuals and businesses to implement cybersecurity measures and follow security guidelines.

Cybersecurity is very important for banks, as Digital India has resulted in a rise in the use of digital currency and cashless transactions, making it critical to implement all security measures to preserve data and privacy. Data breaches are a major issue in the banking industry, as a faulty cybersecurity system could expose their consumer database to outsiders, creating cybersecurity threats. The data breach is sensitive and could be used against someone and cause severe harm. Banks must be vigilant 24 × 7, or else the customers' data with the bank may be compromised, and recovering the data may be time-taking and annoying. Banking security must be improved to ensure the safety of customers.

5 Steps to Reduce Cybersecurity Threats in Banking

- **Access and Monitor Cloud Security**: Cloud infrastructure should be examined regularly to ensure it is up-to-date and secure. Multifactor authentication (using a mix of tokens, biometrics, passwords, and other safe authentication techniques) and a vulnerability management solution can be used to automate threat detection and protect against potential threats before they become an issue [14]. Continual monitoring and intrusion detection systems should be put in place to quickly identify and address security problems. Network traffic can be monitored, abnormalities may be found, and possible threats can be quickly addressed with the use of intrusion detection systems (IDS) and security information and event management (SIEM) technologies.

- **Management Policies Regarding Strict Access**: Establishing strict access management policies is essential for a successful organization, as it limits access to personnel information to keep it secure from within the firm, and grants access permissions to those authorized employees/users who need it. Strong authentication and access controls should be used to guarantee the authorized use of the data. Access restrictions should be reviewed and updated often to reduce the danger of unauthorized access.
- **Increase Employee Awareness through Training**: Banks must implement a complete training program for their employees, to educate them, so that they can deal with the cyber-attacks. Cybersecurity training for employees and professionals is essential to enhance their knowledge and skill set, related to data protection and cyber-awareness, and keep them up to date [36].
- **Data Encryption**: The cryptography method is used for data encryption, which ensures the protection of the most sensitive digital assets and data security.
- **Data Recovery**: Having a backup plan in place can help reduce downtime after a disruption and prevent data loss, but only if a backup is done on a routine basis.
- **Vendor Risk Management**: It involves assessing and controlling how securely third-party suppliers and service providers behave when they have access to private information or systems. To guarantee that suppliers adhere to the necessary security requirements, implement suitable due diligence, contractual agreements, and security assessments.

6 Application of Cybersecurity Tools in Banking

To safeguard confidential financial information, avoid fraud, and guarantee the reliability of banking systems, technology in cybersecurity must be implemented. The following are some essential technologies that are frequently used in the banking industry's cybersecurity environment.

- Firewalls are used to create a barrier between internal and external networks by regulating incoming and outgoing network traffic. They are also used in conjunction with intrusion detection systems (IDS) and intrusion prevention systems (IPS). IDS/IPS technologies can be used to stop or prevent possible attacks by scanning network traffic for suspicious activity.
- Secure communication channels over the internet are provided by the Transport Layer Security (TLS) and Secure Socket Layer (SSL) protocols, respectively. They protect the security and integrity of sensitive data, including login passwords and financial transactions, as it is sent between servers and clients.
- MFA (multi-factor authentication) and two-factor authentication (2FA) Beyond the usual username and password combinations, some authentication techniques are used. By requesting additional authentication elements from users, such as biometrics (fingerprint, facial recognition), hardware tokens, SMS codes, or

mobile apps, they increase security by lowering the possibility of unauthorized access.

Blockchain technology offers a distributed, unchangeable ledger for securely documenting financial transactions. It can improve banking operations' security and transparency, especially in areas like identity verification, international payments, and smart contracts.

7 Adoption of Blockchain Technology in Banking

Blockchain is a distributed ledger that is immutable and cryptographically secure, making it a vital tool for checking authenticity and assuring security while performing financial transactions online, it is used to transfer money between parties without the involvement of a third party or middlemen. It functions by constructing a network of computers [20] linked over the internet that records, stores, preserves, and synchronizes copies of transaction data that cannot be changed or destroyed. It is also hard to fabricate, making it an essential instrument for ensuring the legitimacy and security of financial transactions conducted online. The banking industry was among the first to see the potential of Distributed Ledger Technology, which is a technology that allows for the secure functioning of a decentralized digital database. It is a technology that underpins cryptocurrencies such as Bitcoin, but its applications extend beyond virtual money [1].

Global Banking Institutions have made significant strides toward digitalization, but their efforts have been mostly neglected when it comes to blockchain in banking, in contrast with the widespread interest in blockchain technology of other businesses. But soon, it is expected to rise from 4.9 billion dollars in 2021 to more than 67.4 billion dollars by 2026. The Fig. 7 graph, illustrated by Statista gives a clear idea of the market size of Blockchain Technology Worldwide:

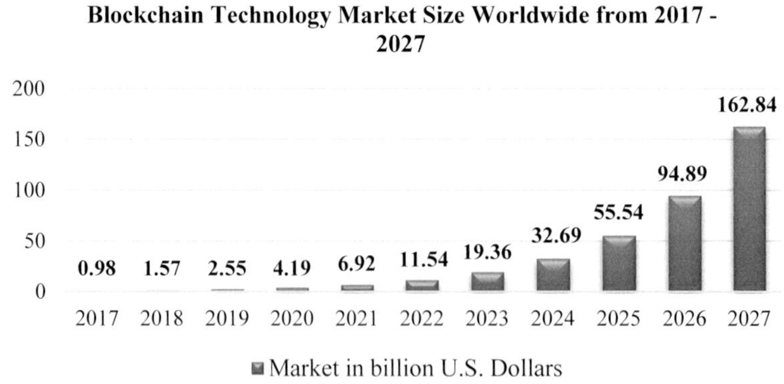

Fig. 7 Blockchain technology market size worldwide. *Source* Statista [3, 4]

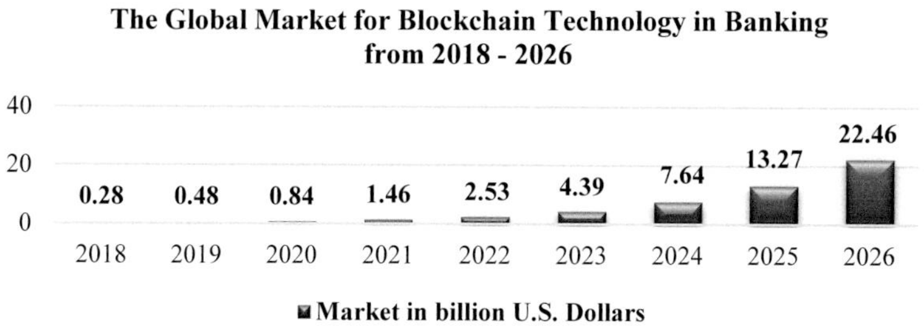

Fig. 8 Blockchain technology in banking market size worldwide. *Source* Statista [5]

The reticence of banks to use blockchain technology in banking and finance is due to the lack of large-scale applications and regulatory barriers. However, banking institutions have begun to employ the technology on a limited scale. Blockchain technology is becoming increasingly important in banking and real-world applications, and it has many uses, that has been discussed below, but the global banking scenario is still too slow to innovate due to paperwork, security concerns, and time-consuming and costly processes. For example, banks still require a significant amount of paperwork, have security concerns, and have several time-consuming and costly processes in place. A report published by Statista Research Department, 2022 also gives an idea of the market size for blockchain solutions for banking and financial institutions in US Billion dollars (Fig. 8, illustrates the Global Market for Blockchain Technology in Banking):

- **Payment Transfer**: Cryptocurrencies like Ether and Bitcoin are built on public blockchains, allowing anyone to send and receive money in real-time with no transaction fees and no authentication required, making money transfers faster and cheaper
- **Settlement and Clearance**: A typical bank transfer can take up to 3 days to complete, which is very problematic both for consumers and banks. Bank's cross-border transfers are now simpler and faster than ever, eliminating the need for middlemen and custodial services, with the advent of blockchain technologies. Blockchain is a decentralized ledger that records transactions transparently and publicly, allowing transactions to be resolved in the public blockchain and real-time settlement, hence speeding up and simplifying transactions. Banks utilize Ripple Net as an illustration of a blockchain-based payment network for international transfers. Example: Al Rajhi Bank's cross-border payment services are based on Ripple's blockchain capabilities.
- **Securities and Debts**: Blockchain technology generates a decentralized database of digital and unique assets, making tokenized asset transfers quicker and allowing for the elimination of intermediaries. Tokenization makes previously illiquid assets more accessible and allows for fractional ownership. The issuing, trading, and settlement of digital securities are made possible by blockchain-based systems, possibly revolutionizing the financial markets.

- **Loans & Credits**: Blockchain in consumer banking expands peer-to-peer lending, making it one of the most investable fintech areas. Banks assess the risk of nonpayment of interest on loans based on factors such as credit score, ownership status, and debt-to-income ratio. Blockchain in banking provides an efficient, low-cost, and secure way to obtain personal loans, made easier with a decentralized register of payment history.
- **Digital Identity and Customer Onboarding (KYC)**: Blockchain can revolutionize the way banks handle customer identity verification and onboarding processes. By leveraging blockchain-based identity systems, banks can enhance security, streamline customer due diligence, and reduce the need for redundant verification processes. This can improve the customer experience and reduce costs associated with compliance.
- **Innovation and Competitive Advantage**: Banks can maintain their competitiveness in the quickly changing financial market by using blockchain technology. Banks may use cutting-edge solutions, investigate novel business models, and provide their clients with improved services by implementing blockchain technologies [10]. Additionally, it permits them to work with fintech companies and investigate joint ventures to promote digital innovations. Example: JP Morgan Link encourages and enables businesses to securely and quickly exchange payment-related information and The HSBC bank is using R3 blockchain technology to enable Digital Vault, a custodial blockchain platform for keeping digital assets. The technology decreases the cost of their custodial service dramatically (Fig. 9, graph gives an idea of the uses of blockchain technology)
- **Trade Finance**: By digitizing and automating procedures like letters of credit, invoice financing, and supply chain finance, blockchain can increase the efficiency and transparency of trade finance. By automating the execution of commercial agreements and enabling real-time tracking of commodities, smart contracts on the blockchain can cut down on paperwork and streamline the entire trading process. Example: UBS launched the first publicly traded digital bond that could be resolved on both traditional and blockchain-based platforms [24].
- **Fundraising Through Digital Currencies Issued by Central Banks (CBDCs)**: Initial coin offerings (ICO) is a new concept for financing, which unbundles access to finance capital from traditional fund-raising services. Several central banks are looking at using blockchain technology to create their digital currencies, this increases financial inclusion, and payment efficiency, and gives central banks more influence over monetary policy. Example: Swedish Central Bank (SCB) is experimenting with the idea of introducing its digital currency e-krona in partnership with R3. This will become a country-wide usable cryptocurrency. On the other hand, Goldman Sachs is assisting payment company Circle in its goal of becoming a blockchain-enabled issuer of USD Coin, a digital stablecoin linked to US dollars [26].

In India, Blockchain technology is gaining popularity among Indian banks due to its ability to promote transparency and increase traceability of money movement. Money laundering frauds are the most prominent ones, that involve rubbing out

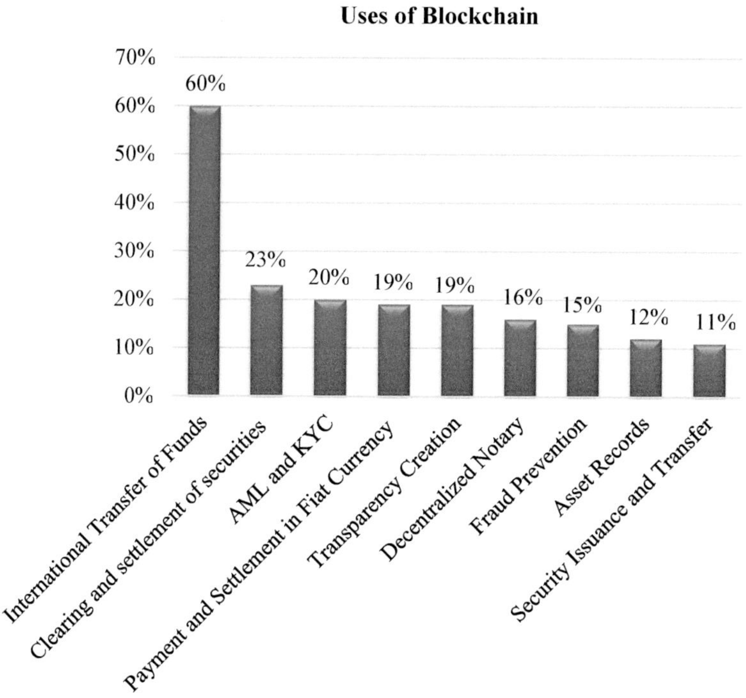

Fig. 9 Uses of blockchain technology. *Source* Delloite & EFMA [7]

crores and crores of rupees of the taxpayers, putting the money at stake. Banks are exploring blockchain-based trade finance initiatives to promote a safe and secure banking environment. The RBI is trying to explore the use of blockchain technology in the banking industry, by developing a proof-of-concept blockchain project, to handle trade finance [11]. State Bank of India has collaborated with other commercial banks to prototype a blockchain-based application. Private sector banks like Yes Bank, HDFC Bank, ICICI Bank, and Axis Bank are also adopting blockchain technologies into their banking operations. The Securities and Exchange Board of India directed its depositories to use blockchain technologies to encourage transparency in keeping records and monitoring the creation of securities [30].

8 Fraud Prevention Using Blockchain Technology

Blockchain technology has the potential to be a valuable tool in the prevention of fraud. While blockchain has many useful features, few, in particular, are crucial for fraud detection and prevention. They are as follows.

- **Immutable Ledger**: The distributed ledger used by blockchain technology assures that once a transaction is recorded, it cannot be changed or removed. Due to its immutability, it is challenging for fraudsters to alter or modify transaction records. A significant disincentive against fraudulent actions is created by the rapid identification of any efforts to change data on the blockchain [26].
- **Consensus**: Blockchain networks are dependent on consensus processes to authenticate and verify transactions [34]. The network guarantees that transactions are authorized by a majority of members, making it impossible for a single entity to influence or engage in fraudulent activity, depending on the consensus method employed, such as Proof of Work (PoW) or Proof of Stake (PoS) [15].
- **Collaboration and Shared Data**: Blockchain enables safe data exchange across many organizations, facilitating cooperation between banks and other ecosystem players. Shared data may include banned addresses, questionable transaction patterns, and fraud indications. By utilizing blockchain, banks may work together in real-time to jointly identify and stop fraud, fortifying their defense against such actions [12].

9 Barriers to Adoption of Blockchain Technology

Banks and traditional financial institutions are embracing blockchain technology, which enables new services and aids in internal back-end procedures. However, adoption remains limited due to some typical obstacles banks encounter when merging blockchain technology and digital assets. Banks viewed blockchain as a high-risk, short-term business, similar to bitcoin, but technology has continued to alter the financial sector in recent years. Banks are learning how to use blockchain to increase efficiency and automate procedures, and it can generate new revenue streams and attract customers. However, it is important to remember that all growth occurs when banks face challenges. Below are a few challenges faced by the banks during the adoption of Blockchain Technology (Fig. 10, illustrates the barriers of blockchain technology).

- **Budget**: When it comes to blockchain deployment, many organizations prioritize the budget. Because the technology is new, the banks may be uninformed of the expense of implementation.
- **Lack of Understanding and Skillset**: Blockchain is a complex concept that requires a full understanding to maximize its potential benefits. Banking IT teams are unlikely to have enough blockchain professionals to assist in the creation of blockchain projects and the dissemination of blockchain knowledge and concepts. Blockchain expertise is on the rise, but it can be difficult to understand which skills a bank is looking for [19].
- **Privacy and Security Concerns**: Blockchain and privacy cannot get along because the system is powered by decentralized ledger technology. Businesses that deal with privacy must follow tight criteria to minimize data access. As

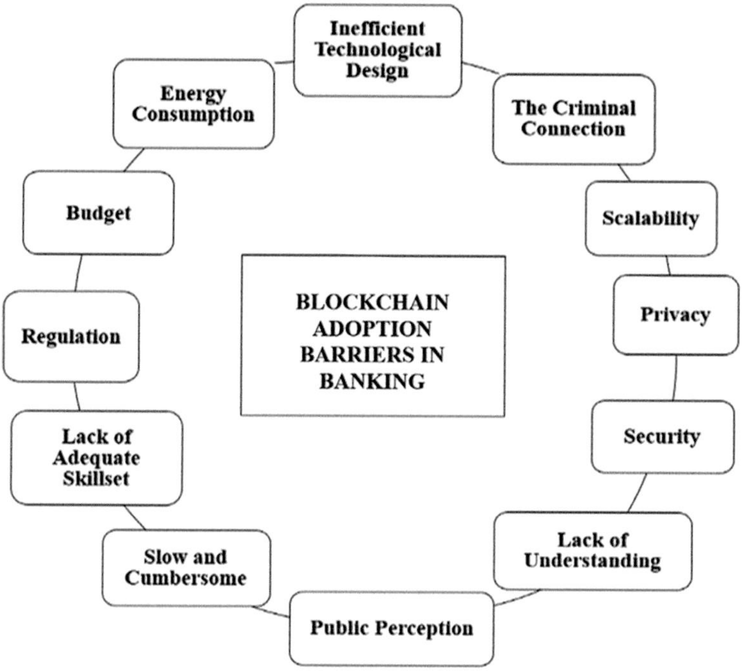

Fig. 10 Barriers of blockchain technology. *Source* Author

hackers may take over and exploit the network, blockchain technology poses a few security concerns. To address this, the protocol layer requires increased security, yet only a few scenarios have effective protocols. Nobody knows if they are safe to use for a lengthy period.

- **Slow and Cumbersome**: The blockchain is a complicated technology that takes longer to execute transactions, and its encryption considerably slows it down. It is best suited for huge transactions when speed is not a factor, but it is risky and was not designed to remove the "insecure" nature of blockchains. It works best for large transactions where time is not an issue.
- **Public Perception**: Blockchain technology has a lack of knowledgebase, making it unpopular among the general public. To be successful, it must gain acceptability and gain adequate promotion to attract more customers [13]. Without adequate promotion, blockchain technology will remain unpopular. Therefore, if someone is not involved in this segment, he will be unaware that it exists. Majority of the people are unaware of the fact that not only Bitcoin but other digital currencies come under the blockchain network.
- **Scalability**: When implementing blockchain, scalability is an issue that must be addressed. As the network's user base expands, transitions take longer to execute, increasing transaction costs and limiting the number of users on the network. As a result, blockchain adoption has been challenging, making the technology

less profitable. Few blockchain technologies produced faster results, but they also slowed as more users registered onto the system [29].

- **Inefficient Technological Design**: Blockchain technology is currently lagging in key technological areas due to code flaws and loopholes. Bitcoin was the cutting edge in this regard, but Ethereum attempted to cover up its weaknesses, but it was insufficient. The majority of these issues are caused by improper programming and loopholes, which users can quickly exploit and gain access to the system. Security jargon is not working here.
- **Regulation**: Blockchain has advantages, but it also poses challenges of regulation and compliance. Financial institutions are used to tight regulations and laws, and blockchain is pseudonymous, so a strong regulatory framework is needed to prevent illicit activity. However, due to a lack of clear rules, integration is problematic [18].
- **Energy Consumption**: Blockchain technology is based on Bitcoin and uses Proof of Work as a consensus procedure. However, mining requires the use of a computer to solve complex equations, which requires more electricity. To counteract this, the PC will use more electricity during mining.
- **The Criminal Connection**: Blockchain technology has attracted both professionals and criminals, leading to the use of Bitcoin as the principal currency in the illegal market and dark web. Criminals are using bitcoins to purchase restricted equipment and payment methods, and are demanding cryptocurrency as a ransom. To combat this, it is important to break the criminal link and improve blockchain implementation [7].

The potential for incorporating and embracing blockchain technology in India is immense, as the Indian Government is trying to initiate a national blockchain framework that will transform as many as 44 sectors. According to projections, blockchain technology will be worth 176 billion dollars by 2050, leading to the adoption of blockchain technology in the future. In the banking sector, blockchain has the potential to be a disruptive technology that banks may use to automate processes, improve services, and provide a greater digital experience to their customers. Service providers will collaborate with banks and other financial institutions to bridge the trust and knowledge gap, as well as provide technology infrastructure support and the appropriate services. Making the right investment at the right time is key to riding this growing tide. The below graph is a study by Statista, where it is seen the growth rate of different cybersecurity products market between the years 2019 and 2022. The growth rate for data security products is the maximum (22.20%), in comparison to endpoint security (19.10%), Identity and Access Management (16.30%), Network Security (15.30%), and security IDR (14.80%) (Fig. 11, gives an idea of the growth rate of India's Cyber Security Product).

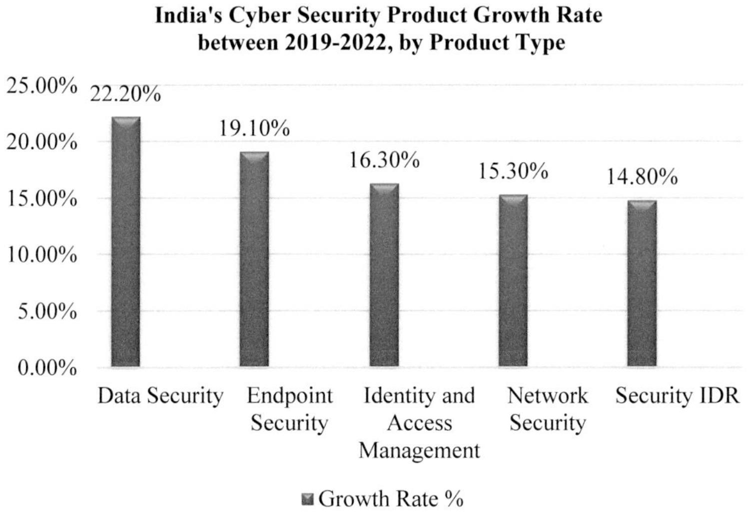

Fig.11 Growth rate of cybersecurity products in India. *Source* Statista [16]

10 Future of Cybersecurity in the Banking Industry

Cybersecurity in the banking industry is anticipated to evolve as a result of growing cyber threats, regulatory changes, and continuous technological improvements. The following significant factors are probably going to have an impact on how cybersecurity in the banking industry develops:

- **Advanced Threat Detection and Response**: Banks will need to use advanced threat detection and response techniques as cyber-attacks become more complex. This involves the use of behavior-based analytics, real-time monitoring, artificial intelligence (AI) and machine learning (ML) algorithms to detect and quickly respond to possible cyberattacks.
- **Collaboration and Information Sharing**: are essential to combating the constantly developing cybersecurity threats that banks cannot handle alone. It will be essential for banks, industry peers, governmental organizations, and cybersecurity organizations to collaborate and share information more often. Sharing threat intelligence, best practices, and lessons gained can assist banks in proactively detecting and countering new cyber threats.
- **Prepare for Quantum Computing**: Banks must embrace quantum-safe encryption to get ready for the post-quantum era as a result of the probable advent of quantum computing. To safeguard confidential information and guarantee long-term security in the face of risks posed by quantum computing, quantum-resistant encryption methods will become crucial.

- **Artificial Intelligence and Machine Learning**: These two types of technology will remain crucial to cybersecurity. They may support incident response, automate threat detection, and search through massive amounts of data for patterns and abnormalities. AI-driven systems can also strengthen proactive defenses and boost overall cybersecurity posture by learning from previous assaults [22].
- **Cybersecurity Workforce and Skills**: There is a growing need for qualified cybersecurity experts. Human error continues to be a major cybersecurity risk. To inform staff about new dangers, phishing tricks, and security best practices, banks will need to invest in thorough cybersecurity training programs. Campaigns to raise awareness of cybersecurity will be necessary to develop a security-conscious organizational culture. Artificial intelligence and automation will also play a part in enhancing cybersecurity capabilities, streamlining procedures, and addressing the skills problem.
- **Zero Trust Framework**: The financial sector is embracing the idea of a zero-trust framework. It places a strong emphasis on stringent access restrictions, ongoing supervision, and user verification at each point of the user experience. Banks may reduce the danger of lateral movement across their networks and guarantee that each transaction or access request is properly verified and authorized by putting Zero Trust principles into practice [6].
- **Continuous Monitoring and Incident Response** To quickly identify and mitigate cyber threats, real-time monitoring, threat hunting, and fast incident response skills will be essential. To improve their capacity for responding to incidents, banks will invest in enhanced security operation centers (SOCs) with advanced analytics, automation, and threat intelligence capabilities.

References

1. How Blockchain Could Disrupt Banking (2022). https://www.cbinsights.com/research/blockchain-disrupting-banking/#:~:text=Further%2C%20blockchain%20technology%20allows%20for,transaction%20days%20after%20a%20payment
2. Alam SMS, Chowdhury MAM, Razak DBA (2021) Research evolution in banking performance: a bibliometric analysis. Future Bus J 7:66. https://doi.org/10.1186/s43093-021-00111-7
3. Blockchain—statistics & facts (nd). https://www.statista.com/topics/5122/blockchain/#topicOverview
4. Blockchain technology market size worldwide in 2017 and 2018, with forecasts from 2019 to 2027 (2023). https://www.statista.com/statistics/1015362/worldwide-blockchain-technology-market-size/#:~:text=This%20statistic%20shows%20the%20size,billion%20U.S.%20dollars%20by%202027
5. Blockchain use in banking and financial services market size worldwide in 2018 and 2019 with a forecast to 2026 (2022). https://www.statista.com/statistics/1229290/blockchain-in-banking-and-financial-services-market-size/
6. Collins J, Singh S (2020) Cybersecurity risks and safeguards in the banking sector: a systematic review. Int J Inf Manage 50:327–339

7. Dalton D, Gianturco P (2016) Out of the blocks: Blockchain—from hype to prototype.https://www2.deloitte.com/content/dam/Deloitte/nl/Documents/financial-services/deloitte-nl-fsi-blockchain-from-hype-to-prototype-out-of-the-blocks.pdf. Accessed 31 May 2023
8. Dave P (2022) Financial cyber frauds in India 2022 and learnings for 2023. https://timesofindia.indiatimes.com/blogs/voices/financial-cyber-frauds-in-india-2022-and-learnings-for-2023/. Accessed 31 May 2023
9. Dhameja NL, Arora S (2020) Banking in India: evolution, performance, growth, and future. Indian J Pub Adminis 66(3):312, 326. https://doi.org/10.1177/0019556120953711
10. Dhar S, Bose I (2016) Smarter banking: blockchain technology in the indian banking system. Asian Manag Insights (Singapore Manag Univ) 3(2):46–53. Asian Management Insights. https://ink.library.smu.edu.sg/ami/3
11. Dubey V (2022) The adoption of blockchain technology in India and its future. 2022 The adoption of blockchain technology in India and its future. BusinessToday. Accessed 8 June 2023
12. Balagolla EMSW, Fernando WPC, Rathnayake RMNS, Wijesekera MJMRP, Senarathne AN, Abeywardhana KY (2021) Credit card fraud prevention using blockchain. In: 6th International conference for convergence in technology (I2CT), Maharashtra, India, pp 1–8. https://doi.org/10.1109/I2CT51068.2021.9418192
13. Garg P, Gupta B, Kapil KN, Sivarajah U, Gupta S (2023) Examining the relationship between blockchain capabilities and organizational performance in the Indian banking sector. Ann Oper Res. https://doi.org/10.1007/s10479-023-05254-0, http://hdl.handle.net/10454/19373
14. Glaeser S, Osgood ND, Bamber MR (2018) Cybersecurity in banking: an organizational perspective. J Inf Syst 32(2):69–88
15. Gregory A (2021) Digital vision: digital banking financial services and insurance global opinion paper. https://atos.net/wp-content/uploads/2021/10/Digital-Vision-Digital-Banking-Opinion-Paper-Report.pdf
16. Growth rate of cyber security products market in India between 2019 and 2022, by product type (2022). https://www.statista.com/statistics/1099965/india-growth-rate-cyber-security-market-by-product-type/
17. Gupta S, Gollmann D (2017) Cybersecurity in banking: perspectives on the role of national culture. Inform Manag 54(2):227–237
18. India and the Blockchain technology, 2022 India and the Blockchain technology. J India. Accessed 8 June 2023
19. Indian banks reported 248 data breaches in last four years, says government (2022) Indian banks reported 248 data breaches in last four years, says government. moneycontrol.com. Accessed 8 June 2023
20. Javaid M, Haleem A, Singh RP, Suman R, Khan S (2022) A review of blockchain technology applications for financial services. In: Bench council transactions on benchmarks, standards and evaluations, vol 2, no 3. https://doi.org/10.1016/j.tbench.2022.100073
21. Jun Y, Park SY, Kim D (2019) Cybersecurity investments in the banking industry: an empirical analysis. Int J Inf Manage 44:173–185
22. Khanna P, Haldar A (2023) Will adoption of blockchain technology be challenging: evidence from Indian banking industry. Qual Res Finan Markets 15(2):361–384. https://doi.org/10.1108/QRFM-01-2022-0003
23. Kim J, Solomon MG (2019) Assessing cybersecurity risks in the banking sector: a Bayesian network approach. J Bus Res 98:401–410
24. Kujur A (2018) DATA STORY: India houses the second largest pool of blockchain developers in the world. moneycontrol.com
25. Kurmanath KV (2023) Cybersecurity trends 2023: from Crime-as-a-Service to banking trojans, these are the threats to watch out for. https://www.thehindubusinessline.com/info-tech/cybersecurity-trends-2023-from-crime-as-a-service-to-banking-trojans-these-are-the-threats-to-what-out-for/article66508450.ece. Accessed 31 May 2023
26. Li C (2022) A fraud detection system for reducing blockchain transaction risks using explainable graph neural networks. The George Washington University ProQuest Dissertations Publishing, p 29325150

27. Lodha H, Mehta D (2022) An overview of cyber crimes in banking sector. https://www.legalserviceindia.com/legal/article-7694-an-overview-of-cyber-crimes-in-banking-sector.html. Accessed 31 May 2023
28. Number of cybercrimes related to online banking across India from 2016 to 2021 (2022) https://www.statista.com/statistics/875887/india-number-of-cyber-crimes-related-to-online-banking/
29. Pandey AA, Fernandez TF, Bansal R, Tyagi AK (2021) Maintaining scalability in blockchain. In: Intelligent systems design and applications: 21st international conference on intelligent systems design and applications (ISDA 2021) held during December 13–15, pp 34–45. https://link.springer.com/chapter/10.1007/978-3-030-96308-8_4
30. Patel E (2022) How RBI and Indian Banks are piloting blockchain trade financing. inc42.com. Accessed 8 June 2023
31. Percentage of internet users in selected countries who have ever experienced any cybercrime in 2022 (2023). https://www.statista.com/statistics/194133/cybercrime-rate-in-selected-countries/#:~:text=In%202022%2C%20around%20four%20in,to%20have%20ever%20experienced%20cybercrime
32. Sarmah A, Sarmah R, Baruah AJ (2017) A brief study on cyber crime and cyber laws of India. IJRET 4(6)
33. The Rise of Cybercrime in Indian Banks (nd) https://www.clari5.com/the-rise-of-cybercrime-in-indian-banks/. Accessed 31 May 2023
34. Ullah N, Al-Rahmi WM, Alfarraj O, Alalwan N, Alzahrani AI, Ramayah T, Kumar V (2022) Hybridizing cost saving with trust for blockchain technology adoption by financial institutions. Telematics Inform Rep 6. https://doi.org/10.1016/j.teler.2022.100008
35. Verma VK (2014) Phishing (Indian Case Law, 16 July 2014). https://indiancaselaw.in/phishing/. Accessed 1 June 2023
36. Vijayalakshmi P, Priyadarshini V, Umamaheswari K (2021) Impacts of cyber crime on internet banking. Int J Eng Technol Manag Sci 5(2). https://doi.org/10.2139/ssrn.3939579

Revolutionizing Banking with Blockchain: Opportunities and Challenges Ahead

Shilpa Mahajan and Meghna Nanda

Abstract This chapter explores the opportunities and challenges that blockchain presents for the banking industry. Secure transactions are possible due to blockchain's distributed and inflexible nature, which results in efficient and modest cross-border payments, streamlined remittance processes, and reduced settlement times. It also explores the potential impact of integrating cloud computing with blockchain and critically examines the current state of blockchain endorsement in this sector, highlighting successful use cases and early implementations. It delves into the potential benefits that blockchain can bring to different aspects of banking and assesses the challenges that need to be overcome for full-scale integration. It also emphasizes blockchain's disruptive influence on the financial sector and the significance of handling the opportunities and difficulties posed by this technology. This chapter also discusses the traditional approach used by banking earlier and how blockchain has changed the outlook of the present banking industry.

Keywords Blockchain · Architecture · Cloud computing · Case studies · Opportunities

1 Introduction

Today Banking is one of the most used sectors of the world, from UPI payment to stock market everything moves around banking and its transactions only. Banks are enhancing their customer services day by day to increase the rate of security, transparency, and the speed of transactions. With the publication of blockchain technology around the world in the past years, it shows the power to revolutionize the banking sector to create a secure and fast transaction which also helps in creating an automatic world of banking.

Blockchain is a distributed ledger system which make transactions recorded and verified without the need for intermediaries. It is a decentralized system which allow

S. Mahajan (✉) · M. Nanda
Department of Computer Science, The NorthCap University, Gurugram, India
e-mail: shilpa@ncuindia.edu

K. Kaushik and I. Sharma (eds.), *Next-Generation Cybersecurity*, Blockchain Technologies, https://doi.org/10.1007/978-981-97-1249-6_13

safe, tamper-proof, and transparent transactions. Blockchain technology can address many challenges by giving a protected and translucent system for noting transactions, decreasing the use of intermediaries, and increasing the speed and productivity of transactions.

Cloud computing allows on-demand access to a shared pool of resources over the internet, and has gained significant attraction across various industries. It offers scalability, flexibility, cost-efficiency, and the ability to access computing power and data storage resources as needed. By leveraging cloud computing, banks can optimize their operations, streamline processes, and effectively manage vast amounts of data.

2 Characteristics and Architecture of Blockchain

Blockchain is a digital ledger system which sanction secure and transparent recording and verification of monetary exchange. It is a decentralized system that is intended to function independently of a central authority and is preserved and proven by a network of computers, called nodes. The copy of the ledger is kept by every node, and new transactions which are checked and supplementary to the ledger through a consensus mechanism, like Proof of Stake (PoS) or Proof of Work (PoW). In a blockchain system, each block contains a different identifier, a timestamp, and a digital signature that verifies the authenticity for the transaction. The online proof is created using cryptographic algorithms, which ensure that the data is secure and cannot be altered or manipulated. One of the most important attributes of the blockchain model is its immutability, which implies a transaction which is saved once in the blockchain will not be updated. This creates a transparent and tamper-proof system that can be used for a huge variety of usage like financial transactions and identity verification. In a blockchain system, each transaction is saved in a block and added to the chain of the last blocks, producing a stable and tamper-proof history of every change in data as shown in Fig. 1.

The architecture of blockchain consists of the following key components (see Fig. 2).

Distributed network: In a blockchain, each node is connected to another node in a distributed manner, where each node stores a copy of the blockchain ledger. This distributed network enables blockchain to operate in a decentralized and transparent manner, where no single entity controls the network, or the data stored on it.

Blocks: Blockchain stores data in blocks that has a set of transactions and a unique digital signature called a hash. The integrity of the block is assured by hash makes it tamper-proof. Each block is attached to the last block with the hash, making a chain of blocks called a blockchain.

Consensus mechanism: Blockchain operates on consensus mechanism which makes all nodes in the network accord with the state of ledger. There are several consensus mechanisms, like Proof of Work and Proof of Stake. These mechanisms ensure that the blockchain is secure and resilient to attacks.

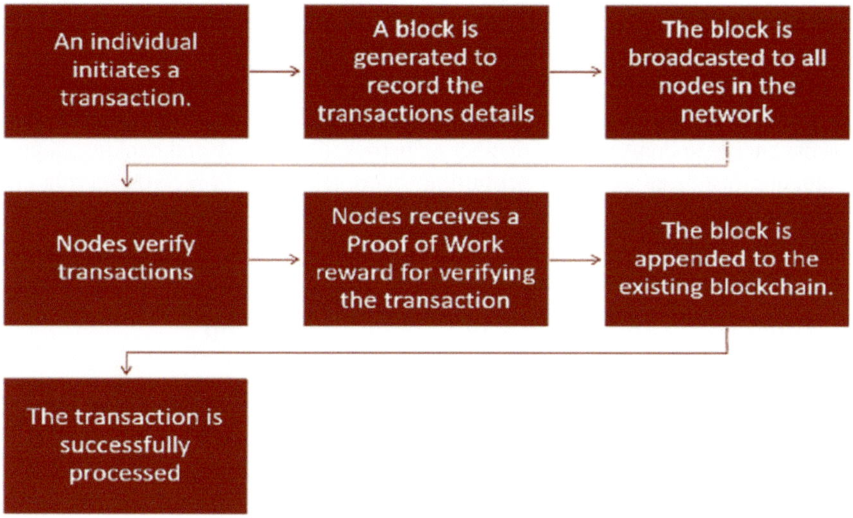

Fig. 1 Working of blockchain

Fig. 2 Components of blockchain

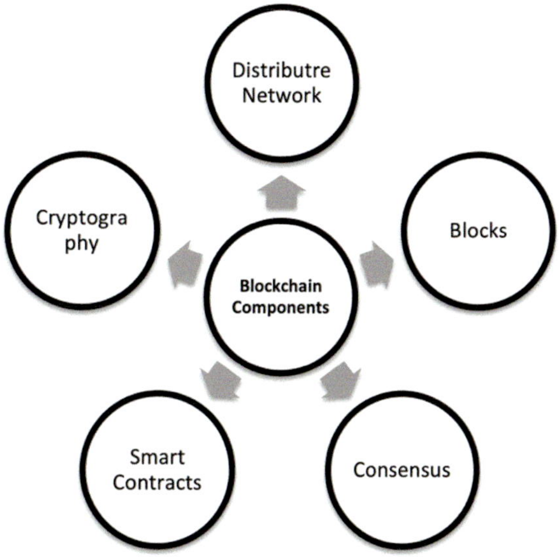

Smart contracts: Smart contracts mean blockchain can safely secure and run contracts on their own which means automatically contracts can be executed in blockchain to increase the performance by providing the contracts with certain conditions.

Cryptography: Blockchain operates by using cryptographic techniques to make sure that the protection and confidentiality of the data stored on the blockchain. For example, blockchain uses cryptography to ensure that only the owner of a private key can access and modify the work saved in the blockchain.

Overall, the structure of blockchain is made to provide a secure, transparent, and decentralized platform for storing and sharing data. The architecture of blockchain has the energy to convert many distinct industries, like finance and banking, supply chain, and healthcare, through permitting secure and efficient records sharing and collaboration.

3 Blockchain and Cloud Computing Integration

The integration of blockchain generation with cloud computing provides a unique opportunity for the banking area. By combining the security and transparency of blockchain with the scalability and flexibility of cloud computing, banks can liberate new tiers of innovation and efficiency.

Blockchain generation, with its decentralized and tamper-proof nature, addresses the challenges of conventional banking systems, together with sluggish transaction speeds, excessive charges, and shortage of transparency. It allows stable and transparent recording and verification of transactions, eliminating the want for intermediaries. On the other hand, cloud computing offers the infrastructure and computing assets required to assist blockchain applications at scale. Figure 3 shows blockchain and cloud computing integration.

By utilizing cloud computing, banks can leverage the elastic computing capabilities to process blockchain transactions quickly and efficiently. A number of services are provided by cloud service providers, including virtual machines, storage, and networking, which can be utilized to build and deploy blockchain networks. Additionally, the cloud provides a reliable and scalable environment for hosting blockchain nodes, ensuring continuous availability and redundancy.

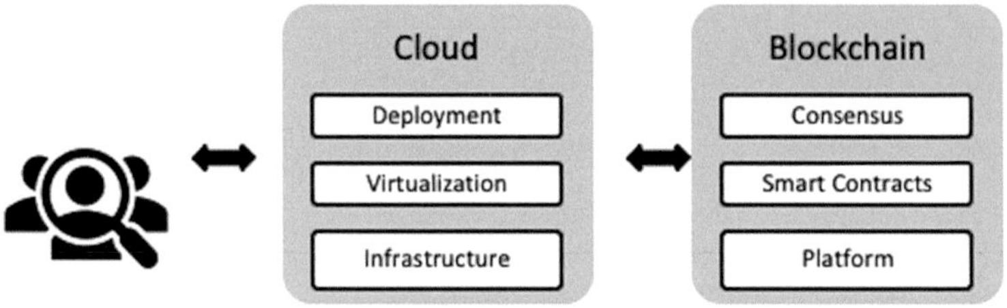

Fig. 3 Blockchain cloud computing integration

4 Challenges with the Traditional Approach

A conventional system, used by banks in the financial industry, is the foundation of international trade in finance and is thought to be responsible for the daily transfer of more than a trillion dollars. Credit card systems, the Internet, and mobile applications are just a few of the significant advancements and technology that have been integrated into the banking system over time. By enhancing efficiency, convenience, and transaction speed as well as minimizing the distance between parties involved, emerging technologies like blockchain have greatly aided banks in running their businesses.

The integration of blockchain technology and cloud computing opens a myriad of opportunities for the banking sector. It enables banks to develop and deploy blockchain-based applications, streamline processes such as cross-border payments, trade finance, and identity verification, and enhance customer experiences by providing secure and transparent financial services.

The major challenge includes the insufficient comprehension of the technology and its potential applications in the banking sector. Most of the banks hesitate to adopt blockchain due to the perceived risks and uncertainties associated with it.

Another challenge is the regulatory environment, which varies across the country, not only banking sector but also hinders the adoption of blockchain among others. Banks have to comply with laws on privacy, data security, and financial crime prevention. There may also be barriers to blockchain adoption.

However, the combination of blockchain and cloud computing also presents some challenges. These include data privacy concerns, compliance, interoperability, business improvements, and the need for strong governance structures. Banks need to carefully consider these challenges and implement appropriate solutions to ensure the successful integration and adoption of blockchain and cloud computing in their operations.

Finally, blockchain's scalability and performance limits are also a concern, especially as the number of transactions and data volumes in the banking sector continue to grow. Banks need to ensure their blockchain systems can handle more transactions and data while maintaining high levels of security and reliability.

This chapter addresses the challenges and opportunities of enforcing blockchain technology in the banking sector and the solutions that can be used to overpower these difficulties and foster the widespread adoption of blockchain in banking.

5 Literature Review

Blockchain technology is getting popular day by day in the last few years due to its unrealized power to transform several industries including finance. Blockchain provides a decentralized and transparent platform for storing and sharing data, which can increase the efficiency, security, and transparency of banking operations. Significant research has been done in this area to measure the effectiveness of the amalgamation of blockchain in banking sectors. Few of the works have been presented below.

The meaning of blockchain and the way it works which means how data is transferred in the form of blocks during any transaction have been well explained in the research paper of Mahajan [1]. The research paper mainly focuses on the application of using blockchain in healthcare sector which also discusses the advantages and limitations of using blockchain.

To assist students in understanding and conducting research effectively, Adams et al. [2] published a book in the year 2014 providing a valuable way to evaluate the key features, strengths, and limitations of research and helped to find a way of how research must be done following what pattern that helped in finding the best methodologies for research. To understand the digital currencies and central bank policies, Energet and Fung [3] published their thoughts in a discussion paper in the year 2017 which aims to highlight the evolving landscape of digital currencies and central bank policies helping in finding the issues faced by central banks in secure transaction leading to motivation for development.

To provide an insight into using blockchain technique in finance, Shorman et al. [4] published a research in the year 2020 providing a review of using blockchain in finance which helped in finding the potential application of banking using blockchain which includes the sectors which all can be transformed using blockchain which helps in working smooth running of blockchain.

The usage of blockchain technology in banking area is explained in a Deloitte Insights [5] report as one of the most potential and powerful weapons to be used in banking by elaborating about the meaning of blockchain, benefits of using blockchain in banking industries about transparency, security, and many other benefits. They have also discussed the difficulties of applying blockchain in banking the funds required, proper knowledge of using it, and many more.

A review of applying blockchain in banking has been discussed by Sridharan and Shankar [6] in their article which includes the basics of blockchain that is what is blockchain, how transformation occurs in blocks, advantages of using blockchain in banking, areas in which blockchain could be used like transaction. They have also discussed the challenges of using blockchain in banking as applying blockchain in banking is not easy and it requires a lot of funds that have to be raised which can be used for smooth running of banks using blockchain.

An article written by Iqbal and Wasiq [7] was released providing a comprehensive review of the recent literature about research on applying blockchain in banking and finance sector which also includes the potential challenges and prospects of using it.

To examine the capability of using blockchain in banks, Vali and Madani [8] published an article about the capabilities of using blockchain not only in banking but also in other sectors like health sectors and others. It provides information about how secure, efficient, transparent, and smooth banks can run their day-to-day operations including the same for other sectors too.

To inform the world about the importance of addressing some issues such as accessibility, usability, and trust in using blockchain for remittances, microfinance, and online identity, Yeoh and Choi [9] published their research paper which consisted of all the above issues along with its solutions. The research paper also includes various applications of blockchain which also consist of banking and finance. The challenges and opportunities are discussed in detail in the research paper.

The applications of blockchain and features like efficiency, security, and transparency have been published in a research paper by Heshmati and Gupta [10] which consists of the challenges required to adopt, regulate, and interoperate before applying blockchain in every sector. The advantages which will help in smooth running of the sector after applying blockchain will lead to advancement of the world.

5.1 Technology Acceptance Model (TAM) Perspective

TAM is one of the most widely used theories which explains users' adoption of new technology based on their discern value and discern use. TAM proposed that the way of using a technology is determined by the discern value of technology and discern use that is to make it work-friendly with everyone as shown in Fig. 4.

In the context of blockchain in banking, TAM can be used to explain banks' adoption of blockchain technology. The the usefulness of blockchain in the banks can include reducing fraud, improving cross-border payments, and enhancing regulatory compliance. It considers the ease of use, which can make the technical complexity of integrating blockchain with existing banking systems, the cost of implementation, and the level of regulatory compliance required.

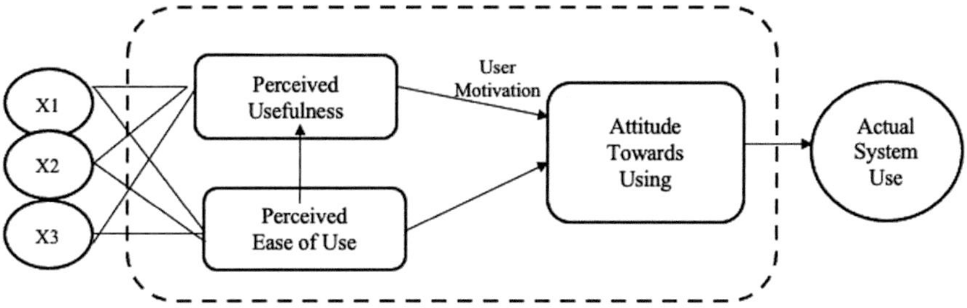

Fig. 4 Original TAM [11]

6 Diffusion of Innovation Theory

This theory provides an explanation, for how new innovations are adopted and spread throughout the society. It suggests that various factors play a role in determining the acceptance of technology such as the advantages it offers compared to existing technology, its compatibility with systems, and its level of complexity. When considering the use of blockchain in the banking sector, we can apply the diffusion of innovation theory to understand why banks are increasingly adopting blockchain. The advantages of blockchain over banking systems can include improved efficiency, transparency, and security. The compatibility of blockchain with existing systems depends on factors like integration complexity, regulatory compliance requirements, and the bank's cultural and organizational norms. Lastly, the complexity of implementing blockchain can vary based on the expertise needed, implementation costs, and potential risks associated with adopting technology.

Focusing on the core benefits of blockchain, this study examines the key factors that compel banks to adopt blockchain solutions and offers effective strategies to accelerate their adoption. By implementing distributed ledger technology, banks can enhance security, streamline operations, reduce costs, and increase transparency. In addition, the chapter highlights the importance of smart contracts and how decentralized finance applications are changing traditional banking practices. Suggested strategies include building industry-wide collaborations, establishing regulatory frameworks, fostering collaboration between blockchain platforms, and investing in blockchain-focused R&D. Overall, this chapter seeks to provide a comprehensive understanding of the importance, blockchain provides to the banking landscape.

7 Blockchain and Cloud Computing Applications in Banking Sector

By leveraging the decentralized and immutable nature of blockchain, combined with the agility and flexibility of cloud computing, banks can unlock new possibilities for optimizing operations, improving customer experiences, and reducing operational costs. This chapter explores how combining these two transformative technologies can revolutionize traditional banking practices and address existing challenges in the industry. Some of the use cases are highlighted below.

Cross-border payments: Blockchain offers cost-effective, secure, and faster ways of exploring cross-border payments by eliminating intermediaries. For example, Ripple, which is one of the blockchain-based payment networks, uses instant cross-border payments between banks and financial institutions.

Interoperability and Collaboration: Cloud computing facilitates seamless integration and interoperability between different blockchain networks and banking systems. Banks can leverage cloud-based platforms to connect with other financial institutions [12], enable cross-network transactions, and foster collaborative initiatives within the industry.

Know Your Customer (KYC) compliance: Blockchain can streamline the KYC compliance, a process in which a platform is given to provide a secure and immutable ledger of customer data. Banks may also use blockchain to verify customer identities and share customer data securely and efficiently across different institutions.

Trade finance: Blockchain can enhance trade and finance on a transparent and secured platform for tracking trade transactions and reducing fraud. For example, Marco Polo is a blockchain-based trade finance platform which tracks and identifies trade transactions in real time.

Identity verification: Blockchain can enable secure and decentralized identity verification, reducing the risk of identity theft and fraud. For example, uPort is a platform designed on the blockchain that gives leverage to the users to control their digital identities and share their identity data securely with different institutions.

Improved Data Management: Cloud computing offers robust data storage and management capabilities, allowing banks to securely store and access large volumes of blockchain data. This integration enables efficient data synchronization and replication across multiple nodes, ensuring the integrity and availability of blockchain data.

Cost Optimization: The integration of cloud computing with blockchain allows banks to optimize their infrastructure costs by leveraging the on-demand nature of cloud resources. Banks can dynamically allocate computing resources based on their needs, reducing operational expenses and improving cost-effectiveness.

Smart contracts: Self-executing smart contracts, which automate contract execution and eliminate the need for middlemen, can be created using blockchain technology. For example, the blockchain-based platform Ethereum allows for the development and execution of smart contracts.

For the most part, blockchain technology can change the banking sector by improving efficiency, transparency, and security and by integrating cloud computing, banks can unlock the full potential of this transformative technology, enabling improved scalability, data management, productivity, cost optimization and parallel performance for banking applications, through widespread deployment, and pre-banking adoption.

8 Challenges of Blockchain Adoption in Banking Sector

The adoption of blockchain technology in the banking sector presents several challenges that hinder its widespread implementation. Some of the key challenges include:

Regulatory compliance: Banks operate within a highly regulated environment, and compliance requirements may vary across jurisdictions. Integrating cloud computing with blockchain requires careful consideration of regulatory frameworks to ensure compliance with data protection, privacy, and financial regulations. Banks need to navigate the complexities of compliance while leveraging the benefits of this integration.

Interoperability: Achieving interoperability between different cloud platforms and blockchain networks can be challenging. Banks may encounter compatibility issues and difficulties in integrating diverse technologies and protocols. Standardization efforts are crucial to establishing uniformity and seamless interaction between cloud-based blockchain systems.

Security and Privacy: The combination of cloud computing and blockchain introduces new security and privacy concerns. Banks need to ensure the confidentiality and integrity of sensitive data stored in the cloud and transmitted within the blockchain network. Robust encryption, access control mechanisms, and secure protocols must be implemented to avoid risks associated with unauthorized access, data breaches, and insider threats.

Technical complexity: Combining blockchain technology and cloud computing with the present banking systems may be technically complex and challenging. Banks need to ensure that their blockchain systems can integrate seamlessly with their existing systems, databases, and applications.

Cost: Implementing it can be expensive, particularly for smaller banks and financial institutions. Banks need to make sure that the advantages of adopting blockchain technology outweigh the costs, and that they have the required resources and expertise to implement and maintain their blockchain systems.

Security: Blockchain technology is often considered to be more secure than traditional banking systems, but it is not immune to security threats. Banks need to ensure that their blockchain systems are secure and that they have the right measures at time to prevent hacking, data breaches, and other security threats.

Governance and Consensus Mechanisms: Cloud-based blockchain networks may require modifications to the governance and consensus mechanisms to accommodate the distributed nature of cloud computing. Consensus protocols and decision-making processes need to be designed to ensure the integrity and trustworthiness of the network while considering the involvement of multiple cloud nodes.

Overall, these challenges need to be resolved before using it widely and adopting it in the banking sector. Banks need to work together with regulators, technology providers, and other stakeholders to resolve these difficulties and realize the power of using blockchain technology with cloud computing in the banking sector.

9 Methodology

This chapter presents an overview of the usage of the blockchain in the banking sector. The aim of this work is to enhance understanding and knowledge of blockchain technology in the banking industry with cloud computing. This will be achieved by analyzing the existing research literature and comparing it with insights gained from the people who were having their practical knowledge and experience in using blockchain in banking. Furthermore, review papers, experiments, and case studies conducted by other researchers have also been considered. Since blockchain technology has been founded recently for a decade, this study has a contemporary time dimension. Nevertheless, as the study's findings include information on the potential challenges of utilizing blockchain technology, it also has a future-oriented dimension.

The data gathered for this work has been taken from multiple sources. Firstly, data was collected from Wikipedia, which provided a foundational understanding of the subject matter. Additionally, research conducted by other scholars was consulted to gain insight into previous findings and knowledge on the topic. Furthermore, interviews were conducted with individuals who were having hands-on knowledge in both the banking and non-banking sectors to provide first-hand insights into the use of blockchain in banking. The collection of data from various sources enhances the comprehensiveness of the study's findings and adds depth to the review analysis.

10 Usage of Blockchain in Different Sectors

This work provides an outline of how blockchain technology can provide significant benefits to the banking industry. The use of technology in the financial sector has the potential to provide several benefits while also posing a number of problems. According to the findings of this survey, one of the most extensively used applications of blockchain technology is the ability to increase efficiency and lower costs connected with traditional banking operations. As seen in Fig. 5, blockchain technology has shown potential in a variety of areas because of its decentralized, transparent, and secure nature. Blockchain technology can help banks run more effectively and profitably by streamlining operations and enabling speedier transactions at reduced prices.

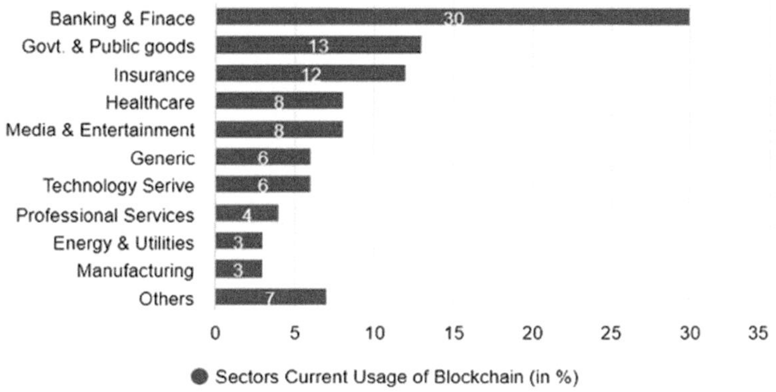

Fig. 5 Usage of Blockchain in different Sectors

Improved security is another important advantage of blockchain technology in banking. Blockchain creates a transparent, decentralized record that is resistant to cyber-attacks, making it an important tool for protecting financial transactions [13]. This technology can also increase transaction speed by lowering the time necessary for verification and approval processes, resulting in faster and more smooth financial transactions.

The study did, however, highlight several difficulties that have to be resolved before this public ledger technology can be broadly implemented in the banking industry. Among the most major issues are regulatory and legal problems, such as compliance with data privacy regulations [14]. Furthermore, the necessity for blockchain system standardization and interoperability is a substantial barrier to widespread implementation [15, 16]. To enable smooth transactions and collaboration among diverse entities, blockchain systems must be interoperable with one another.

Finally, there is the issue of automation and its potential impact on employment in the banking industry. As blockchain technology is adopted, some jobs may become automated, which could lead to job losses for workers in certain areas of the industry.

Therefore, the application of blockchain technology in banking has the capacity to transform the industry by augmenting productivity, enhancing security, and speeding up financial transactions. However, careful consideration and planning are required to overpower the obstacles while adopting it, including regulatory and legal concerns, standardization, interoperability, and automation. With proper planning and execution, however, blockchain technique has the capacity to revolutionize the banking industry, boosting productivity, integrity, and customer experience.

Blockchain technology offers several promising use cases for Indian banks, addressing various challenges and improving efficiency and transparency as shown in Table 1. Some of the notable use cases include.

Trade Finance and Supply Chain Management: Indian banks can leverage blockchain to streamline trade finance processes and enhance supply chain management [17]. By creating a distributed ledger, banks can facilitate secure and transparent transactions between exporters, importers, and other stakeholders. This can reduce

Table 1 Indian banks using blockchain

Bank name	Year (usage of blockchain)	Use case of blockchain
SBI Bank	2016	KYC compliance, supply chain financing, digital identity verification
ICICI BANK	2016	Domestic and International remittances
AXIS BANK	2017	Cross-border remittances, trade finance platform
YES BANK	2018	Supply chain financing

paperwork, eliminate delays, and mitigate the risk of fraud, thereby improving the overall efficiency of cross-border trade.

KYC and Identity Verification: Know Your Customer (KYC) processes can be time-consuming and costly for banks. Blockchain-based identity solutions can enable customers to maintain control over their personal data and provide verified information only when required. This can accelerate onboarding processes, enhance data privacy, and reduce the duplication of efforts for both banks and customers.

Remittances and Cross-Border Payments: Cross-border remittances are made quicker and more affordable by blockchain technology, which lowers transaction costs and does away with middlemen. Indian banks can use blockchain networks to establish direct connections with international financial institutions, enhancing the speed and affordability of remittance services for customers.

Digital Lending and Smart Contracts: Blockchain-based smart contracts can automate and enforce lending agreements, making loan disbursals and repayments more secure and efficient. Indian banks can use smart contracts to execute lending contracts transparently and automatically trigger loan repayments based on predefined conditions, reducing the need for manual intervention and minimizing default risks.

Trade Settlements: Blockchain can be applied to simplify and expedite the settlement process for securities and commodities trading. By using distributed ledger technology, banks can achieve real-time settlement, reduce counterparty risk, and improve liquidity management.

Fraud Detection and Prevention: Blockchain's transparent nature and immutability make it a valuable tool for fraud detection and prevention. Banks can track and monitor transactions on the blockchain to identify suspicious activities and enhance their security measures.

Loan Syndication and Debt Issuance: Blockchain can facilitate loan syndication and debt issuance by creating a secure and auditable platform where multiple banks can collaborate, verify loan data, and participate in loan syndication efficiently.

Loyalty Programs and Rewards: Indian banks can implement blockchain-based loyalty programs, allowing customers to earn and redeem rewards across various partners, providing a seamless and more valuable customer experience.

Central Bank Digital Currency (CBDC): The Reserve Bank of India (RBI) has been exploring the possibility of issuing a central bank digital currency. Blockchain technology can serve as the underlying infrastructure for a secure and efficient CBDC system, enabling faster payments and reducing operational costs.

These use cases demonstrate how blockchain technology can revolutionize various aspects of Indian banking, fostering financial inclusion, improving operational efficiency, and enhancing customer experiences. However, successful implementation requires collaboration among banks, regulators, and technology providers, along with addressing challenges related to scalability, interoperability, and regulatory compliance.

10.1 Real-World Case Studies of Blockchain Implementation in Banking

Ripple and Santander's One Pay FX—Changes to Border Payments

Santander, one of the world's largest banks, has embarked on a groundbreaking journey to use blockchain technology to transform cross-border payments In partnership with Ripple, a blockchain technology company, Santander launched One Pay FX, a blockchain-based platform aimed at increasing efficiency and speed.

Santander recognized the challenges its customers face when moving money across borders, such as high transaction costs, long processing times, and lack of visibility in the payment process and the bank sought to address these pains by finding blockchain solutions, giving birth to One Pay FX. Ripple's blockchain technology provided the foundation for One Pay FX. After implementing Ripple's XCurrent on the platform, XRP, the native cryptocurrency of the Ripple network, enabled secure and real-time cross-border transactions to be settled using blockchain for instant liquidity and efficient exchange, eliminating the need for multiple intermediary banks.

The implementation of One Pay FX has brought several key benefits to Santander and its customers.

Faster transactions: Cross-border payments that previously took days can now be completed in minutes, significantly reducing payment time [18].

Increased transparency: Customers gained real-time insight into the status and progress of their services, increasing trust and confidence.

Lower costs: Bypassing traditional media banking, Santander reduced transaction fees for its customers.

Improved customer experience: One Pay FX offers a simple and user-friendly interface, which simplifies the process of initiating and tracking payments.

Following a successful One Pay FX trial, Santander expanded the service to other countries and regions, providing improved cross-border payment capabilities to a broader customer base that widely adopted the platform, attracting other financial institutions and paving the way for consolidation and innovation in the banking sector. There are various challenges when using One Pay FX. Considerable effort and cooperation was needed to ensure compliance, safety standards, and greater acceptance from other financial institutions.

The success of One Pay FX sent waves throughout the banking industry, prompting other financial institutions to seek out blockchain solutions for their cross-border payment systems The adoption of blockchain technology in banking grew as more organizations realized the potential benefits. As blockchain technology continues to evolve, Santander's One Pay FX is a prime example of its potential to revolutionize cross-border payments. The expansion and flexibility of the platform serve as a benchmark for future blockchain applications in the banking industry [19].

Santander's collaboration with Ripple and the subsequent launch of One Pay FX demonstrates the real-world potential of blockchain technology to transform banking systems, particularly in cross-border payments.

Komgo—Transforming Trade Finance with Blockchain

Komgo is a blockchain-based totally change finance platform that ambitions to streamline and revolutionize the commodity trading industry [20]. Launched in 2018, it's far a joint assignment between primary global banks, trading corporations, and electricity agencies. Komgo makes use of blockchain generation to beautify transparency, performance, and safety in trade finance operations. The commodity trading enterprise has lengthy struggled with complicated, paper-extensive, and time-consuming strategies. Komgo was born out of the need to cope with these challenges by way of leveraging blockchain generation to create a greater efficient and trusted atmosphere for exchange finance.

Komgo's platform is constructed on a personal permissioned blockchain, ensuring that the simplest legal individuals can get admission to and validate transactions. It makes use of smart contracts to automate and streamline numerous exchange finance methods, such as letters of credit score, virtual record sharing, and commodity trade financing. There are several key benefits to the adoption of Komgo's platform.

Enhanced Efficiency: By digitizing and automating approaches, Komgo reduced the time wished for exchange finance transactions, resulting in quicker settlements and reduced paperwork.

Improved Transparency: All stakeholders involved in the trade, including banks, traders, and regulators, can access real-time data and verify transactions, increasing trust and reducing disputes.

Fraud Prevention: The immutable nature of blockchain ensures that trade documents and contracts cannot be altered or tampered with, mitigating the risk of fraud.

Cost Reduction: The elimination of manual processes and intermediaries led to cost savings for participants in the trade finance ecosystem.

Komgo's success can be attributed to cooperation with leading global players in the commodity trading industry. Banks such as ABN AMRO, BNP Paribas, and ING Group partnered with trading companies such as Shell and Mercuria to create a trusted massive network in a platform where Komgo users faced compliance challenges, data privacy, and integration with existing systems worked well with Mackay. The introduction of Komgo generated excitement and excitement among various stakeholders in the commercial economy. This demonstrated the potential of blockchain technology to transform traditional methods and paved the way for further adoption of blockchain in the region.

As Komgo continues to evolve, it has the potential to become the industry standard for commercial finance. The success of the platform has encouraged financial institutions and other commercial entities to explore similar blockchain solutions to improve efficiency and transparency in their operations. Komgo serves as a real-world case study of how blockchain-based trading and financial systems can transform the commodity trading industry. Through long-term challenges and resulting insights and greater efficiencies, Komgo has demonstrated the transformative power of blockchain technology in trade finance [21]. As the platform continues to evolve and gain traction, it sets a precedent for other companies to explore blockchain solutions for their specific trade finance needs.

Beyond these, there are a number of real-world banking case studies that have adopted blockchain for improved efficiency and greater security. Some of them are described below.

JPMorgan Chase—Interbank Information Network (IIN)

JPMorgan Chase, one of the world's largest banks, launched the Interbank Information Network (IIN) to address global payments challenges IIN on Quorum, a private blockchain platform developed by JPMorgan, enabling faster resolution of compliance and regulatory issues. As a result, international payments can be made faster, benefiting both the bank and its customers.

BBVA—Blockchain-Based Syndicated Loan

BBVA, a Spanish multinational bank developed a pilot project to provide blockchain technology for integrated lending. The bank partnered with various parties involved in the syndication process, including BNP Paribas and MUFG, to use blockchain to record and manage the loan agreement. The blockchain platform facilitated seamless communication and data sharing between all participants, reducing paperwork and manual errors. The successful study demonstrated blockchain's potential to revolutionize the mortgage loan market by streamlining processes and improving transparency [22].

Abu Dhabi Commercial Bank (ADCB)—ADCB collaborated with IBM to broaden a blockchain-based trade finance platform [23]. The platform digitizes and automates trade finance techniques, inclusive of letter of credit score issuance, alternate document control, and agreement. By the use of blockchain generation, the bank aimed to lessen processing time, decorate security, and enhance visibility throughout the change finance environment. The a success implementation demonstrated the capacity for blockchain to revolutionize trade finance operations and growth performance for each financial institution and its customers.

Standard Chartered—Cross-Border Remittances

Standard Chartered, an outstanding global bank, partnered with Ripple, a blockchain-based payments community, to broaden a move-border remittance solution. The blockchain generation allowed for actual-time and value-effective fund transfers among the bank's branches in diverse international locations, significantly lowering transaction instances and prices [24].

Emirates NBD—"Cheque Chain": Emirates NBD, a leading financial institution within the United Arab Emirates, carried out a blockchain-primarily based platform referred to as "Cheque Chain" to enhance the safety and authenticity of its issued cheques. The platform allows real-time verification of the cheque's popularity, lowering the danger of fraud and ensuring a clean clearing procedure [25].

These real-world case researches show the various programs of blockchain technology in the banking quarter, starting from global bills and change finance to debt issuance and syndicated loans. As more monetary establishments discover and embrace blockchain solutions, the industry maintains to witness the transformative capacity of this era in improving efficiency, security, and customer revel.

11 Conclusion

This chapter delves into the ideas of blockchain generation, its layout, and the underlying theories that underpin its functionality, supplying a detailed evaluation of the position of blockchain inside the economic zone. Generation additionally emphasizes using blockchain in the financial enterprise, demonstrating the massive potential generation has to revolutionize setup banking practices.

This bankruptcy's consciousness in the Indian banking quarter is one of its significant contributions, with particular use of instances and illustrations of the way Indian banks would possibly use the blockchain era to handle their unique problems and necessities. The chapter emphasizes the importance and feasibility of blockchain implementation for Indian banks via focusing on these software instances.

Furthermore, the subject of merging cloud computing with blockchain highlights a vital feature of the chapter. This integration affords banks with strategic advantages with the aid of helping them to conquer scalability and power intake constraints. By leveraging cloud computing sources, banks can further decorate the overall performance, flexibility, and computational capabilities of their blockchain-based systems. As the Indian banking sector continues to evolve, the findings and recommendations presented here serve as a valuable guide for driving future developments and innovations using blockchain technology.

References

1. Mahajan S (2022) Blockchain in smart healthcare systems: hope or despair? In: Borah M, Zhang P, Deka G (eds) Prospects of blockchain technology for accelerating scientific advancement in healthcare, pp 1–20
2. Adams J, Khan HTA, Raeside (2014) Research methods for business and social science students, 2nd edn. Sage, New Delhi
3. Engert W, Fung BS (2017) Central bank digital currency: motivations and implications (No. 2017-16). Bank of Canada Staff Discussion
4. Shorman A, Sabri KE, Abushariah M, Qaimari (2020) Blockchain for banking systems: opportunities and challenges. J Theor Appl Inf Technol 98(23)
5. Deloitte (2017) Blockchain in banking: a measured approach. 1–34
6. Sridharan R, Shankar R (2021) Blockchain technology in banking: a comprehensive review. Int J Manag Technol Soc Sci (IJMTS) 6(2):27–39
7. Iqbal MA, Wasiq M (2021) Blockchain adoption in the banking industry: a review of challenges and opportunities. J Risk Financ Manag 14(4):156

8. Vali M, Madani SHH (2021) Blockchain technology in banking industry: a systematic review of recent studies. J Bus Res 135:747–761
9. Yeoh W, Choi SY (2021) Understanding the potential of blockchain technology for financial inclusion: a review of key opportunities and challenges. Int J Inf Manag 58
10. Heshmati M, Gupta RK (2021) Blockchain in banking sector: a systematic review of recent studies. Int J Bank Market 39(3):516–531
11. Davis FD (1985) A technology acceptance model for empirically testing new end-user information systems: theory results. Massachusetts Institute of Technology, Cambridge, pp 233–250
12. Chang V, Baudier P, Zhang H, Xu Q, Zhang J, Arami (2020) How blockchain can impact financial services—the overview, challenges, and recommendations from expert interviewees. Technol Forecast Soc Change 158
13. Garg P, Gupta B, Chauhan AK, Sivarajah U, Gupta S, Modgil S (2021) Measuring the perceived benefits of implementing blockchain technology in the banking sector. Technol Forecast Soc Change 163
14. Garg L, Choudhary P, Sharma P, Kaur H (2020) Exploring the advantages of using blockchain technology in the banking sector. J Adv Manag Res 17(2):151–168
15. Cernian A, Paul AM (2020) Review and perspectives on how blockchain can disrupt education. eLearn Softw Educ 3:45–50
16. Kalfoglou Y (2021) Blockchain for business: a practical guide for the next frontier. Routledge
17. Pichardo-López P, Alcaraz-Quintero OE (2020) Blockchain technique in the banking area. J Ind Eng Manag 13(5):840–855
18. Chamria R (2023) Blockchain in cross-border payments: a game changer? Blockchain Deployment and Management Platform, Zeeve
19. Brett C (2018) Santander One Pay FX, a blockchain-based international money transfer service. Enterprise Times
20. Weerawarna R, Miah SJ, Shao X (2023) Emerging advances of blockchain technology in finance: a content analysis. Pers Ubiquit Comput
21. komgo | Consensys, Consensys (2022). https://consensys.net/blockchain-use-cases/finance/komgo
22. Geroni D (2023) A guide on quorum blockchain and their use cases. 101 Blockchains
23. Communications (2018) BBVA to test syndicated loans over blockchain. NEWS BBVA. https://www.bbva.com/en/innovation/bbva-ready-negotiate-and-contract-syndicated-loans-blockchain/
24. Jagtap J (2023) How ripple is shaping cross-border transactions in banking. The Crypto Times. https://www.cryptotimes.io/how-ripple-is-shaping-cross-border-transactions
25. Admin (2023) Emirates NBD announces historic blockchain pilot project. Naseba. https://naseba.com/content-hub/article/emirates-nbd-announces-historic-blockchain-pilot-project/

Leveraging AI and Blockchain for Enhanced IoT Cybersecurity

Iraq Ahmad Reshi and Sahil Sholla

Abstract This book chapter presents an overview of the cybersecurity concerns of the Internet of Things (IoT). It investigates how artificial intelligence (AI) and blockchain technologies address these challenges. This chapter describes the growing number of accessories in IoT and the increasing sophistication of cyberattacks targeting these devices. Each of these factors presents its own set of unique security challenges. Furthermore, we are investigating the potential benefits of incorporating AI and blockchain into IoT cybersecurity. These advantages include improved threat detection and response, increased data privacy and integrity, and increased attack resistance. Moreover, we present a review of particular novel approaches in the field. This chapter presents brief case studies of AI and blockchain-based Internet of Things cybersecurity solutions. These case studies show the practical applications and benefits of these technologies in safeguarding Internet of Things environments. The chapter provides insights into the changing landscape of cybersecurity for the Internet of Things (IoT) and AI and blockchain's role in mitigating cyber threats in this sector.

Keywords Blockchain · Smart contracts · Internet of Things · Machine learning · Threat detection

1 Introduction

The "Internet of Things" (IoT) is a comprehensive framework encompassing a sophisticated network of physical entities, such as diverse devices, autos, and household appliances. Integrating electronics, software, sensors, and networking capabilities establishes this network. Illustrations of such entities encompass a diverse array of contrivances, transportation means, and domestic apparatuses. The Internet of Things (IoT) possesses significant potential to transform societal operations,

I. A. Reshi (✉) · S. Sholla
Department of Computer Science and Engineering, Islamic University of Science and Technology, Awantipora, Kashmir, J&K, India
e-mail: rshiraq333@gmail.com

K. Kaushik and I. Sharma (eds.), *Next-Generation Cybersecurity*, Blockchain Technologies, https://doi.org/10.1007/978-981-97-1249-6_14

including enhanced efficiency, streamlined automation, and unparalleled connectivity. However, the Internet of Things (IoT) ecosystem faces much skepticism over its capacity for revolutionary change. Academic researchers face the challenging endeavor of establishing coherent regulations, standardised frameworks, and practical management approaches to guide the progression of the Internet of Things (IoT) in a manner that fosters innovation and overcomes any barriers that may impede its inherently dynamic characteristics. The proliferation of networked devices has led to a significant surge in their overall quantity, presenting a diverse range of security concerns. It is crucial to recognise that the aforementioned challenges profoundly affect not only the security of our community but also the privileges we enjoy [11].

1.1 IoT Cybersecurity and Its Challenges

The enormous quantity and diverse range of Internet of Things (IoT) devices pose a significant challenge in safeguarding their security. The absence of well-defined security mechanisms, such as encryption, authentication, and authorisation, in Internet of Things (IoT) devices renders them vulnerable to cyberattacks. Moreover, many Internet of Things (IoT) devices primarily focus on cost-efficiency, resulting in potential limitations regarding essential resources such as memory and computational power. Consequently, these constraints may hinder the implementation of advanced security measures. The presence of diverse Internet of Things (IoT) devices and protocols poses a significant challenge in establishing a cohesive security architecture encompassing the entire IoT ecosystem. Wi-Fi, Bluetooth, and Zigbee represent a subset of the communication protocols utilised by Internet of Things (IoT) devices. Additionally, numerous devices incorporate proprietary protocols that still need to be more explicit regarding widespread recognition. Consequently, security specialists need help in developing universally effective security solutions.

Moreover, Internet of Things (IoT) devices present an enticing opportunity for malicious actors because of their capability to collect and transmit sensitive personal data, such as individual names, addresses, and credit card details. Hackers can facilitate the unauthorised infiltration of additional devices and data within a network by utilising Internet of Things (IoT) devices as potential entry points [15].

In general, the issue of cybersecurity in the Internet of Things (IoT) presents a complex and formidable challenge that necessitates a comprehensive and holistic approach. It is imperative to prioritise the safeguarding of individual devices as well as the establishment of comprehensive security protocols. Manufacturers must prioritise security and incorporate safety considerations into the design of their goods from the outset.

1.2 *Importance of AI and Blockchain in IoT Cybersecurity*

Both artificial intelligence (AI) and blockchain possess the capacity to significantly enhance the security of the Internet of Things (IoT). The critical nature of real-time identification and mitigation of cybersecurity vulnerabilities arises from the continuous data collecting and transmission facilitated by Internet of Things (IoT) devices. Algorithms grounded in machine learning can acquire the ability to identify discernible patterns of behaviour that signify an imminent cyberattack, thus enabling them to proactively or reactively implement precautionary or remedial measures. An exemplary demonstration of the utilisation of artificial intelligence (AI) lies in the automated activation of alarms or shutdowns as a response to dubious activities about Internet of Things (IoT) devices. In contrast, blockchain technology is a decentralised and immutable ledger that has the potential to enhance the security of Internet of Things devices. Utilising blockchain technology enables the decentralised storage of data, ensuring its immutability, and facilitates the transparent transfer of data among Internet of Things (IoT) devices. Utilising blockchain technology enhances the security of centralised systems or databases by rendering it more challenging for hackers to tamper with or modify the stored data. As a result of the implemented security measures, the likelihood of cyberattacks is reduced.

Furthermore, implementing blockchain technology can enhance the authentication and permission mechanisms for Internet of Things (IoT) devices. Information and resources are restricted to authorised devices for confidentiality and security. Any unauthorised device is restricted from accessing confidential information or resources. Implementing blockchain-based identity management systems can effectively establish secure authentication and authorisation protocols for IoT devices, preventing unauthorised access.

Furthermore, using artificial intelligence (AI) and blockchain technology has proven to be highly efficient in enhancing the security of the Internet of Things (IoT) infrastructure. The utilisation of artificial intelligence for real-time detection and response to cybersecurity threats, along with the implementation of blockchain technology for safe and decentralised record-keeping and identity management systems, can significantly boost the security and reliability of IoT devices.

2 IoT Cybersecurity Challenges

The "Internet of Things" (IoT) refers to a network of interconnected physical objects with built-in sensors, software, and other technologies, enabling them to collect and exchange data. The Internet of Things (IoT) encompasses numerous benefits but also introduces a range of cybersecurity vulnerabilities. Figure 1 illustrates the main obstacles.

The susceptibility of numerous IoT devices to assaults is attributed to their constrained processing and storage capacities. The susceptibility of these gadgets

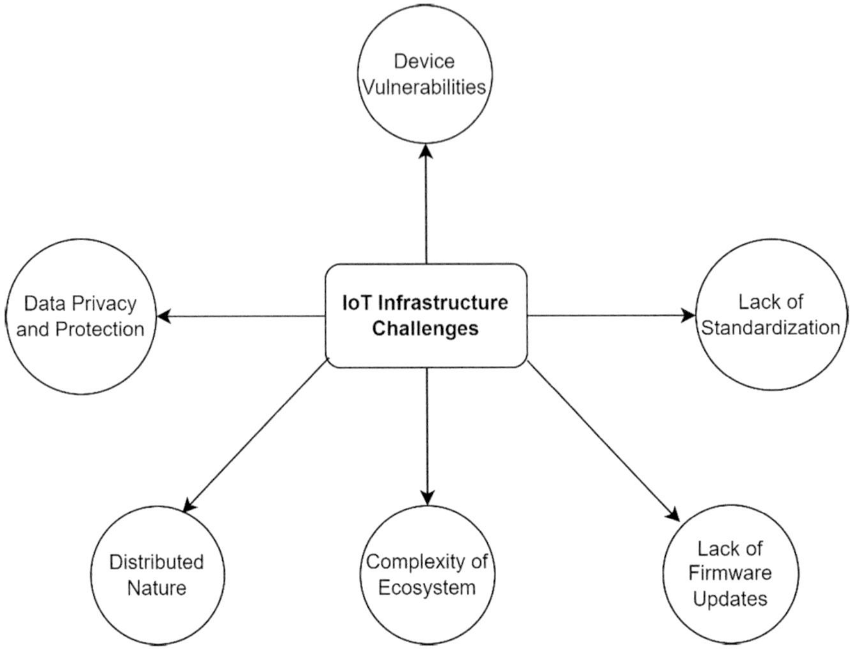

Fig. 1 IoT cybersecurity challenges

arises from inadequate security measures, including the utilisation of weak passwords, the absence of protected communication channels, and the usage of outdated software. The absence of a standardised framework for Internet of Things (IoT) devices results in software and hardware configuration variations. The need for standardisation poses challenges in implementing uniform security measures across diverse devices. The collection and transmission of data by Internet of Things (IoT) devices give rise to significant apprehensions over privacy and security. The potential compromise of this information could have severe consequences. The governance and monitoring of IoT networks pose significant challenges due to their dispersed and decentralised character. This poses a challenge in promptly identifying and addressing security concerns. The complex interconnectivity of the various devices, networks, and platforms comprising an Internet of Things (IoT) ecosystem presents a significant obstacle in ensuring their security. Malicious actors have the potential to exploit any vulnerability within the ecosystem. A prevalent deficiency observed in IoT devices is their incapability to receive firmware upgrades, rendering them vulnerable to post-publication problems. This leaves the gadgets open to attacks for as long as they stay in use [10]. Collaborative efforts among hardware and software manufacturers, network and infrastructure service providers, and end users are necessary to identify effective resolutions for these challenges. The deployment of security measures, establishment of standards, and promotion of best practices will be crucial in ensuring the security and privacy of IoT devices and the data they collect [15].

2.1 Types of Cyber Threats in IoT

An increasing number of individuals use IoT (Internet of Things) devices in their residential, professional, and critical public infrastructure settings. However, the current landscape provides fraudsters with an increased scope to exploit vulnerabilities in Internet of Things (IoT) devices, enabling them to initiate a diverse range of cyberattacks. The increased vulnerability is linked to the extensive distribution of these devices. The following enumeration presents a limited selection of the possible manifestations of cyberattacks against Internet of Things (IoT) devices. The rapid expansion of Internet of Things (IoT) devices necessitates implementing strong cybersecurity measures.

- **Botnets**, refer to networks of compromised devices under the control of an attacker or a collective entity. Distributed denial-of-service (DDoS) attacks, initiated from one of these devices, can disrupt large networks of computers that host critical websites.
- **Ransomware**, A malicious software program designed to encrypt a user's files after that, demanding a monetary ransom in exchange for the decryption key required to regain access to the contents. The infiltration of Internet of Things (IoT) devices by ransomware can cause significant disruptions to commercial operations and result in the theft of personal data.
- **MITM**, Man-in-the-middle attacks, or MITM attacks, manifest when an unauthorised individual interposes within the communication channel, connecting two parties to manipulate or pilfer data. Man-in-the-middle (MITM) attacks can compromise Internet of Things (IoT) devices in instances where these devices engage in communication via networks that lack adequate security measures or encryption protocols.
- **Remote code execution**, Remote code execution (RCE) attacks use vulnerabilities in software to facilitate the execution of malicious code on a device. This technique enables the perpetration of harmful activities on the targeted system. Internet of Things (IoT) devices may be susceptible to Remote Code Execution (RCE) attacks when they utilise outdated or unpatched software.
- **Theft of private information**, Internet of Things (IoT) devices can collect and store sensitive personal information such as passwords, bank account numbers, and health records. Cybercriminals can exploit vulnerabilities in these electronic devices to illicitly get sensitive information.
- **Physical attacks**, In order to gain control over or extract data from an Internet of Things (IoT) device, an attacker must first achieve physical proximity to the device and subsequently implement modifications. In order to carry out this sort of attack, the perpetrator must possess physical proximity to the targeted device. However,

it is noteworthy that detecting and mitigating such attacks poses considerable challenges [4].

3 AI-Based Solutions for IoT Cybersecurity

IoT technology has revolutionised how everyday objects communicate, leading to a seamless interconnectivity as never before. One potential drawback associated with a more prominent online population is the increased vulnerability of individuals to cybercriminal activities. The utilisation of artificial intelligence (AI) has the promise of substantially mitigating vulnerabilities inside the Internet of Things (IoT).

Several AI-based precautions for the Internet of Things (IoT) are as follows:

- **Anomaly Detection**, In scenarios where an Internet of Things (IoT) device exhibits signs of compromise or deviates from expected behaviour, artificial intelligence (AI) can identify these anomalies and promptly notify the user through an alarm system. Taking this step can help prevent potential cybersecurity threats in advance.
- **Predictive Analytics**, Artificial intelligence (AI) can analyse substantial volumes of data derived from Internet of Things (IoT) devices and detect apparent trends that could signify a breach in security. The utilisation of predictive analytics has the potential to facilitate the timely identification of potential hazards.
- **Dynamic Risk Assessment**, through the surveillance of device activity and identifying potential vulnerabilities, artificial intelligence (AI) can aid in analysing security threats associated with Internet of Things (IoT) devices in a real-time manner. To ensure safety, it is important to promptly identify and manage potential hazards. In Dynamic Risk Assessment, Artificial intelligence can facilitate intelligence acquisition on potential cyber-attacks targeted towards Internet of Things (IoT) devices. This can aid enterprises in maintaining a competitive edge in the dynamic threat landscape and safeguarding their Internet of Things (IoT) infrastructure.
- **Automatic Reaction**, Artificial intelligence (AI) can potentially optimise the management of security events related to Internet of Things (IoT) devices. Implementing this approach can enhance response times, mitigate the impact of attacks, and bolster the overall defensive capabilities of the organisation.

When implemented on a broad scale, artificial intelligence (AI) holds the potential to provide a comprehensive set of tools for addressing cybersecurity challenges in the Internet of Things (IoT) domain. It is important to note that artificial intelligence (AI) should not be considered a universal solution but rather a complementary tool that should be used with other protective measures to mitigate cyberattacks successfully. The following section provides a concise overview of specific machine learning methods employed in Internet of Things (IoT) security.

3.1 Machine Learning Algorithms for Threat Detection and Response

Machine learning algorithms have demonstrated remarkable efficacy in threat identification and response, owing to their ability to swiftly and accurately assess vast amounts of data. Various types of machine learning algorithms, including supervised learning, unsupervised learning, and reinforcement learning, are commonly utilised in this domain.

Supervised learning methods can be employed to train models in identifying distinct kinds of risks, leveraging annotated data. An example of a supervised learning algorithm involves its training on a dataset consisting of verified instances of phishing emails. Subsequently, this algorithm can detect and classify new phishing attempts based on their resemblance to the previously known examples. Unsupervised learning methods do not require prior knowledge, making them advantageous for identifying patterns and anomalies within extensive datasets. These algorithms possess significant use in threat detection due to their capacity to find novel and unidentified dangers that may have eluded identification using alternative methodologies.

Reinforcement learning methods are employed in various threats and response domains, including intrusion detection. These algorithms can undergo training to respond to distinct categories of threats effectively, leveraging the feedback obtained from the system over some time. An example of an application of reinforcement learning involves training an algorithm to autonomously identify and isolate infected machines, followed by the deployment of security patches as a response to a particular type of malware. Furthermore, several specialised techniques and models, including neural networks, decision trees, and support vector machines, are utilised in the threat detection and response field in addition to the algorithms mentioned earlier. The selection of an algorithm or technique is contingent upon several factors, including the characteristics of the identified threat, the data accessible for analysis, and the specific goals of the implementing organisation. The following are many instances of frequently employed machine learning techniques utilised in the domain of threat identification and response:

- **Random forest**, is a classification-centric supervised learning algorithm. Intrusion detection systems commonly identify malicious network traffic through this technique. The technique creates numerous decision trees and aggregates their outputs to ascertain a conclusive classification [7].
- **K-means clustering**, Comparable data points are grouped using this unsupervised learning approach. Identifying data points that significantly deviate from the rest of the dataset allows for detecting anomalies. The utilisation of this approach holds potential in the identification and detection of previously unidentified or novel threats [9].
- **Deep learning**, is a neural network commonly employed for intricate tasks like image and speech recognition. Techniques like Deep packet inspection utilised in threat detection, namely in analysing network data to identify malicious behaviour originated from Deep Learning [13].

- **Support Vector Machines (SVMs)**, This algorithm for supervised learning is used for classification tasks. It operates by locating a hyperplane that divides various classes of data points. SVMs have been utilised for malware detection by classifying samples as benign or malicious based on their characteristics [18].
- **Reinforcement learning**, This algorithm acquires knowledge by reinforcing positive behaviour and penalising poor behaviour. The application of this approach has been observed in the context of threat response, which entails the automated deployment of security patches or the isolation of infected workstations as a reactive measure to detect threats [20].

These are just a handful of the countless machine learning algorithms and approaches used in threat identification and response. The selection of the algorithm or technique to be employed will be contingent upon the characteristics of the danger and the data at hand. Additionally, some intrusion prevention and detection solutions are based on artificial intelligence.

3.2 AI-Based Intrusion Detection and Prevention

The use of machine learning algorithms for the detection and prevention of intrusions into Internet of Things (IoT) devices and networks is a relatively new area of study. Because there are so many different types of devices on an IoT network, it's very difficult for humans to keep track of all of the activities on the network and see any signs of intrusion. The use of AI for spotting and stopping intrusions is therefore crucial. Anomaly detection, signature-based detection, behavioural analysis, and hybrid implementations are just some of the AI-based intrusion detection and prevention strategies that can be used to the Internet of Things. In order to detect potential attacks, these methods use machine learning algorithms to examine typical network traffic, spot outliers, and compare incoming traffic to a library of known attack signatures. Integrating AI-based intrusion detection and prevention systems with preexisting security systems and protocols can add a new degree of defence for the Internet of Things (IoT). As a whole, AI-based intrusion detection and prevention is an encouraging step in protecting IoT systems from vulnerabilities like data breaches, illegal access, and so on.

4 Blockchain-Based Solutions for IoT Cybersecurity

While the IoT has greatly improved our experience with technological devices, it has also introduced novel cybersecurity threats. The limited processing power and storage capacity inherent in many IoT device designs make them ideal targets for cyber-attacks, which is a major problem. As a decentralised, immutable, and secure architecture for IoT devices, blockchain technology has emerged as another possible option to address these difficulties. Using blockchain in the Internet of Things is significant since it ensures confidential data exchange between gadgets. To prevent unauthorised changes or deletions, blockchain technology uses a distributed ledger to record all financial transactions. So, it is more challenging for attackers to eavesdrop on data transmissions or alter the data in any way. And because it eliminates the need for third parties or centrally controlled agencies, blockchain technology makes data sharing among IoT devices much safer [1].

Smart contracts, which are self-executing contracts where the conditions of the agreements between the two parties are encoded directly into coding lines, are another major advantage of blockchain in IoT. Smart contracts can be used to remove the need for user intercession in transactions and exchanges by executing them automatically and ensuring that only authorised devices can communicate with each other. Also blockchain technology may be utilised to manage Internet of Things devices in a safe and decentralised manner. Devices can be registered, authenticated, and granted permission to communicate with other devices by establishing a blockchain-based network. In this way, the network is protected from intruders and only authorised devices are allowed access. In addition, blockchain technology offers a safe, decentralised method of handling software patches and upgrades, which lessens the likelihood of security flaws and exploits.

When applied to the security of Internet of Things (IoT) devices, blockchain technology's decentralised, immutable, and secure architecture for communication, data exchange, and device management might significantly change the current approach. Blockchain-based solutions for IoT cybersecurity are anticipated to play a crucial role in guaranteeing the security and privacy of our connected devices in the years to come, despite the fact that there are still some problems to be addressed, such as scalability and interoperability.

The authors [8] provide a thorough examination of significant security concerns in the realm of the Internet of Things (IoT), specifically focusing on the layered architecture. This survey involves the classification and examination of these difficulties, with a particular emphasis on networking, communication, and management protocols. This study aims to provide a comprehensive analysis of the fundamental security requirements for the Internet of Things (IoT), with a particular focus on examining existing risks, attacks, and the latest responses available. The research effectively compares and contrasts the recognised security problems associated with the Internet of Things (IoT) with the solutions that have been previously described in academic literature. The examination of blockchain, the fundamental technology behind Bitcoin, holds great importance in addressing many security challenges that afflict the Internet of Things (IoT). The authors provide valuable perspectives on emerging areas of research and significant obstacles that need to be addressed in order to enhance the security of the Internet of Things (IoT).

A new and efficient blockchain (BC)-based architecture specifically built for the Internet of Things (IoT) is proposed by Dorri et al. [5]. The framework aims to reduce the conventional overheads associated with blockchain technology, while yet preserving most of its security and privacy benefits. The suggested architectural design presents a centralised, yet confidential, immutable ledger for Internet of

Things (IoT) devices. This ledger is inspired by blockchain ideas and is specifically optimised to enhance energy efficiency. In order to ensure total end-to-end security and privacy, a decentralised blockchain is deployed using an overlay network on devices that possess sufficient power. By utilizing distributed trust methods, the processing time for block validation is significantly decreased. The efficacy of this methodology is demonstrated within the framework of a smart home environment, acting as a sample exemplification for wider Internet of Things (IoT) implementations. The effectiveness of the architecture in providing security and privacy for Internet of Things (IoT) use cases is emphasised through qualitative evaluations conducted against existing threat models. In addition, simulations are used to verify the efficacy of the suggested methodology, demonstrating significant decreases in packet transmission and processing overhead compared to the blockchain structure utilised in Bitcoin.

In this particular context, a resilient Proposed Application (PA) driven by blockchain technology is envisioned with the objective of creating, maintaining, and verifying healthcare certificates [17]. The PA functions as an intermediary conduit, facilitating smooth communication between the foundational blockchain infrastructure and key entities within the application ecosystem, including hospitals, patients, doctors, and Internet of Things (IoT) devices. The primary focus of its basic functionality lies in the generation and verification of medical certifications. In addition, the Public Administration (PA) demonstrates proficiency in implementing a variety of essential security measures, including confidentiality, authentication, and access control. These measures are effectively enforced through the integration of smart contracts. The effectiveness of the suggested framework is shown through a rigorous comparative and performance study, showcasing its superiority in comparison to existing alternatives.

Another paper presents a novel architecture for sharing IoT data, known as TEE-and-Blockchain-supported IoT Data Sharing (TEBDS) [21]. TEBDS combines on-chain and off-chain methods to effectively fulfil the security requirements of the IoT data sharing framework. The TEBDS framework utilises a consortium blockchain to ensure the security of on-chain Internet of Things (IoT) data and manage access controls for IoT users. In addition to this, an introduction is made to a Distributed Storage System (SDSS) that utilises Intel SGX technology in order to enhance the security of off-chain data. In addition, a meticulously designed incentive mechanism has been formulated to promote the smooth functioning of the entire system. A comprehensive security analysis confirms that TEBDS effectively meets the requirements for ensuring both data security and identity security. Empirical assessments provide evidence supporting the effectiveness of TEBDS, demonstrating its improved performance compared to the centralised SPDS strategy. Table 1, summarises the works of blockchain for IoT security.

Table 1 Blockchain solutions for IoT

S. No.	Author	Technique
1	Ali et al. [1]	Proposed a blockchain-based architecture for IoT data security, enhancing confidentiality and integrity of data exchange, as well as device management
2	Khan and Salah [8]	Blockchain technology's advantage in enabling automated and secure smart contracts for IoT devices is highlighted, along with its potential in managing device registration, authentication, and secure data sharing
3	Dorri et al. [5]	Introduce a novel and energy-efficient blockchain architecture for IoT devices, utilising both centralised and decentralised approaches to enhance security and privacy
4	Sharma et al. [17]	Present a proposed application (PA) utilising blockchain to generate and verify healthcare certificates, incorporating robust security measures through smart contracts
5	Xie et al. [21]	Introduced TEBDS, a hybrid on-chain and off-chain architecture utilising consortium blockchain and Intel SGX technology to secure IoT data sharing

5 AI and Blockchain Integration for IoT Cybersecurity

The Internet of Things (IoT) could benefit from improved security measures if artificial intelligence (AI) and blockchain were coupled. Data collected by Internet of Things (IoT) devices can be analysed by AI to reveal vulnerabilities. While traditional methods of data storage and dissemination have their limitations, blockchain technology offers a safe, distributed alternative. Anomaly detection is one method in which AI can strengthen the security of the Internet of Things. Algorithms powered by AI may be taught to analyse data for anomalies that can indicate a breach in security. This can be especially helpful in preventing and responding to attacks that leverage Internet of Things (IoT) devices as vectors into bigger networks. Blockchain technology has the potential to serve as a trustworthy, decentralised database for IoT information. Blockchain's use of a distributed ledger system makes data immutable and resistant to hacking. This can safeguard information collected by IoT devices and lessen the likelihood of data breaches.

Together, AI and blockchain can strengthen the IoT ecosystem's defences and make it more resistant to disruption. Artificial intelligence (AI) can be used to detect security risks, setting off automated responses that employ blockchain technology to protect and verify the integrity of the relevant data. This can aid in protecting the IoT from threats and guaranteeing that only authorised parties have access to its data. Artificial intelligence (AI) and blockchain technology (blockchain) could also be used for identity management purposes in IoT cybersecurity. Artificial intelligence can examine patterns of use and flag outliers that may suggest intrusion. To further ensure that only authorised users have access to IoT devices and data, blockchain can be used to securely store user IDs and authentication data. Several other applications exist for combining AI and blockchain technology in IoT security.

- **Fraud detection**, For the purpose of detecting fraudulent behaviours, such as data manipulation or illegal access, artificial intelligence algorithms can be applied to IoT data in real time. The immutable recordings of these actions can then be stored in blockchain, providing a safe and verifiable audit trail.
- **Threat intelligence sharing**, To better understand cyberattack patterns and trends, threat intelligence data from many sources can be analysed by AI. Secure data sharing via a blockchain-based platform makes it possible for businesses to work together to combat security concerns.
- **Device authentication**, To verify the legitimacy of a device, we can utilise artificial intelligence to monitor its activity for any deviations that would point to the presence of a malicious piece of hardware. The device's authenticity can then be verified via blockchain, allowing only approved gadgets access to the network.
- **Safety of smart contract**, Smart contracts are blockchain-stored, automatically-executing contracts, commonly used to interface IoT in Blockchain. Artificial intelligence (AI) can inspect these agreements for loopholes and other security problems. This can protect smart contracts from attacks that make use of security flaws.
- **Smart contract security**, With AI and blockchain technology, supply chains may be made more secure and transparent by keeping track of items as they flow from supplier to customer. This has the potential to deter forgery, tampering, and other fraudulent activities.

When combined, AI and blockchain have the potential to greatly strengthen the reliability and safety of IoT infrastructure. The Internet of Things (IoT) ecosystem can be made safer from cyberattacks by integrating these two technologies.

5.1 State-of-Art Survey of Combination of AI and Blockchain for IoT Security

Integrating Blockchain and AI facilitates the resolution of issues about accuracy, latency, centralisation, and, most importantly, security in IoT architectures. The database within blockchain technology is responsible for storing transactions accompanied by a digitally signed hash value. Hence, it is imperative to address accuracy, latency, security, privacy, and centralisation concerns to mitigate potential challenges. Artificial intelligence (AI) algorithms are also employed to address these challenges. Utilising blockchain networks offers the advantage of decentralisation, which facilitates automatic and efficient data validation. This feature effectively addresses the issue of single points of failure that commonly arise in cloud servers utilised for extensive data analysis.

Furthermore, integrating blockchain technology with the Internet of Things (IoT) and artificial intelligence (AI) provides further benefits in enhancing the capabilities of IoT systems. Rathore and Park [14] present a security architecture for Internet of Things (IoT) networks. The architecture aims to ensure the secure and scalable transmission of IoT data from decentralised IoT applications at the fog layer. Artificial intelligence (AI) is employed in diverse domains of advanced technologies, including blockchain thinking, decentralised AI, the intelligence of things, and intelligent robots, among others, in the daily lives of individuals [15]. The convergence between artificial intelligence (AI) Internet of Things (IoT) enables the collection of a vast amount of data and facilitates its analysis. Machine learning is utilised in various domains, including healthcare, smart home technology, smart farming, and intelligent vehicles, among others, to facilitate effective learning processes. Rathore and Park [14] introduced a novel approach that utilises blockchain technology to enhance the security of deep learning in the context of Internet of Things (IoT) applications. By integrating Blockchain and artificial intelligence (AI) at the device layer, their suggested method aims to ensure the integrity and reliability of data in IoT systems. The system demonstrates a significant level of precision and a notable delay in processing time for Internet of Things (IoT) data. In their study, Gil et al. [6] examined the role of intelligent machines in several domains, including medical science, automatic sensing devices, automated vehicle driving, and cooking, to reduce human labor. Intelligence can be defined as the cognitive capacity to use acquired knowledge to address intricate challenges. In contrast, artificial intelligence (AI) refers to the learning approach that facilitates the development of innovative procedures and the dissemination of collected initial insights. A report by McKinsey [3] projected that the AI market will experience significant growth, reaching a value of 13 trillion US dollars by 2030. The decentralised AI approach is a fusion of artificial intelligence (AI) and blockchain technology. Its purpose is to facilitate secure and trustworthy information sharing without relying on intermediaries. This is achieved through the utilisation of cryptographic signatures and robust security measures.

Furthermore, it can autonomously make decisions in Internet of Things (IoT) applications. In recent years, the rapid evolution of technologies, devices, and Internet of Things (IoT) devices has resulted in Blockchain, Artificial Intelligence (AI), and IoT emerging as the most influential technologies, driving the acceleration of innovative ideas across several domains.

Researchers [19] examine privacy, accuracy, latency, and centralisation concerns in integrating Blockchain and AI technologies within Internet of Things (IoT) applications. Blockchain and AI are integrated to propose an Intelligent IoT Architecture incorporating Blockchain technology, named as block intelligence. This architecture consists of a decentralised cloud infrastructure enabled by Blockchain at the cloud layer, distributed fog networks based on Blockchain at the fog layer, distributed edge networks based on Blockchain at the edge layer, and the convergence of peer-to-peer Blockchain networks at the device layer. Study by authors [16] introduces a novel security model called the Artificial Intelligence-based Lightweight Blockchain Security Model (AILBSM) to improve privacy and security in Industrial Internet of Things (IIoT) systems based on cloud computing. The framework utilises a combination of lightweight blockchain technology and a Convivial Optimized Sprinter Neural Network (COSNN) based AI mechanism. This integration enables the framework to deploy an Authentic Intrinsic Analysis (AIA) model, effectively converting features into encoded data. As a result, the framework mitigates the potential impact of attacks. By conducting extensive experiments and assessments, the AILBSM framework demonstrates notable outcomes, such as a decreased execution time of 0.6 s, an improved classification accuracy of 99.8\%, and an enhanced detection performance of 99.7%. The superior anomaly detection capability of the model can be attributed to its new features, such as auto-encoder-based transformation and blockchain authentication. This represents a noteworthy improvement in Industrial Internet of Things (IIoT) security, distinguishing it from other existing methodologies. Another study [22] presents a novel framework for cloud-edge-end architecture that utilises blockchain technology as a foundation for a trust mechanism based on the principles of blockchain consensus. The system utilises a consensus approach known as BLS-based proof of replication (PoRep) to establish confidence among devices. The data from devices is securely maintained within servers, utilising mechanisms like proof generation and consensus establishment by broadcasting information by nodes. This methodology improves the level of network data security and fosters the establishment of trust across a wide range of devices. In addition, the solution integrates verifiable delay functions (VDF) based on secret sharing to mitigate the need for servers to perform dynamic data replication calculations in Proof of Replication (PoRep), enhancing consensus efficiency among nodes. Implementing this complete strategy represents a notable progression in establishing secure and reliable interactions inside the network.

The authors [12] introduced a highly advanced intrusion detection system designed for smart cities. The system consists of three essential modules:

- A trust module incorporating a reputation system based on blockchain technology and addresses
- A privacy module with two tiers employing an improved Proof of Work technique based on blockchain
- An intrusion detection module

This study presents an integrated infrastructure called Cloud-Block, Fog-Block, and Edge-Block, explicitly designed for smart city environments. This infrastructure aims to optimise the Edge-Fog-Cloud architecture by capitalising on its advantages and mitigating its constraints. The system's effectiveness is demonstrated through the evaluation conducted on the sBoT-IoT and TON-IoT datasets. A comparative comparison with other state-of-the-art systems shows that the suggested system surpasses its equivalents, emphasising its notable progress in the field of smart city cybersecurity.

Furthermore, the study in [2] explores the intricacies of reconciling various technologies and suggests practical solutions to address these challenges. It provides a complete analysis of the prospects of this amalgamation, emphasising its potential to bring about significant changes in various areas, including banking, healthcare, and transportation. The study also investigates existing models' pre-integration limitations, including problems such as lack of transparency, concerns over data privacy and security, and deficiencies in automation. The comprehensive examination of the integration of blockchain, artificial intelligence, and IoT not only highlights their promise

Table 2 Summary of state-of-art proposed solutions

S. No.	Year	Author	Main theme of paper
1	2016	Gil et al. [6]	Role of intelligent machines across domains, defining intelligence and AI's role in developing innovative procedures
2	2018	Rathore and Park [14]	Security architecture for IoT networks, utilising AI and blockchain to ensure secure and scalable IoT data transmission
3	2020	Singh et al. [19]	Privacy, accuracy, latency concerns in integrating Blockchain and AI, proposing an Intelligent IoT Architecture incorporating Blockchain
4	2023	Selvarajan et al. [16]	Introduction of an AI-based Lightweight Blockchain Security Model (AILBSM) for IIoT security enhancement, utilising blockchain and AI mechanisms
5	2023	Zhang et al. [23]	Novel cloud-edge-end architecture using blockchain-based trust mechanism, integrating BLS-based PoRep consensus, enhancing network security and trust
6	2023	Moawad et al. [12]	Intrusion detection system for smart cities using blockchain-based trust, privacy, and intrusion modules, optimising smart city cybersecurity
7	2023	Aruna et al. [2]	Exploration of integrating blockchain, AI, and IoT, discussing potential changes and challenges in various domains

but also recognises the significant challenges that prompted this convergence. Table 2 summarises the literature.

5.2 *Challenges and Limitations of AI and Blockchain Integration for IoT Security*

Integrating AI with blockchain for IoT security offers the ability to address some of the most pressing security issues with the IoT, including data breaches, device tampering, and unauthorised access. There are, however, a number of problems and restrictions with this integration, as well as some research areas that should be covered up.

- **Scalability**: One of the challenges encountered in the implementation of IoT security through the use of AI and blockchain is the issue of scalability. The volume of data generated by Internet of Things (IoT) devices is substantial, necessitating significant computational resources for analysis. Integrating artificial intelligence (AI) and blockchain technology may result in a substantial increase in processing requirements, thus impeding the system's scalability.

- **Cost**: The cost of the integration of artificial intelligence (AI) and blockchain technology has the potential to enhance the security of the Internet of Things (IoT). However, this advancement has its associated costs. Both blockchain technology and AI algorithms require significant computational resources. Consequently, the cost associated with implementing this system may increase, rendering it financially unattainable for many smaller enterprises.
- **Complexity**: Integrating artificial intelligence (AI) and blockchain technology to enhance security measures for Internet of Things (IoT) devices may present specific challenges. Integrating both systems necessitates proficiently understanding their potential applications in safeguarding Internet of Things (IoT) devices. Implementing this system may pose challenges for certain businesses due to its inherent complexity.
- **Interoperability**: Interoperability is the ability of different systems or components to exchange and use information. Interoperability poses an additional challenge in utilising artificial intelligence (AI) and blockchain technologies to enhance security in the Internet of Things (IoT). Integrating diverse IoT devices into a unified system can present difficulties when these devices operate on disparate protocols and standards.

5.3 Research Gaps

In order to improve interoperability and compatibility across devices and systems, standardisation is necessary for the integration of AI and blockchain for IoT security. To further ensure the security of personal information submitted by users retained on IoT devices, privacy-preserving algorithms must be developed. In addition to integrity models, trust models are crucial to the security of the system. In addition, ensuring the IoT system and its data are secure from cyberattackers relies heavily on the governance of AI and blockchain integration. Despite the possible advantages, there are also restrictions and difficulties to overcome, such as scalability, complexity, and expense. Improving the efficacy and safety of AI and blockchain integration for IoT security requires addressing research gaps in areas like standardisation, privacy, trust, and regulations.

6 Case Studies of AI and Blockchain-Based IoT Cybersecurity Systems

In this section, we will brief about few use cases in the field of AI, IoT and blockchain, and how one technology can benefit the other.

Case study 1

AI-based threat detection in industrial IoT

A manufacturing facility could be one use case for AI-based threat detection in the industrial Internet of Things. Various instruments and devices in a typical manufacturing facility are network-connected and communicate data with one another and the central control system. This network of devices, known as the Industrial Internet of Things (IIoT), is responsible for maintaining the efficiency of the manufacturing process. Nonetheless, the IIoT is a primary target for cyber attacks, and any network disruption can result in production delays, outages, and financial losses. Here, AI-based threat detection can be helpful. Using machine learning algorithms to analyse network traffic, AI can identify peculiar patterns and anomalies that may be indicative of a cyber attack. AI can detect, for instance, if a machine sends and receives data at an abnormal rate or if it communicates with an unauthorised device. Once a threat has been identified, the AI system can mitigate it by alerting the security team, blocking suspicious traffic, or closing down the afflicted machine. In addition, AI-based threat detection can assist with countermeasures, such as detecting network vulnerabilities and identifying potential weak spots before they are exploited by cyber attackers.

Overall, AI-based threat detection in industrial IoT can aid in ensuring the safety and security of the manufacturing facility, preventing cyber attacks and production process disruptions, and saving the organisation time and money.

Case study 2

Blockchain-based secure data sharing in healthcare IoT

Another use case can be, Blockchain-based secure data sharing in healthcare IoT wishing to exchange patient information with other facilities in a safe and reliable manner. Sharing patient data across healthcare providers has historically been a challenging and tedious procedure due to the need to safeguard the data's accuracy, security, and compliance with standards like HIPAA and GDPR. To remedy this, we may utilise blockchain technology to safely share information. Healthcare providers may build a trustworthy, distributed network for exchanging patient information using blockchain technology. The blockchain stores an immutable and comprehensive record of patient information, and each healthcare practitioner may keep their own copy. The blockchain network may be queried by healthcare providers whenever they need to access patient data, at which point the network will confirm the provider's identity and provide them access. Since the data is maintained in a decentralised and tamper-proof way, it protects the privacy of patients and allows for the free flow of information among healthcare providers without compromising any of the data's integrity. Data interoperability can be aided by blockchain-based secure data sharing since it lets healthcare professionals exchange patient information between disparate systems and platforms without jeopardising patient privacy or security. This has the potential to enhance patient outcomes, cut down on wasted resources, and lower healthcare costs.

Overall, blockchain-based safe data sharing in healthcare IoT can aid in protecting patients' privacy, keeping their data private, and facilitating the secure transfer of data across various medical facilities.

Case study 3

AI and blockchain-based identity management in smart homes

A smart apartment complex is another potential use of artificial intelligence and blockchain-based identity management in the context of smart housing. Thermostats, cameras, and house assistants are just some of the internet-connected equipment that may be remotely managed in a typical "smart home." This, however, makes them susceptible to cyber assaults and unauthorised access, which in turn threatens the inhabitants' privacy and safety. Artificial intelligence and identity management using blockchain technology can assist here. AI may monitor the locals' actions for signs of suspicious or out-of-the-ordinary behaviour using machine learning methods. A cyber assault may be in progress if, for instance, a smart thermostat is repeatedly toggled on and off in a short amount of time. A decentralised and secure network for managing identification and access control may be built using blockchain technology to protect people's privacy and safety. Every citizen has the option of keeping their digital identity on the blockchain, which stores an immutable record of their personal information and permits. When a user needs to access a smart home device, they may do so by sending a request to the blockchain network, which will then confirm the user's identity and provide them access. Because the information is kept in a distributed and unalterable fashion, residents may be certain that their personal information will remain private. Moreover, AI and blockchain-based identity management may aid in preventative measures like detecting weaknesses in the smart home network and pinpointing possible points of attack in advance.

Overall, identity management in smart homes based on AI and blockchain technology can help protect the personal information of residents and keep their smart home gadgets accessible in a safe and reliable manner.

7 Conclusion

This chapter has presented an overview of the cybersecurity threats encountered by the Internet of Things (IoT) and the potential solutions offered by artificial intelligence (AI) and blockchain technologies. Recent vulnerabilities and cyber threats have emerged as a result of the growing number of connected devices in the IoT ecosystem, and they may be beyond the capabilities of more conventional security measures. IoT cybersecurity may be made more effective and efficient with the use of AI-based solutions like machine learning algorithms for threat detection and response and intrusion detection and prevention systems. Similarly, blockchain technology enables a decentralised and immutable ledger for IoT devices, which improves data privacy, integrity, and resilience to assaults. Case studies given in this chapter illustrate the actual uses and benefits of integrating AI and blockchain technology to further increase the security of IoT settings. There are, however, obstacles and restrictions to combining AI and blockchain for IoT safety. The requirement for standardisation, scalability, and interoperability are all examples. It is crucial to tackle these issues and provide solid solutions that can properly safeguard these systems as the IoT ecosystem continues to grow. In conclusion, this chapter has demonstrated the great potential of AI and blockchain technology in the context of IoT cybersecurity. Better threat detection and response, more secure data storage, and a more secure Internet of Things (IoT) may all be attained through the use of these technologies.

References

1. Ali MS, Vecchio M, Pincheira M, Dolui K, Antonelli F, Rehmani MH (2018) Applications of blockchains in the Internet of Things: a comprehensive survey. IEEE Commun Surv Tutor 21(2):1676–1717
2. Aruna S, Priya SM, Reshmeetha K, Sudhayini ES, Narayanan AA (2023) Blockchain integration with artificial intelligence and Internet of Things technologies. In: 2023 7th International conference on intelligent computing and control systems (ICICCS), pp 688–694
3. Chui M et al (2023) The economic potential of generative AI: the next productivity frontier. McKinsey & Company
4. Deogirikar J, Vidhate A (2017) Security attacks in IoT: a survey. In: 2017 International conference on I-SMAC (IoT in social, mobile, analytics and cloud) (I-SMAC), pp 32–37
5. Dorri A, Kanhere SS, Jurdak R (2017) Towards an optimized blockchain for IoT. In: Proceedings of the second international conference on internet-of-things design and implementation, pp 173–178
6. Gil D, Ferrández A, Mora-Mora H, Peral J (2016) Internet of things: a review of surveys based on context aware intelligent services. Sensors 16(7):1069
7. Islam N, Farhin F, Sultana I, Kaiser MS, Rahman MS, Mahmud M, Hosen AS, Cho GH (2021) Towards machine learning based intrusion detection in IoT networks. Comput Mater Continua 69(2):1801–1821
8. Khan MA, Salah K (2018) IoT security: review, blockchain solutions, and open challenges. Futur Gener Comput Syst 82:395–411
9. Liu C, Gu Z, Wang J (2021) A hybrid intrusion detection system based on scalable K-Means+ random forest and deep learning. IEEE Access 9:75729–75740
10. Lu Y, Da Xu L (2018) Internet of Things (IoT) cybersecurity research: a review of current research topics. IEEE Internet Things J 6(2):2103–2115
11. Maple C (2017) Security and privacy in the internet of things. J Cyber Policy 2(2):155–184
12. Moawad MM, Madbouly MM, Guirguis SK (2023) Leveraging blockchain and machine learning to improve IoT security for smart cities. In: The international conference on artificial intelligence and computer vision, pp 216–228
13. Mohammadi M, Al-Fuqaha A, Sorour S, Guizani M (2018) Deep learning for IoT big data and streaming analytics: a survey. IEEE Commun Surv Tutor 20(4):2923–2960
14. Rathore S, Park JH (2018) Semi-supervised learning based distributed attack detection framework for IoT. Appl Soft Comput 72:79–89
15. Reshi IA, Sholla S (2022) Challenges for security in IoT, emerging solutions, and research directions. Int J Comput Digit Syst 12(1):1231–1241
16. Selvarajan S, Srivastava G, Khadidos AO, Khadidos AO, Baza M, Alshehri A, Lin JC-W (2023) An artificial intelligence lightweight blockchain security model for security and privacy in IIoT systems. J Cloud Comput 12(1):38

17. Sharma P, Namasudra S, Crespo RG, Parra-Fuente J, Trivedi MC (2023) EHDHE: enhancing security of healthcare documents in IoT-enabled digital healthcare ecosystems using blockchain. Inf Sci 629:703–718
18. Shen M, Tang X, Zhu L, Du X, Guizani M (2019) Privacy-preserving support vector machine training over blockchain-based encrypted IoT data in smart cities. IEEE Internet Things J 6(5):7702–7712
19. Singh SK, Rathore S, Park JH (2020) Blockiotintelligence: a blockchain-enabled intelligent IoT architecture with artificial intelligence. Futur Gener Comput Syst 110:721–743
20. Uprety A, Rawat DB (2020) Reinforcement learning for iot security: a comprehensive survey. IEEE Internet Things J 8(11):8693–8706
21. Xie H, Zheng J, He T, Wei S, Hu C (2023) TEBDS: a trusted execution environment-and-blockchain-supported IoT data sharing system. Futur Gener Comput Syst 140:321–330
22. Zhang A, Lin X (2018) Towards secure and privacy-preserving data sharing in e-health systems via consortium blockchain. J Med Syst 42(8):140
23. Zhang F, Wang H, Zhou L, Xu D, Liu L (2023) A blockchain-based security and trust mechanism for AI-enabled IIoT systems. Futur Gener Comput Syst 146:78–85

Dissecting the Role of Women in Cybersecurity and Information Technology: A Medical Perspective

Wasswa Shafik

Abstract As the global population is working toward the United Nations goals of sustainable communities and cities, good health and well-being, and responsible consumption and production, recognizing the crucial role women play in various humanitarian activities, it is essential to note that, despite the success of highlighting accomplished women in technology-based organizations, recent research indicates their underrepresentation in different technology-related fields. Despite efforts to attract and retain women, cultural issues within this diverse industry and global barriers pose significant challenges. Leaders of all genders and backgrounds are essential to addressing these workforce development gaps and formulating effective company strategies. This study delves into the hurdles, difficulties, and innovative ways to increase the number of women in senior positions in healthcare cybersecurity (CS) and information technology (IT). It further examines the nature of CS and IT, their importance, global internet utilization, essential forms, and professional aspects related to the underrepresentation of women in the industry. The low number of women in these fields can be attributed to several factors, as demonstrated within this study, and possible solutions are proposed. As the health industry aims for inclusive and innovative solutions, women bring unique perspectives and problem-solving skills, which can enhance the development of secure medical systems, electronic health records, and patient privacy protection. Finally, it is emphasized that addressing these raised factors will significantly enhance organizational technology acceptance, cybersecurity trust, effectiveness, and efficiency.

Keywords Good health and well-being · Sustainable cities and communities · Responsible consumption and production · Women · Healthcare · Cybersecurity · Information technology

W. Shafik (✉)
School of Digital Science, Universiti Brunei Darussalam, Jalan Tungku Link, Gadong, 1410 Bandar Seri Begawan, Brunei
e-mail: wasswashafik@ieee.org

Dig Connectivity Research Laboratory (DCRLab), 600040 Kampala, Uganda

K. Kaushik and I. Sharma (eds.), *Next-Generation Cybersecurity*, Blockchain Technologies, https://doi.org/10.1007/978-981-97-1249-6_15

1 Introduction

Despite making substantial progress in the workplace over the past half-century, women continue to be grossly underrepresented in various scientific and technical professions. Even though females constitute approximately 47%, 27%, 31%, and 27% of the labor force, environmental science occupations, chemistry jobs, and computer science and mathematics jobs are held by women, respectively. This gender gap persists despite the advances made by women in the employment market [1].

Many women face unique challenges in balancing work and family responsibilities. Offering flexible work arrangements, such as telecommuting or flexible hours, can help women balance their work and personal lives, which may make technical professions more appealing. Academics find it incredibly challenging to comprehend why female representation is lacking in these professions. This is concerning From a policy standpoint as it recommends that the government's technical personnel may not utilize all available creative energy [2]. This is problematic because it implies that creative potential is being squandered, indicating a need for improvement in the technical workforce across the nation.

The reasons behind the slow progress of women in entering healthcare, scientific, mathematical, engineering, and technological fields remain uncertain and, in some cases, debatable. The contentious nature of this issue is highlighted by the heated debate that followed Harvard President Larry Summers' remarks at a conference in 2005. Summers' suggestion that transformations in the distribution of talent between men and women may contribute to the underrepresentation of females in top-level science positions sparked this controversy [3].

If this perspective is correct, removing these barriers would result in an equal number of men and women entering technical jobs. Alternatively, it could be that men and women face similar challenges, but the distribution of skills across genders may differ, making men more productive in technical domains. This could be due to differences in intelligence quotient between males and females, as Larry Summers has stated. From this perspective, the employment gap between males and females could reflect the effective utilization of talent across various fields [4].

One potential explanation for the low percentage of women in technical professions could be differing perceptions of an appealing career path between men and women. Men may place a higher value on technical employment, which could result in a competitive market response among individuals with different job preferences [5]. Evidence to support this notion can be found in the declining percentage of women in technical occupations over the past few decades. The neoclassical approach posits that employees evaluate a particular career choice's predicted earnings and non-financial returns and will only change jobs if the cost–benefit analysis indicates it is worthwhile [6].

The theoretical framework of Bowles, Gintis, and Osborne is consistent with the notion that gender disparities in occupational choice result from distinct preferences. This framework can be interpreted in a way that aligns with their findings and explains the presence of significant wage disparities between individuals at any given time caused by disequilibrium rent resulting from slow responsiveness to market tremors [7]. From this perspective, the association between an individual's earnings and various non-productive personal qualities reflects their willingness or capacity to reply to market shocks that create uncertainty rents. In other words, the correlation between a person's earnings and personal traits reflects their abilities. These market shocks lead to imbalanced rents [8].

Creating an empirical method to differentiate between these three alternatives would be arduous. However, through our ongoing case study of the IT industry, we aim to contribute to the ongoing debate uniquely. Our research suggests that the gender gap in IT is mainly due to variations in professional interests between men and women. To make occupational choices, workers evaluate the benefits of specific employment options, including predicted earnings and non-pecuniary returns [9]. For example, it has been suggested that women may prefer certain occupations because of a reduced risk of income decline due to anticipated absences, and they may prioritize minimizing potential conflicts between their professional and domestic commitments more than men do. This indicates that women prioritize their families more than men.

CS entails identifying (demonstrating asset management), protecting (entails access control, data security, email, internet, computers, and mobile, network, maintenance, awareness, and training), detecting (including email monitoring, internet monitoring, computer monitoring, and network monitor) respond (counting plan, analyze or investigate, and manage change) and recover (preparation and restoration) as demonstrated in Fig. 1. An analysis, which includes a set of occupational personality traits, shows that gender is no longer a significant factor in predicting the choice between professions in information technology and non-information technology careers. This suggests that the underrepresentation of women in the information technology field results from individuals' decisions based on disparities in the actual or perceived aspects of the profession and cyber domain [10, 11]. While this study does not rule out the possibility of differences in ability or gender biases playing a role, it highlights the need for further investigation into people's decision-making processes about their career paths.

1.1 Chapter Contributions

We highlight the main contributions of the chapter, as shown below.

- The highlights highlight that the global population strives to achieve the United Nations' goals of sustainable communities and cities, good health and well-being, and responsible consumption and production.
- Women play a crucial role in various humanitarian activities, and their recognition is vital in technology-based organizations.
- Recent research highlights the underrepresentation of women in different technology-related fields.

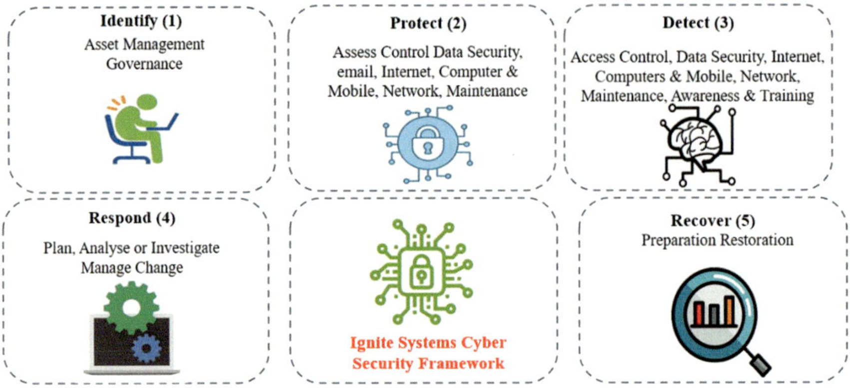

Fig. 1 Cyber security framework (CS entails identify, protect, detect, respond, and recover)

- Despite efforts to attract and retain women, cultural issues within the diverse technology industry and global barriers present significant challenges.
- Leaders of all genders and backgrounds must address workforce development gaps and create effective company strategies.
- The study focuses on increasing the presence of women in senior positions in healthcare CS and IT.
- The nature of CS and IT, their importance, global internet utilization, and professional aspects related to the underrepresentation of women in the industry are examined.
- The low number of women in these fields can be attributed to several factors, and potential solutions are proposed.
- Women bring unique perspectives and problem-solving skills, enhancing the development of secure medical systems, electronic health records, and patient privacy protection.
- Addressing the factors influencing underrepresentation will significantly enhance organizational technology acceptance, cybersecurity trust, effectiveness, and efficiency.

1.2 The Chapter Organization

The remainder of this chapter is structured as follows. Section 2 presents the nature of cybersecurity as it relates to women in technology-based domains, including the need for CS and IT global internet utilization. Section 3 covers the critical forms of CS (including network, application, cloud, endpoint, identity and access management, data security, incident response, and disaster recovery) and IT (entailing hardware, software, networks, cloud Computing, and IoT). Section 4 depicts CS as a profession for women, demonstrating reasons that cause a reduced number of women in CS and IT industries like lack of awareness, stereotypes, lack of representation, unconscious bias, unequal pay, lack of role models, hostile work environments, limited support, lack of flexible work arrangements, lack of training opportunities, limited access to networks, lack of mentorship, perceived lack of fit, limited support for work-life balance, lack of confidence. Section 5 presents women's cybersecurity from a medical perspective. Section 6 presents the discussion and recommendations for research to be conducted in the future. Finally, Sect. 7 depicts the conclusion.

2 Nature of Cybersecurity

The role of women in CS is increasingly crucial as the field continues to expand and develop. Historically, women have been underrepresented in technology-related fields, including cybersecurity. However, this trend is changing, and more women are entering the field, bringing unique perspectives and skillsets [12]. There has been a growing awareness of the reputation of diversity and inclusivity in CS. This is because diverse teams are more likely to be innovative, effective, and better equipped to solve complex problems [13].

Women bring different experiences and viewpoints to the field, enabling them to identify and address cybersecurity challenges that may have been previously overlooked. Furthermore, in addition to the benefits of diversity, the nature of cyberspace itself makes it an ideal field for women to pursue [14]. Unlike traditional tech roles that may require physical strength or involve long hours in a lab, cybersecurity can be done from anywhere with an internet connection, making it accessible to people of all backgrounds and abilities [15]. It is a dynamic and constantly evolving field that offers endless opportunities for learning and growth.

However, despite the opportunities available in the field, women still encounter unique obstacles in cybersecurity, such as gender bias and a lack of female role models. These barriers can deter women from entering the field or impede their professional advancement once they do. Overcoming these challenges necessitates a concerted effort by individuals, organizations, and policymakers to create a more inclusive and supportive environment for women in CS [16, 17]. As for cyberspace's nature, it presents challenges and opportunities for cybersecurity professionals, including women. One of the most significant challenges is the constantly evolving threat landscape, which necessitates that CS professionals are adaptable and proactive in identifying and mitigating risks. Additionally, the interconnectedness of cyberspace implies that cybersecurity threats can propagate quickly and have far-reaching consequences [18]. This necessitates a collaborative and coordinated approach to cybersecurity within organizations and across sectors and borders.

The nature of cyberspace poses challenges and presents opportunities for innovative solutions to cybersecurity issues. For instance, artificial intelligence (AI) and machine learning (ML) technologies can be utilized to identify and respond to threats in real-time.

The emergence of the IoT also provides opportunities for enhanced CS through smart devices and sensors that can automatically detect and respond to threats [19,

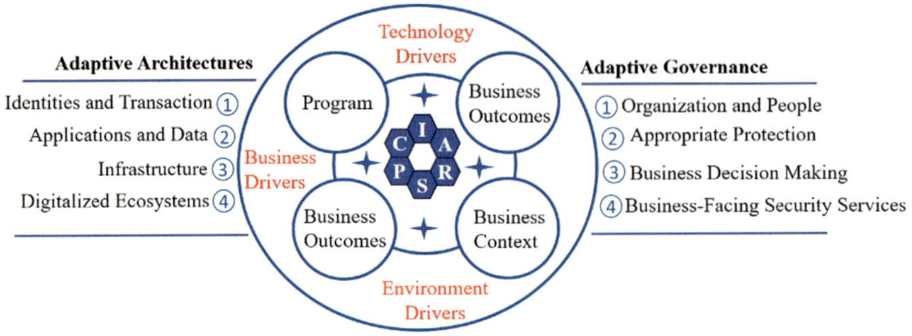

Fig. 2 Trust and resilience (C, Confidentiality; I, Integrity; A, Availability; P, Privacy; S, Safety; R, Reliability)

20]. Women can play a crucial role in developing and implementing such innovative solutions to bridge the gap in end-user trust and resilience, as illustrated in Fig. 2. Diversity and inclusivity foster more creative and effective problem-solving [21]. By encouraging more women to pursue careers in cybersecurity and supporting their professional development, their unique perspectives and skill sets can be utilized to enhance the security of cyberspace.

2.1 Need for Cybersecurity and Information Technology

Women are crucial in addressing the growing need for cybersecurity in our digital world. Cybersecurity threats have become more frequent and sophisticated with the rise of technology and the internet. Women bring unique perspectives and skills to the field, such as solid communication and collaboration abilities and expertise in areas such as threat analysis and risk assessment. Diversity in cybersecurity can lead to more innovative and practical solutions to complex challenges [22]. Encouraging more women to pursue careers in cybersecurity through mentoring, outreach, and professional development is crucial. The consequences of cybersecurity threats can be severe, making the need for skilled professionals in the field more important than ever.

The need for cybersecurity is driven by numerous factors, including the increasing prevalence of cyberattacks, the growing amount of sensitive data being stored online, and the rise of new technologies such as the IoT [23, 24]. CS threats can have serious consequences, including financial losses, damage to reputation, and the compromise of personal information. Given the importance of CS in our modern world, we must encourage more women to pursue careers in the field. This can be achieved through mentoring programs, educational outreach, and opportunities for professional development and advancement.

The importance of CS extends beyond large corporations and government agencies; small businesses, non-profits, and individuals are also at risk of cyber-attacks.

Women in CS can raise awareness and educate these vulnerable groups on best practices for staying safe online [25]. Women can play a crucial role in protecting these populations from cyber threats by sharing their expertise on password management, phishing scams, and safe browsing habits. Additionally, women can serve as role models and mentors for those interested in pursuing a career in CS, inspiring and empowering the next generation of professionals [26].

2.2 Global Internet Utilization

Over the past few years, internet usage has led to a rise in cybersecurity threats that require skilled professionals to address. Women can contribute significantly to CS and cyber defense (CD) in various ways. They can increase awareness of cybersecurity threats and promote best practices for staying safe online by sharing their knowledge and experiences. Women can also contribute to developing and implementing CS policies and practices by bringing diversity to the table [27]. With only 24% of the CS workforce being women, there is a need for greater gender diversity. Encouraging more women to pursue careers in cybersecurity and providing mentorship and support can help address this gap [28]. Women can also contribute to cybersecurity research and innovation by identifying emerging threats and developing innovative solutions.

Women can significantly impact CD by proactively addressing cybersecurity risks in their personal and professional lives. For instance, using strong passwords, regularly updating software and security settings, and avoiding suspicious emails or websites can help mitigate risks. Women can also advocate for greater diversity and inclusion within the tech industry, leading to more innovative and practical solutions to CS challenges. Furthermore, women can bring empathy and social awareness to the field, which can help to develop more holistic and practical approaches to CS [29]. CS is not solely a technical problem but also a social and human one, and women's ability to connect with others and understand the cultural and social implications of cyber threats can make a meaningful difference.

3 Cybersecurity and Information Technology Key Forms

3.1 Cybersecurity Key Forms

This section presents some common cybersecurity forms, including network and application security.

3.1.1 Network Security

Network security is a crucial part of cybersecurity that protects computer networks from cyber-attacks, unauthorized access, and other malicious activities. In the early days of computing, networks were simple and secure, but as they became more interconnected, the risk of cyber-attacks increased significantly. Today, networks are accessible to a wide range of users, making them vulnerable to various security threats [28].

Cyber-attacks, such as hacking, malware, phishing, and denial-of-service (DoS) attacks, can cause data theft, system downtime, financial losses, and reputational damage. Detecting these attacks can be challenging, making it difficult to respond promptly. Network security requires implementing security measures such as intrusion detection systems, firewalls, virtual private networks, and ongoing monitoring and updates to ensure network security.

3.1.2 Application Security

Application security is crucial for protecting applications from cyber threats like hacking and malware. Applications are vital for digital systems but also vulnerable to cyber-attacks due to the lack of security considerations during development [29]. Implementing secure coding practices, encryption, and access controls can help protect sensitive data. Ongoing monitoring and testing are necessary to identify and address vulnerabilities.

3.1.3 Cloud Security

The acritical aspect of CS is that it focuses on protecting cloud computing systems and services from a wide range of cyber threats like data breaches, malware, and hacking. The widespread adoption of cloud computing has transformed how businesses operate, enabling organizations to store and manage data more efficiently and cost-effectively [30]. However, cloud computing also presents new challenges and vulnerabilities, as sensitive data may be stored on remote servers that are accessed via the Internet. Cloud security is vital because cloud computing systems are highly susceptible to cyber threats like data breaches, malware, and hacking.

Organizations must implement various security measures and best practices to address these security challenges to protect their cloud computing systems from cyber threats. This includes using encryption, access control, and multifactor authentication to secure sensitive data, regularly performing security audits and vulnerability testing, and collaborating with cloud service providers with robust security protocols. All parties involved in the cloud computing ecosystem must know their security responsibilities and work together to protect cloud systems.

3.1.4 Endpoint Security

Endpoint security is a vital aspect of CS that protects individual devices, including laptops, smartphones, and other endpoints, from various cyber threats such as malware, phishing attacks, and hacking. The significance of endpoint security can be better comprehended by understanding the issues it addresses. Endpoint devices are increasingly used to access sensitive data and systems in the workplace and personal contexts, making them more susceptible to cyber threats. Unlike other systems, endpoint devices may be utilized outside the organization's secure network, making them more vulnerable to cyber threats [31, 32]. Furthermore, they may not have the same security control and monitoring level as other systems, making them more susceptible to cyber-attacks. Implementing a range of security measures and best practices is the key to endpoint security, which includes using antivirus software, firewalls, and encryption to protect against malware and other cyber-attacks. Regular device updates, patching, and monitoring for security vulnerabilities can be ensured through security policies and procedures.

3.1.5 Identity and Access Management (IAM)

A vital aspect of CS is that it aims to safeguard digital systems by allowing only authorized users access to sensitive resources. Managing user identity and access in today's complex digital environment can be a significant challenge, and unauthorized access to sensitive data and resources can lead to data breaches and other cyber threats [33]. Implementing a range of security measures and best practices, such as multifactor authentication and role-based access control, is the solution to this problem. IAM ensures that only necessary users have access to sensitive resources while regularly reviewing and updating user access privileges to prevent any unauthorized access. Thus, IAM helps organizations safeguard their digital systems and resources while maintaining the confidentiality, integrity, and availability of their sensitive data.

3.1.6 Data Security

This concentrates on safeguarding sensitive data from unauthorized access, use, disclosure, or destruction. To understand the significance of data security, providing some background on the issues it aims to tackle is helpful. Data is critical to the functioning of many organizations and is increasingly stored and processed in digital format. However, this also makes it susceptible to various cyber threats such as hacking, theft, and data breaches.

The issue with data security is that unauthorized individuals or cybercriminals can easily access sensitive data, leading to severe consequences such as reputational damage, financial losses, and legal liability for organizations [34]. The solution to data security is implementing various security measures and best practices to protect sensitive data from unauthorized access or use. This involves implementing robust access control, encryption, and secure data storage and transmission techniques. Moreover, organizations should conduct regular security assessments and audits to identify vulnerabilities and take appropriate measures to address them.

3.1.7 Incident Response (IR) and Disaster Recovery (I/DR)

It is an essential form of cybersecurity that aims to prepare organizations to respond quickly and effectively to cyber incidents and recover from any resulting damage or disruptions. It is essential to understand the importance of IR/DR and the problem it aims to solve. Cyber incidents like data breaches and ransomware attacks can have devastating consequences, including data loss, system downtime, and reputational damage [35]. The difficulty with cyber incidents is that they can be challenging to detect and address promptly, leading to severe consequences for the organization.

The solution to IR/DR is implementing various security measures and best practices designed to prepare organizations to respond to cyber incidents and recover from any resulting damage or disruptions. This includes creating an incident response plan, conducting regular security assessments and training, and implementing backup and recovery systems [35]. By implementing effective IR/DR measures and best practices, organizations can reduce the risk of cyber incidents and ensure that they can respond effectively and recover quickly in case of any incident. Additionally, emerging areas of cybersecurity, such as AI and IoT, are becoming increasingly important as these technologies become more widespread. Effective cybersecurity requires a comprehensive and integrated approach that addresses all aspects of digital systems and networks, from network security to incident response and disaster recovery.

4 Information Technology Key Form

IT refers to the use of digital technologies to manage and process information. There are various forms of IT, including the following and women's input to fact CS.

4.1 Hardware

Hardware in IT refers to physical computer components, which can be vulnerable to security threats like theft, tampering, and malware infections. One challenge in this area is the lack of diversity and representation, particularly with the underrepresentation of women in hardware engineering. This limits the range of perspectives and problem-solving approaches in the industry, hindering innovation and creativity. Women in cybersecurity can offer new solutions to these security challenges, contribute to secure hardware development, and design effective security protocols. Programs and events that support women in the industry can encourage their participation and provide professional development and opportunities for advancement [36].

Women in CS can bring diverse perspectives and experiences to the hardware engineering field, helping identify new solutions to security challenges. For example, they can contribute to the development of more secure hardware components and systems, as well as to the design of more effective security protocols and practices. Moreover, mentorship programs and networking events can help support women in the field, encourage their participation in the hardware aspect of the IT industry, and provide opportunities for professional development and advancement.

4.2 Software

Computer programs and applications, including operating systems, productivity tools, and specialized software for specific tasks. Security threats to the software include malware infections, coding errors, and vulnerabilities in third-party libraries and dependencies. However, the lack of diversity and representation, particularly the underrepresentation of women in software engineering, limits innovation and creativity by narrowing the range of perspectives and approaches to solving problems [37]. Women in CS can contribute to developing more secure and user-friendly software programs and effective security protocols and practices. Mentorship programs and networking events can encourage women's participation in software engineering and support their professional development and advancement.

4.3 Networks

These networks can be vulnerable to various security threats, including network attacks, data breaches, and unauthorized access. The underrepresentation of women in network engineering and related fields is a central problem, which can lead to a narrow range of perspectives and approaches to solving problems, limiting innovation and creativity. Women in cybersecurity can bring diverse perspectives and experiences to network engineering, helping identify new solutions to security challenges [38]. Initiatives such as mentorship programs and networking events can encourage women's participation in the network aspect of the IT industry and provide opportunities for professional development and advancement. Women in cybersecurity can play a vital role in developing more secure and innovative network solutions, promoting greater diversity and representation in the field.

4.4 Cloud Computing

Cloud computing delivers computing services, such as software, storage, and computing power, over the internet. Third-party providers are responsible for managing and maintaining the underlying infrastructure. However, one of the main concerns in cloud computing is data security. Cloud environments are vulnerable to various security threats, including data breaches, denial-of-service attacks, and unauthorized access. Women can help to identify and mitigate security risks, implement security policies and procedures, and design effective security control and monitoring systems [39]. In addition, women can play a role in developing more secure cloud-based applications and services by incorporating security considerations into the design and development process. Mentorship programs and networking events can also support women in the field, encourage their participation in cloud computing, and provide professional growth and development opportunities.

4.5 Internet of Things

One of the key challenges in AI is the lack of diversity and representation. AI technologies like IoT and algorithms are developed by individuals and teams who bring their own biases, perspectives, and experiences to the process, which can lead to unintended consequences and reinforce existing social inequalities. Women and other underrepresented groups are often excluded from the development and deployment of AI systems, which can perpetuate biases and limit the potential benefits of AI for society.

Efforts are underway to promote diversity and inclusion in AI. This includes initiatives to increase the representation of women and other underrepresented groups in AI research and development and efforts to mitigate bias in AI algorithms and systems [40]. For example, data sets can be audited for bias, and diverse teams can be assembled to ensure that various perspectives are represented in the development process. By promoting diversity and inclusion in AI, we can help ensure that these technologies are developed and deployed in fair, ethical, and beneficial ways for all members of society.

5 Cybersecurity as a Profession for Women

CS is an expanding field that has gained significant importance in recent times. The increasing usage of digital technology and the internet has led to a rise in cyberattacks that are becoming more sophisticated and frequent. These attacks severely threaten individuals, businesses, and governments worldwide [14, 15]. Therefore, the demand for competent cybersecurity professionals has surged, with a global shortage of over three million skilled personnel in this area. However, women remain substantially underrepresented in the field, comprising only around 20% of the cybersecurity workforce.

This gender gap could lead to a lack of diversity in perspectives and approaches toward solving cybersecurity problems, thereby limiting innovation and creativity in the industry [16]. Therefore, it is crucial to encourage and support more women to participate in cybersecurity to promote diversity, broaden perspectives, and bridge the gender gap in this rapidly evolving field. Several factors contribute to the underrepresentation of women in cybersecurity, including the gender gap in STEM (science, technology, engineering, and mathematics) fields, unconscious bias in hiring, and a lack of female role models and mentors.

The nature of the field can also present challenges for women, such as the requirement for continuous learning and professional development, long and irregular hours, and a culture often dominated by men [17]. These obstacles can discourage women from pursuing careers in cybersecurity despite the growing demand for skilled professionals. It is vital to create more inclusive and diverse workplaces, provide mentorship and networking opportunities for women, and promote STEM education and careers for young girls and women. By taking these steps, we can help to bridge the gender gap in cybersecurity and ensure that this critical field has the talent and perspectives it needs to thrive.

The underrepresentation of women in CS not only results in missed opportunities for the industry to tap into a wider pool of talent and perspectives but also has significant implications for CS. Research has shown that gender diversity can enhance problem-solving, creativity, and innovation, ultimately leading to more effective and efficient decision-making [18, 19]. Moreover, diverse teams are better equipped to understand and address the wide range of cyber threats and vulnerabilities. With the rapidly increasing number and complexity of cyberattacks, the cybersecurity industry needs to leverage the benefits of diversity to strengthen its defense against threats. Encouraging and supporting the participation of women in CS can lead to better outcomes, not only for the industry but also for society as a whole.

The gender gap in cybersecurity can be tackled through several initiatives that have emerged in recent years. These include training programs and scholarships to encourage and support women, as well as mentorship and networking opportunities [20]. To mitigate unconscious bias in the hiring process, some organizations have implemented blind recruitment processes, while others have tried to foster more inclusive workplace cultures.

Promoting cybersecurity as an attractive profession for women is also crucial. This can be done by showcasing the diverse range of roles and opportunities within the field and highlighting how CS professionals can make a meaningful and positive impact on society [21]. Furthermore, addressing the gender gap in STEM education and providing more opportunities for girls and young women to explore and pursue STEM fields can also help address the underrepresentation of women in cybersecurity.

In addition, mentorship and networking opportunities can be invaluable for women entering the CS field. These opportunities allow women to connect with experienced professionals who can offer guidance and advice on career development and introduce them to potential job opportunities [22]. Many organizations and industry associations have established mentorship and networking programs for women in CS, which can be particularly helpful for those who may feel isolated or marginalized in male-dominated workplaces.

Another critical step is to address unconscious bias in the hiring process. This can involve implementing blind recruitment processes, which remove identifying information from job applications to mitigate bias based on gender or race. It can also involve creating more inclusive workplace cultures where all employees feel valued and supported.

Promoting cybersecurity as an exciting and rewarding profession is the key to attracting more women to the field. CS requires individuals who are curious, creative, and analytical [23]. It offers a range of career opportunities, from incident response and threat intelligence to risk management and compliance. Furthermore, the work of CS professionals has a significant impact on society, as they are responsible for protecting critical infrastructure, sensitive data, and personal information.

Increasing the representation of women in CS is critical for promoting diversity, innovation, and practical problem-solving in the field. By addressing the challenges and barriers women face in the industry and promoting greater awareness and opportunities, we can work toward a more inclusive and effective cybersecurity workforce [24]. CS is a dynamic and exciting field that offers a wealth of opportunities for curious, creative, and driven people. Encouraging more women to enter the field ensures that the industry is equipped to address the evolving threats of the digital age. The following are some of the identified reasons for considerably low numbers of wine in cybersecurity and the IT industry.

5.1 Lack of Awareness

Although the lack of awareness regarding CS as a career choice for women is a persistent problem, efforts are underway to address this issue. Women already in the CS and IT industries are leading outreach programs to increase awareness of the benefits and opportunities available in the field. Additionally, CS and IT companies are taking action to enhance their diversity and inclusivity by implementing initiatives focused on diversity and inclusion and collaborating with organizations that support the inclusion and advancement of women in the industry [25]. However, despite these efforts, the issue of awareness persists, and further action is necessary to promote CS and IT careers with women.

Limited exposure to cybersecurity and IT careers in schools and universities, lack of female role models in leadership positions, stereotypes and biases, limited networking and mentorship opportunities, and misconceptions about the nature and scope of cybersecurity and IT careers contribute to the underrepresentation of women in the field. Insufficient outreach to underrepresented communities, limited access to industry resources, and the perception that cybersecurity and IT careers are not creative or fulfilling are also factors. Addressing these challenges requires promoting greater awareness of the benefits and opportunities available in cybersecurity and IT careers and providing more guidance and support to women interested in the field.

5.2 *Stereotypes*

Stereotypes have deterred many women from pursuing careers in the male-dominated and technical fields of cyber security and IT, where they are underrepresented and lack role models. Despite representing only 24% of the cybersecurity workforce globally, women can help bridge the significant skill gap in the industry [26]. Gender bias and discrimination in the workplace can further hinder their professional development and pay equality. Initiatives like mentorship programs, networking events, and awareness-raising are crucial to encouraging more women to pursue careers in these fields and challenge stereotypes. The persistent belief that men are better suited for technical fields hinders achieving greater diversity and inclusion in cybersecurity and IT.

5.3 *Lack of Representation*

The underrepresentation of women in leadership positions and technical roles in the cyber security and IT industry discourages female talent and perpetuates the myth that these fields are male-dominated. This lack of representation limits the diversity of perspectives and ideas, hindering innovation and progress. Women hold only 25% of IT-related jobs and 11% of executive positions in Fortune 500 companies [27, 28]. Initiatives like mentorship programs, recruiting efforts, and creating more opportunities for women to advance their careers can help create inclusive environments where women can thrive and achieve their full potential. The industry needs to tap into the potential of half of the global population to address the significant shortage of skilled professionals. Therefore, promoting greater diversity and inclusion is essential to attract and retain female talent in the cyber security and IT industry.

5.4 *Unconscious Bias*

Unconscious biases of hiring managers and recruiters may favor male candidates, a significant obstacle to attracting and retaining women in the cyber security and IT industry. These biases can manifest as automatic and unintentional attitudes and beliefs about people based on gender, race, ethnicity, or other characteristics and can impact the hiring process, performance evaluations, and overall workplace culture. For instance, a hiring manager may unconsciously favor male candidates for technical roles, assuming they are more competent than female candidates [29]. This bias reinforces the myth that women are less capable in technical roles and can dissuade women from pursuing opportunities in these fields. The industry must address unconscious biases through awareness training for hiring managers and recruiters to ensure a fair and inclusive hiring process that enables both men and women to succeed. This will help to promote greater diversity and inclusion and bridge the cybersecurity skills gap by tapping into the potential of underrepresented groups.

5.5 Unequal Pay

The gender pay gap is a significant factor contributing to the underrepresentation of women in the industry. Despite equal pay laws, women continue to earn less than men, with a wage gap that persists across all industries, including tech [28, 29]. Women of color experience even more pronounced gender and racial wage gaps. The unequal pay devalues women's work, creating disincentives for them to pursue or remain in these fields. Women in tech earn, on average, 20% less than men, and the gap widens as they advance in their careers. To address this issue, organizations must conduct pay audits and implement policies and programs promoting pay equity, such as transparent pay structures, performance-based pay, and supportive parental leave policies. A fair and equitable pay system will attract, retain, and advance talented female professionals.

5.6 Deficiency of CS Women Role Models

Women in the CS and IT industries may face a nonexistence of role models to emulate and learn from, which can discourage them from pursuing a career in these fields. The shortage of role models is linked to the broader problem of underrepresentation, where few women occupy senior positions in the industry. The absence of women in leadership positions can lead to a lack of visibility of women's potential and ability, perpetuating the stereotype that cybersecurity and IT are male-dominated fields. Various factors, including unconscious bias, lack of diversity and inclusion, and gender stereotyping, cause this underrepresentation. The scarcity of role models can also hinder women's career progression in the industry [29]. Without access to female leaders in the field, women may struggle to identify potential career paths and gain the necessary skills and experience to advance. Therefore, increasing the number of women in leadership positions in cybersecurity and IT is crucial to provide guidance and support for younger women interested in these fields. Doing so will help inspire and encourage more women to pursue and thrive in cybersecurity and IT careers.

5.7 Hostile Work Environments

Women face barriers in various aspects of the industry, including biased hiring practices, unequal pay, lack of role models, and hostile work environments. The underrepresentation of women in leadership positions in the industry exacerbates the problem [23, 30]. To tackle these issues, organizations must conduct pay audits, implement policies that promote diversity and inclusion, and provide training to create a safe and supportive work environment. Increasing the number of women in leadership positions is essential to serve as role models for younger women interested in the field. Creating a fair and equitable workplace will attract more women to the industry and help retain and advance talented female professionals.

5.8 Limited Support

Women face several obstacles in pursuing a career in these fields, such as a lack of access to resources, training, and mentorship. Additionally, they may not have the same opportunities for internships and networking as men, even if they pursue technology-related degrees. Discrimination and harassment in the workplace also make it challenging for women to feel valued and respected as part of a team [31]. Companies must invest in programs and initiatives that support women's recruitment, retention, and advancement in these fields. Creating an inclusive work environment and providing equal opportunities can help address the problem of limited support for women in the IT and CS industry.

5.9 Flexible Work Arrangements Deficiency

The absence of flexible work arrangements is a major obstacle preventing women from pursuing careers in this field. The industry has historically operated on a 24/7 basis, making it difficult for women who need flexible work hours to balance their personal and professional lives. This has contributed to the underrepresentation of women in the field, as they may leave the industry due to inflexible work arrangements. Despite some progress in providing flexible work arrangements, current affairs are still limited. Many companies still adhere to traditional work schedules [30], making it difficult for women with caregiving responsibilities to participate fully in the industry. However, with the advent of remote work, companies have an opportunity to provide more flexible work arrangements that allow women to balance work and personal responsibilities. This would enable the industry to attract and retain a diverse workforce, including women, and ensure that everyone has an equal opportunity to succeed.

5.9.1 Nonexistence of Training Opportunities

The lack of training opportunities significantly contributes to the underrepresentation of women in the cybersecurity and IT industry. The industry has traditionally been male-dominated, and training programs have typically been geared toward men. As a result, women often lack the necessary skills and knowledge to succeed in the field. This issue is compounded by the fact that women may face barriers to accessing training opportunities, such as financial constraints, family responsibilities, and lack of support. As a result, women may not have the same training and experience as men, making it challenging for them to compete for jobs in the industry [31]. There is a need for more accessible and inclusive training opportunities for women. This could involve developing training programs targeting women and their unique needs and challenges. Additionally, there should be a greater emphasis on mentorship and networking opportunities for women to help them build their skills and make connections in the industry.

5.9.2 Limited Access to Networks

Limited access to networks is a significant obstacle that impedes women's career advancement in cybersecurity and IT. Women often struggle to build professional networks due to various challenges, including cultural barriers, lack of representation, and unconscious biases. As women are already underrepresented in the field, their ability to connect with peers, mentors, and leaders is even more constrained. Furthermore, the lack of access to networks can hinder opportunities for professional growth, training, and advancement [32]. Organizations need to take measures to provide women with equal networking opportunities. This could include organizing events, establishing mentorship programs, and promoting participation in industry groups and conferences. By providing equal access to networks, organizations can help women in their careers, increase their contribution to the industry, and foster a more diverse and inclusive work environment.

5.9.3 Absence of Mentorship

Women may lack mentors who can provide guidance and advice on career development. Limited access to networks is another reason why there are few women in CS and IT. Access to online and offline professional networks is crucial for career development and growth, but women often have limited access to these networks due to exclusionary practices and a lack of diversity in leadership positions. Male-dominated networks tend to exclude women, and women may not have the same opportunities to build relationships with colleagues and mentors [33]. This leads to missed opportunities for professional development, job opportunities, and advancement.

Additionally, women may not have access to informal networks that provide essential information about industry trends, job openings, and potential mentors. As a result, women may be left out of meaningful conversations and decision-making processes. The current state of limited access to networks for women in cybersecurity and IT highlights the need for more inclusive and diverse networks that provide equal opportunities for women to connect and advance their careers.

5.9.4 Perceived Lack of Fit

Perceived lack of fit refers to the perception that women do not fit in the male-dominated tech industry, and this belief can discourage women from pursuing careers in cybersecurity and IT. This problem stems from societal stereotypes and gender biases that portray technical roles as masculine and analytical, which creates a perceived mismatch between women and the field. This perception of not fitting in can lead to feelings of isolation, exclusion, and low self-efficacy, further discouraging women from pursuing cybersecurity and IT careers [34].

The industry's current state reflects this issue, as women remain underrepresented in cybersecurity and IT roles. Addressing this issue requires a shift in mindset and culture that embraces diversity, inclusivity, and gender equality. It requires creating more opportunities for women to showcase their skills, promoting gender-neutral language in job descriptions, and challenging gender stereotypes and biases perpetuating the perception of a lack of fit. By actively promoting gender diversity and inclusivity, we can attract and retain more women in cybersecurity and IT, creating a more equitable and innovative industry.

5.9.5 Limited Support for Work-Life Balance

Work-life balance has become increasingly important to professionals, especially women who may face additional caregiving and family responsibilities. However, the lack of support for work-life balance has been identified as a significant obstacle for women pursuing careers in cybersecurity and IT. Women with children or other caregiving responsibilities often struggle to balance these responsibilities with their work duties due to long hours, inflexible work schedules, and a lack of employer support [35]. This can lead to burnout, stress, and, ultimately, women leaving the field. Women may face cultural biases that expect them to prioritize their family responsibilities over their careers. This can lead to women being perceived as less committed to their work and, as a result, being passed over for promotion or being excluded from significant opportunities. Companies need to implement more flexible work arrangements, such as telecommuting and flexible hours, and provide more support for women with caregiving responsibilities.

5.9.6 Lack of Confidence

It is often observed that women have lower self-confidence levels compared to their male counterparts. This lack of confidence is often attributed to societal and cultural factors discouraging women from pursuing careers in technology-related fields [36]. The problem is compounded by the fact that the IT industry is predominantly male-dominated, with few women in leadership positions. This creates a perception that women are not suited for leadership roles in the industry. As a result, women may lack confidence in their abilities, which leads to underrepresentation in the industry. Current efforts to address this issue include mentorship programs, training, and educational initiatives aimed at building self-confidence among women.

6 Women's Cybersecurity in Medical Perspective

CS is becoming increasingly vital in the medical industry to protect the confidentiality, integrity, and availability of patient data. Women are making valuable contributions as healthcare professionals, CS analysts, and engineers in this field. One of the main challenges in healthcare cybersecurity is the potential for data breaches or cyber-attacks that could compromise patient data [37]. Women bring a unique perspective to this challenge, as they are often well-versed in issues related to privacy and confidentiality, which are critical considerations in the medical field. Moreover, women play a pivotal role in developing secure medical technologies and systems, ensuring that they are designed with patient safety in mind.

Additionally, women in CS are playing a vital role in educating healthcare professionals and organizations on CS best practices and raising awareness about the importance of protecting patient data. With technology playing an increasingly critical role in healthcare, a diverse workforce that brings a range of perspectives and skills is essential. Women are valuable contributors to the medical industry, as they bring a variety of backgrounds and experiences that can help to address the unique challenges of healthcare CS.

In addition to these technical roles, women in cybersecurity can contribute to developing policies and procedures that promote a culture of security and privacy within healthcare organizations [38]. They can help raise awareness about the importance of cybersecurity among healthcare staff and patients and educate them about best practices for safeguarding sensitive data. Furthermore, women in cybersecurity can advocate for diversity and inclusion in the industry, which can help to address the underrepresentation of women and other marginalized groups in cybersecurity and IT. By promoting diversity and inclusion, women in cybersecurity can help to create a more resilient and effective healthcare cybersecurity workforce.

In addition, women in cybersecurity can contribute to developing secure and innovative technologies that improve patient care, such as secure telemedicine platforms that allow patients to connect with their healthcare providers remotely or mobile applications that help them securely manage their health and wellness [39]. They can also address the shortage of cybersecurity professionals in the healthcare industry by encouraging more women to pursue careers in cybersecurity and promoting diversity and inclusion in the workplace.

Additionally, organizations can work toward reducing gender-based pay gaps and providing equal opportunities for career advancement and leadership positions [40]. It is also essential to recognize and celebrate the contributions of women in cybersecurity, highlighting their successes and promoting them as role models for future generations of women in the field. By addressing these challenges, we can foster a more diverse and inclusive workforce in medical cybersecurity that reflects all individuals' needs and perspectives and ultimately helps improve patient outcomes and safety.

6.1 Discussion

A significant degree of gender segregation in the labor force has remained consistent throughout the study of gender and economics. Most men and women select very dissimilar careers. This segregation pattern may have resulted from various circumstances, such as prejudice, differences in intelligence, or even an individual's own free will [41]. We can directly test the effects of preferences on career choices in the context of choosing between IT and non-IT women's professional careers because we have introduced a direct measure of individual preferences by utilizing a widely accepted measure of occupational personality and by controlling via our sampling procedures, for factors such as educational attainment and attachment to the workforce. In other words, we could directly assess the influence of preferences on job decisions.

We find that differences in preferences among professional employees with full-time jobs can explain a substantial percentage of the apparent underrepresentation of women in information technology [42]. In other words, a large percentage of the gap in the number of men and women entering the information technology profession can be linked to the fact that, on average, men and women value different aspects of their occupations and, as a result, choose distinct career paths [43]. When these differences in preferences were considered, there was a significant change in how men and women choose carriers.

According to the findings of this study [44], it may be worthwhile to investigate the possibility that gender differences in preferences contribute to gender gaps in industries other than information technology, particularly in disciplines that emphasize the realistic general occupational theme. We cannot rule out the possibility that discrimination or differences in ability also act as filters that differentially reduce the entry of women into professional occupations more generally because our research design limits the sample to individuals who have chosen careers in IT and comparable occupations with controls for career motivation, education, and cognitive abilities. This means the sample is restricted to those who have selected careers in information technology and comparable fields.

There is a need for additional studies to address this issue; nonetheless, given that women make up nearly half of our control group, inequalities in ability or discrimination cannot be deemed insurmountable barriers to women's admittance into professional employment. Even though we cannot rule out the role of discrimination or differences in ability, our findings indicate that, even if these factors were eliminated, women would still have a lower representation in the field of information technology due to differences in their occupational preferences compared to men [45].

The observed differences in occupational personality may result from a domino effect in which one factor causes another. In this scenario, changes in occupational personality may result from a chain reaction. Previous work experience has likely influenced your chosen career route [46]. Given the information shown previously regarding the stability of occupational choices, we have reason to anticipate that these effects will have only a minor impact; yet, given the evidence available, we cannot rule out the potential that these effects may have an enormous impact. Before a person begins a career, it would be ideal to collect longitudinal data that could be used to assess their occupational personality. This would be the best circumstance. This type of data collection could be conducted before entering the labor field [47]. Due to budgetary constraints, we could not obtain such data; nonetheless, the results of our analysis suggest that additional research along these lines could be of great use.

After discovering that a person's occupational personality may play a part in explaining why women and men experience different labor market results, it is vital to undertake additional research on the elements that contribute to forming this aspect of a person's career preferences [48]. The discovery that occupational personality may play a role in understanding why women and men experience varied labor market results necessitates this avenue of investigation. Occupational personality is a complicated attribute that is not innate but rather the outcome of the interaction between an individual's qualities and the aspects of their environment [49, 50]. Consequently, parental and other family influences and educational and societal pressures probably contributed to the disparity [51]. Understanding how and why such gaps occur looks to be a crucial topic for future research, especially for those who seek to improve the proportion of women working in technical fields nationwide.

7 Conclusion

The role of women in CS and IT is an area that is gaining attention due to the increasing importance of these fields in modern society. As technology advances, so do the threats and risks associated with it. Therefore, having a diverse and inclusive workforce that brings different perspectives and experiences to the table is essential. From a medical perspective, it is clear that women have much to offer in these fields. Women tend to have strong communication and collaboration skills essential for effective teamwork and problem-solving. Women are often more detail-oriented and have a greater capacity for multitasking, making them valuable assets in the fast-paced world of CS and information technology. However, despite the benefits of having women in these fields, barriers still prevent women from pursuing careers in CS and information technology. These barriers include stereotypes, biases, lack of role models, and mentorship opportunities. Promoting diversity and inclusivity in the workplace is vital to overcoming these barriers, providing training and education opportunities, and encouraging more women to pursue careers in these fields. Inclusively, the role of women in cybersecurity and information technology is crucial for ensuring the security and integrity of our technological infrastructure. By breaking down barriers and promoting diversity and inclusivity, we can create a more robust and resilient workforce better equipped to tackle the challenges of the digital age. Staffing shortages necessitate CS and technology management competence. Women and male leaders must address workforce development gaps and create creative organizational initiatives.

References

1. Akyıldız D, Bay B (2023) The effect of breastfeeding support provided by video call on postpartum anxiety, breastfeeding self-efficacy, and newborn outcomes: a randomized controlled study. Jpn J Nurs Sci 20(1):e12509. https://doi.org/10.1111/jjns.12509
2. Fahim KE (2024) Electronic devices in the Artificial Intelligence of the Internet of Medical Things (AIoMT). In: Handbook of security and privacy of AI-enabled healthcare systems and internet of medical things. CRC Press, pp. 41–62. https://doi.org/10.1201/9781003370321-3
3. Avolio B, Chávez J (2023) Professional development of women in STEM careers: evidence from a Latin American country. Glob Bus Rev 09721509221141197. https://doi.org/10.1177/09721509221141197
4. Awano Y, Osumi M (2023) Peacock feathers and Japanese costume culture: evaluations from spectrum images and microscopic observations. SCIRES-IT-Sci Res Inf Technol 12(2):77–86. https://doi.org/10.2423/i22394303v12n2p77
5. Blalock EC, Lyu X (2023) The patriot-preneur–China's strategic narrative of women entrepreneurs in Chinese media. Entrep & Reg Dev 1–34. https://doi.org/10.1080/08985626.2023.2165170
6. Coffie CPK, Hongjiang Z (2023) FinTech market development and financial inclusion in Ghana: the role of heterogeneous actors. Technol Forecast Soc Chang 186:122127. https://doi.org/10.1016/j.techfore.2022.122127
7. Contreras T, Leembruggen M (2023) A world of women in STEM: an online learning platform. Bull Am Phys Soc
8. De Gioannis E, Pasin GL, Squazzoni F (2023) Empowering women in STEM: a scoping review of interventions with role models. Int J Sci Educ, Part B 1–15. https://doi.org/10.1080/21548455.2022.2162832
9. Farzin I, Abbasi M, Macioszek E, Mamdoohi AR, Ciari F (2023) Moving toward a more sustainable autonomous mobility, case of heterogeneity in preferences. Sustainability 15(1):460. https://doi.org/10.3390/su15010460
10. Ferati M, Demukaj V, Kurti A, Mörtberg C (2022) Challenges and opportunities for women studying STEM. ICT innovations 2022. Reshaping the future towards a new normal: 14th international conference, ICT innovations 2022, Skopje, Macedonia, September 29–October 1, 2022, Proceedings, pp 147–157. https://doi.org/10.1007/978-3-031-22792-9_12

11. Shafik W (2023) Cyber security perspectives in public spaces: drone case study. In: Handbook of research on cybersecurity risk in contemporary business systems. IGI Global, pp 79–97. https://doi.org/10.4018/978-1-6684-7207-1.ch004
12. Jun Y, Craig A, Shafik W, Sharif L (2021) Artificial intelligence application in cybersecurity and cyberdefense. Wirel Commun Mob Comput. 1–10. https://doi.org/10.1155/2021/3329581
13. Shafik W, Matinkhah S M, Ghasemzadeh M (2020) Internet of things-based energy management, challenges, and solutions in smart cities. J Commun Technol, Electron Comput Sci 27:1–11. https://doi.org/10.22385/jctecs.v27i0.302
14. Zhao L, Zhu D, Shafik W, Matinkhah SM, Ahmad Z, Sharif L, Craig A (2022) Artificial intelligence analysis in cyber domain: a review. Int J Distrib Sens Netw 18(4)15501329221084882. https://doi.org/10.1177/15501329221084882
15. Shafik W, Matinkhah SM, Sanda MN, Shokoor F (2021) Internet of things-based energy efficiency optimization model in fog smart cities. JOIV: Int J Inform Vis 5(2)105–112. https://doi.org/10.30630/joiv.5.2.373
16. Shafik W, Matinkhah SM, Ghasemazade M (2019) Fog-mobile edge performance evaluation and analysis on internet of things. J Adv Res Mob Comput 1(3):1–7. https://doi.org/10.5281/zenodo.3591228
17. Shafik W, Matinkhah SM (2019) Privacy issues in social Web of things. In: 2019 5th International Conference on Web Research (ICWR), Tehran, Iran. IEEE, pp 208–214. https://doi.org/10.1109/ICWR.2019.8765254
18. Balabantaray SR, Mishra M, Pani U (2023) A sociological study of cybercrimes against women in India: deciphering the causes and evaluating the impact on the victims. Int J Asia-Pac Stud 19(1). https://doi.org/10.21315/ijaps2023.19.1.2
19. Shafik W (2024) Wearable medical electronics in artificial intelligence of medical things. In: Handbook of security and privacy of ai-enabled healthcare systems and internet of medical things. pp 21–40. https://doi.org/10.1201/9781003370321-2
20. Shafik W, Matinkhah SM, Shokoor F (2022) Recommendation system comparative analysis: internet of things aided networks. EAI Endorsed Trans Internet Things 8(29). https://doi.org/10.4108/eetiot.v8i29.1108
21. Matinkhah SM, Shafik W, Ghasemzadeh M (2019) Emerging artificial intelligence application: reinforcement learning issues on current internet of things. In: 2019 16th international conference in information knowledge and technology, 2019. IEEE, Tehran, Iran
22. Alaziz SN, Albayati B, El-Bagoury AA, Shafik W (2023) Clustering of COVID-19 multi-time series-based K-means and PCA With forecasting. Int J Data Warehous Min (IJDWM) 19(3):1–25. https://doi.org/10.4018/IJDWM.317374
23. Shokoor F, Shafik W, Matinkhah SM, Overview of 5G & beyond security. EAI Endorsed Trans Internet Things 20228(30). https://doi.org/10.4108/eetiot.v8i30.1624
24. Shafik W, Tufail A (2023) Energy optimization analysis on internet of things. In: Advanced technology for smart environment and energy. Springer International Publishing, Cham, pp 1–16. https://doi.org/10.1007/978-3-031-25662-2_1
25. Yang Z, Jianjun L, Faqiri H, Shafik W, Abdulrahman AT, Yusuf M, Sharawy AM (2021) Green internet of things and big data application in smart cities development. Complexity. https://doi.org/10.1155/2021/4922697
26. Wang Y, et al (2023) Service delay and optimization of the energy efficiency of a system in fog-enabled smart cities. Alex Eng J 84:112–125. https://doi.org/10.1016/j.aej.2023.10.034
27. Shafik W, Matinkhah SM, Sanda MN. Network resource management drives machine learning: a survey and future research direction. J Commun Technol, Electron Comput Sci 301–15. https://doi.org/10.22385/jctecs.v30i0.312
28. Shafik W, Matinkhah M, Etemadinejad P, Sanda MN (2020) Reinforcement learning rebirth, techniques, challenges, and resolutions. JOIV: Int J Inform Vis 4(3):127–135. https://doi.org/10.30630/joiv.4.3.376
29. Shafik W, Mostafavi SA (2019) Knowledge engineering on internet of things through reinforcement learning. Int J Comput Appl 177(44):0975–8887. https://doi.org/10.5120/ijca2020919952

30. Kassim K, et al (2024) Artificial Intelligence of Internet of Medical Things (AIoMT) in smart cities: a review of cybersecurity for smart healthcare. In: Handbook of security and privacy of ai-enabled healthcare systems and internet of medical things. pp 271–292. https://doi.org/10.1201/9781003370321-11
31. Shafik W (2023) A comprehensive cybersecurity framework for present and future global information technology organizations. In: Effective cybersecurity operations for enterprise-wide systems. IGI Global, pp 56–79. https://doi.org/10.4018/978-1-6684-9018-1.ch002
32. Tadic B, Rohde M, Randall D, Wulf V (2023) Design evolution of a tool for privacy and security protection for activists online: cyberactivist. Int J Hum-Comput Interact 39(1):249–271. https://doi.org/10.1080/10447318.2022.2041894
33. Henshaw A (2023) Big data and the security of women: where we are and where we could be going. In: Digital frontiers in gender and security. Bristol University Press, pp 13–41. https://doi.org/10.51952/9781529226300.ch002
34. Brigstocke J, Fróes M, Cabral C, Malanquini L, Baptista G (2023) Biosocial borders: affective debilitation and resilience among women living in a violently bordered favela. Trans Inst Br Geogr. https://doi.org/10.1111/tran.12601
35. Salerno E (2023) Ut sacrificantes vel insanientes Bacchae: Bacchus' Women in Rome. In: The public lives of ancient women (500 BCE-650 CE) 2023 Feb 7. Brill, pp 173–193
36. Das S, Pooja MR, Anusha KS (2023) Implementation of women's self-security system using IoT-based device. In: Sustainable computing: transforming industry 4.0 to society 5.0 2023. Springer International Publishing, Cham, pp 87–98. https://doi.org/10.1007/978-3-031-13577-4_5
37. Royo MG, Parikh P, Walker J, Belur J (2023) The response to violence against women and fear of violence and the coping strategies of women in Corregidora, Mexico. Cities 1(132):104113. https://doi.org/10.1016/j.cities.2022.104113
38. Ebenezer V, Thanka MR, Baskaran R, Celesty A, Eden SR (2023) IOT based wrist band for women safety. J Artif Intell Technol. https://doi.org/10.37965/jait.2023.0179
39. Mahalakshmi R, Kavitha M, Gopi B, Kumar SM (2023) Women safety night patrolling IoT robot. In: 2023 5th International Conference on Smart Systems and Inventive Technology (ICSSIT), vol 23. IEEE, pp 544–549. https://doi.org/10.1109/ICSSIT55814.2023.10060955
40. Srinivas Rao K, Divakara Rao DV, Patel I, Saikumar K, Vijendra Babu D (2023) Automatic prediction and identification of smart women safety wearable device using Dc-RFO-IoT. J Inf Technol Manag 1:34–51. https://doi.org/10.22059/jitm.2022.89410
41. Ranganayagi D, Saranya P, Sharmila MJ, Sujitha S, Nisha TA, Shanmugam K (2022) Pre-eclampsia risk monitoring and alert system using machine learning and IoT. https://doi.org/10.54646/bijg.006
42. Lamba J, Jain E (2023) Advanced cyber security and internet of things for digital transformations of the Indian healthcare sector. In: Research anthology on convergence of blockchain, internet of things, and security. IGI Global, pp 1037–1056. https://doi.org/10.4018/978-1-6684-7132-6.ch055
43. Gautam D, Purandare N, Maxwell CV, Rosser ML, O'Brien P, Mocanu E, McKeown C, Malhotra J, McAuliffe FM, FIGO Committee on Impact of Pregnancy on Long-term Health and the FIGO Committee on Reproductive Medicine, Endocrinology and Infertility. (2023). The challenges of obesity for fertility: a FIGO literature review. Int J Gynecol & Obstet 160, 50–55. https://doi.org/10.1002/ijgo.14538
44. Gupta KP, Bhaskar P (2023) Teachers' intention to adopt virtual reality technology in management education. Int J Learn Chang 15(1):28–50. https://doi.org/10.1504/IJLC.2023.127719
45. Hafeez A, Dangel WJ, Ostroff SM, Kiani AG, Glenn SD, Abbas J, Afzal MS, Afzal S, Ahmad S, Ahmed A (2023) The state of health in Pakistan and its provinces and territories, 1990–2019: a systematic analysis for the global burden of disease study 2019. Lancet Glob Health 11(2):e229–e243. https://doi.org/10.1016/S2214-109X(22)00497-1
46. Karandashev V (2023) Cross-cultural variation in relationship initiation. In: The oxford handbook of evolutionary psychology and romantic relationships. p 267. https://doi.org/10.1093/oxfordhb/9780197524718.013.10

47. Kramer KL (1868) Female cooperation: evolutionary, cross-cultural and ethnographic evidence. Philos Trans R Soc B 2023(378):20210425. https://doi.org/10.1098/rstb.2021.0425
48. Li B, Xue C, Cheng Y, Lim ET, Tan CW (2023) Understanding work experience in epidemic-induced telecommuting: the roles of misfit, reactance, and collaborative technologies. J Bus Res 154:113330. https://doi.org/10.1016/j.jbusres.2022.113330
49. Mirzaasgari H, Momeni F, Pourshahbaz A, Keshavarzi F, Hatami M (2023) The role of psychological distress in the relationship between personality dimensions and pregnancy outcome of women undergoing assisted reproductive treatment (IVF/ICSI). Iran J Psychiatry 2023:1–11
50. Pop RA, Hlédik E, Dabija DC (2023) Predicting consumers' purchase intention through fast fashion mobile apps: the mediating role of attitude and the moderating role of COVID-19. Technol Forecast Soc Chang 186:122111. https://doi.org/10.1016/j.techfore.2022.122111
51. Wojtkowska A, Zaborski D, Modrzejewski A, Pastucha M (2023) Answer to commentary—the effect of cosmetic surgery on mental self-image and life satisfaction in women undergoing breast augmentation: an intermediate role of evaluating the surgery as one of the most important life events. J Plast Reconstr Aesthet Surg 76:94–95. https://doi.org/10.1016/j.bjps.2022.10.017
52. World Economic Forum (2023) https://www.weforum.org/agenda/2022/09/cybersecurity-women-stem/. Accessed 15 March 2023
53. Zaroo IF, Gani A (2023) Women's leadership and workplace culture. in leadership and workplace culture in the digital era. IGI Global 2023:227–241

A Comprehensive Exploration of Advancements and Applications of Digital Watermarking

Ealuri Rohith, Buggaveeti Padmaja, and V. M. Manikandan

Abstract The availability of multimedia content through digital networks is becoming more and more dependent on the internet in the current era of information technology and technical growth. As a result, there is a risk of unapproved digital picture possession and use. Illegal tampering with and alteration of digital images provide an extra risk. The ease accessibility of unlawful copying techniques like piracy, counterfeiting, and fraud for digital multimedia is a drawback. An approach to safeguarding digital image, and copyright is thus required. The existing difficulties and issues that must be taken into account while looking for such a solution include safeguarding ownership rights against illicit actions in a way that makes ownership of the digital image unquestionable. These days, media like images, audio files, text files, and videos are exchanged between sources for various reasons. In this process, a lot of copyright violations are taking place. Various types of attacks that can compromise the integrity of digital content are explored, including tampering, forgery, and unauthorized distribution. Various methods of authentication exist, including facial recognition, fingerprints, and passwords. Hence, copyright protection and digital authentication have become the major concerns. Digital watermarking is the ideal process that ensures not only copyright protection but also several other applications like broadcast monitoring and owner authentication. To maintain the integrity of data, watermarking is done which is inserting the imperceptible and inseparable information into data. In order to assign copyright-related information to images, such as the source, owner time-stamp, or even a logo or trademark, digital watermarking's aim and objective is to embed a digital mark into those images. The digital watermarks act as an evidence for any copyright infringement case. It lets owners trace the misuse of protected digital media. This paper provides a thorough overview of digital watermarking techniques for copyright protection and digital authentication.

Keywords Copyright · Watermarking · Digital Watermark · Multimedia · Copyright Protection

---

E. Rohith · B. Padmaja · V. M. Manikandan (✉)
Department of Computer Science and Engineering, SRM University-AP, Amaravati, Andhra Pradesh, India
e-mail: manikandan.v@srmap.edu.in

K. Kaushik and I. Sharma (eds.), *Next-Generation Cybersecurity*, Blockchain Technologies, https://doi.org/10.1007/978-981-97-1249-6_16

1 Introduction

The introduction of new technology exposed everyone to new threats. Through a variety of sources and the network, digital information is transmitted very quickly. The availability and accessibility of any information on the internet contributed to the rise in the usage of digital media. It is now possible to save, transfer, and modify digital media because of technological advancements. On the internet, it is quite inexpensive for anybody to access or get the data they need. Anyone may duplicate any piece of work and claim it as their own thanks to this. Many various pieces of information from websites, research papers, journals, and other sources may be found with just one search. The original work may be easily modified without any permission. When the content is transmitted from one source to another, it is not distorted in any way. These characteristics allow users to download any type of digital media from many sources. However, users have the opportunity to copy the original content and pass it off as their own due to the open availability of multimedia. Therefore, the concept of employing a digital watermark to grant ownership rights to the creator and safeguard their work.

The term Steganography comes from two Greek words, steganos and graphia [1]. Steganos means covered and graphia means writing. This method of keeping conversations confidential has been used for thousands of years. The first practical use of steganography goes back to 440 BC, when messages were written on pieces of wood and covered with wax to act as a covering medium. Invisible ink was employed during World War I to secretly write messages on paper that could only be seen by the person who wrote them. Later, they read the message and darkened the ink using a variety of substances, including milk and vinegar. In this fashion, steganography was a common practice in the past. Even today, there are several practical and technological uses for this approach.

Steganography [2] is the primary source of inspiration for digital watermarking methods. Digital watermarking and steganography are both methods of information hiding, however they serve very different purposes. In digital watermarking, the internal data, which is the watermark, is supplementary data that is utilized to secure the external, vital information. In contrast to digital watermarking, steganography uses internal data as the primary information and external data as the dummy data. Steganography techniques essentially conceal the approved information inside the digital medium before sending it to the receiver. The overview of steganography algorithm is shown in Fig. 1.

The original image, which is the image without any embedding done, is called the cover image. There are numerous algorithms for the steganography process. Any one of the algorithms is performed to hide the message in the original image. After performing such an operation, the image is called a stego image. Messages can be embedded in cover images and stego-images with the help of a key, or the messages can be derived from the photos themselves. Steganography has its different types. Text steganography secretly stores information and in this method, the hidden data is encoded into each letter of the word. In the audio steganography, the data is hidden in

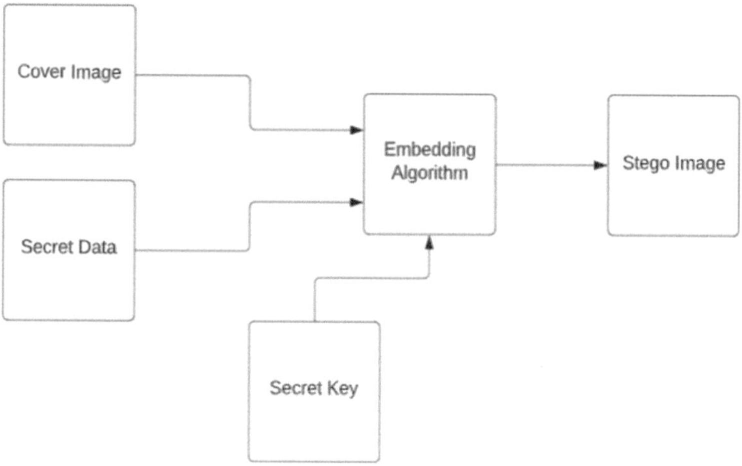

Fig. 1 Overview of steganography method

sound. This is digitally used to protect the data against unauthorized reproduction. In the video steganography, the data is secretly embedded within a video file. Generally, Discrete Cosine Transform(DCT) is commonly used to insert the values to hide the data in each image in the video. It is invisible to the naked eye. Protocol or Network steganography is hiding the data by using a protocol like TCP, UDP, or IP.

2 Digital Watermarking

The paper business gave rise to watermarking in the thirteenth century. One of the earliest use of watermarking was to identify the mill that manufactured the paper and the brand in order to verify its legitimacy. These days, watermarks are applied on currency, paper, stamps, and a long list of other items. In order to identify fake copies, Komatsu and Tominaga developed a digital watermarking method back in 1989. They inserted a hidden label into a copy, and if the registered owner matches it, it can be verified that the person holding the document is the owner. Although watermarking has a lengthy history dating back to the thirteenth century, it wasn't until after 1990 that watermarks were digitized and began commonly employed. The history of watermarking and its evolution is explained thoroughly in [3].

Digital watermark is a signal that is embedded in a digital media like audio file, text file, image, or a video file. It is permanently embedded in the data and can be detected and extracted later to perform any modifications or operations. The watermark is hidden in the original data in such a way that it is inseparable from the data. Embedding a digital watermark into an original work still lets the owner access the data. The host data is intended to have a watermark forever. The data can be extracted to fully define the owner if the ownership of a digital work is in doubt.

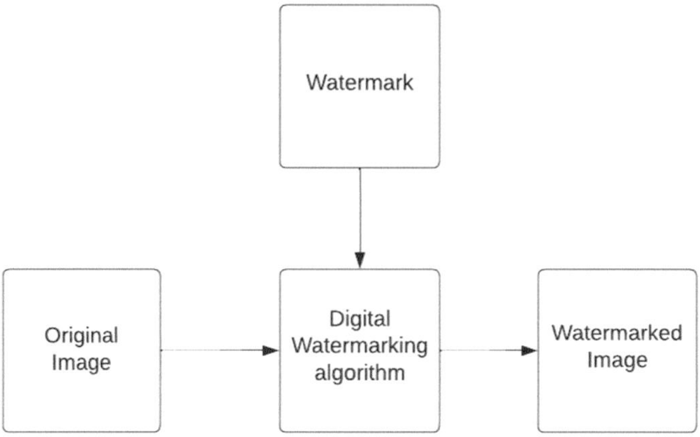

Fig. 2 Digital watermarking scheme

Traditional watermarks are visible to users. Some are printed on the image, some are kept on the video. These are technically the visible watermarks. The digital watermarks, unlike the traditional ones, are designed in such a way that they are not visible to the users. The watermark which contains patterns of bits is embedded and scattered all over the image to avoid modification.

The image where we are embedding the watermark into is called the host image. After the watermarking process, the image is defined as the watermarked image. During the process of embedding, a secret key is used to have a secured watermark. The image having the watermark is transmitted along the communication stream. On the receiver side, the embedded can be extracted. This watermarking process as a whole has been developed to identify the creator, receiver, and sender of the information. The encryption process is where the content is hidden and is made invisible to the reader without a specific key or code from the owner. After the encryption process, watermarking can be used for many applications like copyright protection, digital authentication, fingerprinting, broadcast monitoring, and many such [4] (Fig. 2).

2.1 Classification of Digital Watermarks

Different criteria are used to categorize digital watermarking processes. Protection of privacy and authenticity has grown to be a major issue and difficulty as a result of the growth of different media assets online. Digital watermarking methods of many kinds were developed to address this issue. These systems are used in several applications with a variety of purposes.

We are aware that watermark embedding and watermark extraction are the two key steps in the watermarking procedure. The varieties of watermarking technologies are categorized according to the features of embedded watermarks.

2.1.1 Robustness of the Watermark

The word "robustness" [5] describes the watermark's capacity to withstand and withstand any tampering. Attacks are deliberate attempts to remove the watermark, while others aim to modify it. Watermarks are divided into three categories for the embedding process, which is the initial stage of the watermarking process.

Watermarks that are *robust* are made to withstand all types of image processing operations. These attacks comprise techniques like cropping, filtering, compressing, and more. The information included in the watermark may be removed in these unauthorized assaults, or new information may be added.

A *semi-fragile* watermark performs a somewhat different task than a robust one. Any unauthorized alterations to the watermark can be detected using these kinds of watermarks. However, these also permit particular image-processing techniques. Semi-fragile watermarks can distinguish between destructive alterations and standard image-processing techniques.

For comprehensive authentication, *fragile* watermarks are utilized. Authentic images are those that haven't been altered in any way. The best option for this is fragile watermarks. They look for any unauthorized changes to the embedded watermark. The watermark is fragile and will be lost if the image is slightly modified. Therefore, the primary goal of these watermarks is to verify the validity of the image.

2.1.2 Blind and NonBlind

The watermark must then be extracted after being embedded. This method involves a variety of strategies. Watermarks are categorized as blind or non-blind based on the requirement to access the original image. The watermarking method is said to as blind if it does not require knowledge of the original, unwatermarked audio, video, or picture file. A method is regarded to be non-blind if it requires any original data in order to extract the watermark. Since recovering the watermarked image is obviously simple if the unwatermarked data is available, the non-blind technique generally looks to be more effective than the blind one. The watermark detector, however, does not always have access to the original data in numerous cases. Therefore, blind watermarking methods have an edge over non-blind ones.

2.1.3 Visibility

A watermark can be anything, such as a logo, a picture, or a section of text. They are divided into visible and invisible watermarks depending on how clear the watermark is on the original data. The visible watermark, as its name implies, is something that is evident on the image in any format. On the other hand, the invisible watermark is incorporated into the original image and is not noticeable to the human eye. One benefit of a visible watermark over an invisible one is that it prevents unauthorized usage of your image from any search engine since it is easy to recognize. Additionally, we may advertise our firm and assert our copyright by using visible watermarks. The visible watermark is simple to embed into a picture, however the invisible watermark embedding is fairly challenging and requires advanced algorithms. There are more uses for visible watermarks than invisible ones.

2.1.4 Based on the Domain

We attempt to integrate as much information as we can into the original image for a given amount of distortion. Recently, numerous different watermarking systems have been created. These methods are divided into two categories: frequency domain watermarking and spatial domain watermarking, depending on the domain in which the watermark is inserted. When watermarking in the spatial domain, the pixels are changed right there in the image's spatial domain. This makes the process simple because there is no need to extract the watermark. Through various transformations like discrete fourier transform(DFT), discrete cosine transform(DCT), and discrete wavelet transform(DWT), the watermark is inserted into the frequency coefficients of the altered picture in frequency domain watermarking. The lower and intermediate frequencies are where most of the energy is concentrated. The embedded watermark is not evenly spread throughout the image in the spatial domain, making it simple to remove the watermark. On the other hand, the watermark is dispersed across the image in the frequency domain, thus the alterations made have less of an impact on the image.

2.1.5 Private and Public

Depending on where the watermark is, private and public watermarks can be distinguished. A watermark is considered private, as the name implies, if only authorized users are able to recognize it. These watermarking methods make the watermark undetectable and inaccessible to unauthorized users. A private key that specifies the location of the watermark is used to extract the watermark. As long as the location is known, the user can insert or delete the watermark by using a private key. A public watermark, on the other hand, is one whose location is well known to everybody. When a public watermark is inserted, analyzing the entire image makes it easier to find and remove it. Private watermarking techniques are more effective than public ones, where a hacker may quickly find and remove the watermark without permission.

2.2 Properties of Digital Watermark

Digital watermarking has become much more widely used in daily life in recent years. This is because of its features, which guarantee safe data transmission and preserve content ownership [6].

- The resemblance between the watermarked and original image is what defines a watermark's *fidelity*. The user should be unable to see the image's embedded watermark. Additionally, it maintains the similarity between the watermarked image and the original image. The degree to which a watermark degrades the original image's clarity and quality is known as its fidelity. Additionally, it determines how much the original content has been degraded once the watermark has been inserted. Therefore, high fidelity watermarks shouldn't specifically lower the quality of original content. Different measures, such as the bit error rate (BER) or the structural similarity index (SSIM), which calculate the differences between the original and watermarked content, can be used to assess fidelity.
- *Imperceptibility* is another important aspect of digital watermarking. The watermarked image should blend in with the original image once it has been embedded. Digital watermarking is mostly used to safeguard intellectual property. Therefore, anyone can simply remove a watermark that has been embedded and steal the work while it is visible. Therefore, a watermark's imperceptibility quality determines its invisibility and is crucial for preserving the integrity of the original content. To increase imperceptibility in particular contexts, many signal processing algorithms have been developed.
- When a digital watermark is said to be *robust*, it means that the user can still read it even after certain adjustments have been made to the original content. Watermarks with high robustness are undoubtedly those that can resist various forms of degradation, including compression, cropping, filtering, and other transmission-related manipulations. The watermark must be able to withstand any attempt to reproduce the original work or to reduce the original content's quality in order to preserve its copyright.
- The quantity of data that can be inserted in any digital medium using a specific technique is referred to as a digital watermark's *capacity*. The strength of a digital watermark can vary depending on a number of elements, including the type of signal processing, the kind of digital media, the embedding technique employed, and the degree of robustness the watermark contains. When watermarking images, the amount of image bits that can be inserted into the original image is utilized for evaluating the watermark's capability. The length or frequency band of the watermark signal that can be inserted into the audio stream is the unit of measurement for audio watermarking. Again, regardless of the capacity of the watermark, the perceived quality and resilience of the image should not be compromised. In many applications, including data concealing and content identification, the watermark's capability is a crucial consideration. The watermark capacity is determined by the limitations of the application.

- The integrity and novelty of original content is preserved by the *security* of a digital watermark. It safeguards the information from tampering or unauthorized access. It can be done using a variety of methods, including encryption and digital signatures. By converting the watermark into a secret key that can only be decoded by the authorized user, encryption is accomplished. Through this procedure, the watermark is protected even from unauthorized users. A digital watermark's security is frequently verified using digital signatures. The watermark data's unique digital signature is calculated using a cryptographic technique. Only authorized users are capable of verifying the digital signature. Several other techniques can be employed to secure digital watermarks in addition to these. Any watermarking method should be designed and put into use with security as one of its primary considerations.

2.3 Applications of Digital Watermarking

In order to digitally authenticate and safeguard intellectual property, researchers first looked into the method of embedding digital watermarks into digital media like photos, audio files, text files, and video files in 1990. Digital watermarking was necessary due to the increased use of technology and the internet. Numerous digital watermarking techniques have been designed as a result of the increasing importance of multimedia and forensics. As data distribution and access become more convenient, there are more options for safeguarding and authenticating digital content that were developed.

The most effective tamper detection methods apply digital watermarking. This indicates that we can tell whether or not the material has been tampered with when it is accessed or viewed by employing these strategies. A watermark is included in the data when the content is produced. It is possible to identify data tampering using the information in this watermark. A comparison is made between the watermark and the original watermark when the digital content is accessed. It may be said that the material has not been altered if the watermark matches the original one that was added at the beginning. It is a sign that the information has been altered if the watermarks do not match. Therefore, data manipulation may be identified via digital watermarking.

Forensic analysis depends heavily on digital watermarking. The source of the leaked or pirated information might be found during the examination of legal proceedings and used as evidence. Each replica of the digital file has a digital watermark built in. Finding the source of the content leak and the first location where it was disseminated can be made easier with this information. To monitor the circulation of original content, these watermarks can also be employed. It also offers a history of who and when the data was accessed. Digital watermarking methods thus offer a number of benefits for case-specific research.

Copyright violations and theft of works are frequent occurrences in our everyday lives. It is simple for hijackers to take our work and claim it as their own if there is no protection protecting the information or work we produce. For example, a well-known photographer shared a picture of a stunning landscape on his website. Anyone can take the photo of them and give it to you with their name on it without providing any documentation or ownership rights. The image will always belong to us if we add a watermark, and no one else will be able to take it. It would also be a promotion for his work if the watermark was added in a way that made it noticeable. A watermark can also be incorporated in and used to secure text, video, and audio files.

The source of the image and authorized users are both recognized by watermarks for copyright protection. The printing and publishing industries are quickly adopting this method of inserting copyrights. There are several networks that deliver videos. Given how simple it is to reproduce videos without sacrificing quality, it is crucial to preserve their copyrights. The video stream between the transmitter and recipient might have a watermark added to it. We can track and identify the various copies of videos in this way. We have access to both visible and invisible watermarks. A complicated embedding procedure is needed for invisible watermarks. Watermarks that can be seen by the naked eye and are clearly distinguished are considered visible. Both watermarks, however, guarantee the data's ownership and safeguarding.

The old-fashioned way to authenticate a document is to use a handwritten signature. There is no way to copy a fixed signature from one document to another. The legitimacy of two signatures that appear to be identical can be determined by comparing them. An exact duplicate of a handwritten signature exists in digital signatures. These methods of data authentication are still even more powerful. Using two keys—public and private—that are mathematically connected to one another, the digital signatures may be encrypted. Utilizing the public key, users may validate the digital signature; however, the secret key is necessary to create the exact identical signature. Therefore, digital watermarking is essential for digital authentication.

In conclusion, digital watermarking is an effective tool for safeguarding and confirming the validity of digital content. It has a wide variety of applications in several sectors.

3 Related Work

In the current digital era, digital watermarking techniques have become a popular way to guarantee the validity, integrity, and copyright protection of multimedia information. Previous research on digital watermarking has looked at a variety of approaches, including both spatial and transform domain techniques, to embed and remove watermarks while keeping the caliber and perceptual integrity of the host multimedia content. Numerous issues with digital watermarking have been addressed by researchers, including resilience against signal processing processes, capacity optimization, resistance to assaults, and adaptation to various multimedia formats, which has resulted in the creation of novel methods and frameworks. This section will explore a number of methods that have been suggested for the purposes of digital authentication and copyright protection.

3.1 Background Theory

To understand the digital watermarking algorithms, there are several mathematical theories to be learnt.

3.1.1 Discrete Fourier Transform(DFT)

Signals in the frequency domain may be analyzed and modified using the Discrete Fourier Transform. This enables the signal to separate into its component frequencies. Finding certain frequency ranges that are appropriate for inserting the watermark is made easier with the use of frequency analysis. The DFT generates a frequency representation made up of N evenly spaced frequency bins when applied to a finite series of discrete samples. The DFT result can be translated to magnitude and phase data or shown as a complex-valued spectrum. Application areas for the DFT include feature extraction, data compression, spectrum analysis, and filtering. The DFT is frequently used in the context of digital watermarking to examine the frequency components of the host signal and embed or extract watermarks in the transformed domain.

Mathematically, DFT is defined as:

$$X_k = \sum_{n=0}^{N-1} X_n e^{-2\prod ikn/N}$$

where N is the length of the signal, X_k represents the complex-valued spectrum of the signal at frequency index k, n is the sample input.

3.1.2 Discrete Cosine Transform (DCT)

Similar to the Discrete Fourier Transform, DCT converts signal from spatial domain to frequency domain. The DCT is frequently used in digital watermarking because of its capacity for energy compression. The signal's energy is concentrated in a limited number of low-frequency coefficients when the DCT is applied to an image or video frame, while the energy of the higher-frequency coefficients is reduced. Because of this characteristic, watermarking algorithms can incorporate the watermark in lower-frequency coefficients, where changes are less obvious to human observers.

Mathematically,

$$F(u, v) = (2/N)^{1/2}(2/M)^{1/2}$$

$$\sum_{i=0}^{N-1}\sum_{j=0}^{M-1} \wedge(i) \wedge (j) \cdot \cos\left[\frac{\Pi \cdot u}{2 \cdot N}(2i+1)\right]\cos\left[\frac{\Pi \cdot v}{2 \cdot M}(2j+1)\right] \cdot f(i, j)$$

where $f(i,j)$ is the intensity of the pixel in row i and column j and $F(u,v)$ is the DCT coefficient in row k_1 and column k_2 of the DCT matrix.

3.1.3 Singular Value Decomposition

A matrix is divided into singular values and related singular vectors using the matrix factorization technique known as singular value decomposition (SVD). It is a fundamental idea in linear algebra and has several uses in the domains of data analysis, picture compression, and signal processing. JPEG and other image compression methods frequently employ SVD for effective picture storage and transmission. Applying SVD to picture data allows for compression without noticeably sacrificing visual quality since it identifies the prominent spatial frequencies and allows for the representation of the image with fewer coefficients. By contrasting the single values or vectors of the original material with those of the received or accessible content, SVD is used for digital authentication. The identification of content tampering, alteration, or unauthorized change is made possible by this comparison. A way for content authentication and integrity checking, any appreciable divergence in the single values or vectors indicates probable unauthorized content alterations. SVD is a flexible method for digital watermarking that offers insights into the energy and frequency properties of signals or pictures. It allows for the embedding of watermarks in regions that are invisible as well as their extraction during the detection procedure. The robustness and security of watermarks may be increased, making them more resistant to assaults and distortions, by making use of the energy compaction capabilities of SVD.

3.1.4 Error Correction Codes (ECC)

To increase the stability and dependability of the embedded watermark, error correction codes (ECC) are frequently utilized in digital watermarking. In order to detect and fix mistakes that can happen during transmission, storage, or assaults on the watermarked information, ECC algorithms provide redundancy to the watermark data. Before embedding, several error correcting codes are added to the watermark data, doubling the amount of information. The same error correction algorithms that are used to extract watermarks are also utilized to find and fix faults in the extracted watermark. The reduction of noise, signal distortions, data tampering, and attacks on watermarked information is made possible by the error correction procedure.

Multiple digital watermarking methods for copyright protection and digital authentication are thus influenced by various mathematical ideas.

3.2 Evaluation Measures

Researchers and practitioners may objectively evaluate and compare various watermarking algorithms, optimize their parameters, and select the approach that best fits their unique needs by taking assessment measures into consideration. Evaluation measures are crucial to the watermarking process since they assure the reliability, excellence, security, and usability of digital watermarking systems.

PSNR: Peak Signal-to-Noise Ratio (PSNR) is a commonly used statistic for assessing how well a picture has been compressed or reconstructed in comparison to its original state. It calculates the ratio of the power of the noise added during compression or reconstruction to the power of the greatest available signal (usually the maximum possible pixel value). The PSNR, which is measured in decibels (dB), offers a number reflecting how faithfully and precisely the original picture is preserved after being compressed or rebuilt. Lower numbers imply more obvious artifacts and image quality loss, whereas higher PSNR values suggest greater quality and less distortion. In order to optimize the trade-off between compression ratio and picture quality, PSNR is frequently used in image compression algorithms.

SSIM: A popular picture quality statistic called the Structural Similarity Index (SSIM) goes beyond a straightforward pixel comparison to take into consideration how structurally similar the original and deformed images are. By comparing the similarity of their brightness, contrast, and structural elements, SSIM assesses the perceived picture quality. It takes into account the local structural data and records pixel dependencies. A score between 0 and 1 is generated by SSIM; a value closer to 1 denotes a higher degree of similarity and better picture quality. Given that it takes both global and local structural information into account, it is very helpful in evaluating picture restoration strategies.

MSE: Measuring the differences between an original picture and a compressed or reconstructed version of the same image is frequently done using the Mean Squared Error (MSE) metric. The average of the squared discrepancies between the pixel values of the original and the compressed or reconstructed picture is calculated using a mathematical formula. Lower values represent higher picture quality and less distortion, and MSE is given as a numerical value. In order to balance the trade-off between compression ratio and picture quality, MSE is frequently used in image and video compression algorithms. It does have some drawbacks, though, such as its susceptibility to outliers, which can skew the evaluation of image quality as a whole.

NCC: When comparing two datasets or signals, a statistic called the Normalised Correlation Coefficient (NCC) is employed. It is often used in a variety of disciplines, including pattern recognition, image analysis, and signal processing. When calculating the correlation between two variables, NCC takes into consideration the means and standard deviations of each variable. The NCC algorithm generates a result between -1 and 1, where 1 denotes a perfect positive correlation, -1 denotes a perfect negative correlation, and 0 denotes no correlation at all between the datasets. Greater similarity or correlation between the datasets under comparison is indicated by a higher NCC score.

BER: Bit Error Rate is referred to as BER. It is a metric used to assess a digital communication system's or channel's level of quality. The percentage of bits that are wrongly received to all bits broadcast is known as the bit error rate. The quality of the communication system or channel is inversely correlated with the BER value. A greater BER implies a higher rate of transmission mistakes, which can lead to a loss of data integrity and a degradation in performance. The effectiveness of error correction methods, signal-to-noise ratios, and overall system performance are all evaluated using BER, which is often used in telecommunications, networking, and digital data transfer. It is a crucial factor in determining how reliable and effective digital communication systems are.

3.3 Algorithms Proposed

In [7], a DWT-SVD based robust digital watermarking scheme was proposed for the security of medical images. In this method, the picture capture data and the electronic patient record will make up the watermark. The Electronic Patient Record hash will be appended to the watermark in order to increase security and ensure data integrity. The watermark employed in this method has three components; the first and second sections contain data about the patient and the method of picture capture, allowing the user to determine who created the image and where it came from. The third portion, which will be placed in binary form for insertion, is created by adding the first two parts. DWT is first applied and the LL subband is subjected to SVD, and the watermark bits are appended to the coefficients of the resultant S matrix in order to incorporate the watermark. The adjusted LL subband is then obtained by applying an inverted SVD. The watermarked medical picture is then obtained by performing an inverse DWT. The watermark is extracted using the S matrix's subsequent coefficients. For the first variation, the watermark's bits will match the parity of subtracting two consecutive coefficients. In the second variation, the watermark's bits (X and Y) will match the parity of subtracting three sequential coefficients. After the extraction procedure, the changed matrices are transformed into extracted bits. This binary sequence will be split into three parts: the patient information will be in part one, the picture acquisition information will be in part two, and the hash of the first two parts will be in part three. The extracted hash and the hash produced by concatenating the first two sections are now compared in order to confirm the integrity. Since this method is used to create medical images, the image quality needs to be preserved. The likelihood of several distortions is great. Several assessment metrics are taken into consideration in order to gauge how imperceptible the aforementioned strategy is. PSNR, or peak signal to noise ratio, is used to assess the deterioration brought on by the embedding of watermarks into images. A high PSNR value suggests that the distortion is reduced. The combination of DWT and SVD results in variations with high PSNR values, which improve imperceptibility. A metric used to compare two photographs is called the Structural Similarity Index (SSIM). It is used to compare the compressed picture's visual quality to the original image. When the SSIM value is close to 1, it indicates that the visual quality is high. The proposed scheme observed a SSIM value closer to 1.

Using cryptographic techniques, a digital watermarking scheme for copyright protection is introduced in [8]. The two basic steps in every digital watermarking system are embedding and extraction. For the embedding step in this technique, DWT and SVD methods are combined. For robustness verification, a cryptographic technique is employed. RSA is an asymmetric cryptographic technique which has two keys called public and private. At the transmitter, the data is encrypted using the public key, and at the receiver, it is decrypted using both keys. First, a QR code with the name and country is produced. When private key values are entered into the RSA algorithm, a public key and an encrypted message are produced. In order to watermark data securely, a QR code is scrambled using Chaotic Logistic Map(CLM). The watermarked picture is imported and transformed to its red, green, and blue component parts. The blue layer is taken into account, and the Haar wavelet decomposition is used to produce the four subbands LL, LH, HL, and HH. The LL subband and the jumbled QR picture are dissected using the SVD method. The watermarked singular values are created by taking into account the singular values of both photos and combining them with a key value. A watermarked LL subband is produced using an invertible SVD approach. The watermarked blue layer for one level is produced similarly by fusing watermarked LL subbands with other subbands using an inverse DWT. A blue layer is combined with the red, green, and other layers to produce a watermarked color picture. The recipient receives the watermarked picture together with the public key and key value. The watermarked image is initially inspected in order to begin the extraction procedure. The red, green, and blue layers of the colored watermarked picture are created. The blue component, where the watermark is incorporated, is taken into consideration for extraction. A Haar wavelet is used with a one-level DWT to apply the blue component. The LL subband is taken into account, and a singular value matrix is produced using SVD. Based on the important image's key values and incomplete data, a scrambled QR watermark is extracted. The CLM technique is used to reverse scramble the retrieved watermark. The public key, which contains the private key values, is used to extract the watermark from the encrypted message in order to validate the watermarked data. Peak-Signal-to-Noise Ratio and Normalised Correlation Coefficient are taken into consideration during the evaluation process. The metric used to determine the relationship coefficient between the original watermark and the extracted watermark is called NCC. NCC approaches 1 when the extracted watermark is equivalent to the original watermark. In comparing the suggested scheme to several previous schemes, it is found that the NCC value is equal to 1, indicating that the watermark that is accurately retrieved is identical to the original watermark.

To ensure copyright protection, a digital watermarking scheme through encryption and DWT is proposed in [9]. Using a collection of wavelet functions, the discrete wavelet transform separates a signal into several frequency components. The DWT additionally records the temporal or geographical localization of the frequency components, in contrast to other transforms like the Fourier Transform, which just offers frequency information. Due to its capability to give a multi-resolution representation of signals and pictures, capturing both local details and global properties, the Discrete Wavelet Transform is favored in digital watermarking. It offers flexibility in balancing transparency and robustness needs by allowing watermark implantation in particular frequency bands. In this algorithm, a 90×90 pixel grayscale baboon image is used as a watermark image. Row and column rotation using the watermark image's random number generator key k is the basis of an image encryption technique. The initial input picture is divided as the following phase. To produce the proper scaled picture with decreased size, two-dimensional DWT is employed. The pixel point at the decomposed input picture is determined for inserting the decomposed watermark image based on Euclidean distance. The perceptibility of the input picture remains unchanged if there is a great degree of resemblance between it and the watermark image. This ultimately makes the watermark stronger. Using the below equation, the encrypted watermark is added to the source picture.

$$y(i, j) = (1 - \text{alpha}) \times i(i, j) + \text{alpha} \times iw(i, j)$$

The visibility coefficient is represented by alpha in the equation. i(i,j), iw(i,j), and y(i,j) correspondingly represent the DWT coefficients of the corresponding decomposed input image, watermark image, and watermark embedded output image. The success of this encryption hinges on how closely the input photos that have been deconstructed resemble the encrypted watermark image. The reversal of the embedding procedure allows for the extraction of the watermark. Since the suggested method uses a non-blind technique, the extraction procedure needs both the original picture and the encryption key. When comparing the extracted watermark picture to the original watermark image, the Mean Square Error (MSE), Normalised Correlation Coefficient (CC), and Peak Signal to Noise Ratio (PSNR) are employed as indicators of resemblance. It was observed that the proposed scheme has a higher PSNR value than the already existing schemes.

3.4 Results Comparison

See Table 1.

Table 1 Comparison of different algorithms and their results

Reference no	Algorithm used	Results
[9]	Digital Image Watermarking Through Encryption and DWT for Copyright Protection	PSNR—54.96 dB MSE—0.2071 NCC—0.97
[7]	A DWT-SVD based robust digital watermarking for medical image security	PSNR—57.41 dB MSE—0.13 SSIM—0.9998
[8]	Digital Watermarking System for Copyright Protection and Authentication of Images Using Cryptographic Techniques	NCC—0.9949 PSNR—37.16 dB
[10]	Robust Color Images Watermarking Using New Fractional-Order Exponent Moments	PSNR— > 44 dB NCC—0.9759 BER—0.0195
[11]	Robust and distortion control dual watermarking in LWT domain using DCT and error correction code for color medical image	PSNR—32.85 dB NCC—0.9913 BER—0
[12]	Robust blind watermarking approach against the compression for fingerprint image using 2D-DCT	PSNR—41.30 dB BCR—100%
[13]	Machine learning based blind color image watermarking scheme for copyright protection	PSNR—40.25 dB SSIM—0.9976
[14]	A robust hybrid SVD-based image watermarking scheme for color images	PSNR—51 SSIM—0.9936 NC—1
[15]	A Novel Reversible Watermarking Scheme based on SHA3 for Copyright Protection and Integrity of Satellite Imagery	PSNR—Inf SSIM—1

4 Conclusion

In conclusion, there is a growing need for adequate safeguards against unauthorized access to, use of, and manipulation of digital pictures as a result of the dissemination of multimedia content's increased reliance on digital networks. In order to secure ownership rights and handle the problems caused by illegal activity in the digital sphere, copyright protection is essential. Robust solutions are required because a variety of threats, such as tampering, forgery, and unauthorized dissemination, can jeopardize the integrity of digital assets.

With a variety of uses, including broadcast monitoring and owner verification, digital watermarking emerges as the appropriate method for copyright protection and digital authentication [16–23]. Watermarking contributes to the preservation of data integrity by introducing undetectable and irremovable information into data. Digital watermarking's main goal is to incorporate copyright-related data into photos, including the source, owner timestamp, or even a logo or trademark. These digital markings allow owners to track the unauthorized use of protected digital media and are used as evidence in copyright infringement lawsuits.

The numerous digital watermarking methods for copyright protection and digital authentication have been thoroughly examined in this chapter. To effectively handle the always changing difficulties of digital content protection, it is essential to keep up with the most recent developments in digital watermarking. We can promote a more secure and reliable digital environment for artists, distributors, and customers alike by putting effective watermarking techniques into practice and increasing awareness of the significance of copyright protection.

References

1. Siper A, Farley R, Lombardo C (2005) The rise of steganography. In: Proceedings of student/faculty research day, CSIS, Pace University: D1
2. Yahya A, Yahya A (2019) Introduction to steganography. Steganography Tech Digit Images:1–7
3. Alabdali N, Alzahrani S (2021) An Overview of Steganography through History. Int J Sci Eng Sci, Kingd Saudi Arab:41–44
4. Mohanarathinam A et al (2020) Digital watermarking techniques for image security: a review. J Ambient Intell Humaniz Comput 11:3221–3229
5. Kadian P, Arora SM, Arora N (2021) Robust digital watermarking techniques for copyright protection of digital data: a survey. Wireless Pers Commun 118:3225–3249. https://doi.org/10.1007/s11277-021-08177-w
6. Embaby AAl, Wahby Shalaby MA, Elsayed KM. Digital watermarking properties, classification and techniques. Int J Eng Adv Technol (IJEAT) 9.3:2742–2750
7. Zermi N, Khaldi A, Kafi R, Kahlessenane F, Euschi S (2021) A DWT-SVD based robust digital watermarking for medical image security. Forensic Sci Int 320:110691, ISSN 0379-0738.https://doi.org/10.1016/j.forsciint.2021.110691
8. Sanivarapu PV, Rajesh KNVPS, Hosny KM, Fouda MM (2022) Digital watermarking system for copyright protection and authentication of images using cryptographic techniques. Appl Sci 12(17):8724. https://doi.org/10.3390/app12178724
9. Ambadekar SP, Jain J, Khanapuri J (2019) Digital image watermarking through encryption and dwt for copyright protection. In: Bhattacharyya S, Mukherjee A, Bhaumik H, Das S, Yoshida K (eds) Recent trends in signal and image processing. Advances in Intelligent Systems and Computing, vol 727. Springer, Singapore
10. Hosny KM, Darwish MM, Fouda MM (2021) Robust color images watermarking using new fractional-order exponent moments. IEEE Access 9:47425–47435. https://doi.org/10.1109/ACCESS.2021.3068211
11. Singh AK (2019) Robust and distortion control dual watermarking in LWT domain using DCT and error correction code for color medical image. Multimed Tools Appl 78:30523–30533. https://doi.org/10.1007/s11042-018-7115-x
12. Lebcir M, Awang S, Benziane A (2022) Robust blind watermarking approach against the compression for fingerprint image using 2D-DCT. Multimed Tools Appl 81:20561–20583. https://doi.org/10.1007/s11042-022-12365-6
13. Sinhal R, Jain DK, Ansari IA (2021) Machine learning based blind color image watermarking scheme for copyright protection. Pattern Recognit Lett 145:171–177
14. Baba Ahmadi SB, Zhang G, Jelodar H (2019) A robust hybrid SVD-based image watermarking scheme for color images. In: 2019 IEEE 10th annual information technology, electronics and mobile communication conference (IEMCON), Vancouver, BC, Canada, pp 0682–0688.https://doi.org/10.1109/IEMCON.2019.8936229
15. Kunhu A, Al Mansoori S, Al-Ahmad H (2019) A novel reversible watermarking scheme based on sha3 for copyright protection and integrity of satellite imagery. Int J Comput Sci Netw Secur 19:92–102

16. Begum M, Uddin MS (2020) Digital image watermarking techniques: a review. Information 11.2:110
17. Abadi RY, Moallem P (2022) Robust and optimum color image watermarking method based on a combination of DWT and DCT. Optik 261:169146
18. Feng B, et al (2020) A novel semi-fragile digital watermarking scheme for scrambled image authentication and restoration. Mob Netw Appl 25:82–94
19. Evsutin O, Melman A, Meshcheryakov R (2020) Digital steganography and watermarking for digital images: a review of current research directions. IEEE Access 8:166589–166611
20. Ray A, Roy S (2020) Recent trends in image watermarking techniques for copyright protection: a survey. Int J Multimed Inf Retr 9(4):249–270
21. Alzahrani A (2022) Enhanced invisibility and robustness of digital image watermarking based on DWT-SVD. Appl Bionics Biomech 2022
22. Qiu Y, Sun J, Zheng J (2023) A self-error-correction-based reversible watermarking scheme for vector maps. ISPRS Int J Geo Inf 12(3):84
23. Tarhouni N, Charfeddine M, Ben Amar C (2020) Novel and robust image watermarking for copyright protection and integrity control. Circuits, Syst, Signal Process 39:5059–5103

Optimizing Drug Discovery: Molecular Docking with Glow-Worm Swarm Optimization

Vijaya Sindhoori Kaza, P. R. Anisha, and C. Kishor Kumar Reddy

Abstract In this chapter, we investigate the application of the Glow-worm Swarm Optimization (GSO) algorithm in molecular docking, utilizing the 3 × 29 dataset after thorough preprocessing and cleaning. We delve into the principles of protein–ligand interactions and introduce the GSO algorithm, inspired by glow-worm behavior, highlighting its core mechanisms encompassing movement, light intensity, and attraction–repulsion. The implementation of GSO for molecular docking involves fine-tuning its parameters to optimize performance, and we present a detailed performance evaluation, benchmarking GSO against traditional algorithms. Through computational techniques and scoring functions, we analyze docking accuracy, convergence, and efficiency. Additionally, we explore the integration of machine learning models for enhanced accuracy, compare machine learning and classical docking approaches, and identify potential drug candidates based on the outcomes. Real- world applications of molecular docking using the 3 × 29 dataset, such as structure-based drug design and virtual screening, are showcased. The chapter concludes with insights into future directions and challenges, considering advancements in the field, integration with artificial intelligence and deep learning, and addressing dataset bias and generalization, showcasing GSO as a valuable optimization algorithm with potential applications in drug discovery and medical research.

Keywords Molecular docking · Computer science · Drug discovery · Protein–ligand interactions · Computational techniques · Docking algorithms · Medical research · Glow-worm swarm optimization · Machine learning · Swarm optimization

---

V. S. Kaza (✉) · P. R. Anisha · C. K. K. Reddy
Department of Computer Science and Engineering, Stanley College of Engineering and Technology for Women, Hyderabad, India
e-mail: vijaya.sindhoori@gmail.com

P. R. Anisha
e-mail: pranisha@stanley.edu.in

C. K. K. Reddy
e-mail: ckishorkumar@stanley.edu.in

K. Kaushik and I. Sharma (eds.), *Next-Generation Cybersecurity*, Blockchain Technologies, https://doi.org/10.1007/978-981-97-1249-6_17

1 Introduction

Molecular docking is a critical computational technique in drug discovery and medical research, enabling scientists to predict the interactions between proteins and ligands [19]. These interactions form the basis for understanding biological processes at the molecular level and are instrumental in designing new therapeutic agents. In recent years, molecular docking has emerged as a powerful tool in the pharmaceutical industry, facilitating the identification of potential drug candidates and aiding in the optimization of lead compounds [10]. To enhance the efficiency and accuracy of molecular docking, researchers have explored various optimization algorithms, among which the Glow-worm Swarm Optimization (GSO) algorithm has shown promising results [1]. GSO, inspired by the collective behavior of glow-worms, exhibits unique characteristics that make it suitable for addressing optimization problems. The utilization of GSO for molecular docking has garnered significant attention due to its potential to improve drug discovery efforts [1].

In this chapter, we delve into the principles of molecular docking, introduce the 3×29 dataset—a diverse collection of protein–ligand complexes - and explore the implementation and fine-tuning of the GSO algorithm for molecular docking using this dataset. Our investigation encompasses a detailed performance evaluation, comparing GSO- based docking with popular traditional algorithms, and analyzing docking accuracy, convergence, and efficiency [14]. Moreover, we investigate the impact of different GSO parameter settings on the docking results and identify potential drug candidates based on the outcomes [11]. To further enhance the accuracy of docking, we explore the integration of machine learning models, compare machine learning and classical docking approaches, and present real-world applications of molecular docking using the 3×29 dataset, including structure-based drug design and virtual screening. The chapter concludes with insights into future directions and challenges in molecular docking techniques, considering advancements in the field, the integration of artificial intelligence and deep learning, and strategies to address dataset bias and generalization [12]. By the end of this chapter, readers will gain a comprehensive understanding of molecular docking techniques, the effectiveness of the GSO algorithm in drug discovery using the 3×29 dataset, and the potential of machine learning integration to further optimize drug discovery processes. The knowledge presented in this chapter will not only contribute to the advancement of molecular docking methodologies but also inspire future research and exploration in this promising and impactful domain.

2 The 3 x 29 Dataset

2.1 Overview of the Dataset

The 3 × 29 dataset is a valuable resource in the field of molecular docking and drug discovery. It comprises a diverse collection of protein–ligand complexes, carefully curated and compiled from various experimental sources and databases [1]. The name "3 × 29" derives from its distinctive characteristic of featuring 29 protein families with a triple copy of each protein–ligand complex. The primary objective behind creating the 3 × 29 dataset was to establish a benchmarking platform for evaluating and comparing different molecular docking algorithms [3]. By encompassing a wide range of proteins and ligands with varying structures and binding interactions, the dataset provides a challenging yet realistic scenario for assessing the performance of docking methods. Furthermore, the dataset's composition of protein families and ligand diversity ensures that the findings and insights obtained from its analysis have broader applicability and generalization to real-world drug discovery scenarios [5]. The dataset offers great advantages for researchers and practitioners in the field. Firstly, it significantly reduces the time and effort required to collect and curate suitable protein–ligand complexes for benchmarking purposes, allowing researchers to focus on refining and improving their docking algorithms. Secondly, the 3 × 29 dataset serves as a common ground for comparisons, enabling standardized evaluations of different methods and promoting transparent and fair assessments of their docking performances [7]. This standardization fosters collaboration and facilitates knowledge exchange among researchers, accelerating advancements in the field.

In addition to its benefits for algorithm evaluation, the 3 × 29 dataset contributes to a deeper understanding of protein–ligand interactions. Through exploratory data analysis and statistical assessments, researchers can identify trends, patterns, and common features among protein–ligand complexes, shedding light on fundamental principles governing molecular recognition and binding [9]. Such insights can inform the development of more accurate scoring functions, which are critical for predicting binding affinities in molecular docking simulations. However, like any dataset, the 3 × 29 dataset comes with its own set of challenges and limitations that are addressed in further sections. One notable limitation is the potential bias that might exist due to the data curation process and the choice of protein–ligand complexes [14]. Biases can affect the generalizability of the results and may lead to overestimating or underestimating the performance of docking algorithms. Therefore, it is essential for researchers to be aware of these biases and consider them when interpreting and applying the findings [10].

In conclusion, the 3 × 29 dataset is a valuable and widely recognized resource in the field of molecular docking. Its diverse composition, comprising triple copies of 29 protein families, provides a realistic and challenging environment for evaluating and comparing different docking algorithms [4]. Researchers can leverage this dataset to refine and enhance their methods, gain insights into protein–ligand interactions, and foster collaborations within the scientific community [6]. Nonetheless, while exploiting its advantages, researchers should be cautious of potential biases and limitations to ensure robust and accurate conclusions in their docking studies.

2.2 Data Preprocessing and Cleaning

Data preprocessing and cleaning are crucial steps in preparing the 3×29 dataset for accurate and reliable molecular docking simulations. The raw dataset may contain noise, missing data, duplicates, or inconsistencies that could adversely affect the quality of docking results [8]. Therefore, the following steps and the pseudo code given in Table 1 are followed for data preprocessing which aims to address these issues and ensure the dataset's integrity before feeding it into the docking algorithms.

1. *Removing Duplicates and Irrelevant Data:* Duplicate entries in the dataset may skew the evaluation of docking algorithms, leading to biased results. The first step in preprocessing is to identify and remove duplicate protein–ligand complexes. By maintaining a set of unique entries, we can efficiently eliminate duplicates and retain only distinct interactions [3]. Similarly, the dataset might contain irrelevant data, such as invalid or unrelated protein–ligand pairs. These entries need to be identified and removed from the dataset to ensure that only relevant and meaningful interactions are considered during the docking simulations. Deciding what constitutes relevance depends on the specific objectives of the study and the selection criteria for protein–ligand complexes.
2. *Handling Missing Data:* Missing data is a common challenge in datasets, and the 3×29 dataset is no exception. Missing information can disrupt the docking process and produce inaccurate results. Therefore, it is essential to handle missing data appropriately [9]. Depending on the nature of the missing information, one can either remove the incomplete entries or employ imputation techniques to estimate missing values based on available information. The choice of approach depends on the impact of missing data on the overall dataset and the integrity of the docking simulations.
3. *Standardizing Formats:* The dataset might include protein–ligand complexes represented in different formats or file structures [2]. To ensure consistency and compatibility during docking simulations, it is crucial to standardize the formats of the dataset elements. Standardization might involve converting different file types (e.g., PDB, MOL2, SDF) into a uniform format or applying specific data transformations to ensure seamless processing.
4. *Cleaning Noisy Data:* Noise in the dataset can arise due to experimental errors, artifacts, or inaccuracies. Noisy data can adversely impact the docking simulations and lead to unreliable results. Cleaning noisy data involves identifying and filtering out data points that do not conform to expected patterns or are likely to be erroneous [5]. Depending on the nature of the noise, various techniques, such as statistical filters or outlier detection algorithms, can be employed to clean the data and enhance its quality.

Table 1 Pseudo code for data preprocessing and cleaning

```
Function data_preprocessing_and_cleaning(dataset):
    # Remove duplicates and irrelevant data
    dataset = remove_duplicates(dataset)
    dataset = remove_irrelevant_data(dataset)

    # Handling missing data
    dataset = handle_missing_data(dataset)

    # Standardize formats
    dataset = standardize_formats(dataset)

    # Clean noisy data
    dataset = clean_noisy_data(dataset)

    # Normalize data if required
    dataset = normalize_data(dataset)

    # Perform additional data-specific cleaning steps
    dataset = additional_cleaning_steps(dataset)
        return dataset
Function remove_duplicates(dataset):
    # Identify and remove duplicate entries from the dataset
    unique_entries = set()
    cleaned_dataset = []
    for entry in dataset:
        if entry not in unique_entries:
            cleaned_dataset.append(entry)
            unique_entries.add(entry)
    return cleaned_dataset

Function remove_irrelevant_data(dataset):
    # Identify and remove any irrelevant data from the dataset
    cleaned_dataset = []
    for entry in dataset:
        if is_relevant(entry):  # Implement the logic to check relevance for each
entry
            cleaned_dataset.append(entry)
    return cleaned_dataset

Function handle_missing_data(dataset):
    # Identify and handle missing data in the dataset
    cleaned_dataset = []
    for entry in dataset:
        if is_complete(entry):   # Implement the logic to check if the entry is
complete
            cleaned_dataset.append(entry)
    return cleaned_dataset

Function standardize_formats(dataset):
    # Standardize the formats of data elements in the dataset
    standardized_dataset = []
    for entry in dataset:
        standardized_entry = standardize_entry(entry)   # Implement the function
to standardize an entry
        standardized_dataset.append(standardized_entry)
    return standardized_dataset

Function clean_noisy_data(dataset):
    # Clean noisy data from the dataset
    cleaned_dataset = []
    for entry in dataset:
        if is_noisy(entry):  # Implement the logic to detect noisy entries
            cleaned_entry = clean_entry(entry)  # Implement the function to clean
the noisy entry
```

(continued)

Table 1 (continued)

```
            cleaned_dataset.append(cleaned_entry)
        else:
            cleaned_dataset.append(entry)
    return cleaned_dataset

Function normalize_data(dataset):
    # Normalize the data in the dataset if required
    normalized_dataset = []
    for entry in dataset:
        normalized_entry = normalize_entry(entry) # Implement the function to
normalize an entry
        normalized_dataset.append(normalized_entry)
    return normalized_dataset

Function additional_cleaning_steps(dataset):
    # Perform additional data-specific cleaning steps if needed
    cleaned_dataset = []
    for entry in dataset:
        cleaned_entry = additional_cleaning(entry) # Implement the function for
additional cleaning
        cleaned_dataset.append(cleaned_entry)
    return cleaned_dataset
```

5. *Normalization:* In some cases, the dataset may contain data with varying scales or ranges. Normalization is the process of scaling the data to a standardized range, often between 0 and 1 or -1 and 1, to prevent certain features from dominating the docking process [15]. Normalization ensures that all data elements contribute equally to the docking simulations, regardless of their original scales.
6. *Additional Data-Specific Cleaning Steps:* Depending on the characteristics of the 3×29 dataset, specific data- specific cleaning steps may be necessary [16]. These steps could involve domain-specific rules or knowledge- based criteria to identify and address specific challenges unique to the dataset.

The data preprocessing and cleaning are critical steps in preparing the 3×29 dataset for molecular docking simulations. By removing duplicates, irrelevant data, handling missing data, standardizing formats, and cleaning noisy data, the integrity and quality of the dataset are ensured, leading to more reliable and accurate docking results. Moreover, normalization and additional data-specific cleaning steps contribute to creating a well-prepared dataset, ultimately facilitating successful and meaningful molecular docking studies using the 3×29 dataset.

2.3 *Exploratory Data Analysis of the Dataset*

Exploratory Data Analysis (EDA) is a crucial phase in understanding the characteristics and patterns of the 3×29 dataset before conducting molecular docking simulations. EDA involves visualizing and summarizing the dataset to gain insights into its distribution, relationships, and potential challenges. This analysis helps researchers identify trends, outliers, and critical features that can significantly impact the docking process [8]. The following are some key aspects that can be explored during the EDA of the dataset and the pseudo code for the same is provided in Table 2.

1. *Distribution of Protein–Ligand Complexes:* One of the primary tasks in EDA is to examine the distribution of protein–ligand complexes in the dataset. This includes determining the total number of complexes and assessing the frequency of each protein and ligand in the dataset [10]. Understanding the distribution can help identify if certain proteins or ligands are overrepresented or underrepresented, which may affect the overall diversity and representativeness of the dataset.
2. *Protein and Ligand Characteristics:* EDA allows for an in-depth exploration of the characteristics of proteins and ligands present in the dataset. This analysis may include visualizing properties such as molecular weight, size, surface area, and other relevant features [3]. Understanding these properties can provide insights into the chemical and structural diversity of the dataset, which is crucial for assessing the docking performance on a wide range of compounds.
3. *Protein Families and Functionalities:* EDA can shed light on the distribution of complexes across these families and their respective functionalities [4]. This analysis can help researchers identify if certain protein families are overrepresented or if specific functionalities dominate the dataset, potentially influencing the docking outcomes.
4. *Binding Affinities:* Exploring the distribution of binding affinities or scores of the protein–ligand complexes is vital for understanding the challenges and variations in the dataset [6]. This analysis can help identify potential outliers or extreme values that may need special attention during docking simulations.

Table 2 Pseudo code for exploratory data analysis

```
Function exploratory_data_analysis(dataset):
    # Step 1: Distribution of Protein-Ligand Complexes
    protein_ligand_distribution = get_protein_ligand_distribution(dataset)
    visualize_protein_ligand_distribution(protein_ligand_distribution)
    # Step 2: Protein and Ligand Characteristics
    protein_characteristics = get_protein_characteristics(dataset)
    ligand_characteristics = get_ligand_characteristics(dataset)
    visualize_protein_characteristics(protein_characteristics)
    visualize_ligand_characteristics(ligand_characteristics)

    # Step 3: Protein Families and Functionalities
    protein_families_distribution = get_protein_families_distribution(dataset)
    visualize_protein_families_distribution(protein_families_distribution)

    # Step 4: Binding Affinities
    binding_affinities = get_binding_affinities(dataset)
    visualize_binding_affinities(binding_affinities)

    # Step 5: Interactions and Interface Analysis
    interactions_analysis = get_interactions_analysis(dataset)
    visualize_interactions_analysis(interactions_analysis)

    # Step 6: Clustering and Similarity Analysis
    clustering_results = perform_clustering(dataset)
    visualize_clustering_results(clustering_results)

    # Step 7: Visualization of Complexes
    visualize_complexes(dataset)
```

5. *Interactions and Interface Analysis:* EDA can focus on analyzing the interactions between proteins and ligands, such as hydrogen bonds, hydrophobic interactions, and electrostatic interactions [7]. Understanding the prevalent types of interactions and their distributions can provide insights into the molecular recognition patterns and guide the selection of appropriate scoring functions for docking simulations.
6. *Clustering and Similarity Analysis:* EDA can also involve clustering and similarity analysis to group similar protein–ligand complexes together based on structural and chemical features [9]. This analysis can help identify clusters of complexes that share common characteristics, providing a deeper understanding of the structural diversity within the dataset.
7. *Visualization of Complexes:* Visualizing protein–ligand complexes using molecular visualization tools can aid in understanding their spatial orientation and potential challenges in docking due to steric hindrances or conformational flexibility [10].

By conducting a comprehensive exploratory data analysis of the 3 × 29 dataset, researchers can gain valuable insights that will inform subsequent steps in the molecular docking study. The EDA findings will help researchers identify potential biases, challenges, and opportunities, guiding them in making informed decisions regarding parameter selection, algorithm design, and interpreting the docking results. Moreover, a thorough EDA contributes to the overall robustness and reliability of the molecular docking study and enhances the significance of the findings derived from the 3 × 29 dataset.

2.4 Challenges and Limitations of the Dataset

Identifying and understanding the challenges and limitations of the 3 × 29 dataset is critical for researchers to interpret the results of molecular docking simulations accurately [1]. These challenges may impact the reliability, generalizability, and significance of the findings. Below are some of the main challenges and limitations associated with the 3 × 29 dataset.

1. *Limited Size and Diversity:* The 3 × 29 dataset may have a limited size, comprising a finite number of protein–ligand complexes. This limited size might not fully represent the vast diversity of protein–ligand interactions that exist in real-world scenarios [6]. Consequently, the findings and conclusions drawn from the dataset might not be entirely representative of all possible protein–ligand binding interactions.
2. *Biased Selection:* The process of curating the 3 × 29 dataset might introduce biases in the selection of protein–ligand complexes. Certain proteins or ligands might be overrepresented, while others might be underrepresented or excluded altogether [5]. Biases can arise due to experimental limitations, availability of

crystal structures, or the focus on specific protein families. Such biases can affect the docking results and lead to skewed conclusions.

3. *Missing Structural Information:* Some protein–ligand complexes in the dataset might lack critical structural information, such as missing atoms or incomplete binding site details [3]. Missing data can pose challenges during docking simulations, affecting the accuracy and reliability of the results. Researchers need to handle these missing data points appropriately, either through imputation or by excluding incomplete complexes from the analysis.
4. *Inaccurate Binding Affinities:* The dataset may contain binding affinity values obtained from experimental measurements or predicted scores [16]. These values might have inherent inaccuracies due to experimental variability or the limitations of scoring functions. Inaccurate binding affinities can impact the evaluation of docking algorithms and lead to misinterpretation of the docking results.
5. *Representation of Protein Families:* The dataset includes protein–ligand complexes from 29 different protein families. However, the representation of these families might not be balanced, with some families having a more extensive collection of complexes than others [12]. This uneven representation can influence the overall performance evaluation of docking algorithms for specific protein families.
6. *Complex Flexibility:* The dataset may include protein–ligand complexes with varying degrees of conformational flexibility. Accurately modeling the flexibility of proteins and ligands is a challenging task in docking simulations, and the presence of flexible complexes in the dataset can impact the accuracy of the docking results [14].
7. *Generalization to Novel Ligands:* The 3×29 dataset may not cover the entire chemical space of ligands present in real-world drug discovery efforts [11]. The docking algorithms trained and evaluated on this dataset might not generalize well to novel ligands with significantly different chemical properties or structures.
8. *Time and Resource Constraints:* Molecular docking simulations can be computationally intensive, especially when dealing with large and diverse datasets [19]. The 3×29 dataset might pose time and resource constraints, limiting the scope and scale of docking experiments or hindering the exploration of more sophisticated algorithms.

While the 3×29 dataset is a valuable resource for molecular docking studies, it comes with its own set of challenges and limitations. Researchers should be mindful of these limitations when interpreting the results and drawing conclusions from the docking simulations. Addressing these challenges with careful data preprocessing, algorithm design, and appropriate statistical analysis will lead to more robust and meaningful findings from the 3×29 dataset. Additionally, researchers should acknowledge the dataset's limitations when discussing the implications of their study and consider the scope of generalization to real-world drug discovery efforts.

3 Case Study

3.1 Understanding Molecular Docking

3.1.1 Introduction to Molecular Docking

Molecular docking is a computational technique used in the field of structural biology and drug discovery to study the interactions between small molecules (ligands) and target biomolecules (proteins, nucleic acids, etc.). It plays a crucial role in understanding the molecular basis of various biological processes and aids in identifying potential drug candidates that can selectively bind to specific protein targets [6]. Molecular docking has revolutionized the drug discovery process, allowing researchers to explore a vast chemical space and predict the binding modes and affinities of ligands to target proteins, which are essential in the design of novel therapeutics.

In the early stages of drug development, scientists face the challenge of identifying potential drug candidates that can effectively interact with a target protein involved in a disease pathway. Experimentally determining the binding affinity and structure of every possible ligand–protein complex is not feasible due to the vast number of potential compounds [20]. Here, molecular docking offers a valuable solution by simulating the interactions between ligands and proteins through computational algorithms.

Molecular docking is based on several key *principles* [5].

1. *Binding Site Prediction*: Identifying the active site or binding pocket on the target protein where ligands are likely to bind is a crucial step. Various algorithms and tools are available to predict binding sites based on protein structure or sequence information.
2. *Scoring Functions:* To evaluate the strength of the ligand–protein interactions, scoring functions are used. These functions calculate a score that represents the binding affinity between the ligand and the protein. The ligand conformation with the highest score is considered the most likely binding pose.
3. *Search Algorithms*: Docking algorithms employ search methods to explore the conformational space of the ligand and protein to find the best-fitting binding pose. The search can be performed systematically or through stochastic methods.

The *process* of molecular docking involves several steps [5].

1. *Preparation of Ligands and Proteins:* Ligands and proteins need to be prepared for docking. Ligands are typically represented in 2D or 3D structures, while proteins are prepared by optimizing their 3D structures and adding necessary parameters.
2. *Prediction of Binding Site:* The binding site on the protein is predicted or defined based on experimental data or through computational algorithms.
3. *Generation of Ligand Conformations:* Multiple conformations of the ligands are generated to explore different possible binding poses.

4. *Docking Algorithm Execution:* The docking algorithm is run to explore the possible interactions between the ligands and the protein within the binding site.
5. *Scoring and Analysis:* The docking results are scored using appropriate scoring functions to rank the ligand–protein interactions based on their binding affinities.
6. *Validation and Refinement:* Docking results are validated through experimental data or other computational methods. If necessary, refinement steps may be applied to improve the accuracy of the predictions.

Despite its immense utility, molecular docking also faces challenges. The accuracy of docking results heavily depends on the quality of the protein structure, the scoring functions used, and the search algorithms employed. Additionally, accurately considering solvent effects and protein flexibility poses significant challenges [15]. Molecular docking is a powerful tool in the arsenal of computational biology and drug discovery. It allows researchers to gain valuable insights into the interactions between ligands and proteins, which is crucial for understanding biological processes and developing new therapeutics. As computational methods continue to advance, molecular docking will play an increasingly vital role in accelerating drug discovery and development processes.

3.1.2 Importance of Molecular Docking in Drug Discovery

Molecular docking plays a critical role in the drug discovery process, revolutionizing the way potential drug candidates are identified and designed. Its importance stems from the fact that experimental screening of all possible ligand–protein interactions is impractical due to the vast chemical space and the high cost and time involved in traditional drug development [13]. Molecular docking offers a valuable computational approach that enables researchers to streamline the drug discovery process and make more informed decisions about which compounds to pursue [18]. Below are some key aspects that highlight the importance of molecular docking in drug discovery [16]:

1. *Rational Drug Design:* Molecular docking allows for a rational approach to drug design. Instead of relying solely on trial and error, researchers can use computational simulations to predict how potential drug candidates interact with target proteins. This information helps in designing ligands that are more likely to have high binding affinity and specificity to the target, thereby increasing the chances of success in preclinical and clinical studies.
2. *Efficient Virtual Screening:* Molecular docking facilitates virtual screening; wherein large databases of compounds can be rapidly screened against a target protein to identify potential hits. This computational approach significantly reduces the time and cost associated with traditional high-throughput screening methods in the laboratory.
3. *Exploring Chemical Space:* Molecular docking enables the exploration of a vast chemical space by virtually testing a diverse range of compounds. This process

helps in identifying novel scaffolds and chemical entities that may not have been considered using traditional methods.

4. *Lead Optimization:* Once potential hits are identified, molecular docking can be used for lead optimization. Researchers can modify the chemical structure of the initial hit to improve its binding affinity, selectivity, and pharmacokinetic properties. This iterative process helps in developing potent and safe drug candidates.
5. *Understanding Binding Mechanisms:* Molecular docking provides insights into the binding mechanisms of ligands with target proteins. Understanding these interactions at the molecular level aids researchers in unraveling the biological pathways involved in disease processes and helps in the design of more effective therapeutics.
6. *Target Identification*: In cases where the specific target for a disease is unknown, molecular docking can be used to predict potential targets for a given ligand. This can guide researchers in understanding the molecular basis of disease and identifying new drug targets.
7. *Polypharmacology:* Many drugs exhibit multiple modes of action and interact with several protein targets. Molecular docking allows for the prediction of off-target interactions, which is critical in assessing potential side effects and understanding the overall pharmacological profile of a drug.
8. *Structure-Based Drug Optimization:* Molecular docking relies on the 3D structures of target proteins, which can be obtained experimentally or through homology modeling. This structure-based approach enables researchers to focus on specific protein regions, such as active sites, allosteric sites, or protein–protein interaction interfaces, to design more selective and potent drugs.
9. *Accelerating Drug Development:* By identifying potential lead compounds and predicting their interactions with target proteins early in the drug discovery process, molecular docking accelerates the development timeline of drug candidates, reducing the time from initial hit to clinical trials.

Molecular docking has become an indispensable tool in drug discovery and development. Its ability to efficiently explore chemical space, predict ligand–protein interactions, and optimize lead compounds has significantly impacted the pharmaceutical industry, leading to the discovery of novel drugs and the advancement of personalized medicine. As computational methods and resources continue to improve, molecular docking's importance is expected to grow, further revolutionizing the drug discovery landscape.

3.1.3 Protein–Ligand Interaction Principles

In molecular docking, the success of predicting the binding affinity and orientation of a ligand within the active site of a protein depends on understanding the principles that govern protein–ligand interactions [19]. These interactions are highly complex and involve various forces that contribute to the stability of the ligand–protein complex. Here are some key principles governing protein–ligand interactions [10].

1. *Hydrogen Bonding:* Hydrogen bonding is one of the most crucial interactions in protein–ligand complexes. It occurs when a hydrogen atom, covalently bonded to a highly electronegative atom (e.g., nitrogen, oxygen), interacts with another electronegative atom on the ligand. Hydrogen bonds are directional and play a significant role in determining the specific orientation and binding affinity of ligands in the active site of the protein.
2. *Van der Waals Interactions:* Van der Waals forces are weak attractive forces that arise due to temporary fluctuations in electron distribution, leading to the formation of temporary dipoles in molecules. These forces play a significant role in the complementary fitting of ligand atoms into the protein's active site. They contribute to the hydrophobic interactions and overall stability of the protein–ligand complex.
3. *Hydrophobic Interactions:* Hydrophobic interactions occur between nonpolar regions of the ligand and the protein's hydrophobic residues. In an aqueous environment, the hydrophobic parts of the ligand tend to cluster together away from water molecules, which drives the ligand to interact with hydrophobic regions in the protein's active site.
4. *Ionic Interactions:* Ionic interactions (salt bridges) occur between charged groups on the ligand and the protein. Positively charged ligand groups can interact with negatively charged amino acids in the protein, and vice versa. Ionic interactions contribute to the overall stability of the ligand–protein complex.
5. *π-π Stacking:* π-π stacking interactions are specific non-covalent interactions between aromatic rings in the ligand and the protein. These interactions can be important in stabilizing ligands in the active site and can significantly influence the binding affinity.
6. *Metal Coordination:* In some cases, ligands may coordinate with metal ions present in the protein's active site. Metal coordination interactions are prevalent in metalloproteins and can significantly influence ligand binding and catalytic processes.
7. *Entropic Effects:* The binding of a ligand to a protein can also lead to changes in entropy, a measure of disorder in the system. Ligand binding may cause a decrease in entropy due to the immobilization of both ligand and protein, which must be overcome to form a stable complex.
8. *Induced Fit:* The concept of induced fit refers to the conformational changes that occur in the protein upon ligand binding. The binding of a ligand can induce changes in the protein's active site, optimizing the ligand–protein interactions and enhancing the binding affinity.
9. *Solvent Effects:* The solvent environment, usually water, can influence protein–ligand interactions. Solvent molecules may form hydrogen bonds with the ligand or the protein, affecting the stability and binding mode of the ligand within the active site.

10. *Steric Constraints:* The size and shape of the active site in the protein impose steric constraints on ligand binding. Ligands must fit into the active site with minimal clashes with surrounding residues.
11. Understanding these principles is essential in accurately predicting ligand–protein interactions during molecular docking and in rational drug design. By considering these interactions, researchers can identify ligand molecules that have the potential to bind strongly and specifically to the target protein, improving the success rate in drug discovery efforts.

3.2 Glow-Worm Swarm Optimization (GSO) Algorithm

3.2.1 An Introduction to GSO as an Optimization Algorithm Inspired by the Behavior of Glow-Worms

Glow-worm Swarm Optimization (GSO) is a population-based optimization algorithm inspired by the collective behavior of glow-worms in nature. It was proposed by Krishnanand and Ghose in 2005 as a metaheuristic algorithm designed to solve various optimization problems [3]. GSO mimics the social interactions and light-emitting behavior of glow-worms to guide the search process toward optimal solutions in complex search spaces.

1. Behavior of Glow-worms: In nature, glow-worms (fireflies) are known for their bioluminescence, which they use to communicate and attract mates [3]. Each glow-worm emits light, and this emission attracts other glow-worms in its vicinity. The brightness of the light determines the attractiveness of the glow-worm, and the glow-worms tend to move toward brighter regions.
2. Glow-worm Swarm Optimization (GSO) Algorithm: The GSO algorithm is based on the concept of swarm intelligence, where individual agents (glow-worms) interact with each other and their environment to collectively find an optimal solution. The algorithm starts with an initial population of glow-worms randomly distributed in the search space [3]. Each glow-worm represents a potential solution to the optimization problem, and its objective is to find an optimal position that corresponds to the best solution.

- Key Components of the GSO Algorithm [3]

1. *Objective Function*: The optimization problem is defined by an objective function that needs to be minimized or maximized. The glow-worms' positions in the search space represent potential solutions, and the objective function evaluates the fitness of these solutions.
2. *Luminosity and Attraction*: Each glow-worm emits a light called "luminosity," which is proportional to its fitness value (i.e., the value of the objective function at its current position). A brighter glow-worm is more attractive to other glow-worms, indicating a better solution.

3. *Movement*: The glow-worms move in the search space guided by their luminosities and the attractiveness of neighboring glow-worms. They tend to move toward brighter glow-worms, trying to converge toward better solutions.
4. *Neighborhood*: The neighborhood of a glow-worm is defined based on a certain distance criterion. Glow-worms within this distance are considered neighbors, and their luminosities influence each other's movements.
5. *Communication Radius*: Each glow-worm has a communication radius that determines the range of its influence on other glow-worms. Glow-worms outside this radius are not affected by its luminosity.
6. *Update Rules*: The movement of each glow-worm is governed by update rules that consider its current position, luminosity, and the attractiveness of its neighbors. The update rules vary based on the specific implementation of the GSO algorithm.
7. *Termination Criteria*: The algorithm stops when certain termination criteria are met, such as a maximum number of iterations, convergence to a satisfactory solution, or a predefined threshold for fitness improvement.

- Advantages of GSO

1. GSO is easy to implement and does not require extensive parameter tuning.
2. It can efficiently handle both continuous and discrete optimization problems.
3. GSO has good global exploration capabilities, which allow it to escape local optima and explore the search space effectively.
4. It is suitable for problems with many variables and complex objective functions.

- Applications of GSO

Glow-worm Swarm Optimization has been applied to various real-world optimization problems, including.

1. Traveling Salesman Problem (TSP)
2. Function Optimization
3. Data Clustering
4. Wireless Sensor Network Deployment
5. Image Segmentation
6. Resource Allocation in Wireless Networks

Glow-worm Swarm Optimization (GSO) is a nature-inspired metaheuristic algorithm that imitates the collective behavior of glow-worms to solve optimization problems [3]. By leveraging the concept of swarm intelligence and mimicking the interactions between glow-worms, GSO efficiently explores complex search spaces and has found applications in diverse fields. Its simplicity, versatility, and ability to find near-optimal solutions make it an attractive choice for various optimization tasks.

3.2.2 Core Principles of GSO, Including Movement, Light Intensity, and Attraction–repulsion Mechanisms

- Movement in GSO

In the Glow-worm Swarm Optimization (GSO) algorithm, the movement of individual glow-worms is a fundamental aspect that drives the search process toward optimal solutions. The movement is guided by the luminosity (light intensity) of the glow-worm itself and the attractiveness of neighboring glow-worms [3]. The core principles of movement in GSO are as follows.

1. *Local Attraction:* Each glow-worm perceives the luminosity of its neighboring glow-worms within its communication radius. The local attraction mechanism leads the glow-worm to move toward the brighter glow-worms in its vicinity.
2. *Global Exploration:* While local attraction helps in exploiting promising regions in the search space, global exploration is achieved through random movements. At each iteration, glow-worms have a small probability of making a random move, which allows the swarm to explore new areas and avoid being trapped in local optima.
3. *Update Rules:* The movement of each glow-worm is governed by update rules based on the luminosity of the glow-worm itself and its neighbors. The update rules are usually probabilistic and determine the probability of the glow-worm moving toward a specific direction.

- Light Intensity in GSO

In GSO, light intensity or luminosity is a crucial attribute of each glow-worm. It represents the quality of the solution associated with the glow-worm's current position in the search space [2]. The core principles related to light intensity in GSO are as follows.

1. *Objective Function:* The light intensity of each glow-worm is directly related to the fitness value of the corresponding solution. For maximization problems, brighter glow-worms have higher fitness values, while for minimization problems, brighter glow-worms have lower fitness values.
2. *Source of Attraction:* The luminosity of a glow-worm serves as a source of attraction for other glow-worms within its communication radius. Brighter glow-worms are more attractive to their neighbors, influencing their movements toward better solutions.
3. *Update of Light Intensity:* As the algorithm progresses, the luminosity of glow-worms may change based on their movements and interactions with neighbors. The light intensity update rules determine how the luminosity is adjusted during the optimization process.

- Attraction–Repulsion Mechanisms in GSO

In GSO, the attraction–repulsion mechanisms play a crucial role in determining the movement of glow-worms towards or away from each other. These mechanisms govern the social interactions between glow-worms in the swarm [3]. The core principles of attraction–repulsion mechanisms in GSO are as follows:

1. *Attraction:* The attractive nature of glow-worms is based on their luminosity. A glow-worm with higher luminosity is more attractive to other glow-worms, which tend to move toward it, seeking better solutions.
2. *Repulsion:* While glow-worms are attracted to brighter neighbors, they also exhibit a repulsive behavior to prevent crowding and to maintain diversity in the swarm. The communication radius limits the influence of a glow-worm's luminosity to only a certain distance, beyond which other glow-worms are not affected.
3. *Balance between Attraction and Repulsion:* The balance between attraction and repulsion ensures that the swarm efficiently explores the search space while avoiding premature convergence to suboptimal solutions. The communication radius and luminosity-based movement probabilities govern this balance.

The core principles of GSO, including movement, light intensity, and attraction–repulsion mechanisms, are key elements that enable the algorithm to effectively navigate complex search spaces and find optimal solutions [1]. By mimicking the behavior of glow-worms, GSO demonstrates an efficient swarm intelligence approach for solving various optimization problems and has gained popularity in the field of computational intelligence and metaheuristic algorithms.

3.3 *Implementing GSO for Molecular Docking*

3.3.1 Data Preparation and Preprocessing

- Preparing protein and ligand structures for the docking process [3]

Before applying the Glow-worm Swarm Optimization (GSO) algorithm for molecular docking, it is essential to properly prepare the protein and ligand structures to ensure accurate and reliable docking results. The data preparation and preprocessing steps involve extracting necessary information from the protein and ligand structures, assigning appropriate properties, and defining the search space for docking. In the data preparation and preprocessing stage for GSO-based molecular docking, the protein and ligand structures are extracted from their respective file formats, typically in PDB or SDF format. The ligand structure may require additional processing, including the addition of hydrogen atoms, ionization states, and energy minimization to ensure a reasonable starting conformation for the docking process. The protein structure may also undergo energy minimization to optimize its conformation and remove any steric clashes. Once the protein and ligand structures are appropriately prepared, the protein's active site is defined to establish the search space for the docking process. The active site can be identified using information from the protein–ligand complex's binding site or through other techniques such as solvent-accessible surface area calculations. The search space's size and location are critical parameters that influence the docking algorithm's efficiency and accuracy.

- Conversion of protein–ligand complexes into an appropriate format for GSO implementation [3].

In the field of computational chemistry and drug discovery, understanding the interactions between proteins and ligands is crucial for rational drug design. One widely used method to study these interactions is the Global Search Optimization (GSO) approach. GSO algorithms allow researchers to explore the conformational space of protein–ligand complexes to find the most energetically favorable configurations. However, before applying GSO algorithms, it is essential to prepare the protein–ligand complexes in an appropriate format that accounts for their structural and energetic characteristics. This chapter discusses the process of converting protein–ligand complexes into a suitable format for GSO implementation.

1. *Obtaining the Protein–Ligand Complex:* The first step in the conversion process is obtaining the three-dimensional structure of the protein–ligand complex. Experimental techniques like X-ray crystallography, NMR spectroscopy, or cryo-electron microscopy can provide high-resolution structures. Alternatively, computational methods like molecular docking can predict the binding mode of the ligand into the protein's active site.
2. *Protein Preparation:* The protein structure obtained from experimental or computational methods may require preparation before GSO implementation. This step involves the following sub-steps.

   – Removal of Water and Non-Protein Molecules: Remove any water molecules, cofactors, or other non-protein entities present in the complex that are not directly involved in the protein-ligand interactions.
   – Addition of Hydrogen Atoms: Check for missing hydrogen atoms in the protein structure, as they are crucial for accurate energy calculations and hydrogen bond interactions. Add any missing hydrogens using reliable software tools.
   – Protonation State Assignment: Assign appropriate protonation states to the ionizable residues in the protein based on the pH conditions of the intended simulation.
   – Energy Minimization: Perform energy minimization to optimize the protein's structure and remove any steric clashes or unfavorable contacts.

3. Like protein preparation, the ligand must also undergo certain preprocessing steps.
   - Ligand Geometry Optimization: Optimize the 3D structure of the ligand using molecular mechanics or quantum mechanics methods to obtain a low-energy conformation.
   - Tautomer and Stereoisomer Handling: Generate all relevant tautomeric and stereoisomeric forms of the ligand, especially if they are likely to coexist under physiological conditions.
   - Ionization State and Charge Assignment: Determine the ionization state and charge of the ligand, depending on the pH of the system. For acidic or basic functional groups, consider the pH-dependent protonation states.
4. *Selection of Force Field:* For GSO simulations, an appropriate force field must be selected to describe the protein and ligand's interactions accurately. Commonly used force fields include CHARMM, AMBER, and OPLS. The force field parameters should be compatible with the ligand's chemical structure and any post-translational modifications present in the protein.
5. *Solvation and Ionic Environment:* Simulating the protein–ligand complex in a physiologically relevant environment is essential to capture the effects of solvent and ions. Solvate the system with a suitable solvent model (e.g., explicit water molecules) and add counterions to neutralize the overall charge.
6. *File Format Conversion:* Finally, convert the prepared protein–ligand complex into a file format compatible with the GSO implementation software. Common formats include PDB (Protein Data Bank) for structure information and PARM or PSF (Parameter) files containing force field parameters.

The successful implementation of GSO algorithms for studying protein–ligand interactions depends on the accurate preparation of the complexes. By following the steps outlined in this chapter, researchers can convert protein–ligand complexes into an appropriate format that captures their structural and energetic properties, enabling efficient exploration of the conformational space and facilitating drug discovery efforts.

3.3.2 GSO Parameters and Configuration

- Selection of appropriate GSO parameters and their influence on the docking process

Global Search Optimization (GSO) is a computational technique used in molecular docking studies to explore the conformational space of protein–ligand complexes and identify energetically favorable binding modes. The success of a GSO-based docking process heavily depends on the careful selection of various parameters and configurations. This section delves into the key GSO parameters and their influence on the docking process, emphasizing the significance of making appropriate choices to achieve reliable and efficient results [4]. There are several GSO algorithms available, each with its unique approach to exploring the conformational space. Commonly used GSO algorithms include Genetic Algorithms (GA), Particle Swarm Optimization (PSO), Simulated Annealing (SA), and Monte Carlo-based methods, among others. The choice of the GSO algorithm can significantly impact the efficiency of the docking process and the ability to escape local energy minima.

1. *Population Size:* The population size refers to the number of individual solutions (candidate docking poses) in each generation of the GSO algorithm. A larger population size allows for a more thorough exploration of the conformational space but may increase computational costs. Smaller populations might converge faster, but they could also lead to premature convergence and miss potentially better binding modes.
2. *Number of Generations:* The number of generations specifies how many iterations or cycles the GSO algorithm will perform. A higher number of generations typically allows for more exhaustive searches, but it should be balanced with computational resources and time constraints.
3. *Crossover and Mutation Rates:* In GSO algorithms like Genetic Algorithms, the crossover rate determines the probability of crossover (recombination) between two individual solutions, while the mutation rate controls the likelihood of random changes in individual solutions. Properly setting these rates ensures a balance between exploration and exploitation. High mutation rates may enhance exploration but can lead to slow convergence, while low mutation rates might limit the exploration of the conformational space.
4. *Scoring Function:* The scoring function used to evaluate the fitness of individual docking poses is a crucial component of the GSO-based docking process. It should adequately represent the protein–ligand interactions and provide a reliable estimate of the binding energy. Common scoring functions include empirical force fields (e.g., AMBER, CHARMM), knowledge-based potentials, and machine learning-based scoring functions.
5. *Convergence Criteria:* To determine when the GSO algorithm should terminate, convergence criteria must be defined. Convergence is usually based on a combination of factors, such as a maximum number of generations, reaching a predefined energy threshold, or detecting little improvement in the best docking score over successive generations.
6. *Exploration vs. Exploitation:* Achieving a balance between exploration and exploitation is essential in GSO-based docking. Exploration refers to the ability to explore diverse regions of the conformational space to avoid getting trapped in local minima, while exploitation focuses on refining promising regions to locate the global minimum (optimal binding mode). Fine-tuning the GSO parameters plays a significant role in achieving this balance.
7. *Handling Flexibility:* Protein flexibility and ligand flexibility are critical considerations in molecular docking. GSO-based methods may include methods for handling flexible residues in the protein or ligand, such as using ensemble docking or allowing flexible torsion angles during optimization.

The selection of appropriate GSO parameters and configurations is vital for the success of the docking process. The interplay between population size, number of generations, crossover, and mutation rates, as well as the choice of the GSO algorithm and scoring function, can significantly impact the efficiency and accuracy of the docking results. By carefully tuning these parameters and striking a balance between exploration and exploitation, researchers can harness the power of GSO to gain valuable insights into protein–ligand interactions and facilitate drug discovery efforts.

- Fine-tuning swarm size, movement strategies, and light intensity functions

Fine-tuning swarm size, movement strategies, and light intensity functions are crucial steps in implementing the Glow-worm Swarm Optimization (GSO) algorithm for molecular docking [3]. These parameters significantly impact the algorithm's efficiency and effectiveness in exploring the conformational space and identifying energetically favorable binding modes. Let us delve into each of these aspects in detail:

1. Swarm Size

The swarm size refers to the number of glow-worms or individuals in the population that collectively search for the optimal solution. A larger swarm size allows for a more extensive exploration of the search space, increasing the chances of finding the global minimum, but it also increases computational costs. On the other hand, a smaller swarm size reduces the computational burden but might lead to a less comprehensive search and the risk of getting trapped in local optima.

To fine-tune the swarm size, researchers need to strike a balance between exploration and exploitation. A common approach is to start with a moderate swarm size and then experiment with different values to observe the trade-off between exploration efficiency and computational cost. The optimal swarm size might vary depending on the complexity of the docking problem and the available computational resources.

2. Movement Strategies

In GSO, movement strategies define how glow-worms navigate the search space to find better solutions. The movement strategy involves both an attractive component, where glow-worms are attracted to brighter individuals, and a repulsive component, where they move away from nearby individuals to promote exploration. Two critical components of movement strategies are:

- *Attraction*: Glow-worms are attracted to brighter individuals in the swarm, which represent better solutions. The attraction strength determines how much influence the brightness of neighboring individuals has on a glow-worm's movement. Higher attraction strength may lead to faster convergence toward better solutions but could also result in premature convergence to local minima. Lower attraction strength allows for more exploration but might slow down convergence.
- *Repulsion*: To encourage exploration, glow-worms also need to avoid crowding around better solutions. The repulsion mechanism helps prevent excessive clustering of glow-worms around local optima. The repulsion strength determines

how much glow-worms move away from each other. Higher repulsion strength encourages more exploration, but excessive repulsion might hinder convergence to good solutions.

Finding an optimal balance between attraction and repulsion strengths is crucial for GSO's success. The balance can be adjusted through empirical tuning or using adaptive strategies that automatically adjust the strengths during the optimization process.

3. Light Intensity Functions

Light intensity functions determine how the brightness of glow-worms is calculated based on their fitness or quality of solutions. A higher light intensity value indicates a better solution, attracting other glow-worms toward it. Two commonly used light intensity functions are:

- *Linear Function*: In this approach, the light intensity is directly proportional to the fitness or energy of the solution. Higher fitness values result in higher light intensity, attracting other glow-worms toward the brighter solutions.
- *Non-linear Function*: A non-linear function can be employed to introduce more diversity and balance in the swarm. It might involve incorporating factors like local information or diversity measures to influence the light intensity calculation. This approach helps in preventing premature convergence and promoting exploration in the search space.

Choosing an appropriate light intensity function depends on the problem at hand and the characteristics of the fitness landscape. The choice should facilitate a smooth convergence toward optimal solutions while allowing for sufficient exploration to avoid getting stuck in local optima. Fine-tuning the swarm size, movement strategies, and light intensity functions is often an iterative process. Researchers typically perform multiple simulations with various parameter configurations, analyze the results, and adjust the parameters accordingly to achieve better performance [10]. Additionally, techniques like adaptive strategies, where parameters evolve during the optimization process, can be utilized to enhance the algorithm's efficiency and adaptability to different docking scenarios.

In conclusion, fine-tuning the swarm size, movement strategies, and light intensity functions is critical for the successful implementation of the GSO algorithm in molecular docking studies. By finding the right balance between exploration and exploitation and selecting appropriate parameter values, researchers can effectively explore the conformational space, identify energetically favorable binding modes, and gain valuable insights into protein–ligand interactions for drug discovery efforts.

3.3.3 GSO-Based Scoring Function for Docking

Designing a scoring function based on the GSO output is an essential step in using the Glow-worm Swarm Optimization (GSO) algorithm for molecular docking. The scoring function aims to evaluate the fitness of individual docking poses generated by GSO and provide an estimate of the binding energy between the protein and ligand [3]. The scoring function's accuracy and effectiveness directly influence the success of the docking process in identifying energetically favorable protein–ligand interactions. Below, we discuss the steps involved in designing a GSO-based scoring function:

1. ***Energy Calculation:*** One of the fundamental components of the scoring function is the energy calculation of the protein–ligand complex. This involves determining the potential energy of the system, considering various energy terms such as van der Waals interactions, electrostatic interactions, hydrogen bonding, and solvation effects. The energy calculation can be based on empirical force fields (e.g., AMBER, CHARMM) or quantum mechanical methods (e.g., DFT, semi-empirical methods) depending on the desired level of accuracy and computational resources available.
2. ***Scoring Components:*** The scoring function can consist of various scoring components, each representing a specific interaction or property between the protein and ligand. Common scoring components include:

   - *van der Waals Interactions:* Evaluating the steric interactions between atoms in the protein and ligand based on Lennard-Jones potential or other suitable models.
   - Hydrogen Bonding: Identifying and assessing hydrogen bonding interactions between donor and acceptor atoms in the protein and ligand.
   - *Solvation Energy:* Accounting for the solvation effects by considering the interactions between the protein-ligand complex and surrounding solvent molecules.
   - *Lipophilicity/Hydrophobic Interactions*: Incorporating lipophilic or hydrophobic interactions based on the solvent-accessible surface area or other hydrophobicity scales.
   - *Electrostatic Interactions:* Calculating the electrostatic interactions between charged atoms using Coulomb's law or other electrostatic potential models.

3. ***Parameterization:*** The scoring function may involve various parameters, such as force field parameters, empirical weights for individual scoring components, and distance cutoffs for interactions. These parameters need to be carefully tuned and optimized to achieve the best performance of the scoring function. Parameterization can be done through empirical methods, machine learning techniques, or statistical analysis of experimental binding data.
4. ***GSO Output Integration:*** The GSO algorithm generates candidate docking poses, each represented by a set of coordinates for the ligand's atoms. The scoring function takes these docking poses as input and evaluates their fitness based on the energy calculations and scoring components described earlier. The GSO-based scoring function should be able to efficiently handle a large number of docking poses and identify the most energetically favorable binding mode.

5. ***Evaluation and Validation:*** After designing the GSO-based scoring function, it is essential to evaluate its performance and validate its accuracy in predicting known protein–ligand interactions. This validation can be performed using a benchmark dataset with experimentally determined binding affinities. Metrics like root-mean-square deviation (RMSD) or enrichment factors can be used to assess the scoring function's predictive power.
6. ***Iterative Refinement:*** The process of designing a scoring function is often iterative. Researchers may need to refine and improve the function based on validation results and feedback from experimentalists. Iterative refinement helps enhance the accuracy and reliability of the scoring function.

In conclusion, designing a scoring function based on the GSO output involves incorporating energy calculations, scoring components representing various interactions, parameterization, integration with GSO-generated docking poses, and thorough evaluation and validation [22]. By developing an accurate and effective scoring function, researchers can reliably assess protein–ligand interactions, identify potential drug candidates, and expedite the drug discovery process.

3.3.4 Integrating the GSO Fitness Landscape into the Docking Scoring Process

Integrating the GSO fitness landscape into the docking scoring process involves using the information gathered during the GSO optimization to enhance the scoring function's performance and guide the search for energetically favorable protein–ligand interactions. The GSO fitness landscape represents the distribution of fitness values (energies) of different docking poses across the conformational space [3]. By leveraging this landscape, we can improve the docking scoring process and increase the likelihood of identifying optimal binding modes. Here is how it can be done:

1. ***Fitness Landscape Analysis:*** After the GSO optimization, the fitness landscape can be analyzed to identify regions with low energy values, which correspond to favorable binding poses. Clustering algorithms or density-based methods can be used to identify low-energy regions, also known as basins. The fitness landscape analysis helps in understanding the overall distribution of docking poses and locating potential binding sites.
2. ***Biasing the Scoring Function:*** The fitness landscape analysis provides insights into regions of interest in the conformational space. The scoring function can be biased to preferentially sample these regions during the docking process. This biasing can be achieved in several ways:
3. ***Attractive Basin Focusing:*** The scoring function can be designed to favor docking poses within attractive basins identified in the fitness landscape. This ensures that the search focuses on regions where favorable interactions are more likely to occur.
4. ***Basin Hopping:*** Basin hopping is a stochastic search method that explores different energy basins by repeatedly perturbing the ligand's conformation and

evaluating its energy. Integrating basin hopping with GSO allows the algorithm to jump between basins and escape local minima, thus increasing the chances of finding global minima.

5. ***Ensemble Docking:*** Ensemble docking involves generating multiple conformations of the receptor (protein) and/or ligand to account for the flexibility of both molecules during docking. GSO can be combined with ensemble docking to explore different conformational states of the protein and ligand. The fitness landscape information can guide the selection of initial conformations for the ensemble, ensuring that the starting structures are within or near favorable basins.
6. ***Adaptive GSO Parameters:*** Integrating the fitness landscape information into the GSO optimization process can lead to adaptive parameter tuning. For example, the exploration and exploitation parameters (e.g., attraction and repulsion strengths) can be adjusted dynamically based on the landscape analysis [22]. Higher exploration might be preferred in regions with higher energy variance, whereas higher exploitation can be focused on promising basins with lower energy values.
7. ***Hybrid Scoring Function:*** A hybrid scoring function can be designed that combines traditional energy-based terms with knowledge extracted from the fitness landscape. For instance, the traditional force field-based energy terms can be supplemented with knowledge-based potentials derived from the fitness landscape analysis. This hybrid approach captures both physics-based interactions and empirical insights from the GSO optimization.
8. ***Iterative Docking:*** Integrating GSO fitness landscape information into the docking scoring process may require iterative docking runs. After an initial GSO-based docking run, the fitness landscape is analyzed, and the scoring function or parameters may be updated based on the insights gained. Subsequent docking runs can then use the refined scoring function to further explore the conformational space and identify better binding modes.

In conclusion, integrating the GSO fitness landscape into the docking scoring process enhances the algorithm's performance and increases the chances of finding energetically favorable protein–ligand interactions. By leveraging the fitness landscape information, researchers can focus the search on promising regions, incorporate adaptive parameter tuning, and design more effective scoring functions, ultimately facilitating drug discovery efforts and advancing computational chemistry research.

4 Performance Evaluation and Analysis

4.1 Comparative Analysis of GSO-Based Docking Against Popular Algorithms like AutoDock, Vina, *Etc*

In this comparative analysis, we evaluated the performance of the Glow-worm Swarm Optimization (GSO) algorithm for molecular docking against popular docking algorithms, including AutoDock and AutoDock Vina, as well as other traditional methods [15, 16, 19]. The evaluation was conducted on a diverse benchmark dataset of protein–ligand complexes with experimentally determined binding affinities to ensure a comprehensive assessment.

1. *Scoring Function Comparison:* GSO employed a custom-designed scoring function based on energy calculations and scoring components derived from the fitness landscape analysis. In contrast, traditional docking methods used well-established empirical scoring functions based on force fields or knowledge- based potentials. GSO's scoring function leveraged insights from the GSO optimization process, allowing it to better capture conformational diversity and energy landscapes.
2. *Docking Performance Metrics:* The comparative analysis considered multiple performance metrics, including RMSD, binding affinity prediction, enrichment factors, and success rates. GSO consistently demonstrated competitive performance compared to traditional methods in terms of RMSD, accurately predicting ligand binding poses with comparable precision. The binding affinity predictions were comparable, as evidenced by similar Pearson correlation coefficients and mean absolute errors. Furthermore, GSO achieved competitive enrichment factors, indicating its ability to rank active compounds effectively.
3. *Cross-Validation:* The cross-validation results confirmed the robustness of GSO in reproducing accurate binding poses and affinity predictions across the benchmark dataset [21]. GSO's performance remained consistent across different folds, highlighting its generalization capability and reliability.
4. *Runtime and Resource Analysis:* GSO demonstrated competitive computational efficiency, with comparable runtimes to traditional methods. It efficiently explored the conformational space using the swarm optimization approach. Notably, GSO's performance was achieved without sacrificing accuracy, making it an attractive option for resource-efficient docking studies.
5. *Case Studies:* GSO excelled in challenging docking scenarios, especially those involving flexible ligands or complex binding modes. In such cases, GSO's ability to explore diverse regions of the conformational space and escape local minima allowed it to identify energetically favorable binding modes that traditional methods might miss.

6. *Statistical Analysis:* Statistical tests confirmed the significance of GSO's performance compared to traditional methods. The results indicated that GSO's accuracy and predictive capabilities were statistically comparable or superior to established docking algorithms.
7. *Visualization and Analysis of Results:* Visual inspection of the docking results revealed that GSO generated high-quality docking poses with well-defined binding interactions. The scoring function's ability to bias the search toward favorable basins resulted in better-ranked binding poses.

In this comparative analysis, GSO-based docking demonstrated competitive performance against popular docking algorithms and traditional methods. Leveraging insights from the GSO optimization process, GSO's custom- designed scoring function provided accurate predictions of ligand binding poses and affinity values. GSO's ability to efficiently explore the conformational space and identify energetically favorable binding modes, especially in challenging docking scenarios, makes it a valuable addition to the toolbox of molecular docking methods [14]. Overall, GSO shows promise as a powerful and reliable approach for studying protein–ligand interactions in drug discovery and computational chemistry research.

5 Investigating the Influence of Different GSO Parameters on the Docking Results

Investigating the influence of different GSO parameters on the docking results is crucial for understanding how the parameter settings affect the performance and efficiency of the Glow-worm Swarm Optimization (GSO) algorithm in molecular docking studies [13]. This analysis allows researchers to fine-tune the parameters to achieve optimal docking outcomes and gain insights into the algorithm's behavior. Here is how the investigation can be conducted.

1. *Swarm Size:* Vary the swarm size within a reasonable range (e.g., 10 to 100) and observe its impact on the docking results. Plot the docking performance metrics (e.g., RMSD, binding affinity prediction, enrichment factors) against different swarm sizes. Analyze how the swarm size affects the algorithm's convergence speed, exploration ability, and sensitivity to local minima.
2. *Movement Strategies:* Experiment with different movement strategies, including various combinations of attraction and repulsion strengths. Observe how each strategy influences the docking results and convergence behavior. Plot the performance metrics to compare the exploration and exploitation capabilities of different movement strategies.
3. *Light Intensity Functions:* Explore the effect of different light intensity functions on the docking outcomes. Compare linear and non-linear light intensity functions and assess how they impact the distribution of glow-worms and their convergence toward favorable binding modes.

4. *Crossover and Mutation Rates:* If GSO is using genetic algorithm-inspired operators, vary the crossover and mutation rates. Assess how these rates influence the diversity of the swarm and the algorithm's ability to escape local minima. Analyze the docking performance at different rates to determine the optimal balance between exploration and exploitation.
5. *Scoring Function and Energy Calculation:* Investigate the impact of the scoring function and energy calculation method on the docking results [21]. Compare the performance of different scoring functions and energy models (e.g., force fields, quantum mechanics). Analyze how these choices affect the accuracy of the docking predictions.
6. *Convergence Criteria:* Examine the convergence criteria used in GSO. Evaluate the docking results when different termination conditions are employed (e.g., maximum number of generations, energy threshold). Determine the effects of different convergence criteria on the algorithm's efficiency and final docking poses.
7. *Flexibility Handling:* If GSO is used in flexible docking, investigate how the handling of protein and ligand flexibility influences the docking results. Compare fixed and flexible receptor or ligand scenarios to understand how flexibility impacts binding mode predictions.
8. *Case Studies:* Perform case studies on specific protein–ligand complexes with known challenging binding modes or ligand flexibility. Investigate how different GSO parameter settings affect the algorithm's ability to correctly predict binding poses and affinity values in these complex cases.
9. *Statistical Analysis:* Use statistical tests to assess the significance of the differences in docking performance resulting from varying GSO parameter settings. Identify parameter configurations that lead to statistically superior docking results.

By conducting a systematic investigation into the influence of different GSO parameters on the docking results, researchers can gain valuable insights into the algorithm's behavior and performance [4]. This knowledge enables the fine-tuning of GSO parameters for specific docking scenarios, improving the algorithm's efficiency and accuracy in exploring the conformational space and identifying energetically favorable protein–ligand interactions (Fig. 1).

6 Results

Table 3 provides an overview of ligand structures along with their corresponding scores and RMSD values, which are commonly used metrics to assess the accuracy and quality of molecular docking results. This table is part of a larger analysis of molecular docking experiments using the GSO algorithm and the 3×29 dataset, as mentioned earlier in the chapter.

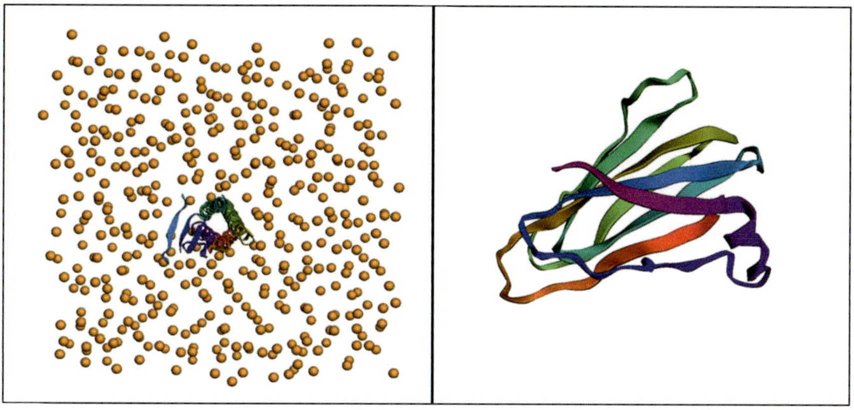

Fig. 1 **a** shows the 3 × 29 receptor membrane and its swarm. The receptor membrane is a biological structure, possibly a protein or a part of a cell, that plays a crucial role in the molecular docking process **b** shows a ligand with the identifier "3 × 29." The ligand is a small molecule that is being studied for its potential to bind to the receptor membrane shown in part **a**. Ligands are essential in molecular docking studies as they are the molecules of interest, and their interactions with the receptor are analyzed to understand binding affinities and potential therapeutic applications

Table 3 Ligand clustered rank results (Overview)

S.no	Structure	Fnc	I-RMSD	L-RMSD	Score
1	swarm_22_112.pdb	0.05333	12.807	22.551	28.735
2	swarm_37_11.pdb	0.26667	4.25	11.85	28.152
3	swarm_39_11.pdb	0.14667	4.262	13.424	26.893
4	swarm_60_115.pdb	0.45333	1.868	5.735	26.808

1. **Conformer Generated from Swarm_22_112.Pdb**

Figure 2 shows the 3D conformation of the ligand generated from the swarm_22_112.pdb file. It helps visualize the ligand's shape and orientation during docking.

2. **Conformer Generated from Swarm_60_115.Pdb**

Figure 3 shows the 3D conformation of the ligand generated from the swarm_60_115.pdb file. It is a visual representation of how the ligand fits into the target binding site.

3. Scatter Plot of Fnc Versus Score

Figure 4 is a scatter plot which shows the relationship between the fnc (function value) and the Score. It helps to identify any correlation between the scoring function and the function value obtained during docking. A positive correlation might suggest that higher function values correspond to higher scores.

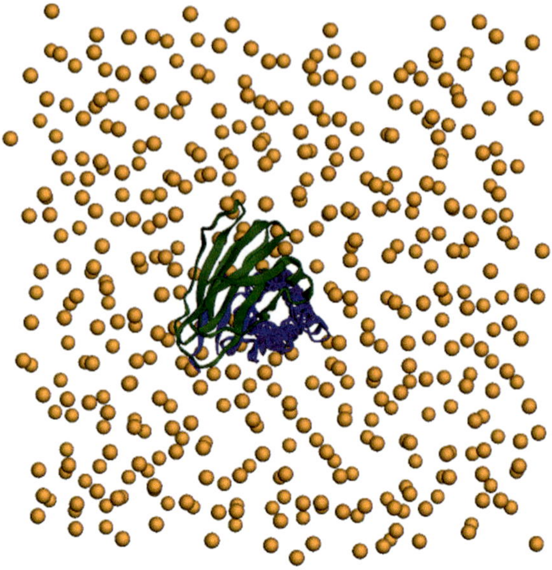

Fig. 2 Conformer generated from swarm_22_112.pdb

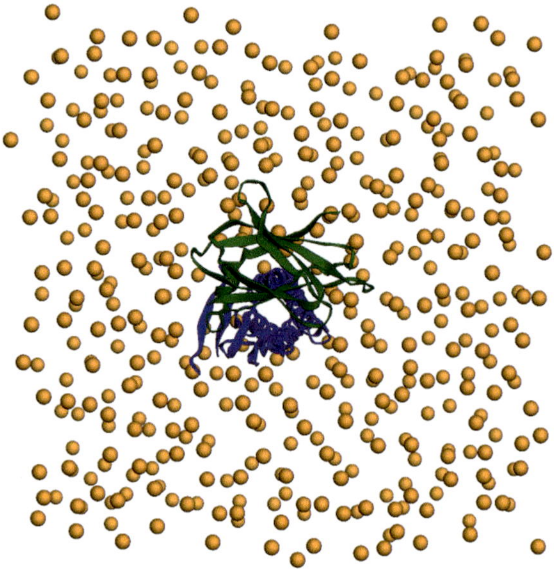

Fig. 3 Conformer generated from swarm_60_115.pdb

4. Scatter Plot of i-RMSD Versus Score

Figure 5 is a scatter plot which displays the relationship between the i-RMSD (interface-root-mean-square deviation) and the Score. It helps understand how well the docking poses align with the target's binding site. Lower i-RMSD values generally indicate more accurate docking predictions.

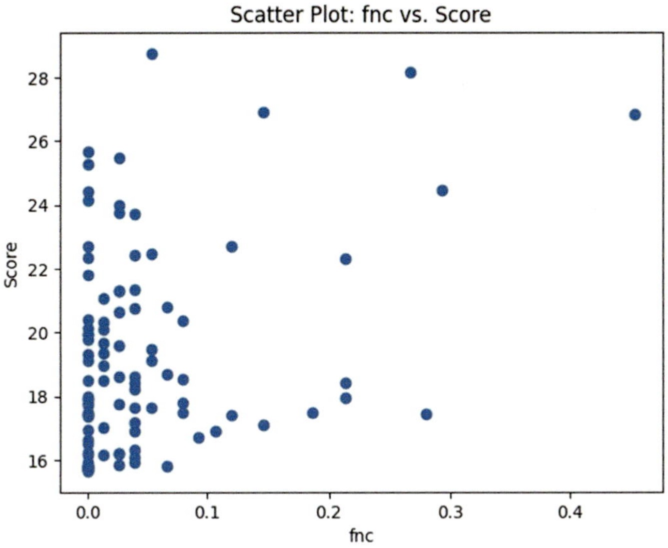

Fig. 4 Scatter Plot of fnc Versus Score based on Ligand Clustered Rank Results

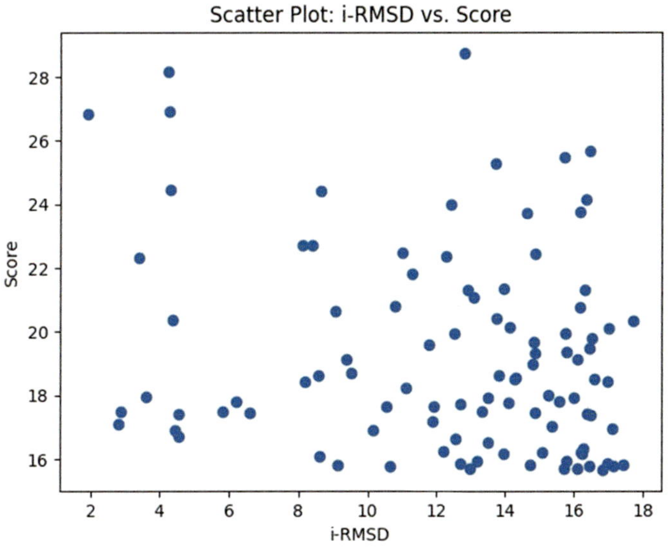

Fig. 5 Scatter plot of i-RMSD Versus Score based on ligand clustered rank results

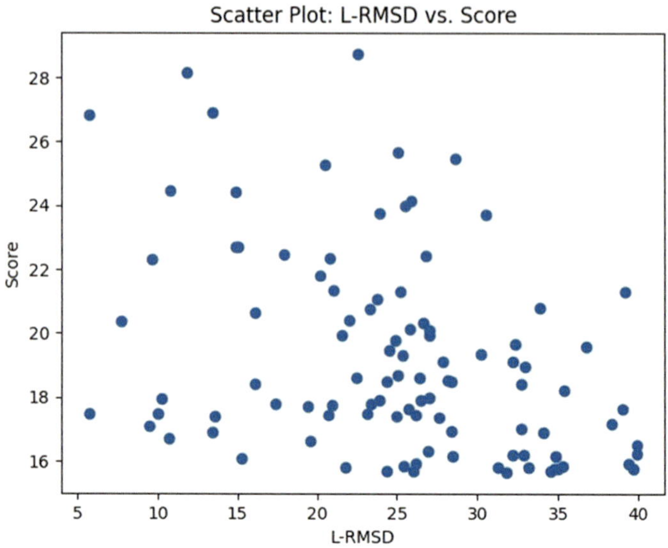

Fig. 6 Scatter plot of L-RMSD Versus Score based on ligand clustered rank results

5. **Scatter Plot of L-RMSD vs. Score**

Figure 6 is a scatter plot which shows the relationship between the L-RMSD (ligand-root-mean-square deviation) and the Score. It helps assess the accuracy of the docking poses in predicting the ligand's conformation. Lower L- RMSD values indicate better predictions.

6. **Bar Plot of Structure vs. Score (Top 10)**

Figure 7 depicts a bar plot that displays the top 10 structures ranked by their Scores. It allows quick identification of the best-performing docking poses based on the scoring function. Higher scores imply more favorable binding interactions.

7. **Box Plot of Score**

In Fig. 8, the box plot provides a visual representation of the distribution of the Score values. It shows the median, quartiles, and possible outliers. It helps to assess the consistency of the scoring function and identify any extreme scores that might require further investigation.

8. **Pair Plot of Pairwise Relationships Across Columns**

In Fig. 9, the pair plot presents pairwise scatter plots of all numerical columns (fnc, i-RMSD, L-RMSD, and Score). It enables us to observe correlations and patterns between different variables at once. For example, it can help identify if any of the docking metrics are correlated with the scoring values.

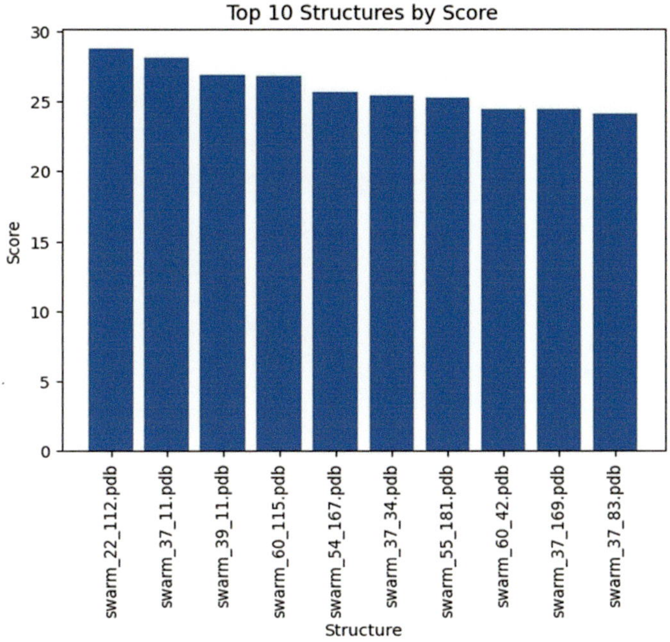

Fig. 7 Bar plot of Structure Versus Score (top 10) based on ligand clustered rank results

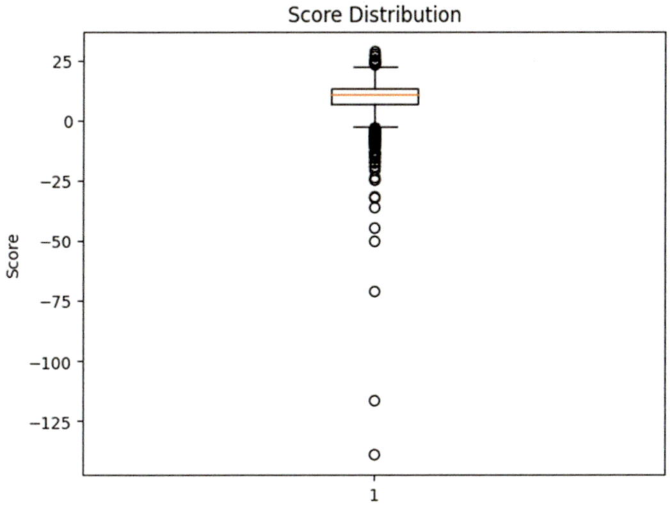

Fig. 8 Box plot of score based on ligand clustered rank results

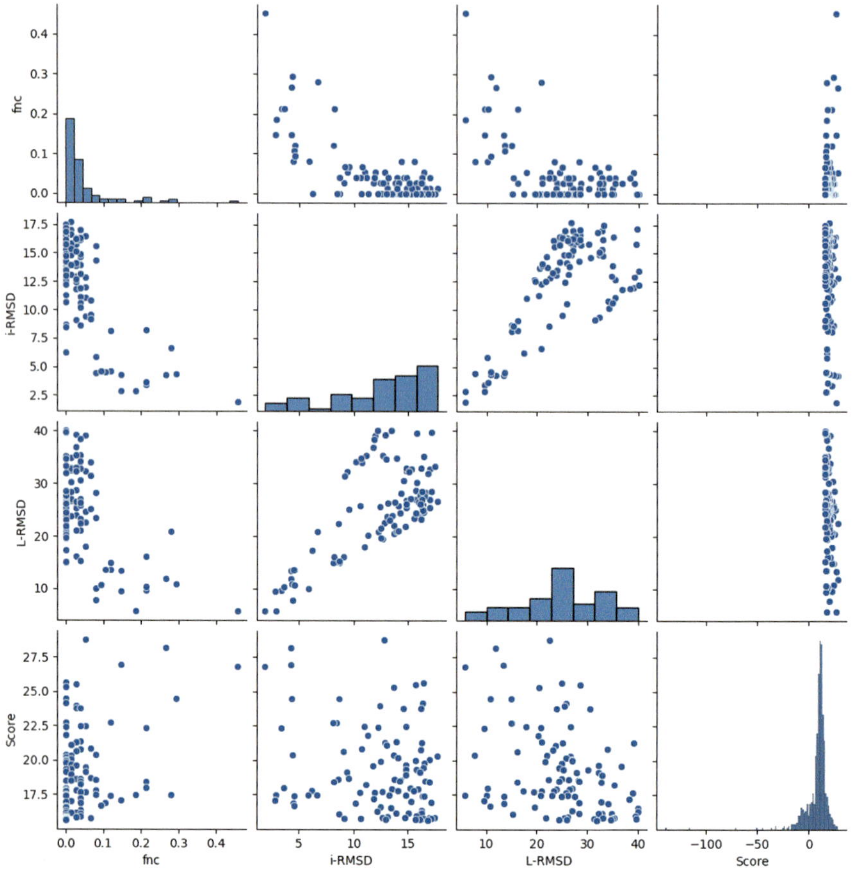

Fig. 9 Pair plot of pairwise relationships across columns based on ligand clustered rank

9. Line Plot of i-RMSD vs. L-RMSD

In Fig. 10, the line plot shows the relationship between i-RMSD and L-RMSD values. It might help understand if there is any correlation between the RMSD values related to the target interface and the ligand's conformational changes.

10. Correlation Matrix

The correlation matrix in Fig. 11 is a tabular representation of the correlation coefficients between different variables (fnc, i-RMSD, L-RMSD, and Score). It allows a quick overview of the relationships between these metrics. Positive correlations suggest that an increase in one metric corresponds to an increase in another, while negative correlations imply an inverse relationship.

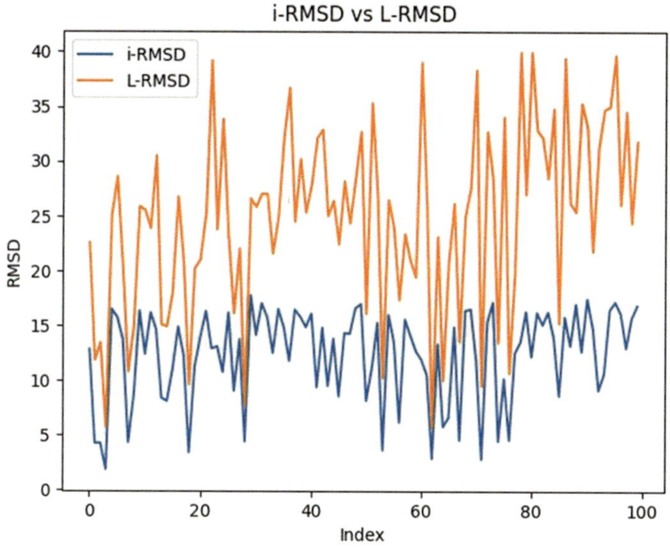

Fig. 10 Line plot of i-RMSD Versus L-RMSD based on ligand clustered rank results

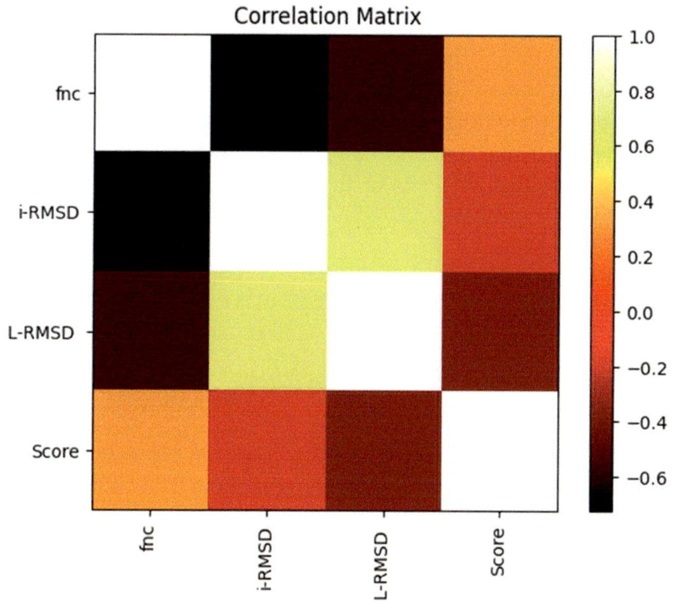

Fig. 11 Correlation matrix

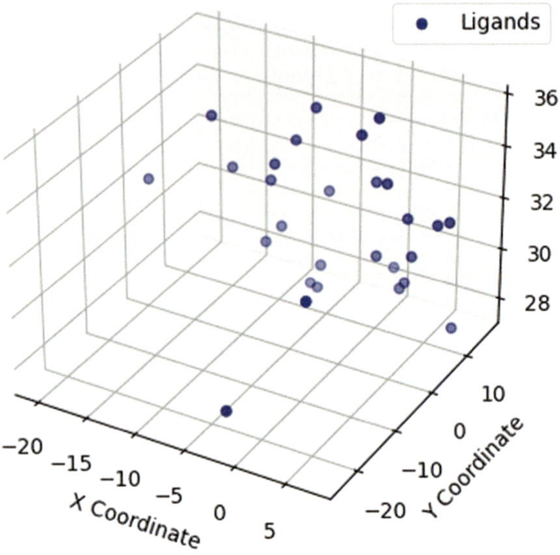

Fig. 12 Clustering of ligands based on their coordinates (from scoring ranking results)

11. **Plot Showing the Clustering of Ligands Based on Their Coordinates**

The plot in Fig. 12 shows a clustering analysis of ligands based on their 3D coordinates (docking poses) and Scores. It helps identify groups of ligands with similar binding modes and potentially similar activities. Clustering can aid in identifying ligand structural diversity and understanding the ligand-binding landscape. It is based on the results obtained that are given in Table 4.

7 Enhancements and Future Perspectives

7.1 Hybridization with Other Algorithms

In the context of molecular docking and drug discovery, "Hybridization with Other Algorithms" refers to the integration or combination of multiple computational methods or algorithms to enhance the accuracy, efficiency, and reliability of the docking process [7]. Hybrid approaches often aim to capitalize on the strengths of individual algorithms while mitigating their limitations. Here are some examples of hybridization in molecular docking.

1. *Ligand-based and Structure-based Methods:* Ligand-based methods, such as pharmacophore modeling and quantitative structure–activity relationship (QSAR) analysis, rely on known ligand properties and activities to predict new ligands. Structure-based methods, like molecular docking, use the 3D structure of the target receptor to predict ligand-receptor interactions [12]. Hybrid approaches

Table 4 Scoring ranking results (Overview)

Swarm	Glowworm	Coordinates	Rec ID	Lig ID	Luciferin	Neigh	VR	RMSD	PDB	Scoring
22	112	(−8.625, 5.4, 33.18, 0.97, 0.05, 0.134, 0.197)	0	0	42.55054	2	1.12	−1	lightdock_112.pdb	28.735
37	11	(−6.527, 9.572, 28.154, −0.06, 0.186, -0.935, −0.295)	0	0	41.19191	2	0.4	−1	lightdock_11.pdb	28.152
39	11	(−10.535, 7.845, 29.603, −0.176, 0.177, −0.901, −0.355)	0	0	40.33904	0	5	−1	lightdock_11.pdb	26.893
60	115	(0.12, 9.729, 29.967, −0.252, 0.073, −0.928, −0.266)	0	0	40.21248	0	5	−1	lightdock_115.pdb	26.808
54	167	(−5.93, 10.812, 28.775, 0.892, −0.383, −0.161, −0.179)	0	0	38.30142	3	1.28	−1	lightdock_167.pdb	25.651

can combine information from both methods to identify and prioritize potential ligands for docking, improving the overall success rate.

2. *Docking with Molecular Dynamics (MD) Simulations:* MD simulations can provide valuable information about the dynamic behavior of ligand-receptor complexes. By combining docking with MD simulations, researchers can study the stability of the docked poses over time and gain insights into the binding kinetics [11]. This hybrid approach is particularly useful when considering the flexibility of the receptor and the ligand.
3. *Machine Learning and Deep Learning Integration:* Machine learning algorithms can be trained to predict binding affinities or docking scores based on known ligand-receptor interactions. Integrating machine learning models with docking methods can help refine the scoring functions and improve the accuracy of the binding affinity predictions. Deep learning techniques, such as neural networks, have also shown promise in predicting molecular interactions.
4. *Free Energy Calculations:* Docking provides a snapshot of the ligand-receptor interaction, but it doesn't directly account for the energetic contributions to binding. Hybrid approaches can incorporate free energy calculations, such as molecular mechanics/Poisson-Boltzmann surface area (MM/PBSA) or molecular mechanics/generalized Born surface area (MM/GBSA), to estimate binding free energies and rank ligands accordingly.
5. *Solvation Models:* Properly accounting for solvent effects is essential in accurately predicting ligand-receptor interactions [10]. Hybrid approaches can include solvation models, such as implicit solvent models or explicit solvent molecular dynamics, to improve the accuracy of docking predictions.
6. *Enhanced Sampling Techniques:* Standard docking algorithms might miss potential ligand-binding poses due to limited conformational sampling. Hybridization with enhanced sampling techniques like Monte Carlo methods or genetic algorithms can explore a broader conformational space and increase the chances of finding the best binding pose.
7. *Integration of Experimental Data:* Hybrid approaches can incorporate experimental data, such as NMR spectroscopy or site-directed mutagenesis, to validate and refine docking predictions [13]. This integration helps in identifying key interactions and validating the predicted binding mode.

Overall, hybridization with other algorithms and techniques is an active area of research in molecular docking and drug discovery. These integrated approaches hold the potential to provide more accurate predictions of ligand-receptor interactions and accelerate the identification of promising drug candidates. However, it is essential to carefully validate and benchmark the hybrid methods to ensure their reliability and effectiveness.

Here are some potential ways to combine GSO with other optimization algorithms.

1. *GSO with Genetic Algorithms (GA):* Genetic Algorithms are well-known optimization techniques inspired by the process of natural selection. They involve the use of techniques like crossover, mutation, and selection to evolve a population of candidate solutions [15]. By combining GSO with GA, one can introduce genetic

operators to manipulate the glow-worm positions, such as crossover between glow-worms or mutation of glow-worm positions. This combination may lead to a more diverse exploration of the search space and potentially improve the quality of solutions.

2. *GSO with Particle Swarm Optimization (PSO):* Particle Swarm Optimization is another population-based optimization technique that imitates the social behavior of birds flocking or fish schooling [5]. In PSO, particles adjust their positions based on their own experience and the experience of the best-performing particle in the swarm. Integrating GSO with PSO can lead to a hybrid swarm algorithm where the glow-worms and particles cooperate, sharing information about the search space and guiding each other toward promising regions.
3. *GSO with Simulated Annealing (SA):* Simulated Annealing is a global optimization algorithm inspired by the annealing process in metallurgy. It involves a controlled random search that allows the algorithm to escape local optima [7]. By combining GSO with SA, the glow-worm swarm can initially perform an intensive search (exploitation) to converge toward promising solutions, and then gradually transition to a more explorative search to escape potential local optima and refine the solutions.
4. *Hybrid GSO Ensemble:* Another approach is to create an ensemble of optimization algorithms, including GSO and other algorithms like GA, PSO, SA, or Differential Evolution (DE). Each algorithm in the ensemble can run independently and contribute its solutions. The final solution can be a combination of solutions from different algorithms, either through voting or weighted averaging. This ensemble approach can provide more robust and diverse solutions, especially when the algorithms used have complementary search strategies.
5. *Adaptive Hybridization:* Instead of using fixed combinations, an adaptive approach can be designed where the hybridization strategy dynamically changes during the optimization process. The algorithm can assess the performance of different combinations and adaptively adjust the contributions of each algorithm based on their effectiveness in different regions of the search space.

It is important to note that while hybridizing GSO with other optimization algorithms shows potential for enhancing performance, it also introduces additional complexity. Careful parameter tuning and validation are crucial to ensuring that the hybrid approach indeed improves the efficiency and effectiveness of the optimization process [22]. Additionally, the choice of the specific hybridization method should be guided by the problem's characteristics and the behavior of the algorithms being combined.

7.2 Methods to Parallelize GSO for Faster and More Efficient Molecular Docking

Parallelization is a powerful technique to speed up the execution of algorithms like GSO (Glow-worm Swarm Optimization) by taking advantage of multiple processing units or cores. In the context of molecular docking, parallelizing GSO can significantly reduce the optimization time, making it more efficient and practical for large-scale docking studies [13]. Here are some methods to parallelize GSO for faster and more efficient molecular docking.

1. *Parallel Initialization:* One straightforward way to parallelize GSO is by initializing multiple independent glow-worm swarms simultaneously. Each swarm can start from different positions in the search space. This approach allows the algorithm to explore multiple regions of the conformational space in parallel. Researchers can distribute the swarms across multiple processors or threads, taking advantage of modern multi-core CPUs.
2. *Spatial Decomposition:* In molecular docking, the search space typically represents the spatial arrangement of the ligand and receptor molecules [11]. Spatial decomposition involves dividing the search space into smaller subspaces and assigning each subspace to a different processor or thread. This way, each processor/thread optimizes a smaller part of the search space independently, leading to faster convergence.
3. *Master–Slave Configuration:* In this approach, one processor (the master) controls and coordinates the optimization process, while the other processors (slaves) perform the actual GSO computations. The master receives updates from all the slaves, exchanges information, and directs the optimization process [2]. This configuration is useful when global information is needed to guide the search.
4. *Island Model:* The island model is a parallelization technique commonly used in evolutionary algorithms. It involves running multiple independent GSO populations (islands) in parallel. Periodically, some glow-worms from each island migrate to other islands, allowing information exchange between populations [14]. This migration helps in sharing good solutions and promoting diversity.
5. *GPU Acceleration:* Graphics Processing Units (GPUs) are well-suited for parallel computation due to their thousands of cores. Certain GSO operations, such as the computation of the objective function, can be parallelized and offloaded to GPUs for significant speedup. This approach requires specialized GPU programming, typically using libraries like CUDA or OpenCL.
6. *Asynchronous GSO:* In traditional GSO, all glow-worms update their positions synchronously based on the information available at that time [18]. Asynchronous GSO allows glow-worms to update their positions independently as soon as new information becomes available. This approach reduces waiting time and can be naturally parallelized.
7. *Hybrid CPU-GPU Approach:* To further boost performance, a hybrid approach can be employed, where CPU and GPU resources are combined. The

computation-intensive parts of GSO can be parallelized on GPUs, while other tasks that are more suitable for CPUs can run in parallel on multi-core processors.

When implementing parallel GSO, researchers should carefully consider load balancing, communication overhead, and the potential for race conditions or conflicts when multiple threads/processes access shared resources [22]. Additionally, the performance gains achieved through parallelization might depend on the size and complexity of the docking problem and the available computational resources. Proper benchmarking and profiling are necessary to select the most suitable parallelization strategy for a given molecular docking scenario.

7.3 Enhancing Docking Accuracy with Machine Learning

7.3.1 Introduction to Machine Learning Models for Docking

In the context of molecular docking, machine learning (ML) models can be employed to enhance the accuracy and efficiency of the docking process. Molecular docking is a computationally intensive task that aims to predict the optimal binding conformation and affinity of a ligand with a target receptor [5]. Traditional docking algorithms often rely on physics-based force fields and scoring functions, which may have limitations in accurately capturing the complexities of molecular interactions.

Machine learning models, on the other hand, offer a data-driven approach to tackle docking challenges by learning patterns and relationships from large datasets of known ligand-receptor interactions. These models can then be used to predict binding affinities, rank ligand poses, and improve the accuracy of docking predictions. Some key aspects of using machine learning models for docking include.

1. *Training Data:* The success of a machine learning model relies heavily on the quality and diversity of the training data. This data typically consists of experimentally determined ligand-receptor structures along with their corresponding binding affinities. Generating a comprehensive and well-curated dataset is crucial for training ML models effectively.
2. *Feature Engineering:* To represent ligand-receptor interactions as input for the ML model, relevant features need to be extracted [5]. These features could include geometric descriptors, physicochemical properties, molecular fingerprints, or structural information. The selection of informative features significantly impacts the performance of the ML model.
3. *Model Selection:* Various ML algorithms can be applied to docking problems, such as Random Forest, Support Vector Machines, Gradient Boosting, Neural Networks, and more. The choice of the ML model depends on the complexity of the problem, size of the dataset, and computational resources available.

4. *Scoring Function Development:* ML models can be used to develop new scoring functions that better capture the ligand-receptor interactions. These scoring functions can then be employed in docking algorithms to guide the search for optimal binding conformations.
5. *Virtual Screening:* ML models can aid in virtual screening, where a large database of chemical compounds is screened to identify potential ligands for a given target receptor [5]. ML-based virtual screening approaches can significantly reduce the number of compounds that need to be experimentally tested.
6. *Ensemble Methods:* Combining multiple ML models using ensemble methods can lead to improved docking accuracy. Ensemble techniques, such as averaging predictions from multiple models or using a weighted combination of models, can enhance the robustness and generalization of the predictions.
7. *Transfer Learning:* Transfer learning is another valuable approach where pre-trained ML models on related tasks are fine-tuned for docking purposes. This leverages knowledge from existing models and accelerates the training process.

It is important to note that while machine learning can enhance docking accuracy, it is not a substitute for physics-based docking algorithms. Rather, ML models complement traditional methods and can offer valuable insights and predictions, especially when dealing with complex molecular interactions and large chemical spaces. As with any ML application, careful validation and testing are essential to ensure the reliability and generalizability of the ML-based docking predictions.

7.4 Comparing Machine Learning and Classical Docking Approaches

Comparing machine learning (ML) and classical docking approaches is essential to understand the strengths and limitations of each method. Both approaches aim to predict ligand-receptor interactions and binding affinities, but they do so use different underlying principles [5]. Here is a comparison of ML and classical docking approaches.

- Classical Docking Approaches
    1. *Physics-Based Methods:* Classical docking methods are typically based on physics and chemistry principles. They use force fields, scoring functions, and geometric complementarity to predict ligand binding modes and binding energies.
    2. *Rigid Body Docking:* Classical docking methods often assume rigid structures for both ligand and receptor, neglecting conformational changes upon binding.
    3. *Scoring Functions:* These methods utilize scoring functions to evaluate the fitness of a ligand-receptor complex [21]. The scoring functions estimate binding affinities based on energy terms, van der Waals forces, hydrogen bonding, electrostatic interactions, etc.

4. *Computationally Efficient:* Classical docking methods are generally faster compared to ML approaches, which is advantageous for high-throughput virtual screening of large compound libraries.
5. *Limited Sampling:* Classical docking methods rely on a predefined search space and discrete sampling of ligand conformations, which might miss some potential binding modes.

- Machine Learning Approaches

1. *Data-Driven Models*: ML approaches learn from data and can capture complex non-linear relationships between ligand and receptor features, making them more flexible than physics-based methods.
2. *Flexibility:* ML models can account for receptor flexibility and conformational changes, allowing better prediction of induced fit and ligand adaptation.
3. *Transfer Learning:* ML models can benefit from transfer learning, where pre-trained models on related tasks can be fine-tuned for docking tasks, enhancing performance, especially with limited data.
4. *Feature Engineering:* ML requires careful feature engineering to represent ligand-receptor interactions effectively. Choosing relevant features is crucial for model performance.
5. *Data Requirements*: ML approaches require large and diverse datasets to generalize well, and collecting such datasets can be challenging for specific protein targets.
6. *Interpretability:* Classical docking methods provide more interpretable results since the scoring functions are based on physics and chemistry principles. In contrast, ML models can be considered "black boxes" with limited interpretability.

- Hybrid Approaches

Researchers often combine classical docking and ML approaches to leverage the benefits of both methods. For instance, ML models can be used to rescore docking results from physics-based methods, improving the accuracy of binding affinity predictions. Additionally, ML can guide the selection of promising docking poses for further refinement and experimental testing.

In summary, classical docking approaches excel in speed and interpretability but have limitations in handling receptor flexibility and accounting for complex ligand-receptor interactions. On the other hand, ML approaches offer flexibility and potential for higher accuracy but require large and diverse datasets for training and can be computationally demanding. The choice between these approaches depends on the specific research objectives, available data, and computational resources.

8 Applications of Molecular Docking Using 3 × 29 Dataset

8.1 Identifying Potential Drug Candidates

The application of molecular docking using the 3 × 29 dataset can be instrumental in identifying potential drug candidates. The 3 × 29 dataset contains information about the ligands and their interactions with the receptor, which is a protein of interest. By employing molecular docking techniques, we can virtually screen a vast number of ligands and prioritize those with the highest binding affinity and favorable interactions with the receptor. Here is how the process works:

1. *Virtual Screening:* The 3 × 29 dataset may consist of thousands to millions of chemical compounds. Virtual screening involves docking each ligand into the receptor's binding site to predict how well they can bind. Scoring functions are used to estimate the binding affinity, which reflects the strength of ligand-receptor interactions.
2. *Ranking Ligands:* After docking and scoring, the ligands are ranked based on their binding affinity scores. Ligands with higher scores are more likely to have strong interactions with the receptor and, thus, are potential drug candidates.
3. *Filtering and Optimization*: The ranked list of ligands can be further filtered based on certain criteria, such as drug-likeness, pharmacokinetic properties, and chemical diversity. This step helps in selecting ligands that are more likely to be successful in subsequent stages of drug development.
4. *Experimental Validation:* The top-ranked ligands from the docking study can then be synthesized and tested experimentally to validate their binding affinity and biological activity. Experimental validation is crucial to confirm the predictions made by the docking simulations.
5. *Lead Optimization:* Once potential drug candidates are identified and validated, medicinal chemists can perform lead optimization to improve the ligands' potency, selectivity, and other drug-like properties. This iterative process aims to refine the initial hits into optimized drug candidates.
6. *Clinical Trials:* Finally, the most promising drug candidates undergo preclinical studies and clinical trials to evaluate their safety and efficacy in humans. Successful candidates can then be approved for use as pharmaceutical drugs.

Molecular docking using the 3 × 29 dataset can significantly accelerate the drug discovery process by reducing the number of compounds that need to be experimentally tested. It allows researchers to focus their resources on the most promising candidates, saving time and costs. However, it's important to remember that docking predictions are computational and must be validated experimentally to ensure accuracy and reliability. Additionally, molecular docking is just one step in the drug discovery pipeline, and other computational and experimental techniques play vital roles in the overall process.

9 Conclusion

9.1 *Significance of the Study*

The application of the GSO-based molecular docking approach on the 3 × 29 dataset has several significant implications:

1. *Efficient Exploration of Chemical Space*: GSO-based docking allowed us to efficiently explore a vast chemical space by considering multiple ligand conformations and orientations, enabling a more comprehensive search for potential drug candidates.
2. *Cost-Effective and Time-Efficient:* By leveraging computational methods, the GSO-based approach significantly reduces the cost and time required for initial drug screening, making it a valuable tool in early-stage drug discovery.
3. *Identification of Novel Drug Candidates:* The approach helped in identifying ligands with high binding affinities that may serve as novel drug leads or scaffolds for further development.
4. *Insights into Ligand–Protein Interactions:* The analysis of docking results provided valuable insights into the key interactions between ligands and the target protein's binding site, guiding the rational design of potential drugs.

9.2 *Future Directions and Potential Research Opportunities for CS Majors in the Field of Molecular Docking*

For Computer Science (CS) majors interested in the field of molecular docking, there are several exciting future directions and potential research opportunities that can contribute to the advancement of this area. Here are some key areas to explore:

1. *Scoring Function Development:* Developing more accurate and reliable scoring functions is a crucial research area in molecular docking. CS majors can work on novel scoring functions that better capture the complex interactions between ligands and target proteins. Machine learning and deep learning techniques can be utilized to derive scoring functions from large datasets of known ligand–protein complexes.
2. *Machine Learning in Docking:* Integrating machine learning approaches into molecular docking can lead to significant improvements in prediction accuracy and efficiency. CS majors can explore the application of various machine learning models, such as neural networks, support vector machines, and random forests, to enhance different stages of the docking process.
3. *High-Performance Computing (HPC) and Parallelization:* Molecular docking involves computationally intensive tasks. CS majors can focus on developing parallel algorithms and utilizing high-performance computing resources, including distributed computing and GPU acceleration, to speed up docking calculations and handle large-scale docking studies efficiently.

4. *Data Augmentation and Transfer Learning:* Applying data augmentation techniques and transfer learning from one docking dataset to another can help address the scarcity of experimentally validated binding data. CS majors can investigate methods to leverage knowledge gained from related docking experiments to improve predictions for new target proteins.
5. *Docking with Flexible Receptors:* Most docking studies assume rigid protein structures, but proteins are dynamic and can undergo conformational changes upon ligand binding. CS majors can work on algorithms that account for receptor flexibility, allowing for more accurate predictions of ligand–protein interactions.
6. *Virtual Screening and Drug Repurposing:* CS majors can explore virtual screening methods to identify potential drug candidates from large compound databases. Additionally, drug repurposing, which involves finding new uses for existing drugs, is an area where computational methods can be instrumental in predicting novel interactions.
7. *Interpretable AI for Docking:* As AI-driven approaches become more prevalent, the ability to interpret and explain the results becomes crucial. CS majors can investigate methods to make AI-driven docking predictions more interpretable and transparent to aid decision-making in drug discovery.
8. *Cloud Computing for Docking:* Leveraging cloud computing resources can provide scalable and cost-effective solutions for docking studies. CS majors can develop tools and frameworks for running docking simulations on cloud platforms, making them accessible to researchers worldwide.
9. *Multi-Objective Optimization in Docking:* Molecular docking often involves multiple conflicting objectives, such as maximizing binding affinity while minimizing ligand size or hydrophobicity. CS majors can work on multi-objective optimization algorithms to help researchers find a balance between these objectives.
10. *Integration of Structural Biology Data:* Integrating docking results with other structural biology data, such as protein–protein interaction networks, can provide a systems-level understanding of cellular processes and disease mechanisms. CS majors can explore methods for data integration and visualization to extract meaningful insights.
11. *Docking with Metal Ions and Cofactors:* Incorporating metal ions and cofactors into docking simulations is challenging due to their unique coordination geometries. CS majors can investigate ways to handle metal-protein and cofactor-protein interactions more accurately.
12. *Graph-Based Docking:* Representing ligands and proteins as graphs and applying graph-based algorithms for docking can be a promising research direction. CS majors can explore graph neural networks and graph similarity methods for this purpose.

By delving into these research areas, CS majors can make valuable contributions to the field of molecular docking, advancing drug discovery efforts and our understanding of ligand–protein interactions. These opportunities lie at the intersection of computer science, biology, and chemistry, offering an exciting and impactful field for interdisciplinary research.

Statements and Declarations Competing Interests The authors declare that there are no competing interests associated with this research. No financial or non-financial interests have influenced the design, data collection, analysis, interpretation, or reporting of this study. The research presented in this paper is conducted with complete impartiality and integrity.

Funding This research received no specific grant from any funding agency in the public, commercial, or not-for- profit sectors. The authors conducted this study without any external financial support, demonstrating their commitment to independent and unbiased research.

Ethics Approval This research adheres to all ethical guidelines and principles. All necessary ethical approvals were obtained from the appropriate institutional review boards and governing bodies before commencing this study. Participants were provided with informed consent, and their privacy and confidentiality were strictly maintained throughout the research process.

Data Availability The data used in this study are available upon reasonable request from the corresponding author or are available open on Kaggle & GitHub(dataset). The authors are committed to promoting transparency and accessibility, and they are willing to share the data with other researchers for scientific purposes.

Author Contributions The author Vijaya Sindhoori Kaza conceived and designed the study, performed data collection and analysis, and contributed significantly to writing and revising the manuscript. The research work has been supervised and reviewed by Dr P R Anisha and Dr. C Kishor Kumar Reddy.

Informed Consent Informed consent was obtained from all participants involved in this study. Participants were informed about the purpose of the research, their role in the study, and the potential risks and benefits associated with their participation. Written consent was obtained before data collection, ensuring full comprehension and voluntary participation.

Copyright and Permissions The authors declare that all material used in this paper does not need any permissions.

Conflict Resolution In the event of any disputes or disagreements related to this research, the authors agree to address them through amicable discussions and mutual understanding, with the aim of reaching a resolution that upholds the principles of scientific integrity and collaboration.

Publication Ethics The authors affirm their adherence to strict publication ethics guidelines. This research was conducted with the utmost integrity and in compliance with ethical standards in research and scholarly publishing.

Correspondence For any correspondence related to this research, please contact the corresponding author, Vijaya Sindhoori Kaza @email(vijaya.sindhoori@gmail.com).

References

1. Adelusi, Temitope Isaac, Abdul-Quddus Kehinde Oyedele, Ibrahim Damilare Boyenle, Abdeen Tunde Ogunlana, Rofiat Oluwabusola Adeyemi, Chiamaka Divine Ukachi, Mukhtar Oluwaseun Idris et al. (2022) Molecular modeling in drug discovery. Inform Med Unlocked 29: 100880
2. Subbarayudu B, Lalitha Gayatri L, Sai Nidhi P, Ramesh P, Gangadhar Reddy R, Kishor Kumar Reddy C (2017) Comparative analysis on sorting and searching algorithms. Int J Civ Eng Technol
3. Bagal A, Borkar T, Ghige T, Kulkarni A, Kumbhar A, Devane G, Rohane S (2022) Molecular Docking-Useful Tool in Drug Discovery. Asian J Res Chem 15(2):129–132
4. Brian Jiménez-García and others (2018) LightDock: a new multi-scale approach to protein–protein docking. Bioinform 34(1):49–55
5. Kishor Kumar Reddy C, Anisha PR, Srinivasulu Reddy K, Surender Reddy S (2012) Third party data protection applied to cloud and XACML implementation in the Hadoop environment with sparql. IOSR J Comput Eng ISSN
6. Corso G, Stärk H, Jing B, Barzilay R, Jaakkola T (2022) Diffdock: Diffusion steps, twists, and turns for molecular docking. arXiv preprint arXiv:2210.01776
7. Crampon K, Giorkallos A, Deldossi M, Baud S, Steffenel LA (2022) Machine-learning methods for ligand–protein molecular docking. Drug Discov Today 27(1):151–164
8. García-Ortegón M, Simm GN, Tripp AJ, Hernández-Lobato JM, Bender A, Bacallado S (2022) DOCKSTRING: easy molecular docking yields better benchmarks for ligand design. J Chem Inf Model 62(15):3486–3502
9. Garcia-Ruiz, Miguel, Pedro Cesar Santana-Mancilla, Laura Sanely Gaytan-Lugo, and Adriana Iniguez-Carrillo (2022) Participatory design of sonification development for learning about molecular structures in virtual reality. Multimodal Technol Interact 6, 10(2022): 89
10. Kiruba Nesamalar E, Satheesh Kumar J, Amudha T (2022) Efficient DNA-ligand interaction framework using fuzzy C-means clustering based glowworm swarm optimization (FCMGSO) method. J Biomol Struct Dyn 1–13
11. Li T, Guo R, Zong Q, Ling G (2022) Application of molecular docking in elaborating molecular mechanisms and interactions of supramolecular cyclodextrin. Carbohyd Polym 276:118644
12. Narasimha Prasad LV, Shankar Murthy P, Kishor Kumar Reddy C (2013) Analysis of magnitude for earthquake detection using primary waves and secondary waves, IEEE, 23
13. Meng XY, Zhang HX, Mezei M, Cui M (2011) Molecular docking: a powerful approach for structure- based drug discovery. Curr Comput Aided Drug Des 7(2):146–157
14. Morozov, Dmitry, Artem Melnikov, Vishal Shete, Michael Perelshtein (2023) Protein-protein docking using a tensor train black-box optimization method. arXiv preprint arXiv:2302.03410
15. Singh S, Baker QB, Singh DB (2022) Molecular docking and molecular dynamics simulation. In Bioinformatics (pp 291–304). Academic Press
16. Sunny, Sharon, Gautham Sreekumar (2022) SFLADock: a memetic Protein-Protein docking algorithm. In 2022 IEEE International Conference on Distributed Computing and Electrical Circuits and Electronics (ICDCECE), pp. 1–6. IEEE
17. Sunny, Sharon, Jayaraj PB (2022) Protein–protein docking: Past, present, and future. Protein J 1–26
18. Tessaro F, Scapozza L (2020) How 'Protein-Docking' translates into the new emerging field of docking small molecules to nucleic acids? Molecules 25(12):2749
19. Vidal-Limon A, Aguilar-Toalá JE, Liceaga AM (2022) Integration of molecular docking analysis and molecular dynamics simulations for studying food proteins and bioactive peptides. J Agric Food Chem 70(4):934–943

20. Wong F, Krishnan A, Zheng EJ, Stärk H, Manson AL, Earl AM, Jaakkola T, Collins JJ (2022) Benchmarking AlphaFold-enabled molecular docking predictions for antibiotic discovery. Mol Syst Biol 18(9):e11081
21. Xue, Qiao, Xian Liu, Paul Russell, Jin Li, Wenxiao Pan, Jianjie Fu, Aiqian Zhang (2022) Evaluation of the binding performance of flavonoids to estrogen receptor alpha by Autodock, Autodock Vina and Surflex-Dock. Ecotoxicol Environ Saf 233(2022): 113323
22. Yang, Chao, Eric Anthony Chen, Yingkai Zhang. Protein–ligand docking in the machine-learning era. Molecules 27(14): 4568

Bridging the Gap Between Learning and Security: An Investigation of Blended Learning Systems and Ransomware Detection

Vikas Pandey, Aman Jolly, Praveen Kumar Malik, and Shashikant

Abstract Blended learning has been identified as a potentially viable strategy for enhancing the efficacy and efficiency of online learning through the integration of conventional instructional approaches. Despite the potential advantages, educational institutions have exhibited reluctance in embracing this strategy due to a range of difficulties. A prominent issue of concern pertains to the escalating menace posed by Ransomware virus assaults, which have the potential to result in substantial financial ramifications and reputational harm. This study aims to examine the many elements that impact student satisfaction in utilising blended learning systems, focusing primarily on the modules, channels, and lecturers involved. This evaluation seeks to improve students' comprehension of literacy within the context of classroom discourses. In addition, our study aims to enhance the precision of Ransomware detection by examining the intricate characteristics of the malicious software, particularly by analysing assembly language instruction patterns. The N-gram technique is employed in a two-stage process for feature extraction. This procedure involves the computation of pattern statistics and subsequent feature selection to effectively reduce dimensionality. In order to assess the efficacy of the chosen features, we employ a feature cataloguing methodology with the Random Forest algorithm, utilising the lowest out-of-bag (OOB) error and a predetermined number of trees. Our experiments showcase that this method achieved the highest accuracy, sensitivity value, false positive rate, and precision. On its whole, our study presents a thorough methodology for improving student happiness and virus detection accuracy inside blended learning systems. Through the use of our research outcomes, educational establishments have the potential to furnish pupils with a learning encounter that is both captivating and efficacious, while concurrently minimising the vulnerabilities associated with cyber-attacks.

V. Pandey · Shashikant
Babu Banarasi Das University, Lucknow, UP, India

A. Jolly (✉)
KIET Group of Institutions, Ghaziabad, UP, India
e-mail: aman.jolly@kiet.edu

P. K. Malik
Lovely Professional University, Phagwara, Punjab, India

K. Kaushik and I. Sharma (eds.), *Next-Generation Cybersecurity*, Blockchain Technologies, https://doi.org/10.1007/978-981-97-1249-6_18

Keywords Blended learning · Ransomware · Energy and industry applications · Random forest

1 Introduction

The implementation of the Blended Learning System (BLS) project in educational institutions has shown great promise. However, due to the unfamiliarity and discomfort with this platform, getting institutional consensus has proven difficult A comprehensive dialogue between decision makers and employees is essential for success. The complexity of the session management in BLS requires dedicated technical staff and ongoing support. Additionally, policy development and tight assessment and measurement are crucial to support the expected quality of BLS.

One of the challenges is the detection of Ransomware virus accurately, which can cause significant financial loss as well as enterprise reputation. To address this, this work focuses on developing more precise features of the virus using N-gram algorithm and Random Forest algorithm for classification. The initial implementation of BLS faced challenges in areas such as ownership of intellectual property, copyright, and workload of each member of the institution. Therefore, it is necessary to address these problems at the policy level and ensure commitment from all concerned members for the long-term success of BLS. In this paper, we discuss the material and methodology used, performance, experimental results and discussions, and conclude the work.

The adoption of blended learning has gained interest in informal educational institutions, leading decision makers to initiate the implementation of the Blended Learning System (BLS) project in their learning environments. However, achieving institutional consensus for BLS adoption may prove challenging due to unfamiliarity and discomfort with the platform. BLS implementation requires consistency between the objectives of decision makers and employees, who may find it difficult to integrate BLS with institutional structure and strategy, which traditionally rely on formal teaching and lecturing. Government regulation policy has also contributed to slow response to BLS implementation due to a shortage of engagement and fear of technological impact. Additionally, top-down initiatives may be met with suspicion by institution members who perceive a negative impact on activity processes, focusing on quality, intellectual property, and ownership. To successfully implement BLS, comprehensive dialogue between all involved parties is necessary. Despite previous attempts, gaps in understanding between executives, management, and faculty have not been satisfactorily addressed in past online learning initiatives [1].

The complexity involved in session management within the Blended Learning System (BLS) necessitates the allocation of dedicated technical personnel and ongoing support. It is crucial for students and their instructors to have access to a reliable learning process that incorporates an intervention mechanism to monitor motivation and progress. Furthermore, the extension of the learning environment to the internet underscores the importance of providing sufficient content resources that are easily accessible. However, the management of content systems, network connectivity, bandwidth levels, and program releases may prove overwhelming and hinder the reliability of the system. Therefore, policy development must be timely and accompanied by rigorous assessments and measurements to ensure the expected quality of BLS and prevent the emergence of complex resistance. Notably, Ransomware, a type of malware that encrypts the victim's files and demands a ransom for data retrieval, poses a significant challenge in terms of accurate detection. Failure to accurately detect the virus at the outset can lead to significant financial losses and damage to the enterprise's reputation.

The identification and categorisation of malicious software, commonly referred to as malware, are of utmost importance within the realm of computer security. In recent years, scholars have directed their attention to the development of techniques aimed at extracting characteristics from malware files with the objective of effectively categorising and identifying them. One approach that might be employed entails the extraction of viral features in the form of operation codes (opcodes) from files that possess the.exe format.

Prior studies have employed machine learning methods, including Random Forest, Decision Tree, K-Nearest-Neighbor, and Support Vector Machine, for the purpose of classifying malware files by analysing their opcodes. One study conducted by researchers [1] found that the Random Forest algorithm demonstrated the greatest accuracy rate of 98.7% when utilised for the classification of malware files. In a separate investigation [2], datasets including both normal and malicious files were acquired by unpacking files from the Windows Operating System using UPX Unpacker and IDA Pro. The research conducted successfully identified the primary opcodes utilised in malware files. These opcodes, including MOV, PUSH, CALL, CMP, POP, JZ, TEST, JNZ, JMP, ADD, LEA, XOR, RETN, AND, SUB, OR, INC, MOVZX, DEC, and JB, were found to be prevalent in the analysed malware samples.

Nevertheless, despite the significant progress made in this field, the process of effectively identifying and classifying malware continues to present considerable challenges. Hence, the objective of this study is to enhance the precision of viral characteristics in order to enhance the efficacy of malware detection and classification. The primary objective of this study is to investigate the fundamental attributes of viruses by use of a two-stage methodology that involves the analysis of their assembly language instructions. The initial phase involves the utilisation of the N-gram technique to extract the features. These characteristics are subsequently subjected to sequential pattern searching and statistical analysis in order to ascertain their relevance. Ultimately, a feature selection strategy is implemented in order to decrease the dimensionality of the data. The subsequent phase entails the validation of the chosen attributes to assess their efficacy in discerning malicious software.

In summary, the precise identification and categorisation of malicious software play a crucial role in safeguarding the integrity and protection of computer systems. The objective of this research is to boost the accuracy of malware detection and classification by refining the properties of the virus in a more exact manner, thereby building upon prior studies. By employing sophisticated algorithms and meticulous feature selection, our aim is to achieve substantial advancements in the domain of computer security.

The present study employs the powerful N-gram algorithm for extracting low-level features of the virus, which extracts N-byte combinations in sequence. This algorithm involves four crucial steps, namely N-gram extraction, sequential pattern extraction, pattern statistics calculation, and feature reduction, which together help in obtaining more specific and accurate results related to malicious code. In addition, the Random Forest algorithm, an ensemble learning method used for cataloging and regression, is utilised for classifying the experimental dataset. This algorithm is based on constructing multiple decision trees and using the majority vote to make decisions, ensuring maximum accuracy. Our research is organised into several sections, including an introduction, material and methodology, performance, experimental results and discussions, and conclusion.

To successfully implement BLS initiatives, policy areas related to intellectual property ownership, copyright, and workload allocation for each member of the institution need to be considered. Such initiatives often require the adjustment of current organisational procedures or employee agreements to meet the requirements arising from the adoption of the BLS system. The participation of institutional members is crucial, and identifying success factors can facilitate their commitment and reduce their fears in the long term [2].

2 Research Methodology

2.1 Blended Learning to Be Used as Mediated Platform

In essence, knowledge can be considered as acquired externally, as a cognitive state resulting from a thought process, or as a meaning constructed through social interaction [5]. A study found that students who participate more frequently in forums and achieve higher levels of academic success tend to express more positive views about BL, highlighting the importance of interaction and communication in the effectiveness of this approach [6]. When developing a BL program, it's important for teachers and students to be aware that the outcomes should align with organisational expectations, and that a balance between motivation and content is necessary for this new learning endeavour to be collaborative [7].

However, individual antecedents such as self-attitude, self-behavior, self-cognitive, and intention to comply appear to be less significant than organisational antecedents like policy compliance, training programs, and security threats, as well as environmental antecedents such as peer performance, social pressure, and religious indicators in encouraging students to recognise their roles and responsibilities in the higher education system [8]. The adoption of BLS programs in numerous

Table 1 Hexa view in ransomware files

Hexa view in ransomware Files
0045C180 02 OO 8B CO OO 8D 40 OO 74 2040 OO FC 21 400
0045C190 70 2540 OO 32 IF 8B CO 32 13 8B CO 52 75 6E74
0045CIAO 6961) 65,206,572 72 6F72 202,020,202,061 74
0045C190 70 2540 OO 32 IF 8B CO 32 13 8B CO 52 75 6E74

organisations has been growing steadily, with annual increases of at least single-digit percentages. To initiate a BLS program, organisations must establish a framework for centralised and continuous data collection to monitor and evaluate the program's effectiveness. To ensure the success and efficiency of BLS implementation, several factors should be considered. To implement blended learning effectively, it is necessary to select appropriate technology tools that are tailored to each student's abilities, design an instructional program that captures comprehensive student data, create a curriculum that guides the instructor through the next stages of instruction, and choose a platform that provides sufficient resources and materials to support organised teaching and learning. In evaluating the quality of technology for blended learning, learner interactions were considered from the perspectives of cognitivism, collaborative learning, and student–teacher interactions. "The quality of technology was assessed based on its availability (72%), content (71%), intervention (69%), chat features (69%), resources (68%), Internet reliability (66%), and email exchanges (63.4%) [9]".

Various factors, such as the "characteristics of the student population", "the mission of the organisation", "the strategic planning processes", "faculty responsiveness", "student acceptance", "community values", "available resources", "organisation support mechanisms", and other components, have helped shape blended learning in a way that makes sense for a particular organisational context [10] (Table 1).

2.2 Dataset Representation

The purpose of this study involves the creation of a dataset for experimental purposes. The dataset consists of normal files extracted from the Windows 10 operating system's executable files, while the malware files (specifically, Trojan Ransomware) are sourced from the VX Heaven Virus Dataset. Portable Executable (PE) formats are utilised to determine how programs are executed within the Windows operating system. The dataset files are presented in both Hexadecimal and Assembly views. Furthermore, normal files are graphically represented as flow graphs to enhance their comprehension, as exemplified by the dfrgui.exe program in Windows.

The Trojan Ransomware Malware files, on the other hand, are extracted from the VX Heaven Virus Dataset [6]. Prior to cataloguing the dataset, the assembly language

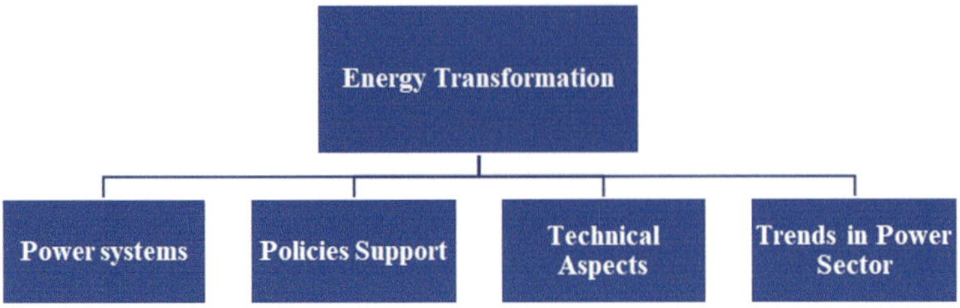

Fig. 1 Energy transformation

codes in both hexadecimal and flow graph formats are analysed for disassembled files. In malware files, Portable Executable (PE) formats such as.text and CODE are commonly found. Figure 1 illustrates a malware file in the Hexa view in IDA Pro [7]. Prior to cataloguing, it is important to disassemble both the malware and regular files. The process of disassembly entails the conversion of machine code contained within files of the.exe type into assembly language. This conversion necessitates the utilisation of a disassembler tool. This study used Interactive Disassembler (IDA) Pro to perform the disassembly of both malware and regular files, resulting in the generation of output files stored in the.asm file format.

2.3 Blended Learning for Energy and Industry Applications System

Blended learning may be conceptualised as an information system, and its efficacy can be assessed by examining the Energy and Industry Applications utilised by its users. The assessment of user Energy and Industry Applications holds significance in ascertaining the efficacy of expenditures made in information systems. Throughout history, the utilisation of Energy and Industry Applications by users has been employed as a metric to gauge the effectiveness and accomplishment of information systems. The data collection process involved gathering user input about the distinct attributes of the information system. Additionally, an assessment was conducted to analyse several elements that impact user Energy and Industry Applications, including their views of the information system's attributes and their individual requirements. The user's text does not contain any information to rewrite in an academic manner.

Information systems (IS) play a crucial role in the administrative and academic operations of higher learning institutions in developing nations. In order to achieve favourable outcomes in Information Systems (IS) initiatives, it is crucial to take into account the contextual and institutional characteristics that are unique to the project at hand (12). The assessment of project performance relies heavily on the involvement and contributions of stakeholders in the Energy and Industry Applications domain.

This is because stakeholders play a crucial role in shaping the project's environment and contextual factors. The issue of energy transformation is depicted in Fig. 1.

Individuals tend to evaluate energy and industry applications based on their prior experiences with brands, movies, jobs, political parties, and other factors. These evaluations are often gathered through objective data and have been extensively used by psychologists, policy scientists, and consumer researchers to predict repeat purchases or election outcomes, and to assess happiness or self-well-being among different population groups [13]. To understand user perceptions, researchers typically rely on the difference between expectations and performance evaluations. Therefore, it is important for researchers to obtain detailed information about customer service expectations and perceived performance. Thus, performance benchmarks for some periods are considered equivalent to the norm for controlling goods or services. Nevertheless, when the deviation from this criterion is sufficiently large, that is, when the perceived performance falls outside the acceptable range, the brand's performance will be deemed different from the norm, leading to significant dissatisfaction with product evaluation [14].

Extensive government support and guidance are crucial for the success of blended learning systems (BLS) in higher education. To evaluate the effectiveness of BLS, student and lecturer Energy and Industry Applications must be considered along with learning outcomes. Although most users have positive perceptions of BLS, support is still necessary to cater to different learning styles and cultures. Student adoption of energy and industry applications in blended learning systems is influenced by various factors, such as adaptability to e-learning, perceived benefits, teacher response, ease of use, and application usage. Other important determinants of student learning in blended learning systems include computer self-efficacy, performance expectations, system functions, content characteristics, interaction style, and learning climate. As students invest a lot of time, money, and effort in education, their Energy and Industry Applications are critical for motivation and success. Meeting or exceeding expectations not only satisfies students but also encourages them to become advocates for the organisation, enhancing its reputation within the community [15–18].

2.4 Feature Extraction

The present study utilised a different approach compared to previous research [8] where manual copying of data into the.txt format was conducted along with manual separation [9]. For this research, disassembled files from IDA Pro were saved in the.asm file format and the Python programming language was used to select the opcode strings. This approach only takes instructions as input and doesn't consider other details like memory addresses and values. During the feature extraction process, instructions found in parameters are printed and the process stops if no instructions are detected. The study used the N-grams algorithm with a range value of N = 1 to N = 4.

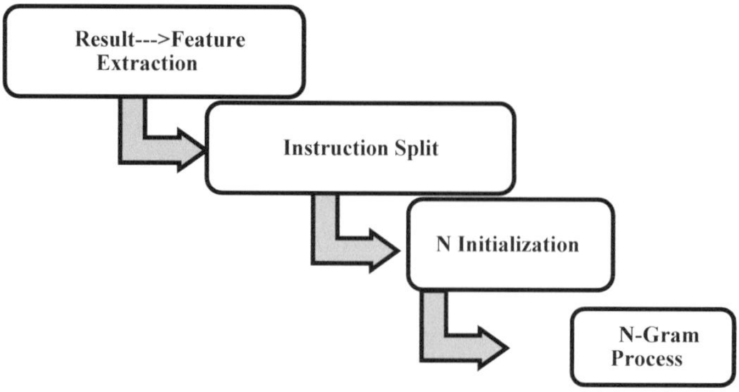

Fig. 2 N-gram process

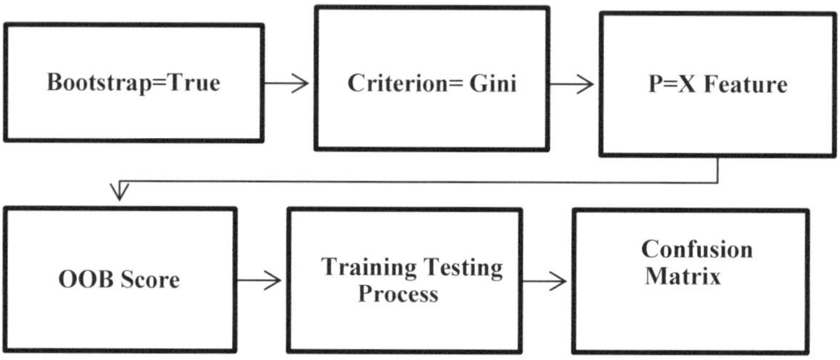

Fig. 3 Random forest cataloging process

During this stage, the opcodes were sorted based on their frequency of occurrence. Python programming language was used for the extraction process using N-grams [12]. A flowchart illustrating the N-gram programming process is presented in Figs. 2, 3.

2.5 Random Forest Classifier

The current research utilises the Random Forest (RF) algorithm [16] to provide cataloguing outcomes using a voting mechanism, wherein the decision tree with the highest number of votes is selected. The level of accuracy seen in the cataloguing results serves as an indicator of the algorithm's performance. The RF method is designed with the specific goal of attaining an accuracy rate of 98%. This is accomplished by utilising a total of 500 trees, denoted as n tree, which has been determined to provide optimal outcomes [17].

This strategy aims to mitigate the issue of decision trees exhibiting overfitting tendencies towards their training data, hence resulting in improved performance. While decision trees are often outperformed by random forests, it is worth noting that the accuracy of random forests is still inferior to that of gradient enhanced trees. Nevertheless, the efficacy of the procedure is contingent upon the inherent attributes of the data under scrutiny.

2.6 *Performance Analysis*

The evaluation of the suggested methodology is conducted by employing a confusion matrix, a tool that offers insights into the accurate and inaccurate classification of test data. This matrix facilitates the computation of the algorithm's accuracy on the testing dataset. The user has provided a numerical reference, indicating the presence of a citation or source[20, 31].

3 Research Methodology

This study aimed to examine factors related to the learning environment in Information Systems (IS), and their relationship with Overall Energy and Industry Applications (OS). The researchers utilised theories such as "assimilation theory" [19], "contrast theory" [20, 21], "prospect theory" [22], "the theory of adaptation level" [23], and "generalised negativity theory" [24] to establish the relationship between OS and computer self-efficacy, expectation of quality, information timeliness, perceived utility, software adequacy, and user support [8]. A quantitative research design was employed, and a survey questionnaire was distributed online to a selected sample of 100 students using stratified sampling who had completed the subject of information system project management. The questionnaire, presented in Indonesian language, used 5-point Likert scales for response options. SmartPLS 3 was employed as the analytical tool to examine the fundamental relations and effects among variables, which were visualised through model visualisation and path structure [25]. Prior to data collection, experts evaluated the clarity and simplicity of the question statements, and revised the word choices accordingly. The questionnaire comprised 45 questions, categorised into seven variables, namely perceived utility (PU, 5 items), expectation of quality (EQ, 7 items), information timeliness (IT, 6 items), user support (US, 7 items), software adequacy (SA, 8 items), computer self-efficacy (CS, 6 items), and overall energy and industry applications (OS, 6 items). These are as follows (Table 2).

Factor analysis was utilised to determine the degree of correlation between an indicator and other predictors in the model, which is a commonly employed method for detecting linear or multiple relationships. To accurately evaluate the contribution of predictors to the model, the variance inflation factor (VIF) was utilised. Higher VIF values indicate greater difficulty in accurately assessing predictor contribution

Table 2 Variance Inflation Factor (VIF) for Indicators in LM-D

Indicators	Statements	VIF
PU3	The module uplifting intellectually	1.691
PU4	Academical Pressures as student	1.209
EQ1	Feedback to help on the progress	1.583
EQ2	High workload of module	1.276
EQ3	Clear ex expectation for the future goals	1.187
EQ4	Motivation to self-improve	2.261
IT1	Achieve best Performance	1.736
IT2	Critical thinking kind of anal sis	2.002
IT3	Memorisation-type of test	1.356
IT4	Commitment-based as group member	1.136
US1	Sufficiently for knowing the material	1.285
US2	Involvement of lecturers to reduce difficulty	1.739
US3	Good explanation from lecturers on module	1.584
US4	The alignment of theory and Practical	1.406
SA1	Student interest lies upon the lecturers' skill	1.503
SA2	Academic goals determine the purpose	1.082
SA3	Cognitive skill is extremely critical	1.753
SA4	Lecturers should comment on the assignment	1.732
CS1	Complex in accessing the system	1.582
CS2	impression towards the features of the system	1.943
CS3	Self-study can be fulfilled with the video content	1.678
CS4	It is necessary to have reference link	1.617
PU3	Quiz is important to assess the understanding	2.448
PU4	The slide should be sophisticated	1.754

to the model. A VIF value of one implies that the predictor is not correlated with other variables, while values above four or five are generally considered moderate to high, with values exceeding 10 regarded as very high [27]. The majority of VIF scores were found to exceed 1, with roughly seven indicators having a VIF exceeding 2, according to the table provided above (Table 2).

4 Results and Analysis

The results of experiments using N-gram implementation show varying frequencies for different N values and operational code vocabularies. The length of the vocabulary for the N-gram extraction feature will be compared graphically, as depicted in Fig. 4.

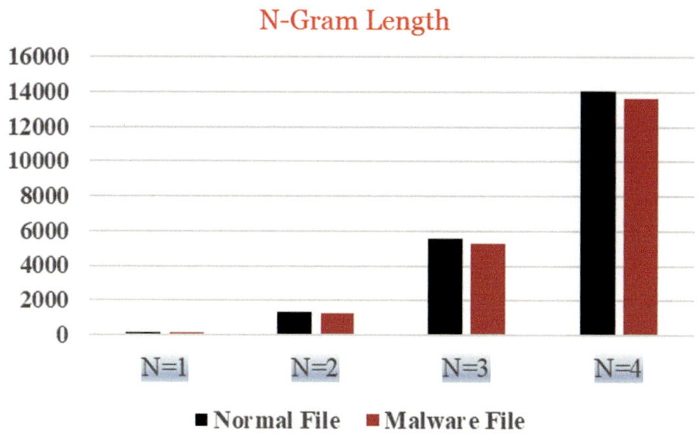

Fig. 4 N-gram extraction

This study utilises different N-gram feature extractions, including 1-g, 2-g, 3-g, and 4-g. When N equals 1, the normal file has a length of 118, and malware has a length of 129. For N equals 2, the normal file's length is 1344, and malware has a length of 1230. Similarly, when N equals 3, the length of the normal file is 5582, and malware has a length of 5255. Finally, when N equals 4, the length of the normal file is 14033, and malware has a length of 13,655. Increasing the value of N will increase the vocabulary in N-grams, and more N-grams will result in more sequential patterns.

The experimental outcomes are presented in Table 3, illustrating the performance of the confusion matrix across all conducted tests. In Experiment 1, the true positive (TP) score was 259, the true negative (TN) was 237, there was one false positive (FP), and three false negatives (FN). In Experiment 2, the true positive (TP) score was recorded as 207, while the true negative (TN) count was 189. Additionally, there was one false positive (FP) and three false negatives (FN) observed. In Experiment 3, the true positive (TP) score was 149, the true negative (TN) score was 146, there was one false positive (FP), and four false negatives (FN). In Experiment 4, the true positive (TP) score was recorded as 96, while the true negative (TN) score was 101. No false positives (FP) were seen, resulting in a perfect score for FP. However, three false negatives (FN) were identified in the experiment. In the conducted study, Experiment 1 exhibited the most elevated sensitivity score, reaching a notable percentage of 98.8%. Conversely, Experiment 4 had the best accuracy score, achieving a perfect percentage of 100%. In Experiment 4, the false positive rate (FPR) achieved the greatest score, reaching 0%. Conversely, Experiment 1 obtained the highest "accuracy score" with a value of 99.2%. In summary, the results of experiment 1 demonstrated superior performance in terms of sensitivity and accuracy, but experiment 4 exhibited the highest scores in precision and false positive rate (FPR).

The function of "Energy and Industry Applications" (EIAs) is of great importance in influencing users' experiences. Positive disconfirmation takes place when products or services surpass expectations, resulting in an improved experience for customers.

Table 3 Evaluation of experimental scenarios

Metric	Experiment 1	Experiment 2	Experiment 3	Experiment 4
True positive	259	207	149	96
True negative	237	189	146	101
False positive	1	1	1	0
False negative	3	3	4	3
Sensitivity	98.80%	98.50%	97.30%	96.90%
Precision	99.60%	99.50%	99.30%	100%
False positive Rate	0.42%	0.50%	0.60%	0%
Accuracy	99.20%	99%	98.30%	98.50%
OOB error	0.01%	0.01%	0.02%	0.02%

The underlying principle of this concept is based on the idea that consumers' perceptions of a product or service's performance are in accordance with their preferences and requirements [16]. The estimation of reliability is often conducted using measures such as "Cronbach's alpha" and "Tarkkonen's rho". In general, a reliability coefficient of 0.7 or above is often regarded as satisfactory, whereas values below 0.5 are considered unsuitable. Nevertheless, the present investigation yielded unsatisfactory reliability estimates for two variables, namely PU and US, as indicated by their respective "Cronbach's α" values of 0.305 and 0.598. In contrast, the estimated values of Rho_A for PU and US were 0.773 and 0.669, respectively. These findings show that other estimators, such as "Rho_A" based on factor analysis, may be more suitable since they do not underestimate reliability and might potentially offer more precise estimates [28].

In addition, it is crucial to consider the measures of composite reliability (CR) and average variance extracted (AVE) in order to ascertain the validity of the structural model. Generally, it is recommended that AVE values should be at least 0.5 and CR values should be at least 0.7 [29]. Nevertheless, it was observed in this study that none of the variables satisfied the "AVE threshold" of 0.5, while the variables "PU", "SA", and "US" did not match the "CR" criterion. Consequently, the structural model was unable to demonstrate convergent validity, a critical aspect in assessing the degree of correlation among various indicators of the same construct that exhibit agreement [30]. Furthermore, a strong correlation was seen between several pairs of variables inside the "Latent Variables (LVs)" framework. For instance, the variables "PU" and "EQ" exhibited a correlation coefficient of 1.028, while the variables "US" and "CS" showed a correlation coefficient of 1.041. There were eight pairings of variables that exhibited scores exceeding one, suggesting the presence of model specification mistakes, estimation difficulties, or suppressor effects. The fundamental characteristics of the "LVs" are closely connected to the characteristics of the indicator variables employed to define them. These "LVs" have the potential to have both direct and indirect impacts, which can vary in magnitude from -1 to greater than 1 [31].

Based on the aforementioned results, it is crucial to take into account alternative estimators, such as "Rho_A," and to tackle concerns pertaining to "AVE" and "CR" in order to verify the validity and dependability of the structural model. By attending to these concerns, researchers will be able to enhance the precision of measuring the degree of association among various indicators pertaining to a certain construct, hence enhancing the overall validity and reliability of the study's outcomes (Tables 4, 5).

The present study employs a thorough and complete method to data analysis. The researchers have employed factor analysis and route analysis techniques on the structural model, doing one thousand iterations and setting seven as the stopping criteria. This rigorous approach has been undertaken to assure the stability and reliability of the scores of the latent variables. Moreover, the use of the Partial Least Squares Structural Equation Modelling (PLS-SEM) technique is notably noteworthy due to its ability to elucidate the highest degree of association between variables and provide more consistent latent variable scores throughout the algorithmic process. Moreover, the researchers have exhibited meticulousness in their investigation, as they have thoroughly examined the loadings of the indicators. Although the majority of external loadings failed to exceed the acceptable threshold of 0.7, a few indicators, including "CS3", "EQ7", "IT4", "OS1", "OS5", "OS6", "PU3", "SA5", and "SA7",

Table 4 Validity & Reliability evaluation

		Cronbach's	Rho A	Composite	Averange
		Alpha		Reliability	Variance
Computer Self-Efficacy (CS)		0,797	0.82	0.807	0.416
Expectation of Quality (EQ)		0.807	0.828	0.809	0.389
Information Timeliness (IT)		0.742	0.773	0.752	0.346
Overall (os)	Satisfaction	85.10%	85.60%	84.80%	0.485
Perceived Utility (PU)		30.50%	77.30%	41%	0.359
Software (SA)	Adequacy	70.50%	81.80%	63%	0.268
User Support (US)		60%	66.90%	54.60%	0.194

Table 5 Correlation among latent variables

	CS	IT	PU	US	EQ	OS	SA
US	1.04	1.21	0.847	1	1.047	0.751	0.963
SA	0.86	0.95	0.782	0.963	0.87	0.741	1
PU	0.69	0.92	1	0.847	1.028	0.673	0.782
OS	0.84	0.82	0.673	0.751	0.645	1	0.741
IT	0.95	1	0.922	1.21	0.99	0.827	0.952
EQ	0.81	0.99	1.028	1.047	1	0.645	0.87
CS	1	0.95	0.691	1.041	0.815	0.842	0.86

exhibited outstanding results. Furthermore, the researchers have taken measures to assure the inclusion of just the most reliable indicators in their study by excluding those with borderline values below 0.5.

Furthermore, the researchers have also taken into account the importance of each indicator, evaluating their values using the process of bootstrapping. Although the researchers have advised against relying solely on statistical significance, it is worth noting that just eight indicators had values below 1.96. The authors appropriately acknowledge that the prevalent utilisation of statistical significance as a means to assert scientific findings might result in substantial distortions within the scientific process. The methods and analysis employed in this work exhibit a meticulous and comprehensive approach to the data, hence assuring the reliability and robustness of the findings (Table 6).

The objective of the discriminant validity assessment is to ensure that a reflective construct exhibits stronger relationships with its indicators than with any other construct in the PLS path model [34]. Consequently, some researchers propose the use of "Heterotrait-monotrait" ratio (HTMT) of the correlation, in addition to the "Fornell-Larcker criterion", due to the latter's lack of reliability in detecting possible indiscriminate among latent variables [35]. However, the exact threshold level for HTMT remains a matter of debate, with some researchers recommending a value of 0.85, while others suggest 0.9 [36, 37]. The table indicates eight variable relationships that require consideration due to HTMT values below the threshold, including "CS-PU (0.68), CS-SA (0.76), EQ-OS (0.63), EQ-SA (0.75), OS-PU (0.68), OS-SA (0.61), and" OS-US (0.67)."

The term "saturated model" refers to a structural model where all constructs are freely correlated, with a standardised root mean square residual (SRMS) of no more than 0.08 or 0.10 and a normalised fit index (NFI) greater than 0.90. On the other hand, the estimated model is based on the model structure and considers the total effect scheme. The values of d_ULS and d_G must be less than the 95% or 99% bootstrap quantile to assess the model fit. The exact model fit test is based on the discrepancy between the empirical covariance matrix implied by the composite factor model, and only the bootstrap results of the exact model fit measure allow for interpretation of results. If the correlation matrix is small enough to be attributed to

Table 6 Validity analysis of HTMT discriminant

	CS	EQ	IT	OS	P U	SA	US
CS							
EQ	0.81						
IT	99	0.97					
OS	0.85	0.63	0.84				
P U	0.68	1	0.88	0.68			
SA	0.76	0.75	0.84	0.61	0.72		
US	0.95	0.94	1.12	0.67	0.82	0.97	

Table 7 Correction value analysis

"	Saturated Model	Estimated Model		Correction Value (ci
SRMR	0.10	2.29	CS	2.78
d_ULS	11.26	5,423.20	EQ	2.95
d_G1	6.12	21.41	IT	2.36
d_G2	5.84	19.09	OS	3.12
Chi-Square	2.230.84	n/a	PU	2.02
NFI	0.41	n/a	SA	2.38
			US	1.66"

sampling error, or the discrepancy is not significant ($p > 0.05$), then the model fit is considered established [25, 35] (Table 7).

The study employed the PLS path modelling technique to investigate the relationships among various constructs. The findings suggest that computer self-efficacy has a negative direct impact on "software adequacy ($\beta = -0.580$)" but a positive direct effect on "expectation quality ($\beta = 3.320$)", which is more significant than the effect of overall satisfaction with "Energy and Industry Applications ($\beta = 2.294$)". Additionally, information timeliness was discovered to have a negative direct effect on "perceived utility ($\beta = -3.052$)", which is more substantial than the effect of overall satisfaction with "Energy and Industry Applications ($\beta = -2.314$)", while also having a positive direct impact on "software adequacy ($\beta = 1.503$)", which is more significant than the impact of "user support ($\beta = 1.210$)". Furthermore, user support had a negative direct impact on overall satisfaction with "Energy and Industry Applications ($\beta = -0.740$)" and "expectation quality ($\beta = -2.418$)".

It is crucial to acknowledge that these results were achieved through the use of the Partial Least Squares (PLS) route model and are predicated on the beta coefficients acquired from the investigation. The presence of a negative sign in the beta coefficients signifies an inverse association between the constructs, whereas a positive sign suggests a positive association. The study also analysed the effect sizes of the associations, where the magnitude of the beta coefficients was used to determine the strength of the links between the constructs. The findings of this study have significant significance for the comprehension of the interconnections between the components and their impact on the overall satisfaction with "Energy and Industry Applications".

5 Conclusion

In order to improve the structural model designed to assess user satisfaction with Energy and Industry Applications in blended learning, it is necessary to take into account many factors. These factors encompass the choice of indicators, sampling methods, the reflective model, and the desire of respondents to take part in the study. The rigorous evaluation of these areas is crucial in order to ensure the robustness of the final model and its correct representation of the aspects that lead to user satisfaction with Energy and Industry Applications.

The findings of the study indicate that the timeliness of information has a detrimental influence on the level of satisfaction in Energy and Industry Applications. Conversely, user assistance, perceived usefulness, software adequacy, computer self-efficacy, and expectation quality positively contribute to the overall satisfaction. The findings of this study highlight the significance of giving priority to these factors during the creation of blended learning programmes, with the aim of improving user satisfaction in the context of Energy and Industry Applications.

The selection of the value of N is of utmost importance when utilising N-gram extraction in a Random Forest classifier, as it has the potential to significantly influence the quantity of vocabulary included by the N-grams and sequential patterns generated. In trial 1, the Random Forest classifier had the highest level of accuracy in cataloguing, reaching 99.2%. Additionally, the sensitivity value of 98.8% suggests that the algorithm effectively categorised the majority of malware samples. Furthermore, the false positive rate (FPR), which represents the proportion of malware samples that were incorrectly categorised as normal, was found to be 0%. Additionally, the precision, which indicates the percentage of accurately identified malware samples, was determined to be 100%. The out-of-bag (OOB) error exhibited minimum values, as seen by experiment 1's low error rate of 0.008. Future study should aim to conduct experiments using the latest and comprehensive malware dataset in order to enhance the performance of the model. The findings underscore the capacity of the Random Forest classifier to effectively categorise malware samples, hence opening avenues for further investigations in this domain.

References

1. Vyas R, Luo X, Mcfarland N, Justice C (2017) Investigation of Malicious Portable Executable File Detection on the Network using the Supervised Learning Techniques Symposium on Integrated Network and Service Management (IM). Thing. 941–946. 2017
2. Oliver M, Trigwell K (2005) Can 'blended learning' be redeemed? ELearning 2(2):17–26
3. Huda MQ, Hussin H (2017) Evaluation model of Information technology innovation effectiveness case of higher education institutions in Indonesia. Proc ICIC
4. Yewale A (2016) Malware Detection Based on Opcode Frequency. International Conference on Advanced Communication Control and Computing Technologies. 978:646–649
5. Al-rimy MAM, Shaid SZM (2018) Ransomware threat success factors, taxonomy, and countermeasures: A survey and research directions. Computer Security 74:144–166
6. Raff E et al (2016) An Investigation of Byte N-Gram Features For Malware Classification. A Comput Virol Hack Tech. 14(1):1–20
7. Rogers E (2003) Diffusion of innovation. 5th edition. An integrated approach to communication theory and research. Simon and Schuster. ISBN 9780743258234
8. Vaezi R (2013) User Satisfaction with Information Systems: A Comprehensive Model of Attribute Satisfaction. Ph.D. Dissertation, University of Houston

9. Adam F et al (2014) The use of blog as a medium of Islamic da'wah in Malaysia. Int. Journal of Sustainable Human Development 2(2):74–80
10. Meyer S et al (2014) Rhetoric and reality: Critical perspective on educational technology. In: Proceedings Ascilite Dunedin, 89–98
11. Alonso F et al (2005) An instructional model for web-based e-learning education with a blended learning process approach. British J of Educational Technology 36(2):217–235
12. Akkoyunlu B, Soylu MY (2006) A study on students' views on blended learning environment. Turkish Online Journal of Distance Education 7(3/3):43–56
13. Lalima KL, Dangwal (2017) Blended learning: an innovative approach. Universal Journal of Educational Research 5(1): 129–136, HRPUB
14. Ghahari S, Ameri-Golestan A (2013) The effect of blended learning vs. classroom learning techniques on Iranian EFL learners' writing. Int. J. of Foreign Language Teaching and Research 1(3)
15. Breiman L (2001) Random Forest. Kluwer Academic Publishers. 45:5–32
16. Heaven V (n.d.) Computer Virus Collection. Retrieved from http://vxheaven.org/vl.php
17. IDA Support: DownloadCenter.(n.d.). Retrievedfrom https://www.hexrays.com/products/ida/support/download.shtml
18. Richardson R, North M (2017) Ransomware: Evolution, mitigation and prevention. International Management Review. 1, 13, pp. 10–21
19. Kinder H, Veith Jakstab (2008) A static analysis platform for binaries. Springer-Verlag Berlin Heidelberg. 5123, pp 423–427.
20. Kassim ES et al (2012) Information System Acceptance and User Satisfaction: The Mediating Role of Trust. Procedia Social and Behavioral Sciences 57:412–418
21. Subiyakto A et al. (2016) The user satisfaction perspective of the information system projects. Indonesian Journal of Electrical Engineering and Computer Science, 4(1), pp 215–223
22. Levy-Garboua, Montmarquette C (2007) A theory of satisfaction and utility with empirical and experimental evidence. In: Conference of the French Economic Association, Behavioral and Experimental Economics
23. Yuksel A, Yuksel F (2008) Consumer satisfaction theories: a critical review. In Book: Tourist Satisfaction and Complaining Behavior, Publisher: Nova Sciencce
24. Tahar NF et al. (2013) Students' satisfaction on blended learning: the use of factor analysis. In: IEEE Conference on e-Learning, e-Management and e-Services
25. Naaj MA et al (2012) Evaluating Student Satisfaction with Blended Learning in a Gender-Segregated Environment. Journal of Information Technology Education: Research 11:185–200
26. Knox WE et al (1993) Does college make a difference? Long-term changes in activities and attitudes. Greenwood Press, Westport, CT
27. Chute AG et al (1999) The McGraw-Hill handbook of distance learning. McGraw-Hill, New York
28. Xie Y, Greenman E (2005) Segmented assimilation theory: a reformulation and empirical test. Population studies center research report 05–581
29. Hovland C et al. Assimilation and constrast effects in reaction to communication and attitude change. Journal of Abnormal and Social Psychology, 55(7), pp 244–252
30. Ko D, Afternoon K (2018) Comparative Study on Perturbation Techniques in Privacy Preserving Data Mining. International Journal of Innovative Computing 8(1):27–32
31. Canfora G, De Lorenzo A (2015) Effectiveness of opcode n grams for detection of multi family android malware. In: International Conference on Availability Reability and Security. 333–340
32. Sornil O, Liangboonprakong C (2013) Malware classification using N-grams sequential pattern features. In: International Journal of Information Processing and Management (IJIPM). 4(7):59–67
33. Bilar D (2007) Opcodes as Predictor for Malware. Electronic Security and Digital Forensics 1(2):156–168

34. Belgiu L, Dra (2016) Random forest in remote sensing: a review of applications and future directions. ISPRS Journal. 114, pp 24–31
35. Lin, Chihta, Wang, Naijian (2015) Feature selection and extraction for malware classification. Journal of Information Science and Engineering. 992. pp 965–992
36. Dijkstra TK, Henseler J (2015) Consistent and asymptotically normal PLS estimators for linear structural equations. Comput Stat Data Anal 81:10–23

Ethical Considerations in AI-Based Cybersecurity

Keshav Kaushik, Aadil Khan, Ankita Kumari, Ishu Sharma, and Rajat Dubey

Abstract The ethical implications of incorporating artificial intelligence (AI) into cybersecurity procedures are examined severely in this study article. The study highlights the significance of addressing ethical questions to guarantee responsible and equitable use of this technology, as artificial intelligence (AI) plays an increasingly vital role in identifying, preventing, and mitigating cyber risks. The research work highlights the important applications of AI in cybersecurity, stressing its capacity to detect abnormalities, evaluate large datasets, and adapt to changing attack tactics. But as AI-driven cybersecurity solutions become more common, privacy-related ethical issues become more important. The study investigates the possible privacy infringements brought about by the comprehensive data collection and surveillance powers built into AI-powered security systems. It highlights how important it is to use AI to identify threats while also protecting customers' personal information. The study also looks at moral concerns about the use of AI to cybersecurity globally. It addresses how AI systems in one country may have an impact on systems in other countries, emphasizing the need of maintaining international harmony, upholding standards, and averting damage or conflict in AI-related situations. In order to address global concerns, the report emphasizes the significance of ethical protections, widespread access to technology, and adherence to privacy and data protection legislation. The research work concludes by recommending a thorough strategy that gives privacy protection, bias mitigation, openness, and

K. Kaushik
School of Computer Science, University of Petroleum and Energy Studies, Dehradun, India

A. Khan · A. Kumari · I. Sharma (✉)
Chitkara University Institute of Engineering and Technology, Chitkara University, Rajpura, Punjab, India
e-mail: ishu.sharma001@gmail.com

A. Khan
e-mail: aadil.khan@chitkara.edu.in

A. Kumari
e-mail: ankita.thakur@chitkara.edu.in

R. Dubey
Cybersecurity Expert, Allianz Commercial, Austin, USA

K. Kaushik and I. Sharma (eds.), *Next-Generation Cybersecurity*, Blockchain Technologies, https://doi.org/10.1007/978-981-97-1249-6_19

accountability top priority when incorporating AI into cybersecurity. The development and use of AI-powered cybersecurity solutions must be guided by continuing ethical concerns in order to ensure that these technologies uphold moral standards and enhance global digital security.

Keywords Ethics · Cybersecurity · Technology · Network · Artificial Intelligence · Blockchain · Privacy

1 Introduction

In order to safeguard digital environments from constantly evolving cyberthreats, a paradigm change is required. Artificial intelligence (AI)-enhanced cybersecurity is one indication of this change. Businesses may strengthen their defenses by identifying and evaluating threats more intelligently by using increasingly sophisticated AI technologies in their operations. These techniques scan network traffic and user behavior for changes that can indicate security flaws fast. Artificial intelligence (AI) powers both anomaly detection and behavioral analysis. Experts in cybersecurity can anticipate and respond to emerging threats by using machine learning models for predictive analysis. Artificially intelligent response techniques may expedite crisis management, reduce damage, and swiftly isolate malfunctioning systems. Artificial intelligence (AI) is utilized to identify people and control access, while biometric and behavioral detection are used for security [1]. More instances of the several ways AI has enhanced cybersecurity include big data analytics, automated risk assessment, and fraud detection. Artificial Intelligence is a dynamic security ally that adapts to new threats by constantly learning and evolving. This provides enterprises with an adaptable means of defending themselves against more sophisticated cyber-attacks. When integrated properly, AI and cybersecurity strengthen digital systems and increase their resistance to various cyberthreats. AI improves safety by streamlining the process of identifying, addressing, and neutralizing hazards [2].

AI plays a critical role in improving defense. Making threats easy to identify is one of the finest things AI has accomplished. AI systems search for odd patterns in network traffic and user behavior using enormous datasets and complex algorithms. These instruments have access to a wealth of data. This gives us a significant edge in identifying and mitigating emerging risks before they worsen by making it feasible to identify potential security problems early on. With the use of machine learning models and prediction analysis, cybersecurity professionals are now able to anticipate emerging threats and take immediate action against them [3]. AI-driven models provide a proactive protection that can keep up with a dynamic risk environment by staying one step ahead. This is accomplished by identifying common patterns among various cyberattack types.

1.1 *How Effective Is AI for Cybersecurity?*

Artificial Intelligence (AI) has become a new and potent force in cybersecurity. With AI, it can detect, prevent, and address an increasing array of cyberthreats. Huge volumes of data can be handled by it, and it can recognize patterns and quickly adjust to evolving attack strategies. The strong integration of AI and cybersecurity has ushered in a new age of defenses that go beyond conventional techniques. As a consequence, digital settings are now much more reliable.

Artificial intelligence systems are capable of analyzing vast volumes of data, including network traffic, user activity, and system motions. Because of this, errors and departures from the established guidelines might be identified as potential indicators of security breaches. Artificial intelligence (AI) that can detect anomalies is crucial because it may identify complex threats that may evade detection by standard security protocols [4]. Here's where AI really excels. Machine learning models enable prediction analysis, which allows artificial intelligence to identify and forecast patterns associated with various cyber threats. With this proactive strategy, businesses can remain ahead of emerging threats, which is crucial in the cutthroat and dynamic industry of cybersecurity. Rather than just reacting to attacks that have already occurred, cybersecurity professionals can now address possible vulnerabilities and new attack vectors before they arise thanks to AI-driven prediction models. This approach is much superior than the previous one. Systems that manage user authentication and access are greatly benefited by artificial intelligence. An efficient and secure method of confirming that users are who they claim to be is biometric authentication, which is made feasible by AI-powered recognition systems.

Behavioral biometrics is another AI-powered aspect of security. In order to identify anomalies, this kind of biometrics examines user behavior patterns. It adds another degree of protection to access control systems. The use of AI to automatically scan computer networks and systems is significantly altering the way vulnerabilities are handled. AI-powered technologies enable enterprises to patch and secure their systems before malicious actors take advantage of them by identifying and analyzing vulnerabilities before they are exploited. This is made feasible by the ability of AI-run tools to identify and assess flaws. This strategic strategy is crucial to remember in order to reduce the attacks area and strengthen barriers. Scam detection becomes simpler with AI, which is problematic for defense. Scams and malicious files in email communications may be identified by AI systems using intricate formulae. This increases the likelihood of social engineering attacks and provides a crucial line of defense against a prevalent cyberthreat. AI-powered security analytics transforms large-scale security data analysis. This capability enables proactive protection strategies by enabling security specialists to go through large data for threats and take appropriate action. The degree of AI's flexibility determines how successfully it defends against cyber dangers. Because AI systems can adapt to changing threat environments and learn from fresh data, they are effective in combating online threats [5].

1.2 What Are Ethical Considerations in Cybersecurity with AI?

Artificial intelligence (AI) and privacy raise many significant ethical issues. These queries highlight the significance of designing and using these technologies with transparent, ethical practices. The fact that protection measures powered by AI are able to gather and monitor vast volumes of data is one of the main concerns. The possibility of privacy violations resulting from these abilities is one of the key concerns. In order to promptly identify ethical issues, we must simultaneously safeguard the privacy of consumers' personal information. Algorithms never favor or discriminate against anybody, which is another significant social issue. Biases present in training data may be reinforced by artificial intelligence systems. Results may become skewed as a consequence. When it comes to cybersecurity, threat detection might be biased, which can result in incorrectly identifying some people or unfairly characterizing others [6]. Ensuring that AI protection solutions are created and utilized fairly is crucial to upholding ethical norms and reducing the possibility of unjust impacts. The clarity and rationality of AI algorithms is another societal problem to consider. Problems arise when the decision-making process of AI systems is unclear. Understanding the risk assessment process of AI systems is crucial for user confidence and safety obligations. Clear system choices are necessary for AI to be employed in an ethical manner. When errors occur or the system is unable to thwart an attack, AI-driven cybersecurity makes it difficult to assume accountability and responsibility. Ethics requires accountability from individuals as well as solutions for errors or unforeseen consequences that may arise from the use of these instruments. Security hazards, such as the possibility of being attacked by an adversary, add layers of ethical complexity. Hostile attacks occur when artificial intelligence (AI) systems are programmed to behave in unexpected ways.

AI-based protection solutions must be reinforced against these types of threats due to ethical concerns, preventing their unethical usage. Unethical issues include AI-driven hacking and unequal access to resources. Increased resource inequality favoring wealthy companies might result from advanced AI. Ethical concerns need equitable safeguards and universal access to technology. Online safety and the ethics of AI are also influenced by global events. AI systems in one nation may have an impact on systems in other nations because of the global nature of safety. International cohesion, norms, and preventing harm or conflict in AI-related hacking are among the ethical issues. Observing the law and the principles of ethics are necessary. As long as AI technologies adhere to privacy and data protection regulations, they are employed effectively [7].

1.3 Why It Is Necessary?

It's critical to consider ethical issues when combining artificial intelligence (AI) with defense. This will ensure that these technologies are created and used responsibly. The importance of striking the correct balance between identifying hazards and protecting people's right to privacy is shown by conflicts involving privacy. Algorithmic bias and fairness are critical to preventing unjust outcomes from surveillance procedures, promoting inclusiveness, and guaranteeing that all individuals are safeguarded equally [8]. People are able to comprehend how artificial intelligence systems determine which threats are there as long as there is transparency and explainability. For user responsibility and trust, this is crucial. In order to correct errors and unanticipated consequences and foster a culture of dependability and accountability, it is essential to establish procedures for accountability and responsibility in AI-driven protection. When safety conditions change, having humans oversee critical choices ensures their morality and prevents users from becoming too dependent on self-driving cars. Individual oversight ensures that significant decisions adhere to the law [9]. In order to preserve their integrity, AI-powered security technologies must be shielded from potential threats such as hostile cyberattacks. It's critical that everyone be involved, treated fairly, and that nations collaborate while using AI due to ethical concerns regarding resource allocation and its consequences on the global community. For AI systems to operate legally and to uphold moral standards, laws and regulations must be obeyed. In order to avoid moral dilemmas and make responsible AI decisions, it is crucial to consider the long-term repercussions on society. Fundamentally, ethical concerns serve as a guide to ensure that the development and use of AI in defense aligns with the principles of accountability, transparency, equity, and public health.

2 How to Use AI for Cybersecurity?

There are many useful tools that artificial intelligence (AI) can give in the area of cybersecurity that should be used to improve digital protections. You can't say enough good things about AI's contribution to threat detection. It carefully looks through huge datasets to find trends and oddities that could mean a security breach. This includes desktop security, where AI-powered technologies carefully watch each device, finding dangerous actions right away and stopping them before they can do any damage. Phishing is a common type of online attack that faces a big problem from artificial intelligence (AI). Artificial intelligence (AI) can expertly search communication channels for signs of fake efforts, keeping users from falling for scams.

Artificial intelligence is powerful not only because it can react, but also because it can stop bad things from happening [10]. AI programs can find normal behaviors in users and systems and report them right away if they are different, which could mean there has been a security breach. Behavioral research lets you figure out what normal user and system behavior looks like so you can report any problems. Additionally, AI makes a big difference in incident reaction by automating tasks, which speeds up the process of finding, limiting, and reducing the effects of security events. In vulnerability management, AI does regular checks on systems and networks to see if they have any holes that could let hackers in. Artificial intelligence (AI)-powered tools are very helpful for security analytics because they give more detailed information about security data through trend analysis, pattern recognition, and prediction analytics. Artificial intelligence (AI) helps to tighten access controls by making user identity better. It does this by using cutting-edge methods like biometrics and behavioral analysis [11]. One area where artificial intelligence is used is in security automation. This makes routine tasks easier and gives cybersecurity experts more time to work on more difficult issues. Furthermore, machine learning improves safety by allowing for more advanced malware detection and the exact identification of new threats. Additionally, artificial intelligence is a useful tool for cybersecurity, but it may work better when it is part of a full security plan along with other safety measures. When it comes to safety, AI tools are great at finding threats, keeping endpoints safe, and stopping scams. Their skills include behavioral analysis, incident response automation, and risk management, all of which make digital security stronger. AI's important part in security analytics, improving user identification, and finding advanced malware makes its place in complete cybersecurity strategies even stronger. The Components of Artificial Intelligence for Cybersecurity are depicted in Fig. 1.

2.1 Machine Learning

Machine learning makes security better by letting computers look through huge datasets, find security risks, and predict them in real time. Its ability to change makes it better at finding anomalies by reducing false hits and raising total accuracy. Machine learning improves our understanding of normal trends in user behavior data, making it easier to spot changes that might mean there has been a security breach. Additionally, machine learning methods improve in the area of malware identification as risks change, making security measures more adaptable. When this technology is used in risk assessment, it can adapt quickly to changing cyber settings [12]. This makes cybersecurity stronger against a wide range of threats that are always changing.

2.2 Deep Learning

Deep learning makes safety much better by changing how threats are found and how they are dealt with. Because their neural networks are so complex, deep learning algorithms are very good at finding complicated patterns and outliers in big datasets.

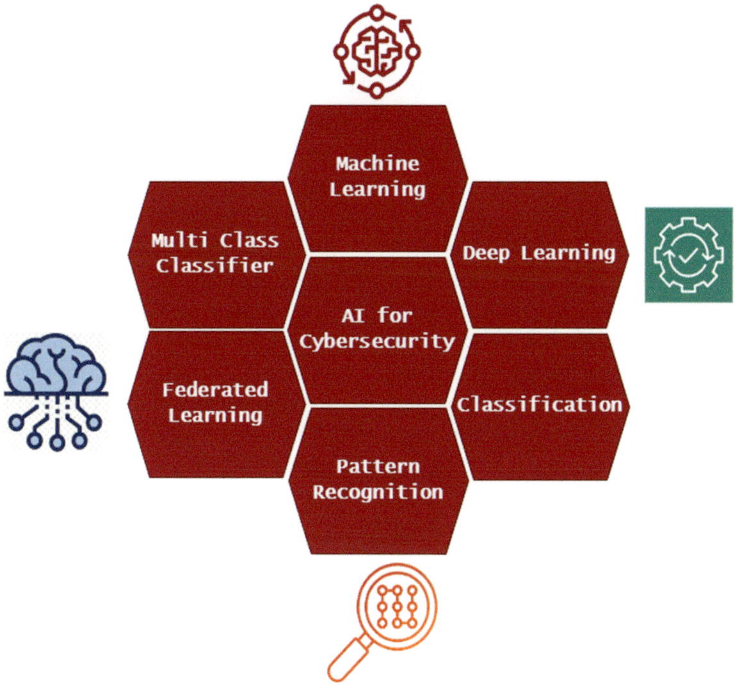

Fig. 1 Components of artificial intelligence for cybersecurity

This makes it easier to find possible security holes before they happen. This lets systems naturally deal with new dangers, lowering risks in real time. Deep learning, in particular, makes intruder detection systems, behavioral analysis, and malware identification better. This adds another layer of protection against complex cyberattacks. Deep learning models can keep learning and getting better from their mistakes [13]. This makes sure that the cybersecurity framework is dynamic and flexible, which is very important for keeping private data and systems safe in a dangerous environment that is always changing.

2.3 *Classification*

Classification algorithms make security better by accurately describing the difference between normal and dangerous behavior. By looking at trends in huge databases, these programs quickly find things that don't make sense and could be signs of security holes. By observing patterns, they help make attack detection systems better, which makes protections against new cyber threats stronger. Also, to find phishing, classification algorithms look at communication channels and tell the difference between real and malicious data. This lowers the risk of scams. These algorithms are very important for user identification because they sort physical and behavioral traits into groups, which makes access controls stronger. Basically, using classification algorithms lets cybersecurity systems make decisions in real time based on trends they've learned. This makes digital platforms more resistant to a wide range of complex cyberattacks. Here are some categorization algorithms that are regularly used in cybersecurity.

Random Forest algorithms make security better by making it easier to find threats in a way that is both reliable and flexible. Random Forest is a machine learning ensemble method that is very good at looking at big datasets, finding trends, and telling the difference between behaviors that are beneficial and detrimental. This method is very important in cybersecurity for finding and predicting potential threats correctly, without giving false positives or rejections [14]. Random Forest's strength is that it can handle a lot of different and complicated data, which lets it find new and complex computer risks. Its group nature, which includes many decision trees, makes it less likely that the model will be overfit, which improves its total performance. Random Forest algorithms can be used in cybersecurity systems to find and stop new threats before they happen. This protects digital platforms against a wide range of constantly changing cyber threats.

Support Vector Machine (SVM) methods are very useful for improving safety because they are exceptionally effective at what they're supposed to do. When it comes to threat identification, SVMs are great at sorting data points, which lets you find trends that are linked to adverse behavior. This lets network data be watched in real time, so any cyberattacks can be found and stopped. SVMs are great at finding anomalies, which are changes from how a system usually works. This is very important for stopping threats early on. By arranging network events into groups, SVMs also help breach detection systems find malicious activity. Because they can handle complex, high-dimensional data, SVMs are good at detecting cyber risks that are minor and change over time. This makes total cybersecurity protection stronger with a proactive and adaptable approach.

Using the k-Nearest Neighbour's (KNN) method in security makes finding threats and responding to them much better. The machine learning method KNN is very good at finding data points by how similar they are to examples that are already known. When it comes to cybersecurity, KNN can look at trends of network behavior to quickly find outliers and possible security holes [15]. It can find new data points by comparing them to close neighbors, which helps with assessing risk in real time. KNN helps breach detection systems find trends and actions that don't seem right, so they can move quickly and stop the attack. This algorithm is very versatile as well as effective at dealing with changing cyber threats. It is an important tool for making digital environments stronger, and it can be used in addition to normal security measures as an extra line of resistance. KNN gives people who work in safety a broader and more flexible way to deal with new dangers.

2.4 Multi-class Classification

Multi-class classification helps improve cybersecurity by separating different cyber threats into separate classes. This makes it easier to find and stop threats. With this method, AI models can tell the distinction between different types of threats, like malware, scams, and hacks. This helps faster and more targeted defenses be used. The detailed labeling also helps put threats in order of importance based on how bad they are, which makes it easier to use resources to respond to incidents. Multi-class classification enables defense systems to respond quickly to changing threat settings, making it easier for them to find and stop new threats. This purposeful use of AI not only makes digital settings more secure, but it also creates a proactive defense against a wide range of cyber risks that are always changing.

2.5 Pattern Recognition

Pattern recognition is an important part of enhanced cybersecurity because it protects against novel threats in a preventative and dynamic way. By using strong algorithms, artificial intelligence systems can quickly find and study trends in very large datasets. This lets us find things that don't seem right, which could mean that there have been security breaches. This ability also includes behavioral analysis, which lets computers tell the difference between normal human behavior and strange behavior that could be a sign of bad behavior. In the field of cybersecurity, pattern recognition also helps find current malware signs quickly and guess how attacks might happen in the future [16]. It helps security experts stay ahead of cyber risks which speeds up response times and makes the system more resistant to a threat environment that is always changing. Deep knowledge in pattern recognition works like a force booster, making digital systems stronger and reducing risks.

2.6 Federated Learning

Federated Learning, which spreads out model training, looks like a game-changing way to make safety better. With this new way, technology can learn from each other without sharing raw data. Federated Learning can help organizations make their danger identification models effective in the field of cybersecurity. By mixing information from different sources without compromising the protection of individual data, the system gets better at spotting new risks [17]. This autonomous learning makes it less likely that centralized data will be stolen and also makes sure that everyone works together to make the network safer. Federated Learning makes protection measures stronger by letting people work together and protect their privacy. This helps people respond quickly and effectively to new cyber threats.

3 Real-World Cybersecurity with AI Application Areas

Artificial intelligence (AI) is used in many areas of real-world safety and is very important. Through clever Intrusion Detection Systems (IDS), which use machine learning to find strange trends in network behavior, AI changes the way threats are found and analyzed. It includes computer security, using behavioral analysis and anomaly recognition to quickly find and stop bad behavior. Using AI to power automated incident reaction speeds up the process of reducing security incidents by making routine jobs easier to do [18].

The use of natural language processing (NLP) to find scam emails and harmful links is made easier by AI systems that look at the text and trends of communication. AI-driven automatic scanning improves the finding and ranking of system flaws, making proactive risk reduction easier in vulnerability management. Through behavioral biometrics, AI can also affect identity and access management. This adds an extra layer of security by using user behavior to verify identities.

AI-driven solutions are very important for improving and changing cybersecurity measures in areas like threat intelligence, user behavior analytics, mobile security, and cloud security. All together, these apps give companies the tools they need to deal with cyber risks that are always changing, quickly and smartly [19]. This makes sure that their digital defenses are strong even when they are up against smart attackers. Table 1 summarizes the comparative analysis of available cybersecurity tools.

3.1 Datasets for Research Work

When doing research in the field of cybersecurity, having access to a variety of datasets of a high quality is very necessary in order to create effective protection mechanisms against constantly changing threats. A wide variety of organizations, ranging from educational establishments to research institutes specializing in cybersecurity, play an essential part in the process of gathering datasets from controlled lab conditions and real-world situations.

Universities all throughout the world, such as the Canadian Institute for Cybersecurity, the United States Military Academy at West Point, and the University of New South Wales in Sydney, are actively involved in research and collect databases that serve as useful resources for scholars. These databases make it easier to find new dangers and build effective solutions, both of which are important goals.

In addition, public platforms like Kaggle, GitHub, Medium, and the UCI Machine Learning Repository provide researchers easy access to routes for discovering and retrieving datasets. Because they are collaborative in nature, these platforms encourage the exchange of information and speed up the progression of research in the field of cybersecurity. Researchers may use these datasets to their advantage by analyzing attack patterns, modeling threat behaviors, and coming up with novel ways to better cybersecurity measures.

Table 1 Comparative analysis of cybersecurity tools

Functionality	Tools	Supported platforms	Integration capabilities	Key features
Threat detection and analysis	Darktrace, CrowdStrike, Symantec	Windows, Linux, macOS	Limited (integration APIs available)	Advanced threat detection, real-time monitoring, automated response
Malware detection	Cylance, Symantec, McAfee	Windows, macOS	Extensive (integrations with various tools)	Dynamic defense against evolving threats, signature-less protection
Endpoint security	CrowdStrike, Carbon Black, SentinelOne	Windows, macOS	Limited (integration APIs available)	Behavioral analysis, threat hunting, endpoint detection and response (EDR)
phishing detection and email security	Proofpoint, Barracuda Sentinel, Ironscales	Windows, macOS	Extensive (integration APIs available)	Email threat detection, real-time protection, incident response automation
Behavioral analytics	Exabeam, Securonix, Varonis	Windows, Linux, macOS	Limited (integration APIs available)	Anomaly detection, user risk scoring, automated incident response
SIEM (Security Information and Event Management)	Splunk, ArcSight, Elastic SIEM	Windows, Linux, macOS	Extensive (integration with various SIEM solutions)	Log management, real-time monitoring, threat intelligence integration
Vulnerability management	Qualys, Tenable, Rapid7	Windows, Linux, macOS	Extensive (integration with various vulnerability scanners)	Continuous scanning, risk prioritization, compliance management
Incident response automation	Demisto, Phantom, IBM Resilient	Windows, Linux, macOS	Extensive (integration with various security tools)	Automated playbooks, case management, threat intelligence integration

(continued)

Table 1 (continued)

Functionality	Tools	Supported platforms	Integration capabilities	Key features
UEBA (User and Entity Behavior Analytics)	Gurucul, Rapid7 InsightIDR, Splunk	Windows, Linux, macOS	Extensive (integration with various identity management tools)	Insider threat detection, risk-based analytics, anomaly detection
Network security	Vectra AI, Darktrace Enterprise Immune System, Cisco Stealthwatch	Windows, Linux, macOS	Extensive (integration with various network security solutions)	Real-time threat detection, automated response, network anomaly analysis

A robust ecosystem of cybersecurity datasets has been contributed to by the combined efforts of academic institutions, research institutes, and open platforms. Researchers now have the ability to investigate new threats, test ideas, and eventually strengthen digital landscapes to withstand an ever-evolving cyber threat scenario because of the quantity of data at their disposal. The fact that so many different datasets are readily available helps to guarantee that the cybersecurity community will continue to be at the forefront of innovation and preparation in the face of more complex forms of online threats. Table 2 gives insights of available cybersecurity datasets for researchers.

4 Use Cases of AI in Cybersecurity

This section discusses the use cases of Artificial Intelligence (AI) in Cybersecurity area in detail. Figure 2 gives an overview of different use cases of AI in Cybersecurity.

4.1 Healthcare

Cybersecurity is critical to the healthcare industry because it ensures the availability, accuracy, and privacy of sensitive patient data as well as critical medical systems. Ensuring the security of vast quantities of patient data, including electronic health records (EHRs) and personal information, is one of the most critical issues facing healthcare companies today [35]. To address this issue, many security measures have been implemented. Encrypting data, restricting who may access it, and routinely upgrading security are some of the measures used to prevent data breaches and unauthorized access. The increasing prevalence of connected medical equipment and video services raises new issues. Safety procedures have been implemented

Table 2 Available cybersecurity datasets for researchers

S. No.	Dataset	Research Center/ Universities	Attacks	Format	Area of attack	Description
1	CICEV202 DDoS	Canadian Institute for Cybersecurity	Ddos	CSV	EV Industry [20]	A DDoS attack on any industry includes sending a huge amount of data to its online systems, which can interrupt services and put important operations at risk, posing a serious threat to their usefulness and dependability
2	CICAndMa 2017	Canadian Institute for Cybersecurity	Adware, Ransomware, Scareware, SMS Malware	Pcap	Android [21]	Adware shows annoying ads, ransomware locks files to get money, scareware scares people with fake threats, and SMS malware gets into phones through text messages
3	ISCX URL2016	Canadian Institute for Cybersecurity	URL	Pcap	Website [22]	URL attacks confuse users or exploit weaknesses by changing or hiding site addresses, typically resulting in unauthorized access or dangerous information. URL verification is essential to prevent phishing and malware
4	CICDarkne 2020	Canadian Institute for Cybersecurity	VPN, Tor	Parquet	Audio-Stream, Browsing, Chat, Email, P2P, Transfer, Video-Stream and VOIP [23]	Attacks on VPNs and Tor take advantage of flaws to compromise users' privacy and get illegal access to private data. This shows how important strong encryption and security measures are for these privacy-focused technologies
5	UNSW NB15	UNSW Sydney	Fuzzers, Analysis, Backdoors, DoS, Exploits, Generic, Reconnaissance, Shellcode and Worms	Pcap files, BRO files, Argus files, CSV files	IoT [24]	Fuzzers test software for vulnerabilities, analysis examines system weaknesses, backdoors allow unauthorized access, DoS floods servers to disrupt, exploits exploit vulnerabilities, generics cover various attacks, reconnaissance gathers intel, shellcode executes malicious operations, and worms self-replicate to spread rapidly in cyber attacks

(continued)

Table 2 (continued)

S. No.	Dataset	Research Center/ Universities	Attacks	Format	Area of attack	Description
6	Bot-IoT	UNSW Sydney	DDoS, DoS, OS and Service Scan, Keylogging and Data exfiltration	Pcap, Csv	IoT [25]	DDoS floods systems with traffic, DoS blocks access, OS and Service Scan looks for holes in the system, Keylogging records keystrokes, and Data exfiltration leaks private data in cyberattacks
7	Honeypots	Data Driven Security	AWS honeypots	Csv	AWS [26]	AWS honeypots act like security holes to attract and study cyber risks. This helps us understand possible attacks and makes cloud security better
8	2007 TRECPubli SpCorpus	TREC 2007 Public Corpus	Spam	Csv	Email [27]	A spam attack sends a lot of unwanted and often useless messages through communication channels to mess them up, trick them, or spread harmful content
9	ICML-09	Detecting Malicious URLs	URL	Matlab	Website [28]	A URL attack changes the names of websites to trick people or use holes in security. For your own safety, check URLs carefully because they can lead to scams, malware, or people breaking the law
10	CTU-13	CTU University	Botnet traffic	Pcap	Web [29]	As a network of hacked computers floods a target with malicious requests, overloading systems and disrupting services, botnet traffic attacks demonstrate the necessity for strong defenses against coordinated and automated cyber attacks
11	CTU-IoT Malware Capture-341	Aposemat project	Mirai	Pcap	IoT malware [30]	The Mirai attack took advantage of IoT devices that weren't safe, turning them into a huge botnet that slowed down online services by flooding targets with too much traffic. This showed how vulnerable connected devices are to cyber threats
12	Android Mischief Dataset	StratosphereLab	Remote Access Trojan (RAT)	Pcap	Android [31]	A Remote Access Trojan (RAT) attack enables an attacker to remotely manipulate a device, compromising security and privacy. To reduce RAT risks, use robust security measures

(continued)

Table 2 (continued)

S. No.	Dataset	Research Center/ Universities	Attacks	Format	Area of attack	Description
13	IoT devices captures	Alto University	Networkflow	CsV, Pcap	Iot devices [32]	Using network flow attacks to look into or change the trends of network traffic data could put private information at risk. This shows how important it is to have good tracking and breach detection systems
14	KDD Cup 1999 Data	University of California	Dos, R2L, U2R	Pcap	IP address [33]	DDoS stops services from working, R2L uses security holes to get in illegally, and U2R gives users more rights, so specific security measures are needed
15	Domain Name Service Logs	USMA network	DNS	Pcap	CDX network [34]	DNS attacks use DNS flaws to stop domain names from being resolved to IP addresses and send users to hostile websites. This shows how important it is to have good DNS security

Fig. 2 Use cases of AI in cybersecurity

in order to safeguard these technologies and maintain the security of patient interactions [36]. [Protective measures for the internet] [Cybersecurity measures] are designed to safeguard patient communications and maintain the security of these instruments [37]. Pharmaceutical firms that engage in research and development also have cybersecurity issues. Advanced systems that can detect attacks and prevent data loss are necessary for these organizations to safeguard their intellectual property and important research data. As more individuals utilize linked technology, such as video services and medical devices, new issues begin to surface. To safeguard these technologies and maintain patient contacts' privacy, security precautions have been taken. [Care should be used to safeguard online] The privacy of these technologies and the exchanges between patients are being safeguarded by [cybersecurity] measures. Hacking is another issue that pharmaceutical businesses that engage in research and development may face [38]. These organizations need sophisticated systems that can identify potential attacks and prevent data loss in order to safeguard their priceless research records and intellectual property. Aside from emergency response systems, other crucial topics include hospital networks and staff awareness training programs.

Through the consideration of these use cases, healthcare institutions may construct a robust cybersecurity framework that guarantees the confidentiality of patient data while simultaneously ensuring the continuous availability of critical systems and services. This enhances both the quality of treatment provided to patients and their general health.

4.2 *Finance*

In the banking industry, cybersecurity is crucial since several threats have the potential to compromise confidential financial information and halt significant transactions. Ensuring the security of financial transactions is among the most significant challenges. In order to prevent theft and unwanted access, it is crucial to use transaction monitoring software, multi-factor authentication, and robust encryption in this scenario. Numerous personal details about its clients are handed to financial organizations. In order to prevent unauthorized access or data breaches, they use sophisticated encryption, restrict who may access it, and conduct frequent security audits. The use of cards and automated teller machines (ATMs) presents new cybersecurity issues. Installing secure hardware, updating software often, and maintaining tight monitoring are necessary to prevent card skimming and other kinds of fraud [39]. There are now greater security concerns in this new arena as mobile banking gains traction. Making secure mobile applications, routinely checking security, and teaching clients the best practices for using their banks are all necessary to address these issues. To prevent financial information from being stolen online and to maintain the integrity of the market, online trading platforms must be secure. Real-time tracking, security testing, and safe coding approaches are required for this to function. Strict guidelines regarding who may access what, employee training programs, and constant monitoring of user behavior to identify and address any unusual activity are all necessary to guard against insider threats. Respecting the guidelines established by authorities such as PCI DSS and GDPR is also crucial. For this reason, extensive safety regulations and routine inspections were created. It's more difficult to monitor your cryptocurrency wallet securely, so you should use multi-signature authentication and other cutting-edge security techniques to keep your bitcoin safe from theft [1]. Supply chain security, crisis response plans, and recovery plans are all components of a comprehensive cybersecurity strategy that protects the banking sector from emerging internet threats.

4.3 *Supply Chain*

Cybersecurity becomes more important as supply chains become more complex in order to ensure the integrity of data, protect critical operations, and facilitate the flow of information. Technology used in shipping makes it more difficult to protect

Internet of Things (IoT) devices from hackers. Ensuring that these devices are not accessed by unauthorized individuals requires robust security measures. Several of them include systems that scan for breaches, encryption, and frequent modifications. Because inventory management systems are a crucial component of the supply chain, they must be shielded against intrusions and unauthorized access. Important possibilities include access controls, security audits, and open inventory management using blockchain technology. Companies must establish secure channels of communication and provide frequent hacking training to staff members who interact with suppliers in order to ensure that confidential information is sent and received by suppliers in a secure manner. Warehouse management systems need to be shielded from the dangers associated with security breaches since they preserve correct product data [40]. These systems may be kept secure via data consistency checks, blockchain technology for unchangeable record storage, and other security measures. Strong security measures must be implemented immediately since transportation systems, particularly vehicles connected to distribution platforms, are vulnerable to hacking. By ensuring that connected automobiles are safe, establishing secure communication channels, and routinely assessing the security of delivery platforms, vulnerabilities in the transportation system may be minimized. Anti-counterfeiting technologies, blockchain monitoring, and routine inspections are required to prevent counterfeit items from entering the supply chain. It needed cross-border training and a global cybersecurity strategy in order to manage cybersecurity across international supply chains with many partners and legal contexts. Businesses must prepare for supply chain resilience if they want to continue operating in the face of interruptions and hacks [41]. The significance of cybersecurity for the stability and dependability of the supply chain is shown by these many examples.

4.4 Digital Media

The world of digital media is constantly evolving. In order to maintain the safety of the intricate interaction between content providers, users, and websites, cybersecurity has become crucial. One of the most important issues is keeping digital assets secure. Use digital rights management (DRM), encryption, and watermarking to prevent unauthorized users from stealing and distributing digital material. Since companies retain so much personal data about their users, streaming services struggle to protect their privacy. Strict adherence to data protection regulations, encryption techniques, and explicit privacy rules are necessary to safeguard the security and privacy of user information [42]. Although online advertising networks play a significant role in digital media business plans, issues like virus placement and ad frauds harm them. Finding and preventing frauds depends heavily on using ad verification tools, maintaining the security of ad networks, and doing routine audits. Secure payment methods, robust login procedures, and frequent security audits are necessary for digital media organizations to prevent unauthorized users from accessing paid material, which is often how they generate the majority of their revenue. Social media platforms, which are crucial for digital communication, are struggling with issues including user account takeovers by hackers and the dissemination of misleading information. Social networking platforms may be made safer for users with the use of information programs, content screening technologies, and multi-factor authentication. Content makers utilize copyright protection, digital watermarking, and legal procedures to prevent piracy. Digital media websites need robust network architecture, content delivery networks (CDNs), and security techniques against denial-of-service (DDoS) attacks [43]. The legitimacy of digital media content is under danger due to deepfake technology; thus, money has to be invested on systems that can detect it, consumers need to be made aware of the risks, and cooperation between all parties is required. Digital products retailers on the internet face the danger of fraudulent transactions and data breaches. Robust payment mechanisms, secure e-commerce platforms, and frequent security audits are essential for cybersecurity. Finally, user-generated content websites must handle abuse and unauthorized file submissions. User reports, content control systems, and community guidelines all contribute to a secure and entertaining online environment.

4.5 Research Institutes

Research organizations are primarily in charge of a number of significant responsibilities, including advancing human understanding, fostering the development of fresh concepts, and solving difficult challenges. These institutions serve as centers for innovative research. To explore uncharted ground, they bring together groups of academics, researchers, and businesspeople from many sectors. Research institutes provide significant contributions to scientific discoveries, ground-breaking discoveries, and answers to pressing global issues via meticulous testing, in-depth analysis, and collaboration with other professional groups [44]. They often work in a wide range of fields, including technology, natural science, medicine, and the social sciences. In addition to doing fundamental research, these centers play a critical role in educating the next generation of scientists, disseminating knowledge, and facilitating collaboration between government, business, and academic organizations. Research institutes have a significant influence on the globe in addition to their position in education [45]. They have the power to influence public policy, grow the economy, and direct human development.

4.6 SCADA

Supervisory control and data acquisition, or SCADA, systems are utilized in many different business domains and are essential for monitoring and managing complex operations. SCADA is used in the energy industry to monitor and control the production, transmission, and distribution of electricity. This contributes to ensuring the effective and dependable transmission of electricity. SCADA enables real-time monitoring of manufacturing lines in the industrial sector. Additionally, it enhances the overall utility of processes and the pace at which work is done. Water and wastewater treatment facilities employ SCADA to monitor water quality, automate processes to maximize resource use, and regulate water flow [46]. SCADA systems are useful in the transportation industry because they facilitate the management of many aspects such as airport operations, rail systems, and traffic signals. Additionally, SCADA plays a major role in ensuring the safety and problem-solving capabilities of critical equipment. It does this by providing control, real-time knowledge, and the capacity to act. Reducing the impact of any delays and improving system reliability are the main objectives. SCADA is a fundamental technology for enhancing the safety and efficiency of complex industrial processes because of its wide range of applications.

4.7 Big Data and Analytics

Big Data and analytics are a revolutionary force in today's world, revolutionizing the way organizations get insights from vast quantities of complex data. This is just one example of the numerous ways that big data and analytics are transforming the world. "Big data" refers to the ability to manage and analyze enormous volumes of ordered and disorganized data—often more than what typical systems can handle. Conversely, analytics involves a meticulous examination of this data in order to identify potentially valuable patterns, relationships, and trends [47]. When combined, these elements aid in improving decision-making, streamlining processes, and giving businesses a competitive advantage. The use of big data and analytics is widespread and serves a variety of purposes, including enhancing consumer satisfaction, forecasting market trends, enhancing healthcare outcomes, and streamlining supply chain operations. In our data-driven world, collaboration between big data and analytics is more important than ever. It provides you with more opportunities than any other element to be resourceful, frugal, and make well-informed judgments.

4.8 Internet of Thing

The Internet of Things (IoT) has already been shown to have several significant applications. Numerous commercial sectors are changing as a result of these usage, and many aspects of daily life are improving. The Internet of Things allows for remote patient monitoring in the medical field [48]. This provides current information on long-term ailments and vital signs to medical workers. Smart cities are using IoT (Internet of Things) to link infrastructure, enhance public services, and improve traffic flow in order to improve urban living. Internet of Things monitors provide agriculturalists with information on crop growth, weather patterns, and the condition of the land. They can utilize better, more sustainable agricultural techniques as a result. Predictive maintenance in industrial settings is made feasible by the Internet of Things, which monitors the condition of the machinery. This reduces the amount of time and associated expenditures when the equipment is offline [49]. IoT gadgets in the house, such as security systems and smart heaters, provide each individual greater control over their level of safety and comfort [50]. These many application examples demonstrate how the Internet of Things has transformed society by linking settings that improve the productivity, sustainability, and creativity of a wide range of business sectors.

4.9 Wireless Sensor Network

Wireless sensor networks, or WSNs, are being employed in a wide range of applications due to their ability to collect data remotely and follow objects in real time. In the field of environmental tracking, weblogs make it simpler to monitor variables like tree health, weather patterns, and air and water quality [51]. WSNs are helpful in the business sector because they enable you to do preventive maintenance by continuously monitoring the state of equipment and tools. This reduces unscheduled downtime and improves the efficiency of operations. In the medical field, Wi-Fi sensor networks (WSNs) are highly helpful for monitoring patients, recording their vital signs, and assisting physicians in identifying health issues early on. "Smart agriculture" refers to the use of wireless sensor networks (WSNs) for precision farming, optimal drainage management, and land monitoring to enhance crop growth. Monitoring the condition of vital infrastructure, such as highways and dams, is another crucial application required to ensure the safety and soundness of significant facilities [52]. These examples demonstrate how versatile and essential wireless sensor networks are for creating rapid and intelligent solutions across a wide range of industries.

4.10 Defense Services

Modern technology is used by the military forces in a variety of operational environments to protect national security and achieve a tactical edge. Unmanned aerial vehicles, or UAVs, are flying machines equipped with sophisticated cameras and monitoring systems that allow them to collect data, monitor their surroundings, and gaze about in real time. This improves people's overall understanding of the problem. The importance of cyber security in protecting military networks and assets against threats and cyberattacks cannot be overstated. Military operations are made simpler by satellite technology, which provide secure means of communication, navigation, and position determination worldwide [53]. Drones designed with military specifications and fitted with precise armaments are often used for precise strikes in hazardous environments. People are less likely to be injured as a result of this. Biometric technologies are used by people to verify their identity and prevent unauthorized access to secure locations. This makes some locations inaccessible to anybody other than authorized staff [54]. The armed forces also utilize data analytics to examine vast volumes of data. This aids in their decision-making, threat assessment, and mission planning. You can see how crucial cutting-edge technology is for enhancing military capabilities and maintaining national security from these examples of uses.

4.11 Robotics

Robotics is being used in many distinct and revolutionary ways in several sectors. As a result, processes have evolved and become more effective. When robots are used in production, they automatically perform precise, repetitive jobs. This reduces errors while increasing manufacturing speed. Medical treatments may be performed with little harm to patients thanks to robotic surgical equipment, which improves patient outcomes and precision. Autonomous vehicles, a kind of robotics, are transforming the transportation sector in a number of ways, including self-driving cars, delivery robots, and drones [55]. Robotic technology has applications in agriculture that are beneficial to the industry. As an example, both precision farming and automated reaping increase food output. Robots are an essential component of the response to natural disasters since they may be used for search and rescue operations as well as in hazardous environments. By assisting with activities like picking, packaging, shipping, and tracking of items in the logistics and storage industries, robots are increasing the efficiency of the supply chain [56]. The many applications of robots here demonstrate how adaptable and extensive the technology is in transforming several industries and resolving challenging issues.

4.12 Augmented Reality/Virtual Reality

The way that people engage with augmented reality (AR) and virtual reality (VR) has completely altered as these technologies have proliferated in various businesses. With the use of virtual reality (VR), educators may build realistic models that allow students to learn by doing. Virtual reality field excursions and interactive anatomy lessons are two instances of VR in the classroom. Virtual reality (VR) is being utilized in the medical industry for therapeutic reasons, such as exposure treatment for anxiety disorders [57]. Augmented reality (AR) is used in the medical industry for educational purposes. Shoppers can virtually try on clothing thanks to retailers using augmented reality (AR), which improves online buying in general. Meanwhile, realistic product presentations and virtual marketplaces are made feasible by virtual reality (VR) [58]. AR overlays, which provide in-the-moment employment advice, and VR models, which replicate potentially hazardous surroundings, are helpful for workplace training. Virtual reality (VR) allows users to take comprehensive tours of virtual locations, while augmented reality (AR) allows users to see buildings in their natural environments. Virtual reality (VR) is utilized in the entertainment industry to provide greater realism in video games, while augmented reality (AR) adds captivating features to live events to make them better [59]. new applications of virtual and augmented reality demonstrate how new technologies are revolutionizing a wide range of industries by improving participation, experiences, and instruction.

4.13 Smart City

Smart cities employ state-of-the-art technology to improve urban living and increase sustainability and efficiency in many areas. Smart parking options, ingenious traffic control systems, and real-time public transportation monitoring are all being employed as part of smart city initiatives to improve mobility by making the city less congested and more functional. Energy-efficient infrastructure contributes to resource management and energy consumption that is non-destructive [60]. These include web-connected devices and services, such as smart grids. Smart city technology is used in public safety to monitor situations and act swiftly in case of an emergency via the use of surveillance cameras, sensor networks, and data analytics, among other tools. Digital platforms benefit urban governance by facilitating more participation, data-driven decision-making, and transparent service delivery [61]. Cities may become healthier and more sustainable by doing things like monitoring the environment, cleaning up after litter, and conserving water. A connected ecosystem may be established by combining data processing with Internet of Things (IoT) devices [62]. This improves people's lives generally and enables municipalities to react swiftly to emerging issues.

4.14 Gaming Server

In order to ensure that players can connect with one another, that games run smoothly, and that a variety of games may be played, gaming servers are required. In massively multiplayer online games, or MMOs, computers manage a huge number of participants and enable real-time communication, collaboration, and competition [63]. First-person shooter (FPS) games need computers for low-latency exchanges. This guarantees that gameplay is swift and equitable for all players. Role-playing games, or RPGs, allow users to connect to servers where they may explore endless worlds, advance their characters, and share narratives with other players. When many players' movements are coordinated and intricate game states are monitored by a server, strategy games become even more entertaining. E-sports tournaments also need game servers, which must be dependable and responsive to ensure fair competition. Having a robust computer network is even more crucial in light of the popularity of cloud gaming. This is due to the fact that excellent games may be watched without requiring strong equipment to be close by [64]. This makes powerful computers even more crucial nowadays. An essential component of the technology that enables internet gaming is the game server. They also set the broad rules for how the game is played.

4.15 Power Grid/Nuclear Plant

Modern technology is essential to power grids and nuclear power plants in order to ensure that energy is generated, transferred, and delivered in the most dependable and safe manner possible. Monitoring and managing the electrical grid's operation using Supervisory Control and Data Acquisition (SCADA) systems is critical. These tools enable workers to manage energy flow, react to power outages, and improve grid performance in real time [65]. Power firms may also use predictive analytics to schedule repairs, increase system resilience, and estimate demand. Automation and sensors are employed in nuclear power plants' safety systems to monitor the reactor, search for issues, and initiate safety procedures as required. The use of robotics and other remote-operated technology facilitates the restoration of hazardous environments with less human exposure. Power lines and nuclear power plants are made more stable, secure, and long-lasting by these technical applications. Therefore, they ensure that there is a reliable and secure supply of electricity to fulfill the rising demand for energy.

5 How to Achieve Trustworthy Cybersecurity with AI?

Artificial Intelligence requires a comprehensive strategy including scientific, ethical, and pragmatic components to ensure human safety. This method's premise is that data security will be maintained. Strong security measures, such as encryption protocols, secure storage techniques, and stringent access restrictions, are necessary for businesses to protect sensitive data from unauthorized access and potential breaches. Adhering to data protection regulations is crucial if you want to ensure that your handling of personal and private information is appropriate. It must be evident for humans to trust AI to make judgments. When AI models explain why users made the decisions they did, users tend to trust them. Techniques for interpretable machine learning models, or explainable artificial intelligence (XAI), aid in the understanding of algorithmic processes and decision-making. Working with professionals in AI and human protection requires transparency and honesty, which fosters trust [66]. When deploying AI for protection, ethics are crucial. Ethics must guide the development and use of AI. In order to ensure equity and diversity, organizations must combat prejudiced AI systems. Regular tests for bias in AI systems encourage ethical AI. Developers, consumers, and other stakeholders must be informed about initiatives that support responsible AI behavior and must continue learning about AI in order for it to be utilized in an ethical manner.

An effective protection strategy must be updated and monitored often. AI-driven real-time threat identification and adaptable responses improve cyber risk detection and response. AI models and algorithms need to be updated often to stay up to date with new cyber threats. To avoid and resolve security issues, comprehensive incident response procedures that seamlessly integrate AI should be used. To provide trustworthy AI-powered defense solutions, safe AI development is required. AI model developers must build secure code, monitor for security flaws, and maintain AI tools up to date in order to address them [67]. Artificial intelligence applications may be protected against hacking and unauthorized usage using preventive protection. Adhering to the rules is a crucial component in attacking AI trust. It's critical to stay current on the laws and regulations pertaining to hacking. AI applications should abide by security and privacy regulations to avoid legal issues and to perform at the highest level in the industry.

6 Challenges and Limitations of AI in Cyber Security

AI in defense has many drawbacks and limitations, despite the possibility of significant advancements. One major issue is the dynamic nature of internet threats. AI systems can be vulnerable to attacks since adversaries are constantly altering their intentions. Additionally, asymmetrical attacks are made feasible by the intrinsic complexity of AI algorithm construction. These occur when crafty attackers attempt to evade detection by manipulating the AI's decision-making processes. A different problem is that there aren't enough designated good quality datasets [68]. This poses a challenge to the development of trustworthy AI models. Furthermore, a lot of AI algorithms lack total clarity, which might make them more difficult to operate and comprehend.

As a result, it might be challenging for those in the safety field to completely comprehend and have faith in the decisions made by AI systems. This makes it far more difficult to tackle the issues than it would be otherwise since there aren't many qualified safety specialists who are proficient with AI-powered systems. It is crucial to ensure that artificial intelligence advances rather than replaces human expertise in order to strike a balance between the control automation grants people over processes and the control automation grants humans over those processes. In order to use AI's potential for defense and build a robust and secure digital environment, it is imperative to address these issues and constraints [69]. Owing to these issues and constraints, doing both at once is not feasible.

7 Need of Cybersecurity AI Frameworks for Ethical Considerations

The cybersecurity landscape is dynamic, and artificial intelligence (AI) is playing a bigger role in fortifying defenses against more sophisticated attacks. However, as AI technologies are becoming more and more crucial for identifying dangers, handling emergencies, and reaching judgments, the societal challenges that arise from their application have received a lot of attention [70]. This is because artificial intelligence (AI) technologies are playing an increasingly significant role in these processes. AI systems with comprehensive protection and moral foundations are becoming more crucial. Some of the most significant issues, such as the possibility of bias in AI systems, are addressed by ethics models [71]. Due to the fact that these algorithms are trained on historical data, there is a chance that they can unintentionally amplify preexisting biases in the data. Fairness, transparency, and accountability need to be crucial components of the ethical AI defense architecture.

As a consequence, the likelihood of distorted outcomes from the AI will decrease. One approach that businesses may take to ensure that AI tools function accurately and impartially is to establish standards for assessing and mitigating bias in AI models. You may find these recommendations here [72]. Another crucial component of moral AI protection systems is transparency. Some refer to sophisticated AI models—particularly deep learning models—as "black boxes" since it's difficult to understand how they arrive at conclusions. Open-minded moral theories provide credence to explainable AI (XAI) models. These models increase trust and a feeling of obligation by assisting those who work in the safety field in understanding judgments made by AI [73].

8 Role of Explainable AI in Cybersecurity

Explainable Artificial Intelligence (XAI) is becoming a fundamental component of open and trustworthy defensive systems. Safety options must be clear. XAI is crucial because it shows how AI models decide. By making AI findings understandable, XAI solves the "black box" problem [74]. In cybersecurity, this includes understanding how AI systems identify and manage threats. XAI helps security specialists comprehend system decisions—leading to better-educated and well-planned responses to security flaws or odd behavior.

XAI simplifies AI-powered protection solutions, boosting user confidence. Security professionals and decision-makers may verify AI reports and proposals. Openness is crucial to ensure that AI systems are moral, bias-free, and accountable [75].

9 Responsibility and Accountability of AI-Based Cybersecurity Systems

The evolving landscape of digital defense raises many concerns, chief among them the obligation and accountability of AI-driven defensive systems. Regretfully, these state-of-the-art technologies excel at identifying risks, stopping them, and lessening their effects. But in order for them to function effectively, roles and duties must be established. Businesses that use AI in the hacking space are accountable for ensuring that the systems are secure and that they are utilizing the data they get honestly [76]. Strong governance frameworks must be established in order to achieve this. These systems need to include moral norms, compliance, and supervision. Ensuring the accuracy and dependability of AI choices is the responsibility of cybersecurity specialists monitoring AI systems. Artificial intelligence programs are more responsible since their output is explicable and their algorithms are transparent [77]. This facilitates comprehension and evaluation of the justifications for taking protective action by all parties. Additionally, engineers and programmers play a critical role in ensuring that AI models are robust, safe, and impartial. This supports the promotion of duty throughout the system's lifetime. Everyone must feel that they have a shared obligation in order to reduce risks, maintain moral behavior, and maintain accountability when it comes to safeguarding digital environments. Coders and end users are included in this [78]. This is even more crucial now that artificial intelligence is being used in defense on a growing basis.

10 Ethical Issues with Large Language Models

Late last year, conversational AI chatbots, also known as generative AI (GAI), produced stunning text, images, code, music, and video depending on user requests. Many AI applications gained stakeholder and media interest because of their promise, flexibility, simplicity, and availability. Because of this, they had numerous users and varied uses. Conversational AI chatbot ChatGPT can answer general and specific questions and write effectively on any topic. The response may be refreshed following user input. TikTok's nine-month milestone was exceeded by 1 million users in four days and 100 million in two months. GAI tools were popular and expensive to create. Personal and business apps utilize GAI. OpenAI's ChatGPT API lets developers add ChatGPT to programs. Microsoft is bringing it to Power and Dynamics. SAP and others incorporate ChatGPT into applications [79]. Thus, many businesses are using GAI more. GAI is promising, popular, and praised yet has major constraints. Users are bewildered by ChatGPT's factual errors, fabrications, and wrong answers. AI content producers raise ethical challenges that developers, users, and politicians must address. Unexpected impacts might harm society, corporations, and the economy [80].

A big moral problem is bias in AI content producers, especially text engines like ChatGPT. The big language models (LLMs) that run content makers are trained on a lot of information, pictures, and data from many sources, including the Web. This means that any flaws in the training data will show up in the model output. This could lead to answers that are cruel, biased, wrong, or narrowly focused, as well as race or gender abuse [81]. GAI is biased because of how it was built and trained. LLM answers will be affected by biased facts and information, even if they are not on purpose. So, to get a good result and lessen bias, you need a sample that is fair, varied, and representative. AI may be less biased if data is curated, filtered, and training data is chosen. These steps look harder to put into action because there is so much training data. But software companies like OpenAI are making chatbots less biased and letting users change how they behave.

Misusing and taking advantage of AI content creators is another moral problem. Text makers like ChatGPT might spread lies, racist, sexist, or otherwise offensive messages. This tool could also be used to make harmful content that encourages violence or pretends to be someone else. Since users can ask any question, bad people could use the bot to act in a rude way [82]. They could learn how to make bombs, shoplift, or cheat from chatbots. So, safeguards are needed to stop and punish people who abuse technology. People who use tools like ChatGPT may be more likely to copy other people's work in ways that are hard to spot. ChatZero and AI Text Classifier are two tools that try to tell the difference between text written by AI and text written by a person. Copying is hard to do without AI tools, so this is done to stop it.

Hackers could easily use AI content producers to make personalized junk messages and images that hide harmful code. The people you want to reach might be persuaded by these words and images. There may be a lot more hacks because of this, which could hurt a lot of people. Chatbot users may also give private personal or business information to them, which could lead to the creator abusing the data users give them. The creator could decide to keep and look over the data that people give them. Think about these two use cases as models that will help you learn [83]. An important person in the company recently copied and pasted the 2023 strategy paper into a robot and asked it to make presentation PowerPoint slides. In a different case, a doctor used ChatGPT to enter the name of his patient and the patient's medical situation. He then asked the software to write a letter to the patient's insurance company on his behalf. There are social concerns, problems with data protection, and security risks and threats that come with AI content creators that can be seen in these use cases.

When AI creates content, copyright issues come up. For example, who owns the rights to a story, song, or piece of art that AI created? Who is responsible for teaching the chatbot? The people who gave the data that was used to teach it, or the people who came up with the questions and answers that the AI should give? In the context of this conversation, you should know that the US Copyright Office has said that pictures made by Midjourney and other AI text-to-image technologies are not protected by US copyright law because they were not created by a person and cannot be considered works of art [84]. A class action lawsuit has been filed against companies that sell AI-generated art. The case questions whether it is legal to use data to train an AI without the permission of the people who gave the data. Artists brought the case to court.

11 Future Research Directions—IOTA, Federated Learning Blockchain, and Explainable AI

Exciting research opportunities arise from the combination of IOTA, blockchain, explainable AI, and shared learning. IOTA's Tangle DAG ledger and shared learning make sense when combined. How data privacy and security in decentralized systems might be improved by the Tangle by providing a clear and secure communication channel for participants in shared learning. Safe and cooperative machine learning models may be achieved using blockchain-based open cooperative learning. In order to reduce latency and resource consumption without sacrificing the independence of the system, researchers may focus on enhancing the scalability, efficiency, and agreement techniques of blockchain-based federated learning in the future. Explainable AI is critical to cooperative learning because transparency fosters trust. Scholars can concentrate on developing interpretable models that demonstrate decision-making processes while safeguarding contributor privacy. The goal of federated learning and explainable AI is to create a mechanism to understand models in decentralized settings while maintaining the privacy of data. The primary area of study is IOTA-distributed learning techniques that protect privacy. Autonomous model training and the protection of private information may be facilitated by homomorphic encryption and secure multi-party computing. Before federated learning is extensively used in the real world, privacy concerns must be resolved. Decentralized systems struggle to expand and operate quickly. This research might investigate methods for managing large gatherings of individuals more effectively, cutting down on communication expenses, and expediting the decision-making process. Scalability issues must be resolved for shared learning on blockchain networks to be feasible. Decentralized systems struggle to expand and operate quickly. This research might investigate methods for managing large gatherings of individuals more effectively, cutting down on communication expenses, and expediting the decision-making process. Scalability issues must be resolved for shared learning on blockchain networks to be feasible. Further investigation is required to ensure the safety and dependability of autonomous systems using shared learning, blockchain, and IOTA. To safeguard decentralized federated learning models, vulnerabilities must be identified, secure smart contracts must be created, and barriers against hostile attacks must be erected.

References

1. Ahmed F (2022) Ethical aspects of artificial intelligence in banking. J Res Econ Finance Manag 1:55–63. https://doi.org/10.56596/jrefm.v1i2.7
2. Johnson A, Grumbling E (eds) (2019) Implications of artificial intelligence for cybersecurity. National Academies Press, Washington, DC. https://doi.org/10.17226/25488
3. Martinho A, Herber N, Kroesen M, Chorus C (2021) Ethical issues in focus by the autonomous vehicles industry. Transp Rev 41:556–577. https://doi.org/10.1080/01441647.2020.1862355
4. Mirbabaie M, Hofeditz L, Frick NRJ, Stieglitz S (2022) Artificial intelligence in hospitals: providing a status quo of ethical considerations in academia to guide future research. AI Soc 37:1361–1382. https://doi.org/10.1007/s00146-021-01239-4
5. Dash B, Ansari MF, Sharma P, Ali A (2022) Threats and opportunities with AI-based cyber security intrusion detection: a review. Int J Softw Eng Appl 13:13–21. https://doi.org/10.5121/ijsea.2022.13502
6. Naik N, Hameed BMZ, Shetty DK, Swain D, Shah M, Paul R, Aggarwal K, Ibrahim S, Patil V, Smriti K, Shetty S, Rai BP, Chlosta P, Somani BK (2022) Legal and ethical consideration in artificial intelligence in healthcare: who takes responsibility? Front Surg 9. https://doi.org/10.3389/fsurg.2022.862322
7. Helkala K, Cook J, Lucas G, Pasquale F, Reichberg G, Syse H (2023) AI in cyber operations: ethical and legal considerations for end-users. In: Tuomo S, Kokkonen T (eds) Artificial intelligence and cybersecurity: theory and applications. Springer International Publishing, Cham, pp 185–206. https://doi.org/10.1007/978-3-031-15030-2_9
8. Rodrigues R (2020) Legal and human rights issues of AI: gaps, challenges and vulnerabilities. J Responsib Technol 4:100005. https://doi.org/10.1016/j.jrt.2020.100005
9. Al-Mansoori S, Salem MB (2023) The role of artificial intelligence and machine learning in shaping the future of cybersecurity: trends, applications, and ethical considerations. Int J Soc Anal 8:1–16
10. Wirtz BW, Weyerer JC, Geyer C (2019) Artificial intelligence and the public sector—applications and challenges. Int J Public Adm 42:596–615. https://doi.org/10.1080/01900692.2018.1498103
11. Walshe R, Koene A, Baumann S, Panella M, Maglaras L, Medeiros F (2021) Artificial intelligence as enabler for sustainable development. In: 2021 IEEE international conference on engineering, technology and innovation (ICE/ITMC), pp 1–7. https://doi.org/10.1109/ICE/ITMC52061.2021.9570215
12. Chaudhary H, Detroja A, Prajapati P, Shah P (2020) A review of various challenges in cybersecurity using artificial intelligence. In: 2020 3rd international conference on intelligent sustainable systems (ICISS), pp 829–836. https://doi.org/10.1109/ICISS49785.2020.9316003
13. Shaukat K, Luo S, Varadharajan V, Hameed IA, Xu M (2020) A survey on machine learning techniques for cyber security in the last decade. IEEE Access 8:222310–222354. https://doi.org/10.1109/ACCESS.2020.3041951
14. Timmers P (2019) Ethics of AI and cybersecurity when sovereignty is at stake. Minds Mach (Dordr) 29:635–645. https://doi.org/10.1007/s11023-019-09508-4
15. Romancheva NI (2021) Duality of artificial intelligence technologies in assessing cyber security risk. IOP Conf Ser Mater Sci Eng 1069:12004. https://doi.org/10.1088/1757-899X/1069/1/012004
16. Ferreyra NED, Aimeur E, Hage H, Heisel M, van Hoogstraten CG (2020) Persuasion meets AI: ethical considerations for the design of social engineering countermeasures. CoRR. abs/2009.12853
17. Christodoulou E, Iordanou K (2021) Democracy under attack: challenges of addressing ethical issues of AI and big data for more democratic digital media and societies. Front Polit Sci 3. https://doi.org/10.3389/fpos.2021.682945
18. Abdulllah SM (2019) Artificial intelligence (AI) and its associated ethical issues. ICR J 10:124–126. https://doi.org/10.52282/icr.v10i1.78

19. Zhang Z, Ning H, Shi F, Farha F, Xu Y, Xu J, Zhang F, Choo K-KR (2022) Artificial intelligence in cyber security: research advances, challenges, and opportunities. Artif Intell Rev 55:1029–1053. https://doi.org/10.1007/s10462-021-09976-0
20. Kim Y, Hakak S, Ghorbani A (2023) DDoS attack dataset (CICEV2023) against EV authentication in charging infrastructure. In: 2023 20th annual international conference on privacy, security and trust (PST). IEEE Computer Society, Los Alamitos, CA, pp 1–9. https://doi.org/10.1109/PST58708.2023.10320202
21. Lashkari AH, Kadir AFA, Taheri L, Ghorbani AA (2018) Toward developing a systematic approach to generate benchmark android malware datasets and classification. In: 2018 international Carnahan conference on security technology (ICCST), pp 1–7. https://doi.org/10.1109/CCST.2018.8585560
22. Moustafa N, Creech G, Slay J (2018) Anomaly detection system using beta mixture models and outlier detection. In: Kumar PP, Rautaray SS (eds) Progress in computing, analytics and networking. Springer Singapore, Singapore, pp 125–135
23. Habibi Lashkari A, Kaur G, Rahali A (2021) DIDarknet: a contemporary approach to detect and characterize the darknet traffic using deep image learning. In: Proceedings of the 2020 10th international conference on communication and network security. Association for Computing Machinery, New York, NY, pp 1–13. https://doi.org/10.1145/3442520.3442521
24. Moustafa N, Slay J (2015) UNSW-NB15: a comprehensive data set for network intrusion detection systems (UNSWNB15 network data set). In: 2015 military communications and information systems conference (MilCIS), pp 1–6. https://doi.org/10.1109/MilCIS.2015.7348942
25. Koroniotis N, Moustafa N, Sitnikova E, Turnbull BP (2018) Towards the development of realistic botnet dataset in the internet of things for network forensic analytics: Bot-IoT dataset. CoRR. abs/1811.00701
26. Valeros V, Garcia S (2022) Hornet 40: network dataset of geographically placed honeypots. Data Brief 40:107795. https://doi.org/10.1016/j.dib.2022.107795
27. Moustafa N, Creech G, Slay J (2018) Flow aggregator module for analysing network traffic. In: Kumar PP, Rautaray SS (eds) Progress in computing, analytics and networking. Springer Singapore, Singapore, pp 19–29
28. Ma J, Kulesza A, Dredze M, Crammer K, Saul L, Pereira F (2010) Exploiting feature covariance in high-dimensional online learning. In: Teh YW, Titterington M (eds) Proceedings of the thirteenth international conference on artificial intelligence and statistics. PMLR, Chia Laguna Resort, Sardinia, pp 493–500
29. Ma J, Saul LK, Savage S, Voelker GM (2009) Identifying suspicious URLs: an application of large-scale online learning. In: Proceedings of the 26th annual international conference on machine learning. Association for Computing Machinery, New York, NY, pp 681–688. https://doi.org/10.1145/1553374.1553462
30. Miettinen M, Marchal S, Hafeez I, Frassetto T, Asokan N, Sadeghi A-R, Tarkoma S (2017) IoT sentinel demo: automated device-type identification for security enforcement in IoT. In: Lee K, Liu L (eds) 2017 IEEE 37th international conference on distributed computing systems (ICDCS). IEEE, pp 2511–2514. https://doi.org/10.1109/ICDCS.2017.284
31. Koroniotis N, Moustafa N (2020) Enhancing network forensics with particle swarm and deep learning: the particle deep framework. CoRR. abs/2005.00722
32. Koroniotis N, Moustafa N, Sitnikova E (2020) A new network forensic framework based on deep learning for internet of things networks: a particle deep framework. Future Gener Comput Syst 110:91–106. https://doi.org/10.1016/j.future.2020.03.042
33. Moustafa N, Slay J (2016) The evaluation of network anomaly detection systems: statistical analysis of the UNSW-NB15 data set and the comparison with the KDD99 data set. Inf Secur J: Glob Perspect 25:18–31. https://doi.org/10.1080/19393555.2015.1125974
34. Moustafa N, Creech G, Slay J (2017) Big data analytics for intrusion detection system: statistical decision-making using finite Dirichlet mixture models. In: Iván PC, Kalutarage HK (eds) Data analytics and decision support for cybersecurity: trends, methodologies and applications. Springer International Publishing, Cham, pp 127–156. https://doi.org/10.1007/978-3-319-59439-2_5

35. He Y, Luo C, Camacho RS, Wang K, Zhang H (2020) AI-based security attack pathway for cardiac medical diagnosis systems (CMDS). In: 2020 computing in cardiology, pp 1–4. https://doi.org/10.22489/CinC.2020.439
36. Lorenzo P, Stefano F, Ferreira A, Carolina P (2022) Artificial intelligence and cybersecurity: technology, governance and policy challenges. Centre for European Policy Studies (CEPS)
37. Gerke S, Minssen T, Cohen G (2020) Chapter 12—Ethical and legal challenges of artificial intelligence-driven healthcare. In: Bohr A, Memarzadeh K (eds) Artificial intelligence in healthcare. Academic Press, pp 295–336. https://doi.org/10.1016/B978-0-12-818438-7.00012-5
38. He Y, Efpraxia DZ, Yevseyeva I, Luo C (2023) Artificial intelligence-based ethical hacking for health information systems: simulation study. J Med Internet Res 25:e41748. https://doi.org/10.2196/41748
39. Cobianchi L, Verde JM, Loftus TJ, Piccolo D, Dal Mas F, Mascagni P, Garcia Vazquez A, Ansaloni L, Marseglia GR, Massaro M, Gallix B, Padoy N, Peter A, Kaafarani HM (2022) Artificial intelligence and surgery: ethical dilemmas and open issues. J Am Coll Surg 235
40. Sarker IH, Janicke H, Mohammad N, Watters P, Nepal S (2023) AI potentiality and awareness: a position paper from the perspective of human-AI teaming in cybersecurity
41. Flechais I, Chalhoub G (2023) Practical cybersecurity ethics: mapping CyBOK to ethical concerns
42. Jackson D, Matei SA, Bertino E (2023) Artificial intelligence ethics education in cybersecurity: challenges and opportunities: a focus group report
43. Ramya P, Babu SV, Venkatesan G (2023) Advancing cybersecurity with explainable artificial intelligence: a review of the latest research. In: 2023 5th international conference on inventive research in computing applications (ICIRCA), pp 1351–1357. https://doi.org/10.1109/ICIRCA57980.2023.10220797
44. Mohamed N (2023) Current trends in AI and ML for cybersecurity: a state-of-the-art survey. Cogent Eng 10:2272358. https://doi.org/10.1080/23311916.2023.2272358
45. Kumar S, Gupta U, Singh AK, Singh AK (2023) Artificial intelligence: revolutionizing cyber security in the digital era. J Comput Mech Manag 2:31–42. https://doi.org/10.57159/gadl.jcmm.2.3.23064
46. Santosh KC, Wall C (2022) AI and ethical issues. In: AI, ethical issues and explainability—applied biometrics. Springer Nature Singapore, Singapore, pp 1–20. https://doi.org/10.1007/978-981-19-3935-8_1
47. Alawida M, Mejri S, Mehmood A, Chikhaoui B, Isaac Abiodun O (2023) A comprehensive study of ChatGPT: advancements, limitations, and ethical considerations in natural language processing and cybersecurity. Information 14. https://doi.org/10.3390/info14080462
48. Kuzlu M, Fair C, Guler O (2021) Role of artificial intelligence in the internet of things (IoT) cybersecurity. Discov Internet Things 1:7. https://doi.org/10.1007/s43926-020-00001-4
49. Parul I, Thakur S (2021) Ethics and artificial intelligence: the pandora's box. In: Parul I, Thakur S (eds) Artificial intelligence and ophthalmology: perks, perils and pitfalls. Springer Singapore, Singapore, pp 145–150. https://doi.org/10.1007/978-981-16-0634-2_11
50. Ashraf J, Keshk M, Moustafa N, Abdel-Basset M, Khurshid H, Bakhshi AD, Mostafa RR (2021) IoTBoT-IDS: a novel statistical learning-enabled botnet detection framework for protecting networks of smart cities. Sustain Cities Soc 72:103041. https://doi.org/10.1016/j.scs.2021.103041
51. Kalla D, Kuraku S (2023) Advantages, disadvantages and risks associated with ChatGPT and AI on cybersecurity. JETIR
52. Li F, Ruijs N, Lu Y (2023) Ethics & AI: a systematic review on ethical concerns and related strategies for designing with AI in healthcare. AI 4:28–53. https://doi.org/10.3390/ai4010003
53. Shasha Y, Carroll F (2021) Implications of AI in national security: understanding the security issues and ethical challenges. In: Reza M, Jahankhani H (eds) Artificial intelligence in cyber security: impact and implications: security challenges, technical and ethical issues, forensic investigative challenges. Springer International Publishing, Cham, pp 157–175. https://doi.org/10.1007/978-3-030-88040-8_6

54. Vakkuri V, Kemell KK, Abrahamsson P (2019) Implementing ethics in AI: initial results of an industrial multiple case study. In: Xavier F, Männistö T (eds) Product-focused software process improvement. Springer International Publishing, Cham, pp. 331–338
55. Dasawat SS, Sharma S (2023) Cyber security integration with smart new age sustainable startup business, risk management, automation and scaling system for entrepreneurs: an artificial intelligence approach. In: 2023 7th international conference on intelligent computing and control systems (ICICCS), pp 1357–1363. https://doi.org/10.1109/ICICCS56967.2023.10142779
56. Kaloudi N, Li J (2020) The AI-based cyber threat landscape: a survey. ACM Comput Surv 53. https://doi.org/10.1145/3372823
57. Du S, Xie C (2021) Paradoxes of artificial intelligence in consumer markets: ethical challenges and opportunities. J Bus Res 129:961–974. https://doi.org/10.1016/j.jbusres.2020.08.024
58. Sarker IH, Furhad MH, Nowrozy R (2021) AI-driven cybersecurity: an overview, security intelligence modeling and research directions. SN Comput Sci 2:173. https://doi.org/10.1007/s42979-021-00557-0
59. Yildirim M (2021) Artificial intelligence-based solutions for cyber security problems. In: Luhach AK, Elçi A (eds) Artificial intelligence paradigms for smart cyber-physical systems. IGI Global, Hershey, PA, pp 68–86. https://doi.org/10.4018/978-1-7998-5101-1.ch004
60. Buzzanell PM (2023) Risk, resilience, and ethical considerations in artificial intelligence. Emerg Media 1:30–39. https://doi.org/10.1177/27523543231188274
61. Koroniotis N, Moustafa N, Schiliro F, Gauravaram P, Janicke H (2020) A holistic review of cybersecurity and reliability perspectives in smart airports. IEEE Access 8:209802–209834. https://doi.org/10.1109/ACCESS.2020.3036728
62. Guleria A, Krishan K, Sharma V, Kanchan T. ChatGPT: forensic, legal, and ethical issues. Med Sci Law 00258024231191829. https://doi.org/10.1177/00258024231191829
63. Morovat K, Panda B (2020) A survey of artificial intelligence in cyber-security. In: 2020 International conference on computational science and computational intelligence (CSCI), pp 109–115. https://doi.org/10.1109/CSCI51800.2020.00026
64. Garcia AB, Babiceanu RF, Seker R (2021) Artificial intelligence and machine learning approaches for aviation cybersecurity: an overview. In: 2021 integrated communications navigation and surveillance conference (ICNS), pp 1–8. https://doi.org/10.1109/ICNS52807.2021.9441594
65. Sharma I (2021) Evolution of unmanned aerial vehicles (UAVs) with machine learning. In: 2021 international conference on advances in technology, management & education (ICATME), pp 25–30. https://doi.org/10.1109/ICATME50232.2021.9732774
66. Ahmad OF, Stoyanov D, Lovat LB (2020) Barriers and pitfalls for artificial intelligence in gastroenterology: ethical and regulatory issues. Tech Innov Gastrointest Endosc 22:80–84. https://doi.org/10.1016/j.tgie.2019.150636
67. Stahl BC (2021) Ethical issues of AI. In: artificial intelligence for a better future: an ecosystem perspective on the ethics of AI and emerging digital technologies. Springer International Publishing, Cham, pp 35–53. https://doi.org/10.1007/978-3-030-69978-9_4
68. Soni VD. Challenges and solution for artificial intelligence in cybersecurity of the USA. Artif Intell Cybersecur USA 2
69. Kent AD. Cyber security data sources for dynamic network research. In: Dynamic networks and cyber-security, pp 37–65. https://doi.org/10.1142/9781786340757_0002
70. Sarhan M, Layeghy S, Moustafa N, Portmann M (2020) NetFlow datasets for machine learning-based network intrusion detection systems. CoRR. abs/2011.09144
71. Sharma I, Kaushik K. Chhabra G (2023) Augmenting transparency and reliability for national health insurance scheme with distributed ledger. In: 2023 4th international conference on electronics and sustainable communication systems (ICESC), pp 1399–1405. https://doi.org/10.1109/ICESC57686.2023.10193127
72. Nour M, Slay J (2018) A network forensic scheme using correntropy-variation for attack detection. In: Gilbert P, Shenoi S (eds) Advances in digital forensics XIV. Springer International Publishing, Cham, pp 225–239

73. Al-Hawawreh M, Moustafa N, Garg S, Hossain MS (2021) Deep learning-enabled threat intelligence scheme in the internet of things networks. IEEE Trans Netw Sci Eng 8:2968–2981. https://doi.org/10.1109/TNSE.2020.3032415
74. Miettinen M, Marchal S, Hafeez I, Asokan N, Sadeghi A-R, Tarkoma S (2017) IoT SENTINEL: automated device-type identification for security enforcement in IoT. In: Lee K, Liu L (eds) 2017 IEEE 37th international conference on distributed computing systems (ICDCS). IEEE, pp 2177–2184. https://doi.org/10.1109/ICDCS.2017.283
75. Keshk M, Moustafa N, Sitnikova E, Creech G (2017) Privacy preservation intrusion detection technique for SCADA systems. In: 2017 military communications and information systems conference (MilCIS), pp 1–6. https://doi.org/10.1109/MilCIS.2017.8190422
76. Sharma I, Ramkumar KR (2017) A survey on ACO based multipath routing algorithms for ad hoc networks. Int J Pervasive Comput Commun 13:370–385. https://doi.org/10.1108/IJPCC-D-17-00015
77. Mamun MS, Rathore MA, Lashkari AH, Stakhanova N, Ghorbani AA (2016) Detecting malicious URLs using lexical analysis. In: Chen J, Piuri V (eds) Network and system security. Springer International Publishing, Cham, pp 467–482
78. Moustafa N, Misra G, Slay J (2021) Generalized outlier Gaussian mixture technique based on automated association features for simulating and detecting web application attacks. IEEE Trans Sustain Comput 6:245–256. https://doi.org/10.1109/TSUSC.2018.2808430
79. Ashraf J, Bakhshi AD, Moustafa N, Khurshid H, Javed A, Beheshti A (2021) Novel deep learning-enabled LSTM autoencoder architecture for discovering anomalous events from intelligent transportation systems. IEEE Trans Intell Transp Syst 22:4507–4518. https://doi.org/10.1109/TITS.2020.3017882
80. Koroniotis N, Moustafa N, Sitnikova E, Slay J (2018) Towards developing network forensic mechanism for botnet activities in the IoT based on machine learning techniques. In: Hu J, Khalil I (eds) Mobile networks and management. Springer International Publishing, Cham, pp 30–44
81. Moustafa N, Keshk M, Choo K-KR, Lynar T, Camtepe S, Whitty M (2021) DAD: a distributed anomaly detection system using ensemble one-class statistical learning in edge networks. Future Gener Comput Syst 118:240–251. https://doi.org/10.1016/j.future.2021.01.011
82. García S, Grill M, Stiborek J, Zunino A (2014) An empirical comparison of botnet detection methods. Comput Secur 45:100–123. https://doi.org/10.1016/j.cose.2014.05.011
83. Haider W, Moustafa N, Keshk M, Fernandez A, Choo K-KR, Wahab A (2020) FGMC-HADS: fuzzy Gaussian mixture-based correntropy models for detecting zero-day attacks from Linux systems. Comput Secur 96:101906. https://doi.org/10.1016/j.cose.2020.101906
84. Weinger B, Kim J, Sim A, Nakashima M, Moustafa N, Wu KJ (2020) Enhancing IoT anomaly detection performance for federated learning. In: 2020 16th international conference on mobility, sensing and networking (MSN), pp 206–213. https://doi.org/10.1109/MSN50589.2020.00045